Quantification of Biophysical Parameters in Medical Imaging

Ingolf Sack • Tobias Schaeffter
Editors

Quantification of Biophysical Parameters in Medical Imaging

 Springer

Editors
Ingolf Sack
Department of Radiology
Humboldt University
of Berlin Charité University Hospital
Berlin
Germany

Tobias Schaeffter
Medical Physics
and Metrological Information Technology
Physikalisch-Technische Bundesanstalt
Berlin
Germany

ISBN 978-3-030-09755-4   ISBN 978-3-319-65924-4   (eBook)
https://doi.org/10.1007/978-3-319-65924-4

Printed on acid-free paper

This Springer imprint is published by Springer Nature
The registered company is Springer International Publishing AG
The registered company address is: Gewerbestrasse 11, 6330 Cham, Switzerland

Acknowledgements

The authors of this book wish to thank the many colleagues and collaborators from whom they received continuous support, inspiration, scientific advice, and positive criticism. The endeavor of writing a book like this would not have been possible without prospering communication and collaboration across the established borders of disciplines and institutions. Many researchers involved in writing chapters of this book have ties to both clinical research and basic science and dedicate themselves to obtaining a higher precision, improved accuracy, and better understanding of clinical imaging markers. This spirit has evolved in the new research training group BIOQIC, Biophysical Quantitative Imaging towards Clinical Diagnosis, which is funded by the German Research Foundation (DFG GRK2260). The authors are grateful to the DFG and their reviewers for having been granted this unique opportunity to establish such an interdisciplinary training program in the important fields of biophysics, medical imaging technologies, and clinical radiology as covered by this book.

Contents

Introduction: Medical Imaging for the Quantitative Measurement of Biophysical Parameters

Ingolf Sack and Tobias Schaeffter

Abstract

Medical imaging is the backbone of modern clinical diagnosis with the vast majority of images obtained by high-end tomographic modalities such as computed tomography (CT), magnetic resonance imaging (MRI), 3D ultrasound (US), and emission tomography (PET, SPECT). However, the current radiological practice is based on visual inspection of images and most diseases are rated in qualitative terms, which results in limited comparability and accuracy of image-based diagnostic decisions. Therefore, there is a need for quantitative imaging technologies for the assessment of biophysical tissue parameters. This book focuses on imaging approaches to obtain quantitative information of fluid transport, soft tissue mechanics, and tissue structures; and gives examples on clinical applications that benefit from quantitative imaging.

Medical imaging is the backbone of modern clinical diagnosis. More than 5 billion imaging investigations are performed worldwide each year [1, 2]. The vast majority of radiological imaging procedures are based on high-end tomographic modalities such as computed tomography (CT), magnetic resonance imaging (MRI), 3D ultrasound (US), and emission tomography (PET, SPECT) [3]. Medical imaging is a prospering market with estimated sales of $31.9 billion in 2019 alone in the United

I. Sack (✉)
Department of Radiology, Charité - Universitätsmedizin Berlin, Berlin, Germany
e-mail: ingolf.sack@charite.de

T. Schaeffter
Medical Physics and Metrological Information Technology,
Physikalisch-Technische Bundesanstalt, Berlin, Germany
e-mail: tobias.schaeffter@ptb.de

© Springer International Publishing AG 2018
I. Sack, T. Schaeffter (eds.), *Quantification of Biophysical Parameters in Medical Imaging*, https://doi.org/10.1007/978-3-319-65924-4_1

States.[1] The growing demand for high-end imaging systems reflects advancing technologies, aging demographic trends, evolving epidemiological patterns, and changing patient care strategies. Notwithstanding this success, medical imaging has not yet fulfilled all expectations about its reproducibility, consistency, and accuracy.

1.1   Most Clinical Imaging Examinations Are Still Nonquantitative

Medical imaging modalities today can acquire morphological information with high spatial resolution in a relatively short time. This technical progress has led to the current radiological practice to make decisions based on visual inspection of images and to rate diseases in qualitative terms like "enlarged," "small," or "enhanced." Such qualitative information often suffers from the limited comparability among readers, modalities, platforms, and scan protocols.

Virtually none of the currently used diagnostic information is related to the biophysical properties of the affected tissue. The lack of established quantitative and biophysics-based medical imaging markers has wide implications for clinical practice. Today, radiologists need many years of training to master the interpretation of subtle morphometric variations. Although principally feasible using current imaging methods, quantitative staging of many diseases such as tumors, hepatic fibrosis, neurodegeneration, and cardiovascular pressure usually requires invasive interventional methods and/or histological examination for final validation.

For example, in cardiovascular disease diagnosis and therapeutic decisions are based on vascular anatomical features, such as the diameter of stenotic vessels, rather than on the physiological relevance of the stenotic vessels for the heart or brain. The COURAGE trial [4] showed no significant outcome difference in coronary catheterization in comparison with standard medical therapy alone when vascular anatomical features were used. However, it was shown in a sub-study that certain patients can benefit from such expensive intervention, when the blood flow is measured in tissue (called perfusion) as a biophysical property [5]. Also in cancer new imaging parameters are required for early diagnosis and teatment of tumors. There is a strong trend to expand the current metrics, which have been based solely on size, like the World Health Organization (WHO) criteria and the Response Evaluation Criteria in Solid Tumors (RECIST) [6, 7].

Both applications also require medical imaging to asess therapy effects non-invasively. However, such serial imaging is often hampered by a low reproducibility due to technical variations and inter-observer variability. The same applies to multi-center studies, where different scanners and interobserver variability add to the issues. For these reasons, the demand for quantitative imaging technologies has been identified as the major new challenge for the advancement of medical imaging over the next decade [8].

[1] www.freedoniagroup.com.

1.2 Toward Quantitative Medical Imaging

The mission to *improve the value and practicality of quantitative imaging biomarkers* has emerged in worldwide alliances such as QIBA (Quantitative Imaging Biomarkers Alliance[2]) by RSNA, the Quantitative Imaging Network (QIN[3]) of the National Cancer Institute in the United States, or EIBALL (European Imaging Biomarkers Alliance) by ESR [9], which are all committed to making medical imaging a more quantitative science. In parallel, graduate training programs emerge which specifically focus on quantitative medical imaging. One activity is BIOQIC—Biophysical Quantitative Imaging towards Clinical diagnosis [10]—a graduate school centered in Berlin, Germany, and funded by the German Research Foundation (DFG), which addresses the investigation and clinical application of new quantitative imaging approaches. Most of the authors of this book are involved in BIOQIC activities—from defining clinical needs to current research and teaching activities.

1.3 About the Book

This book has the mission of teaching medical imaging sciences to interdisciplinary scientists and PhD students with emphasis on quantitative biophysical tissue parameters. These parameters include transport-related parameters, such as flow velocity; tracer kinetics; diffusion; perfusion; and mechanical properties, like stiffness, elasticity, and viscoelasticity, and structure-related properties like cell density, heterogeneity, and anisotropy. Many of these properties also influence each other, e.g., the anisotropy of tissue microstructure strongly affects the amount and direction of diffusion. This book emphasizes imaging research in *modality- and system-independent* biophysical properties. This biophysics-based view on quantitative medical imaging is complementary to the worldwide research effort in quantitative parameter mapping of modality-related parameters such as relaxation times in MRI or CT attenuation coefficients. Figure 1.1 illustrates the primary parameter classes addressed by biophysical quantitative imaging, that is, *fluid transport, soft tissue mechanics,* and *tissue structures*. The close interrelation of soft tissue structures including vascular components and cellular and extracellular networks at multiple scales naturally yields an overlap in the prescribed parameter categories. For example, the crosslinking of extracellular matrix influences the macroscopic mechanical response of the tissue. Another example is the microstructures such as microvessels and interstitial spaces, which significantly control the fluid transport properties in the tissue. Chapter 2 outlines the theoretical foundations for the parameter categories as illustrated in Fig. 1.1. The interrelation between fluid transport, structures, and mechanical properties in biological tissues is of high relevance for basic research and clinical applications in imaging sciences.

---

[2] http://www.rsna.org/QIBA/.

[3] https://imaging.cancer.gov/informatics/qin.htm.

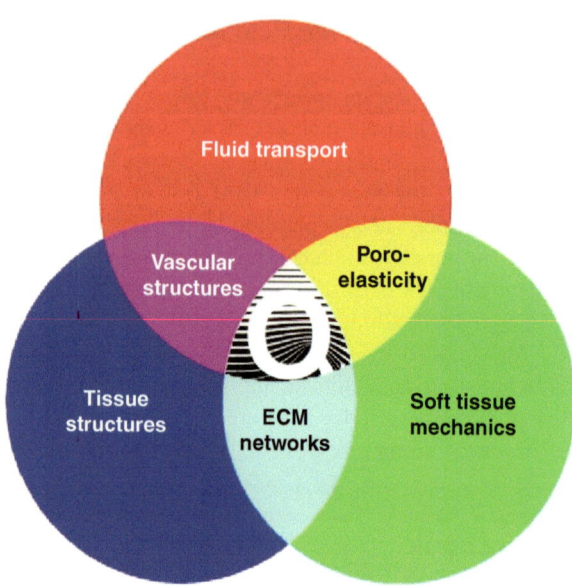

Fig. 1.1 Categories of biophysical medical imaging parameters to which this book is targeted. With fluid transport, we specify the phenomenon of water mobility and blood transport in biological tissues and within the body occurring at different scales. Soft tissue mechanics summarizes all parameters related to mechanical deformation, stress, and pressure within biological matter. Tissue structures specifies quantitative imaging markers sensitive to the constitution of a tissue on the microlevel, including extracellular matrix (ECM) network structures, cell density and the physical forces acting on cells, as well as microvascular structures (e.g., microvessel density, bifurcation index, fractal dimension)

The book is organized in three parts: The first part focuses on the biological and physical fundamentals of quantitative imaging by covering transport equations (Chap. 2), mathematical modeling of fluid motion at different scales (Chaps. 3 and 4), as well as biophysical properties and molecular signatures of cells and extracellular matrix (Chaps. 5 and 6). The second part of this book is dedicated to medical imaging modalities from mathematical foundations (Chap. 7) and data sampling strategies in magnetic resonance imaging (MRI) (Chap. 8) to novel imaging approaches including Chemical Exchange Saturation Transfer (CEST) MRI (Chap. 10) and photoacoustic tomography (Chap. 13). The third part of this book presents applications of new medical imaging approaches in clinical research, from measuring different tissue properties in cancer and cardiovascular disease to the assessment of perfusion in the brain and in the heart using different imaging modalities (Chaps. 15–23).

1.4   Quantitative Imaging Versus Biological Variability

The advancement of imaging modalities over the past few years has increased the precision with which details of structure and function of living tissues can be captured. However, such higher precision without understanding the underlying biological principles does not automatically yield higher diagnostic accuracy. Instead, biological diversity and physiological variability naturally limit the level of accuracy with which an imaging marker can classify diseases or predict therapy responses and ultimately clinical outcome. Many medical imaging technologies have recently encountered the limits of biological variability. An example here is the limit of spatial resolution in MRI. Ultrahigh-field MRI as described in Chap. 15 has enabled examinations of human tissue with a spatial resolution much below a millimeter voxel size. However, high-resolution 3D brain imaging examinations can take up more than 10 min during which pulsation in the order of a millimeter causes blurring of signals if acquired unsynchronized to the heartbeat. Thus, high spatial resolution requires that data are accumulated within intermediary phases of relatively still tissue. Moreover, physiological effects can affect accurate measurement of imaging markers and are often summarized in "physiological noise." For example, measuring in vivo tissue stiffness by elastography by techniques outlined in Chaps. 12, 19, and 20 is influenced by blood perfusion. This biophysical property can be measured by different imaging modalities and requires mathematical models (Chap. 3) to estimate the quantitative value of perfusion (Chap. 22 and 23). The measurement of perfusion is also important in positron emission tomography (PET) studies, in which innovative imaging tracers (Chap. 11) are used. High perfusion can result in high uptake of specific tracers without the abundance of specific targets in that region. Therefore, the measurement of perfusion can help researchers to disentangle flow-related effects for the better characterization of tumors (Chap. 18). Combining the physiological information acquired by PET with in vivo mechanical tissue properties could open a new window into quantitative medical imaging of tissue function. In general, gathering information from different modalities may help us to better understand the structure and function of biological tissue in vivo. Consequently, there is a trend to multimodality examinations with combined systems like SPECT-CT, PET-CT, and PET-MR. How this knowledge gained by different medical imaging technologies can be translated into clinical value is part of ongoing research activities—some of them are covered by this book.

1.5   Clinical Focus

Methods described in this book are related to a variety of diseases with focus on *cardiovascular diseases* and *cancer* which represent two of the most prevalent groups of diseases in Western countries. Imaging methods for the detection, staging, and quantification of cardiovascular diseases and tumors include molecular probes for MRI (Chaps. 6, 10, and 16), radionuclide tracers for PET and SPECT (Chaps. 11, 18, and 21), flow, perfusion and diffusion measurement by MRI (Chaps. 9, 17, and 22), as well as MRI- and US-based elastography (Chaps. 12, 19, and 20). The signals measured by these methods often represent with overlapping signatures due to the same physical principles acting on signal emitting tracers or contrast manipulating agents in the tissue. At the same time, the different capabilities of medical imaging devices can capture physiological signals with different sensitivities and different temporal and spatial resolution providing complementary information. Such multiparametric data sets offer new insights into pathological processes than possible by each modality alone. It is important to understand the underlying system-independent imaging markers that affect the different imaging readouts today. Due to the limited broadness of an introductory book as the present is, it can only cover a fraction of imaging concepts. Instead, this book aims at the basic concepts of quantitative medical imaging and their clinical perspectives. The future will show the value of quantitative imaging to reduce variability across devices, patient groups, and time.

References

1. Owens B. Scans: enhanced medical vision. Nature. 2013;502(7473):S82–3.
2. Roobottom CA, Mitchell G, Morgan-Hughes G. Radiation-reduction strategies in cardiac computed tomographic angiography. Clin Radiol. 2010;65(11):859–67.
3. Gwynne P. Next-generation scans: seeing into the future. Nature. 2013;502(7473):S96–7.
4. Boden WE, O'Rourke RA, Teo KK, Hartigan PM, Maron DJ, Kostuk WJ, Knudtson M, Dada M, Casperson P, Harris CL, Chaitman BR, Shaw L, Gosselin G, Nawaz S, Title LM, Gau G, Blaustein AS, Booth DC, Bates ER, Spertus JA, Berman DS, Mancini GB, Weintraub WS. Optimal medical therapy with or without PCI for stable coronary disease. N Engl J Med. 2007;356(15):1503–16.
5. Shaw LJ, Berman DS, Maron DJ, Mancini GB, Hayes SW, Hartigan PM, Weintraub WS, O'Rourke RA, Dada M, Spertus JA, Chaitman BR, Friedman J, Slomka P, Heller GV, Germano G, Gosselin G, Berger P, Kostuk WJ, Schwartz RG, Knudtson M, Veledar E, Bates ER, McCallister B, Teo KK, Boden WE. Optimal medical therapy with or without percutaneous coronary intervention to reduce ischemic burden: results from the Clinical Outcomes Utilizing Revascularization and Aggressive Drug Evaluation (COURAGE) trial nuclear substudy. Circulation. 2008;117(10):1283–91.
6. Moertel CG, Hanley JA. The effect of measuring error on the results of therapeutic trials in advanced cancer. Cancer. 1976;38(1):388–94.
7. Therasse P, Arbuck SG, Eisenhauer EA, Wanders J, Kaplan RS, Rubinstein L, Verweij J, Van Glabbeke M, van Oosterom AT, Christian MC, Gwyther SG. New guidelines to evaluate the response to treatment in solid tumors. J Natl Cancer Inst. 2000;92(3):205–16.
8. Kessler LG, Barnhart HX, Buckler AJ, Choudhury KR, Kondratovich MV, Toledano A, Guimaraes AR, Filice R, Zhang Z, Sullivan DC. The emerging science of quantitative imaging biomarkers terminology and definitions for scientific studies and regulatory submissions. Stat Methods Med Res. 2015;24(1):9–26.
9. Trattnig S. The shift in paradigm to precision medicine in imaging: international initiatives for the promotion of imaging biomarkers. In: Martí-Bonmatí L, Alberich-Bayarri A, editors. Imaging biomarkers. Cham: Springer; 2017.
10. Sack I. Forschungsförderung – BIOQIC – neue generation von bildgebungsspezialisten. RöFo. 2017;188(1):86.

Part I
Biological and Physical Fundamentals

The Fundamentals of Transport in Living Tissues Quantified by Medical Imaging Technologies

2

Sebastian Hirsch, Tobias Schaeffter, and Ingolf Sack

Abstract
Physiology is the science of the mechanical, physical, bioelectrical, and bio-chemical functions of living systems. All physiological processes are based on physical and biochemical principles. Quantitative medical imaging exploits these principles to measure parameters of those processes noninvasively in vivo. Parameters measured by quantitative medical imaging have to be in agreement with values that would be obtained from standardized measurements from physics or material sciences, if these were applicable for living tissues. Technical advancements have led to the emergence of various methods for quantifying bio-physical and constitutive tissue parameters. This chapter focuses on quantitative medical imaging of physiological processes that are related to different types of physical transport mechanisms. More specifically, we will show that continuity of mass and energy can be interpreted as overarching principles that govern seemingly unrelated modes of energy or mass transport. For this, the derived transport equations will be reviewed from the perspective of medical imaging modalities such as magnetic resonance imaging (MRI), positron emission tomography (PET), or ultrasound with a focus on water diffusion, blood perfusion, fluid flow, and mechanical wave propagation.

S. Hirsch
Department of Medical Informatics, Charité – Universitätsmedizin Berlin, Berlin, Germany

T. Schaeffter
Medical Physics and Metrological Information Technology Physikalisch-Technische Bundesanstalt, Berlin, Germany

I. Sack (✉)
Department of Radiology, Charité – Universitätsmedizin Berlin, Berlin, Germany
e-mail: ingolf.sack@charite.de

© Springer International Publishing AG 2018
I. Sack, T. Schaeffter (eds.), *Quantification of Biophysical Parameters in Medical Imaging*, https://doi.org/10.1007/978-3-319-65924-4_2

9

Notation
Attention is paid to keep the mathematical notation as consistent and self-explanatory as possible. The following conventions were used:

- Matrices, tensors, and vectors are denoted by bold, upright Latin or Greek letters: $\mathbf{a} = \mathbf{C} \cdot \mathbf{b}$.
- The dot operator "$\cdot$" is used exclusively to denote the scalar product of two vectors or tensorial expressions leading always to a reduction: $a = \mathbf{b} \cdot \mathbf{c} \equiv \sum_i b_i c_i$.
- Matrix multiplication which preserves or increases the rank of a tensor is denoted by two adjacent tensorial symbols: $\mathbf{a} = \mathbf{bc}$.
- Where appropriate, temporal derivatives are marked by a dot and spatial derivatives by an apostrophe: $\dfrac{\partial f}{\partial t} = \dot{f}$ and $\dfrac{\partial f}{\partial r} = f'$.
- Summation over an index is always specified explicitly by the sum symbol.
- Only indices or exponents that take numerical values are printed in italics; super- or subscripts that serve as a specification are always printed upright: x_i, a^b, and u_{shear}.
- The complex unit is always represented by an upright letter to make it distinguishable from an index: $\sqrt{-1} = i \leftrightarrow u_i, i \in \mathbb{N}$.
- Fourier transform is denoted either by FT or a diacritic symbol: $FT(\mathbf{u}) \equiv \tilde{\mathbf{u}}$.

List of Symbols
The following table lists commonly used mathematical symbols. Note that some symbols are reused in a local context to denote something else, but this will always be explained in the text.

Roman symbols	
$\mathbf{B}$, BF, b	Magnetic field, blood flow, b-value
$\mathbf{C}$, C, c	Elasticity tensor, particle concentration, wave speed
D	Diffusion constant
E	Energy
$\mathbf{G}$, g, G^*	Linear magnetic field gradient, gradient amplitude, complex shear modulus
i, (i)	Imaginary unit, index of incident wave
K	Compression modulus
$\mathbf{m}$, m	Magnetic moment, mass
N	Gaussian distribution function
$\mathbf{P}$	Flux vector
R, (r), $\mathbf{r}$, r	Reflection coefficient, index of reflected wave, position vector, traveled distance
S	MRI signal
T, t, (t)	Transmission coefficient, time, index of transmitted waves

u	Particle deflection
v	Deflection velocity
x, y, z	Spatial coordinates
Greek symbols	
α	Springpot interpolation parameter
δ	Dirac delta distribution, loss angle
ε	Strain tensor
η	Viscosity
φ	Spin phase
γ	Gyromagnetic ratio
κ	Lumped springpot modulus
λ	Perfusion partition coefficient
μ	Shear modulus
Θ	Heaviside function
ρ	Mass density
σ	Stress tensor
ς	Compression viscosity
$\tau, (\tau_\Delta)$	Time delay, application time of MRI gradient (delay between MRI gradients)
ω	Angular frequency
Mathematical operators	
$\nabla, \nabla\times, \nabla\cdot$	Gradient operator (vector), curl, divergence
Δ	Laplacian operator
FT	Fourier transform
1	Identity operator

2.1   Introduction

This chapter pursues two objectives. Firstly, transport phenomena in the human body will be discussed from a physics point of view. Secondly, we will explain how medical imaging—especially magnetic resonance imaging (MRI)—can be used to assess and quantify these transport phenomena.

For the physics part, we follow a route that is different from most textbook discussions of waves, diffusion, and flow, which are based on Newton's law. Instead, we will show that the continuity equation and the associated concept of conservation of mass and energy can be understood as the overarching principles applying to all these transport mechanisms.

Physiology is the science of the mechanical, physical, bioelectrical, and biochemical functions of living systems. Physiology is closely related to anatomy, the study of form, and both are studied in tandem as part of a medical curriculum. Biomedical imaging allows noninvasive assessment of anatomy and physiology, i.e., form and function. Physiology can be assessed in more detail by quantitative imaging techniques when the underlying physical principles are known. In particular, the transport of particles, mass, or energy is fundamental to many living processes. In fact, all vital functions depend on the transport of blood, oxygen, nutrients, metabolites, or electrical signals. All transport mechanisms have in common that they involve a *local density* (or concentration) of a substance (such as nutrients) or energy and *flux*, which describes the motion of the said quantity. The fundamental relationship between density and flux is defined by the continuity equation [1]:

$$\frac{\partial \rho}{\partial t} = -\nabla \cdot \mathbf{P}, \tag{2.1}$$

which states that any influx ($\mathbf{P}$) of energy or mass into a volume has to be balanced by a change in energy or mass density (ρ) within that volume. $\nabla \cdot \mathbf{P}$ denotes the divergence of the flux, which quantifies the local source or sink density of the flux field. ∇ is a vector operator whose components are first-order spatial derivative operators. Equation (2.1) means that mass or energy can neither be created nor destroyed. Furthermore, the continuity equation states that matter is conserved *locally*. This is a strong conservation statement and the foundation of many quantitative imaging techniques, which account for mass and energy within a given volume. $\mathbf{P}$ is the flux vector whose components describe the flow of mass or energy with velocity $\mathbf{v}$ through a surface element of unit area perpendicular to $\mathbf{v}$:

$$\mathbf{P} = \rho\,\mathbf{v}. \tag{2.2}$$

Equations (2.1) and (2.2) state that flow occurs along the negative density gradient, meaning that flow is induced from sources ($\nabla \cdot \mathbf{P} > 0$) to sinks ($\nabla \cdot \mathbf{P} < 0$). Furthermore, according to Eq. (2.1), wherever the divergence of flux ($\nabla \cdot \mathbf{P}$) is distinct from zero, mass or energy density has to change.

Transport phenomena quantified by medical imaging technologies include many scales and various physical principles. Based on the continuity equation (2.1), we will derive the governing (equilibrium) equations and measured quantities for the following transport mechanisms:

- Diffusion
- Energy transport by mechanical waves
- Fluid flow
- Perfusion

The complexity of biological structures naturally leads to overlapping physical representations of these mechanisms. This motivates our intention to review the underlying principles from the perspective of medical imaging, which, however, often means that we must make assumptions and introduce simplifications in order to treat the basic equations in a straightforward manner. The following sections therefore briefly compile the basic equations starting with Eq. (2.1) to facilitate the comparison of the physical concepts of quantitative medical imaging with a focus on mass and energy transport.

2.2   Diffusion

Diffusion is ubiquitous in nature and related to any type of stochastic motion of molecules and particles driven by Brownian motion. Diffusion is temperature dependent, since the average kinetic energy per particle increases with temperature, so that particles can travel farther within a given time span when the temperature is higher [2, 3].

In tissues, many vital transport mechanisms are governed by diffusion; and detecting changes in diffusion can indicate diseases with high sensitivity (see Chap. 17). Therefore, measurement of diffusion of water or tracers by medical imaging modalities has been part of clinical routine for a long time. In our generalized treatment of transport physics, we derive the equilibrium equation for diffusion from the continuity equation using Fick's first law. Fick's law states that, in equilibrium, the steady-state flux of particles occurs along the negative density gradient (or concentration gradient), i.e., from high to low densities, and with a velocity that is proportional to the density gradient:

$$\mathbf{P} = -D\nabla\rho. \tag{2.3}$$

D is called the *diffusion coefficient* and is expressed in the unit of area per unit time, typically mm^2/s in diffusion-weighted MRI. D quantifies how easily a particle can move in the direction of flux vector $\mathbf{P}$ according to Fick's law. Combining Eqs. (2.1) and (2.3) yields

$$\frac{\partial\rho}{\partial t} = D\Delta\rho, \tag{2.4}$$

which constitutes the diffusion equation in three dimensions. Δ is the 3D Laplace operator as obtained by the product of gradient vectors: $\Delta = \nabla\cdot\nabla = \dfrac{\partial^2}{\partial x^2} + \dfrac{\partial^2}{\partial y^2} + \dfrac{\partial^2}{\partial z^2}$.

The 3D diffusion equation is satisfied by a normal Gaussian distribution of probability N of the particles having traveled a distance r [4]:

$$\rho(r,t) \propto N(r,t) = \frac{1}{(4\pi Dt)^{\frac{3}{2}}} e^{-\frac{r^2}{4Dt}}. \tag{2.5}$$

This Gaussian distribution has standard deviation $\sqrt{2Dt}$ and variance $2Dt = \langle\mathbf{r}^2\rangle$. The relationship $t \propto \langle\mathbf{r}^2\rangle$ applies to *regular diffusion*. Equation (2.5) highlights the probabilistic nature of diffusion: at any time t, particle density ρ is distributed over a continuum of radii r. This means that particle velocities and thus kinetic energies are within the whole range of values from zero to infinity (in a nonrelativistic model for the relationship between speed and kinetic energy). Consequently, already at the beginning of the diffusion process, the probability of finding diffusing particles at arbitrarily large distances is larger than zero. As the diffusion process continues, the particle cloud spreads out, and the probability of finding particles at large distances increases, while the concentration at the source decreases. This process is thermodynamically driven by an increase in entropy due to the increasing disorder in the system. Thus, passive diffusion is a spontaneous process and does not require input of energy. Section 7.3 explains how diffusion can be measured by MRI along with classifications of nonregular diffusion processes as detected by MRI in biological tissues.

It is important to note that the term "density" or "concentration," as used in the context of Eqs. (2.3)–(2.5), does not have to be an actual mass density but can rather refer to the "density of specifically labeled particles." The meaning of this statement becomes clearer when analyzing diffusion in a glass of pure water. Ignoring gravitational effects, the density of water is equal everywhere, so that the density gradient in Eq. (2.3) vanishes and thus the flux becomes zero. However, this is not physically correct. The underlying process is called "self-diffusion." It can be understood by assuming that the water molecules in each small volume element are assigned a specific label and that the label is unique for each volume element. In that case, "density" refers to the concentration of water molecules carrying one specific label. All volume elements act as the sources of many separate diffusion processes running in parallel. Initially, all molecules with a specific label are contained in the same volume element. As time passes, through random motion, molecules leave their original volume element and diffuse into other volume elements. Thus, the concentration of molecules with a specific label broadens over time, with the concentration being represented by a Gaussian function as described above (Fig. 2.1). In the long-time limit, all volume elements contain water molecules from all other volume elements, i.e., a uniform mixture of labels. However, since the whole process consists of water molecules moving around, the bulk density remains constant. In the context of diffusion MRI, the labeling of water molecules is achieved by imposing a position-dependent spin phase onto every spin ensemble, as shall be described in Sect. 2.7.3.

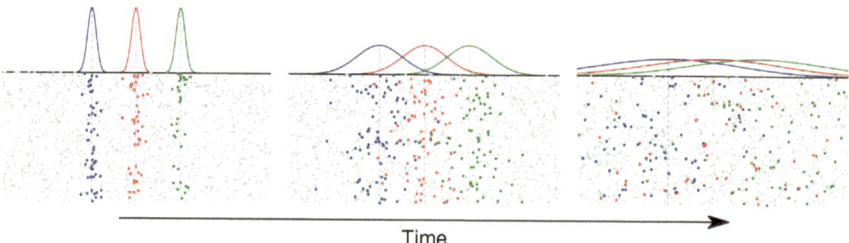

Time

Fig. 2.1 Illustration of one-dimensional self-diffusion. Each dot represents a particle (e.g., a water molecule). Initially, particles in three volume elements are labeled, indicated red, green, and blue. Over time, the particles diffuse along the horizontal axis, as can be seen from the broadening of the corresponding probability distributions. Nevertheless, the bulk density remains unchanged, since unlabeled particles compensate the motion of the labeled particles. The dashed lines indicate the positions of the original labeling

2.3   Wave Transport

A number of imaging modalities including ultrasound, elastography, and photo-acoustic tomography exploit the propagation of classical waves for generating image contrast. Mechanical waves can be understood as the propagation of a distortion of the mechanical equilibrium state through time and space [2]. Wave propagation is associated with transport of energy by local particle deflections, i.e., by local mass displacements around equilibrium position in a static reference coordinate system (see Fig. 2.2). The deformation of an object can be described in terms of the

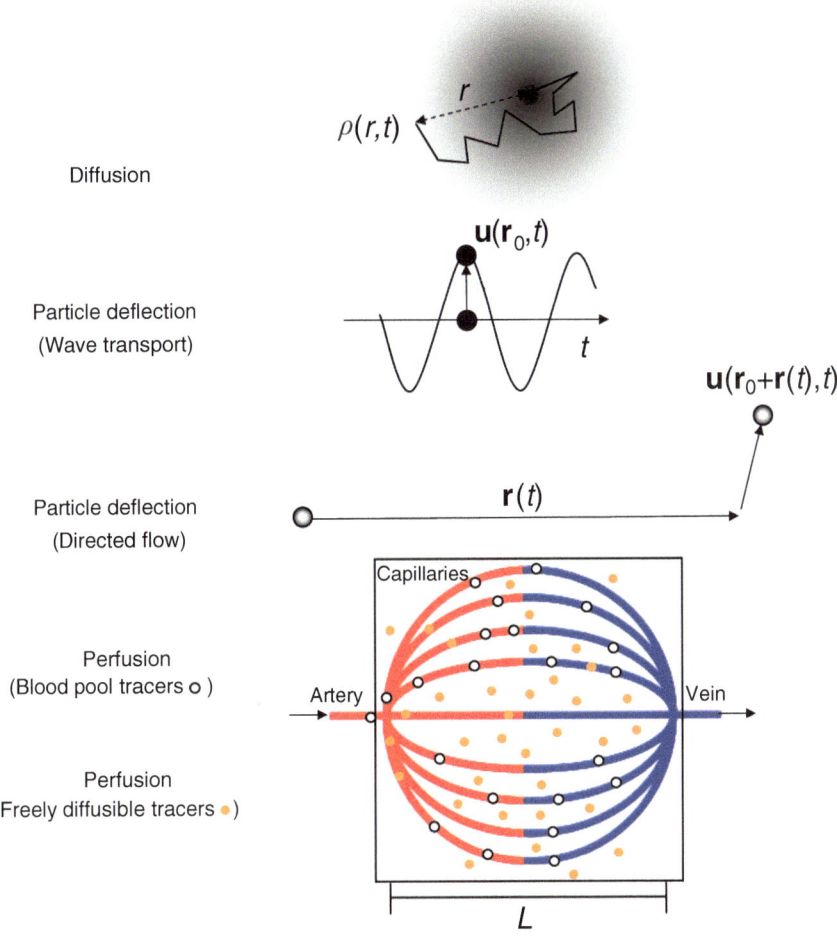

Fig. 2.2 Notation of transport phenomena addressed in this chapter

displacement field **u** in the sense that every point **r** is shifted to a different position **r**+**u**(**r**, t). This displacement field **u** is the dynamic parameter behind ultrasound-based medical imaging or elastography. Before we begin the discussion of mechanical wave propagation and how it is related to the continuity equation, we will first introduce elementary terminology, namely, *stress* and *strain*. More details can be found in standard textbooks of elasticity theory and continuum mechanics such as [5] and [6].

2.3.1 Strain, Stress, and Linear Elasticity Coefficients

Strain describes the elastic deformation a body has undergone upon exertion of a force [6]. Strain is usually measured as displacement **u** relative to the size Δ**r** of the deformed body. A compact tensor notation of **u** over Δ**r** is given by the infinitesimal linear strain tensor ε with components

$$\varepsilon_{ij} = \frac{1}{2}\left(\frac{\partial u_i}{\partial r_j} + \frac{\partial u_j}{\partial r_i}\right) = \frac{1}{2}\left(\nabla\mathbf{u} + \left(\nabla\mathbf{u}\right)^T\right). \tag{2.6}$$

∇**u** is the Jacobian of the displacement field, i.e., the matrix that includes all spatial derivatives, $\left(\nabla\mathbf{u}\right)_{ij} = \frac{\partial u_i}{\partial r_j}$. ε is a symmetric (i.e., $\varepsilon_{ij} = \varepsilon_{ji}$) tensor of rank 2 which can be expressed as a 3×3 matrix. Due to its symmetry, only six of its nine entries are independent. Any strain in a material is caused by a force **F** acting on a surface element A_j. $j \in \{1, 2, 3\}$ denotes the direction of surface normal vectors along the Cartesian coordinates (see Fig. 2.3). Force per area defines stress, i.e.,

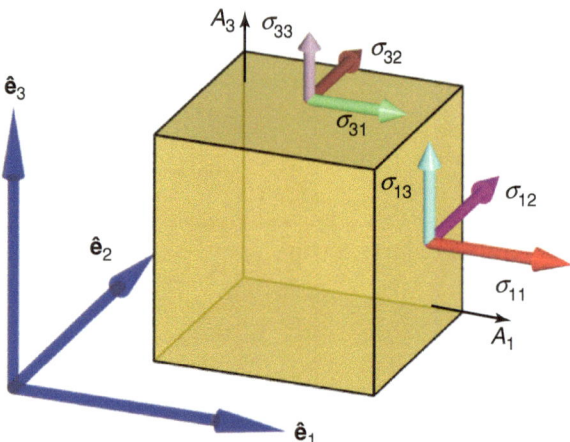

Fig. 2.3 Notation of the stress tensor elements which are defined by the directions of acting forces and surface normals A_j

$$\sigma_{ij} = \frac{F_i}{A_j}, \tag{2.7}$$

which is a tensor of rank 2, analogous to strain. The diagonal components of the stress tensor $\boldsymbol{\sigma}$, σ_{ii}, act orthogonally on the three surfaces A_i. The sum $\sum_i \sigma_{ii}$ is proportional to the negative pressure acting against volumetric changes, i.e., compression or dilatation. The off-diagonal entries σ_{ij}, $i \neq j$, characterize stresses tangential to their respective surfaces. Tangential stresses exert a torque on the volume element, causing shear deformation while preserving volume. If the resultant torque of all stresses does not vanish, the cube will rotate. Therefore, the necessary and sufficient condition for a static (equilibrium) configuration is that all tangential (shear) stress components cancel each other, which is fulfilled if $\sigma_{ij} = \sigma_{ji}$. Thus, the stress tensor possesses the same symmetry as the strain tensor.

The relationship between stress and strain is, in linear approximation, characterized by a constant rank-four tensor $\mathbf{C}$ which contains all the elastic coefficients necessary to describe the stress throughout a material that causes or is caused by a given strain:

$$\boldsymbol{\sigma} = \mathbf{C} \cdot \boldsymbol{\varepsilon} \; (\text{tensor notation}), \tag{2.8}$$

$$\sigma_{ij} = \sum_{k=1}^{3}\sum_{l=1}^{3} C_{ijkl}\varepsilon_{kl} \; (\text{component notation}). \tag{2.9}$$

This is Hooke's law—the most fundamental constitutive law in material science, which accounts for a spring-like elastic response of a solid in the limit of small deformations. Since $\mathbf{C}$ is a fourth-order tensor, it formally has $3^4 = 81$ elements, of which only 21 are independent because of symmetry in the case of the most general anisotropic solid. Applications of Hooke's law to biological materials usually require far fewer coefficients. In a transversely isotropic elastic material (featuring a single plane of isotropy and a principal axis orthogonal to that plane), $\mathbf{C}$ has only five independent coefficients, whereas under perfect isotropy, only two independent coefficients exist [6]. There are several ways to parameterize an isotropic material in terms of pairs of elastic moduli such as shear modulus μ and compression modulus K, Young's modulus E and Poisson's ratio ν or both Lamé coefficients (the second of which again being the shear modulus μ). All combinations are equivalent and can be converted to any other combination. As an example, Hooke's law of an isotropic material expressed in terms of compression modulus K and shear modulus μ is

$$\boldsymbol{\sigma} = \left(K - \frac{2}{3}\mu \right)(\nabla \cdot \mathbf{u})\mathbf{I} + \mu\,\boldsymbol{\varepsilon}, \tag{2.10}$$

where $\mathbf{I}$ is the 3×3 unit matrix and $(\nabla \cdot \mathbf{u})$ signifies volumetric strain. Due to the high water content, soft biological tissues normally have much higher compression moduli than shear moduli, since water has a very high compression modulus of approximately 2.2 GPa but can undergo shear deformation with no resistive response at all (see Table 12.1 in Chap. 12).

2.3.2 Mechanical Waves

The equilibrium equation of classical wave fields $\mathbf{u}$ is obtained from the continuity of dynamic strain energy. Hence, we need to account for kinetic energy density E_{kin} and potential energy density E_{pot} of the displacement field $\mathbf{u}$ in order to assess total energy density $E = E_{kin} + E_{pot}$. Similar to the kinetic energy of a mass point with velocity $\dot{u}$ ($E_{kin} = \frac{1}{2} m \dot{u}^2$), the kinetic energy density of a coherent wave field $\mathbf{u}$ can be expressed as

$$E_{kin} = \frac{1}{2} \rho \dot{\mathbf{u}} \cdot \dot{\mathbf{u}} = \rho \frac{1}{2} \sum_{i=1}^{3} \dot{u}_i^2. \tag{2.11}$$

To proceed with potential energy, we need the stress-strain relation that we introduced in the previous section.

2.3.2.1 From the Continuity of Dynamic Strain Energy to the Navier Equation

Hooke's law is directly linked to the internal strain energy (potential energy) density of an elastically deformed body by [5]

$$E_{pot} = \frac{1}{2} \sigma \cdot \varepsilon = \frac{1}{2} \sum_{i=1}^{3} \sum_{j=1}^{3} \sum_{k=1}^{3} \sum_{l=1}^{3} C_{ijkl} \varepsilon_{ij} \varepsilon_{kl}. \tag{2.12}$$

The total energy density, $E = E_{kin} + E_{pot}$, is the preserved quantity of interest for wave propagation. We therefore formulate the continuity equation (2.1) in terms of energy density:

$$\frac{\partial E}{\partial t} = \frac{\partial \left(E_{kin} + E_{pot} \right)}{\partial t} = -\nabla \cdot \mathbf{P}. \tag{2.13}$$

The left-hand side of Eq. (2.13) can be expanded by inserting E_{kin} from Eq. (2.11) and E_{pot} from Eq. (2.12):

$$\frac{\partial E}{\partial t} = \rho \underbrace{\frac{1}{2} \frac{\partial}{\partial t} \left(\frac{\partial \mathbf{u}}{\partial t} \cdot \frac{\partial \mathbf{u}}{\partial t} \right)}_{\rho \ddot{\mathbf{u}} \cdot \dot{\mathbf{u}}} + \underbrace{\frac{1}{2} \frac{\partial}{\partial t} \left(\mathbf{C} \cdot \varepsilon \cdot \varepsilon \right)}_{\sigma \cdot \dot{\varepsilon}}. \tag{2.14}$$

The right-hand side of Eq. (2.13) denotes the flux of strain energy into a volume element. The flux of strain energy is stress multiplied by deflection velocity as given in [7]

$$\mathbf{P} = -\sigma \cdot \dot{\mathbf{u}}$$
$$\Rightarrow \nabla \cdot \mathbf{P} = \left(\nabla \cdot \sigma \right) \cdot \dot{\mathbf{u}} + \sigma \cdot \nabla \dot{\mathbf{u}}. \tag{2.15}$$

Combining Eq. (2.14) with Eq. (2.15) leads to

$$\rho \ddot{\mathbf{u}} \cdot \dot{\mathbf{u}} + \boldsymbol{\sigma} \cdot \dot{\boldsymbol{\varepsilon}} = \left(\nabla \cdot \boldsymbol{\sigma} \right) \cdot \dot{\mathbf{u}} + \boldsymbol{\sigma} \cdot \nabla \dot{\mathbf{u}}$$

$$\Rightarrow \rho \ddot{\mathbf{u}} \cdot \dot{\mathbf{u}} = \left(\nabla \cdot \boldsymbol{\sigma} \right) \cdot \dot{\mathbf{u}}.$$

(2.16)

In the above equation, we exploited the fact that $\boldsymbol{\sigma} \cdot \left(\nabla \dot{\mathbf{u}} - \dot{\boldsymbol{\varepsilon}} \right) = 0$ due to the tensorial symmetries of $\boldsymbol{\sigma}$ and $\boldsymbol{\varepsilon}$. Equation (2.16) agrees with Newton's second law, which can be expressed as

$$\rho \ddot{\mathbf{u}} = \nabla \cdot \boldsymbol{\sigma}$$

(2.17)

and which also constitutes the equilibrium equation for elastic waves. Equation (2.17) can be expanded for isotropic materials in terms of compression modulus K and shear modulus μ as defined in Eq. (2.10):

$$\rho \ddot{\mathbf{u}} = \left(K + \frac{4}{3} \mu \right) \nabla \left(\nabla \cdot \mathbf{u} \right) - \mu \nabla \times \left(\nabla \times \mathbf{u} \right).$$

(2.18)

In the literature, Eq. (2.18) is often referred to as Navier equation for isotropic, homogeneous solids. It has the characteristic form of a wave equation, relating a second-order temporal derivative with second-order spatial derivatives. For simplicity we neglected all spatial variations of coefficients K and μ, leading us to the assumption of zero spatial derivatives of the elastic coefficients. This *assumption of local homogeneity* is applied in many treatments of the wave equation in wave-inversion-based imaging modalities. Equation (2.18) illustrates that the full displacement field is a superposition of two separate fields: a compression field ($\nabla \cdot \mathbf{u}$) and a shear field ($\nabla \times \mathbf{u}$). Each term represents decoupled and independent types of motion, which can be separated into two independent wave equations by virtue of Helmholtz decomposition (see Chap. 4, Eqs. (4.8)–(4.10).) [5]. Applying the divergence operator or curl operator to Eq. (2.18) yields two separate equations for compression waves and shear waves, respectively:

$$\rho \left(\nabla \cdot \ddot{\mathbf{u}} \right) = \left(K + \frac{4}{3} \mu \right) \Delta \left(\nabla \cdot \mathbf{u} \right),$$

(2.19)

$$\rho \left(\nabla \times \ddot{\mathbf{u}} \right) = \mu \; \Delta \left(\nabla \times \mathbf{u} \right).$$

(2.20)

Equation (2.19) is the compression wave equation, which is a scalar equation for waves polarized parallel to the propagation direction. Equation (2.20) is the shear wave equation for waves with polarization perpendicular to the direction of travel. Both equations are satisfied by a plane wave

$$\mathbf{u} \left(\mathbf{r}, t \right) = \mathbf{u}_0 e^{i \omega \left(\frac{1}{c} \mathbf{n} \cdot \mathbf{r} - t \right)}$$

(2.21)

with $\mathbf{n}$ being the wave normal vector. c denotes the wave speed, which is

$$c = \sqrt{\frac{K + \frac{4}{3}\mu}{\rho}} \quad \text{for compression waves and} \qquad (2.22)$$

$$c = \sqrt{\frac{\mu}{\rho}} \quad \text{for shear waves.} \qquad (2.23)$$

Elastography measures the shear modulus of soft tissues by stimulating shear waves and using medical imaging such as ultrasound [8] or MRI [9]: to detect them. The investigated tissues are often considered incompressible, since compression modulus K is several orders of magnitude larger than shear modulus μ. However, true incompressibility implies both vanishing volumetric strain ($\nabla \cdot \mathbf{u} \to 0$) and infinitely high compression modulus ($K \to \infty$). Even though in incompressible media neither of the two parameters alone can be measured with sufficient precision, their product is finite and is identified as isotropic pressure:

$$p = -K(\nabla \cdot \mathbf{u}) = -\frac{1}{3}(\sigma_{11} + \sigma_{22} + \sigma_{33}). \qquad (2.24)$$

Note that, due to the high content of water in biological soft tissues (>75%), the compression modulus does not vary much when the material behaves monophasically. Monophasic properties imply that all constituents of the tissue, including solid and fluid compartments, move in synchrony as one field. Biological soft tissues normally behave monophasically at high stimulation frequencies as in sonography (on the order of MHz). For this reason, the compression modulus measured by ultrasound can be approximated by the speed of sound in water (~1500 m/s), which corresponds to a compression wavelength of 1.5 mm at 1 MHz frequency. On the other hand, the frequency range of elastography, which is between 10 and 100 Hz in clinical applications, results in compression wavelengths on the order of 15–150 m, entailing a pressure gradient which does not vary in space, that is, $\nabla p = (K + 4/3\mu)\nabla(\nabla \cdot \mathbf{u}) = const$. Hence, Eq. (2.18) can be simplified for incompressible media to

$$\rho\ddot{\mathbf{u}} = \mu\,\Delta\mathbf{u} + const \qquad (2.25)$$

by using the vector calculus identity $\Delta\mathbf{u} = \nabla(\nabla \cdot \mathbf{u}) - \nabla \times (\nabla \times \mathbf{u})$. Time-harmonic elastography usually solves Eq. (2.25) for shear modulus μ by direct inversion, phase gradient methods, or local frequency estimation [10] after suppressing the offset pressure term using spatial filters or the curl operator as in Eq. (2.20) [11]. In contrast, without assuming a constant pressure gradient, multiparameter recovery can yield both Lamé coefficients by either variational approaches [12–14] or direct inversion [15, 16].

Poroelastography explicitly aims at quantifying tissue pressure based on the hypothesis that the compression modulus in multiphasic media at low stimulation frequencies is much lower than predicted by ultrasound. A good example is the brain: the speed of sound (compression waves) in ultrasound imaging of brain matter is, as in any other soft tissues, on the order of 1500 m/s, corresponding to a compression modulus of 2.2 GPa. In contrast, arterial pulsation causes cyclic brain expansion to an order of tens of milliliters, which can be observed by MRI as movement of tissue boundaries. The estimated compression modulus of the brain in this quasi-static dynamic range is on the order of only 26 kPa [17] highlighting the relevance of multiphasic models for understanding tissue compression in the static and quasi-static limit. Further details and applications are discussed in Chaps. 4 and 20.

2.4   Wave Scattering and Diffusive Waves

So far we have considered only unscattered plane waves which propagate along a straight line between two points. However, in heterogeneous media such as biological tissues, waves are typically scattered at elastic discontinuities, giving rise to a "smeared" wave intensity within the region through which the wave has passed. When traversing a scattering medium, the propagating wave front continuously loses amplitude since, at each single scattering event, wave intensity is split between the transmitted and reflected waves. Moreover, wave coherence in a global ensemble of stochastically scattered small wave packages is lost, causing a decrease in average wave amplitude. The most fundamental principle of wave scattering is that a wave changes direction when it hits an interface. This means that part of the wave is reflected while another part is transmitted through the interface. The two parts of the wave travel with different amplitudes into different directions. The change in amplitudes of scattered waves in 1D is derived in the following.

2.4.1   Wave Scattering at Planar Interfaces

The most basic wave scattering scenario is a plane incident wave orthogonally hitting an interface between two media with different elastic properties. In this scenario, the normal vector of the interface is parallel to the wave normal vector $\mathbf{n}$, and the polarization of the incident wave is preserved for the reflected and transmitted waves. We will therefore ignore the polarization vector in Eq. (2.21) and instead discuss a scalar 1D plane wave of amplitude u_0. Coordinate r identifies the position along the propagation direction of the wave. The elastic discontinuity is located at $r=0$, dividing the medium into two compartments, compartment 1 for $r<0$ and compartment 2 for $r>0$. Incident and reflected waves traverse 1, while the transmitted wave propagates through 2 (Fig. 2.4). We first define the transmission and reflection coefficients, T and R, as the amplitude of the transmitted ($u_{0(t)}$) and reflected ($u_{0(r)}$) wave at the boundary, normalized to the incident wave amplitude ($u_{0(i)}$):

$$T = \frac{u_{0(t)}}{u_{0(i)}}, \tag{2.26}$$

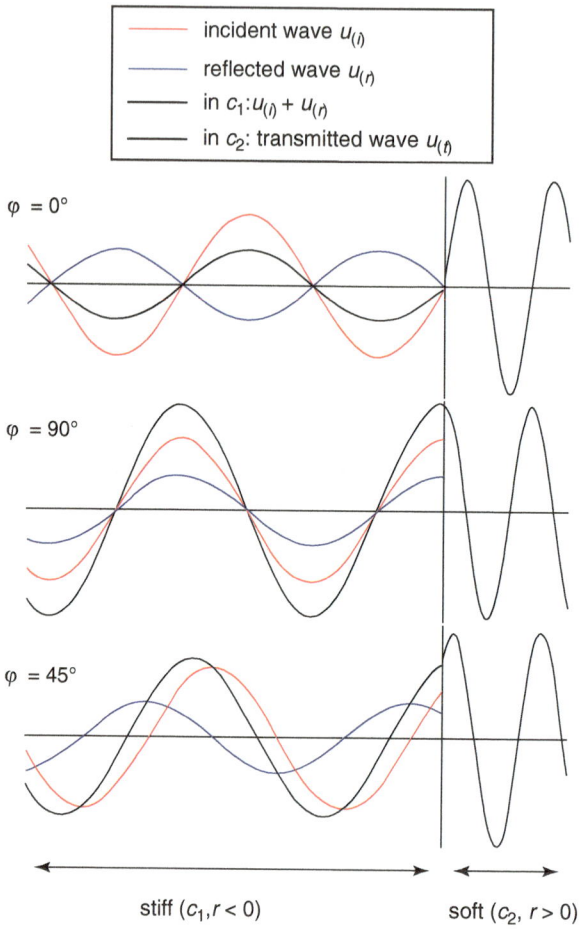

Fig. 2.4 1D wave scattering at an elastic interface at $r = 0$. The wave amplitudes on both sides of the interface are considered to be equal as implied by the boundary condition of Eq. (2.28) (nonslip condition). Shown are three different phases ϕ of the waves with respect to $r = 0$

$$R = \frac{u_{0(r)}}{u_{0(i)}}. \tag{2.27}$$

We assume nonslip conditions, i.e., 1 and 2 are in close contact, so that wave amplitudes in both compartments are equal at $r = 0$, which yields

$$u_{0(i)} + u_{0(r)} = u_{0(t)}$$
$$T - R = 1. \tag{2.28}$$

Furthermore, conservation of energy requires a continuous flux of wave energy through the interface in the steady state. Therefore, energy inflow equals energy outflow or

$$\left\langle \mathbf{P}_{(i)} \right\rangle = \left\langle \mathbf{P}_{(r)} \right\rangle + \left\langle \mathbf{P}_{(t)} \right\rangle. \tag{2.29}$$

$\left\langle \mathbf{P}_{(i)} \right\rangle$, $\left\langle \mathbf{P}_{(r)} \right\rangle$, and $\left\langle \mathbf{P}_{(t)} \right\rangle$ denote the time average of harmonic steady-state energy fluxes through the interface of incident, reflected, and transmitted waves, respectively. The flux vectors are obtained by combining Eqs. (2.15) and (2.21), such that

$$\left\langle P \right\rangle = \rho c^2 \frac{\partial u}{\partial r} \frac{\partial u}{\partial t} \quad \text{with} \quad u = \mathrm{Re}\left[u_0 e^{i\omega\left(\frac{1}{c}r - t\right)} \right]. \tag{2.30}$$

Since energy is a scalar, real-valued quantity, we accounted for the real part of the complex wave (Re) hitting the interface at $r = 0$ (thus $\left\langle P \right\rangle$ denotes here the scalar flux amplitude). From the above equation, one obtains, for the average energy of each wave component,

$$\left\langle P_{(i)} \right\rangle = \frac{1}{2} \rho_1 c_1 u_{0(i)}^2, \tag{2.31}$$

$$\left\langle P_{(r)} \right\rangle = \frac{1}{2} \rho_1 c_1 u_{0(r)}^2, \tag{2.32}$$

$$\left\langle P_{(t)} \right\rangle = \frac{1}{2} \rho_2 c_2 u_{0(t)}^2, \tag{2.33}$$

with indices 1 and 2 for the respective compartments. Combining Eqs. (2.28)–(2.33) and solving for T and R results in the well-known scattering amplitude coefficients in 1D for perpendicular incidence:

$$T = \frac{2 c_1 \rho_1}{c_1 \rho_1 + c_2 \rho_2} \quad \left(\text{transmitted wave amplitude}\right), \tag{2.34}$$

$$R = \frac{c_1 \rho_1 - c_2 \rho_2}{c_1 \rho_1 + c_2 \rho_2} \quad \left(\text{reflected wave amplitude}\right). \tag{2.35}$$

When multiple interfaces (scatterers) are present and $R \neq 0$, wave amplitudes behind the wave front become speckled through constructive and destructive interferences of incident and reflected waves. In 2D and 3D, T and R depend on the direction cosines of incident and transmitted waves relative to the interface, rendering the amplitudes of the transmitted and reflected wave front direction dependent [18].

2.4.2 Stochastic Wave Scattering

Turning back to biological tissues characterized by a stochastic distribution of elastic discontinuities, we will further consider the superposition of many wave packages with individual effective propagation path lengths r. In this model, we assume a stochastic distribution of small scattering inclusions. If scatterer density is high enough, every voxel will contain a superposition of multiply scattered and transmitted partial waves with no well-defined phase relation between them. In other words, the wave amplitude in every voxel is defined by the interference of several waves with random amplitudes and phases.

Similar to diffusion, r would then represent the effective distance a wave package has reached. Hence, we find a distribution of effective wave speeds c defined by the effective distance r divided by waiting time t. However, contrary to diffusion, there is a maximum c-value which corresponds to the wave speed of the unscattered wave [11]. A wave front is defined by the most advanced wave components, since—as previously said—any scattering event is a deflection from a straight line of propagation and hence the effective distance the wave has traveled from the source is reduced, corresponding to a reduced effective wave speed. In other words, the wave front contains only the components that were transmitted at each single scattering event with relative amplitudes T. Assuming a uniform distribution of solid scatterers throughout the sample and thus a constant time interval τ_s between two scattering events, we can define the decrease in amplitude $u_0(t)$ by

$$u_0\left(t+\tau_s\right) = T\ u_0\left(t\right). \tag{2.36}$$

For sufficiently small intervals τ_s, one can represent $u_0(t+\tau_s)$ by the first order of the Taylor series around t:

$$u_0\left(t+\tau_s\right) \approx u_0\left(t\right) + \dot{u}_0\left(t\right)\ \tau_s. \tag{2.37}$$

Combining the right-hand sides of these two equations yields

$$\dot{u}_0\left(t\right) = \frac{T-1}{\tau_s} u_0\left(t\right)$$

$$\Rightarrow u_0\left(t\right) = u_0\left(t=0\right)\ e^{\frac{1-T}{\tau_s}t}. \tag{2.38}$$

The unscattered wave front therefore decays exponentially with time at a rate of $-(1-T)/\tau_s = R/\tau_s$ and, since propagation velocity is constant, exponentially with the distance from the source. $u_0(t=0)$ is the wave amplitude at the source, which is switched on at $t=0$. Note that the wave amplitude of a wave front decreases even in nonabsorbing materials. However, the intensity of scattered waves is not lost but concentrated in the area "behind" the wave front. In this region, superposition of multiply scattered waves occurs, giving rise to constructive and destructive interferences, which appear in ultrasound images as speckles. Overall wave intensity (energy) in our object is not affected by elastic scattering as long as no viscous damping occurs. Viscous damping, similar to diffusion, is a spontaneous irreversible process characterized by an increase in entropy. In contrast, elastic wave transport is reversible, which is reflected by the symmetry of the wave equation (second-order derivatives in time and space). As a consequence, time reversal of propagating waves is feasible and has already been used in medical imaging and therapeutic applications of ultrasound [19]. In contrast, photoacoustic tomography as detailed in Chap. 13 basically relies on optical energy absorption but is influenced by optical scattering processes. Despite the principal differences between diffusion equation and wave equation, both can be mathematically combined to obtain a very compact and elegant description of dispersive wave transport as outlined in the following section.

2.4.3   Wave Diffusion and Scale-Free Viscoelastic Properties

The diffusion equation (2.4) and the wave equation as given in Eq. (2.19) or Eq. (2.20) can be regarded as just two special cases of a more general description of wave diffusion. An equilibrium equation which fulfills the two limits of diffusion and unattenuated waves is obtained utilizing a fractional derivative operator α:

$$\rho \frac{\partial^{2-\alpha} \mathbf{u}}{\partial t^{2-\alpha}} = \kappa \Delta \mathbf{u}, \text{ with } 0 < \alpha \le 1. \tag{2.39}$$

$\alpha = 1$ and $\kappa = \rho D$ apply for pure diffusive particle motion, while $\alpha = 0$ and $\kappa = \rho c^2$ for unattenuated waves. A general treatment of fractional derivative operators is beyond the scope of this review. We instead refer to the literature of fractional calculus for general problems in diffusion and viscoelasticity such as [20]. For harmonic functions, the Weyl definition of the fractional derivative operator yields the following linear relationship between the Fourier transform (FT) and derivatives [21]:

$$FT\left[\frac{\partial^{\alpha} f(t)}{\partial t^{\alpha}}\right] = (i\omega)^{\alpha} FT\left[f(t)\right]. \tag{2.40}$$

Thus, for time-harmonic waves in the frequency domain $\tilde{\mathbf{u}}(\omega)$, Eq. (2.39), the following wave equation is obtained:

$$\rho FT\left[\frac{\partial^{2-\alpha} \mathbf{u}(t)}{\partial t^{2-\alpha}}\right] \Rightarrow \rho (i\omega)^{2-\alpha} \tilde{\mathbf{u}}(\omega) = -\frac{\rho \omega^2}{(i\omega)^{\alpha}} \tilde{\mathbf{u}}(\omega). \tag{2.41}$$

Correspondingly, the Laplacian of a time-harmonic wave as defined in Eq. (2.21) gives

$$\kappa \ FT\big[\Delta \mathbf{u}(t)\big] \Rightarrow -\kappa \frac{\omega^2}{c^2}\tilde{\mathbf{u}}(\omega). \tag{2.42}$$

Combining Eqs. (2.41) and (2.42) yields

$$\rho c^2 = \kappa\left(\mathrm{i}\omega\right)^{\alpha} \equiv G^*(\omega). \tag{2.43}$$

We here defined the complex-valued modulus $G^*(\omega)$, the real part of which is the storage modulus that governs the elastic properties of a material, while the imaginary part is the loss modulus that quantifies viscous properties. Equation (2.43) illustrates that transition from non-damped waves to diffusive waves is associated with loss of wave energy due to loss of phase coherence, which is equivalent to viscous attenuation as usually described in viscoelastic theory by viscoelastic models based on springs and dashpots. The combination of the two as defined in Eq. (2.43) is the *springpot*. Notably, the springpot is a power law with the same exponent α for both the real and imaginary parts of $G^*(\omega)$, meaning that the ratio between loss and storage properties is constant over frequency. If a material shows a constant ratio between viscous and elastic properties over a wide range of frequencies, the material is considered to be scale-free, meaning that viscoelastic properties measured at the macroscopic scale of typical medical imaging resolutions can be directly translated into much smaller dimensions, such as the cellular scale, by means of a hierarchical network with only two parameters κ and α [22]. The springpot predicts a constant loss angle δ, which quantifies the ratio between viscous and elastic properties:

$$\delta = \arctan\left(\frac{\mathrm{Im}\big[G^*(\omega)\big]}{\mathrm{Re}\big[G^*(\omega)\big]}\right) = \alpha \frac{\pi}{2}. \tag{2.44}$$

Note that κ in Eq. (2.43) has a cumbersome dimension, which depends on the value of α. Therefore, in the literature, κ is usually converted to an elastic modulus by $\kappa = \mu^{1-\alpha}\eta^{\alpha}$ comprising a shear modulus μ and shear viscosity η. To derive μ from κ requires assumptions on η. In the literature, η is often set to unity, to 3.7 Pa·s for brain tissue, or 7.3 Pa·s for liver tissue [23]. Springpot-based viscoelastic dispersion functions are shown in comparison to other two-parameter models of viscoelasticity in Fig. 2.5. Experimental data demonstrate that the springpot is widely applicable to describe viscoelastic properties of biological soft matter [22, 23]. It is an intriguing observation that we arrived at the springpot-related properties of biological tissues starting with the diffusion-wave equation (Eq. (2.39)). This further highlights that a mixed description of coherent and incoherent wave transport phenomena applies to biological soft tissues.

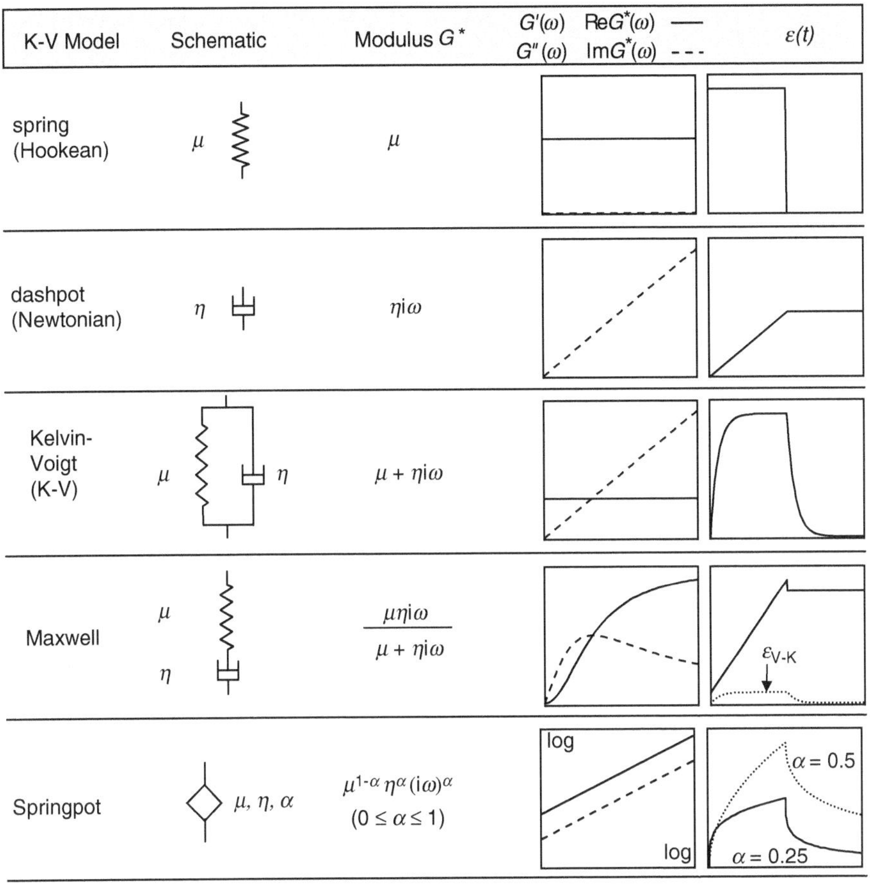

K-V Model	Schematic	Modulus G^*	$G'(\omega)$ $\mathrm{Re}G^*(\omega)$ — $G''(\omega)$ $\mathrm{Im}G^*(\omega)$ - - -	$\varepsilon(t)$
spring (Hookean)	μ	μ		
dashpot (Newtonian)	η	$\eta i\omega$		
Kelvin-Voigt (K-V)	μ η	$\mu + \eta i\omega$		
Maxwell	μ η	$\dfrac{\mu\eta i\omega}{\mu + \eta i\omega}$		$\varepsilon_{\text{V-K}}$
Springpot	μ, η, α	$\mu^{1-\alpha}\eta^{\alpha}(i\omega)^{\alpha}$ $(0 \leq \alpha \leq 1)$	log ... log	$\alpha = 0.5$ $\alpha = 0.25$

Fig. 2.5 Basic elements, spring and dashpot, which model the elastic and viscous response of a material and are used to assemble two-parameter viscoelastic models as sketched in the second column. Complex modulus G^* for the shown models is analytically given in the third column and schematically plotted in the fourth column over angular frequency (G' and G'' denote the storage and loss modulus, respectively). "log" refers to double logarithmic plots to better illustrate the power law behavior of the springpot. The rightmost column shows strain $\varepsilon(t)$ on a time axis in response to a boxcar stress spanning the first half of the time axis

2.5   Flow-Related Transport

Flow comprises many length scales in living tissue, ranging from large-distance-directed blood flow to microperfusion through stochastically distributed capillaries, where flow is similar to diffusive motion. The mathematical description of flow phenomena is similar to that of waves; however, since flow involves transport of mass rather than energy, convection has to be taken into account. Most fluids cannot store shear strain energy due to their property of moving along an acting stress without generating a resistive force. Therefore, shear stress in fluids exists only in its viscous form, which is proportional to strain rate. In the previous section, we already introduced viscoelasticity based on the two-parameter springpot model. The two other models with only two independent parameters are the Voigt and Maxwell models, which represent parallel and serial arrangements of a spring- and a dashpot, respectively [11]. In the Voigt model, total stress is the sum of two partial stresses corresponding to spring and dashpot, whereas in the Maxwell model, total strain is the sum of the two partial strains. A simple extension of Hooke's law given in Eq. (2.8) to include a viscosity tensor is

$$\sigma = \mathbf{C}_{\text{elastic}} \cdot \boldsymbol{\varepsilon} + \mathbf{C}_{\text{viscous}} \cdot \dot{\boldsymbol{\varepsilon}}. \tag{2.45}$$

$\dot{\boldsymbol{\varepsilon}}$ is the strain-rate tensor. As stated before, since fluids do not respond elastically to shear deformation, all shear-related entries of $\mathbf{C}_{\text{elastic}}$ are zero. However, it is important to note that even purely viscous materials with respect to shear can still respond elastically to compression, resulting in an isotropic pressure p. Furthermore, viscosity induces a rate-dependent resistance to shear deformation, even if there is no elastic response. We therefore reduce the elastic stress to its compression elements ($\nabla \cdot \mathbf{u}$), while the full viscous stress tensor is retained [24]:

$$\sigma = \left[\left(K - \frac{2}{3}\mu \right)(\nabla \cdot \mathbf{u}) + \left(\varsigma - \frac{2}{3}\eta \right)(\nabla \cdot \dot{\mathbf{u}}) \right] \mathbf{I} + \eta \dot{\boldsymbol{\varepsilon}}. \tag{2.46}$$

ς denotes compression viscosity, whereas η is shear viscosity (both in units Pa·s). By addressing motion of incompressible fluids such as blood, we can collapse both divergence terms in Eq. (2.46) into a single pressure parameter similar to pressure in incompressible elastic solids:

$$\sigma = -p\mathbf{I} + \eta \dot{\boldsymbol{\varepsilon}}. \tag{2.47}$$

Here, p is a lumped pressure parameter that combines volumetric strain and volumetric strain rate. Both volumetric strain terms are evanescent in the limit of incompressibility, while compression modulus K and compression viscosity ς become infinitely large, resulting in p as a nonkinetic stress quantity which depends neither explicitly on strain nor strain rate. However, since p can vary in space, this parameter remains part of the equilibrium equation as shown below.

So far, we have accounted for viscous damping and have eliminated the shear strain from $\boldsymbol{\sigma}$. A further distinction between solid and fluid motion is the coordinate system relative to which displacements are measured. In a solid medium, each particle is very much restricted in its range of motion and can only undergo very small deflections from its equilibrium position. The strain field is therefore a measure of the deflection of each mass point, as seen from the static coordinate system that defines the equilibrium state. This point of view is also referred to as the *Eulerian* description. In fluids, on the other hand, mass can be deflected by an arbitrary amplitude from its original position due to the lack of a restoring force. Particle deflections are therefore not small but can accumulate in an unconstrained fashion over time. It is therefore more common to use the *Lagrangian* description, which moves along the trajectory $\mathbf{r}(t)$ of the flowing medium [25]. Consequently, a fluid displacement field $\mathbf{u}$ can be parameterized by $\mathbf{u}(\mathbf{r}(t),t)$, which is the motion around equilibrium position plus motion of the coordinate system. In this case, the temporal derivative has to be calculated as an absolute derivative (also called *material derivative*) [1]:

$$\frac{d\mathbf{u}}{dt} = \frac{\partial \mathbf{u}}{\partial t} + \frac{\partial \mathbf{r}}{\partial t} \cdot \nabla \mathbf{u} = \left(\frac{\partial}{\partial t} + \mathbf{v} \cdot \nabla \right) \mathbf{u}. \tag{2.48}$$

$\mathbf{v} = \dfrac{\partial \mathbf{r}}{\partial t}$ denotes the velocity of the material's coordinate system in which any local property change occurs. In contrast to the Eulerian description that only quantifies displacement from a reference position, as caused by propagating (pressure) waves, Eq. (2.48) incorporates the flow-related aspect of convection, which comprises the trajectory of the flow as well as boundary effects, such as the acceleration of flow in a funnel. The parameter measured in flow imaging is flow velocity $\mathbf{v}$. The equilibrium equation of a flow field $\mathbf{v}$ is obtained from the continuity equation (2.1) similar to Newton's second law, which, above, led us to Navier equation (2.16). Combining the material derivative from Eq. (2.48) with Eq. (2.16) leads to

$$\rho \left(\dot{\mathbf{v}} + \mathbf{v} \cdot \nabla \mathbf{v} \right) = \nabla \cdot \sigma. \tag{2.49}$$

The Navier-Stokes equation for incompressible fluids is readily obtained by inserting fluid stress tensor σ of Eq. (2.47) into Eq. (2.49):

$$\rho \left(\dot{\mathbf{v}} + \mathbf{v} \cdot \nabla \mathbf{v} \right) = -\nabla p + \eta \Delta \mathbf{v}. \tag{2.50}$$

The Navier-Stokes equation plays a central role in medical imaging of fluid dynamics since it fully describes the evolution of flow stream lines depending on the geometry and elasticity of the vessel pipelines as well as the distribution of pressure within the cardiovascular system. Flow imaging by Doppler ultrasound is of utmost importance for the clinical diagnosis of vascular dysfunctions in many organs [26]. Phase-contrast (PC) MRI has extended the knowledge of blood flow in the cardiovascular system in health and disease [27]. The strength of flow PC-MRI is to provide three-dimensional vector fields of $\mathbf{v}$ (and acceleration terms $\dot{\mathbf{v}}$) with good spatiotemporal resolution, enabling researchers to model flow by implementing different boundary conditions into the Navier-Stokes equation and to solve it by numerical methods. There is no general solution to the Navier-Stokes equation due to the convective term $(\mathbf{v} \cdot \nabla \mathbf{v})$ on the left-hand side of Eq. (2.50), which makes it nonlinear and difficult to handle without a priori assumptions on boundary values. As an example, if an incompressible fluid flowing through a thick pipe is funneled into a pipe of smaller diameter, this can only be achieved if the fluid is accelerated. Hence, there is an acceleration term that is caused by the geometric properties of the boundaries, which cannot be derived from the acting forces alone. In contrast, waves can be focused or defocused (e.g., by the source geometry) without affecting their propagation velocity but only their amplitudes. Further details on the numerical treatment of the Navier-Stokes equation for cardiovascular MRI as well as PC-MRI experiments are provided in Chaps. 3 and 9.

2.6 Perfusion

Perfusion is blood flow at the capillary level and is closely related to the delivery of oxygen and other nutrients to the tissue [28]. Hence, perfusion is linked to the metabolism of the tissue, which can change tremendously in the presence of diseases such as tumors or ischemia. For this reason, perfusion has been used as a clinical imaging marker in CT, MRI, sonography, PET, and SPECT for a long time. Perfusion can be measured using either exogenously administrated tracers or intrinsically labeled particles based on their change in concentration within the targeted tissue. This means that perfusion measurement can be seen as the measurement of the concentration $C(t, \mathbf{r})$ of any kind of labeled particles over time. Common labeling methods include radioactive tracers (PET, SPECT), magnetically labeled proton spins (arterial spin labeling MRI, ASL-MRI), paramagnetic contrast agents (dynamic susceptibility contrast MRI), radiation-absorbing agents (CT), or contrast agents based on gas bubbles (sonography). During the first phase of particle inflow, the tracer concentration rises with arrival in the arteries, which is followed by a second phase during which the tracer concentration decreases due to washout. Mechanisms for tracer washout include diffusion from the vascular bed into the tissue, venous drainage, and a combination of both. Perfusion falls into the range between flow and diffusion: on the one hand, tracers are delivered by the instreaming blood, which is a coherent transport mechanism. On the other hand, random distribution of the transport pipelines or diffusion of the tracers from the blood into the tissue can result in a diffusive decay of the measured signals. Hence, we have to account for both tracer density ρ and blood velocity $\mathbf{v}$ variation in space and time, resulting in the perfusion flux vector

$$\mathbf{P}(\mathbf{r},t) = \rho(\mathbf{r},t)\mathbf{v}(\mathbf{r},t). \tag{2.51}$$

The divergence of the flux in the continuity equation (2.1) readily follows as

$$\frac{\partial \rho}{\partial t} = -(\nabla \rho) \cdot \mathbf{v} - \rho \nabla \cdot \mathbf{v}. \tag{2.52}$$

We note that the divergence of the flow field in the blood is nearly zero ($\nabla \cdot \mathbf{v} = 0$) due to incompressibility as discussed above.

For the following discussion, we assume that uptake of nutrients and tracers occurs only at the smallest scale of vascular hierarchy, marking the transition from the arterial to the venous tree. We denote the "thickness" of this zone by L (see Fig. 2.2) and approximate the concentration gradient with a finite difference. We also substitute density ρ in Eq. (2.52) with a concentration $C(t, \mathbf{r})$ [29]:

$$\frac{\partial C}{\partial t} = -\nabla C \cdot \mathbf{v} = \frac{C_a - C_v}{L} \cdot \frac{L}{\Delta t}. \tag{2.53}$$

Here, the spatial gradient of $C(t, \mathbf{r})$ is approximated by the difference between local concentration of tracers in arteries and veins, $C_a(t)$ and $C_v(t)$, respectively; hence $C(t, \mathbf{r}) \Rightarrow C(t)$. This implicitly assumes that the concentration predominantly changes along the direction of the path from arteries to veins. Δt is the time that the tracer requires to transit the exchange zone. Unlike flow imaging, which quantifies blood flow velocities, perfusion usually measures blood flow (BF) as a scalar parameter, which is blood volume (ml) per transport time (minutes) normalized to tissue mass (100 g). However, while the decay rate ($1/\tau$) of the tracer signal is the actual measure of perfusion, the blood volume per tissue mass is normally not known. Instead, assumptions have to be made to calibrate the initial rate of perfusion to BF [30]. τ is identified as the mean transit time, which is the time a certain volume of blood (BV) spends in the capillary circulation, i.e., $\tau = BV/BF$. Using τ we can rewrite Eq. (2.53) to obtain the equilibrium equation of perfusion:

$$\frac{d\bar{C}(t)}{dt} = \frac{1}{\tau}\left(C_a(t) - C_v(t)\right). \tag{2.54}$$

$\bar{C}(t)$ is the tracer concentration averaged over the exchange region. A common assumption is $C_v(t) = \lambda \bar{C}(t)$, with $\lambda < 1$ [29]. Partition coefficient λ quantifies the retention of a tracer by the tissue. Delay of washout can be effected by various mechanisms including diffusion of the tracers from the blood into the tissue (diffusible tracers) or by circulation inside the blood pool (blood pool agent). The differential equation

$$\frac{d\bar{C}(t)}{dt} = \frac{1}{\tau}\left(C_a(t) - \lambda \bar{C}(t)\right) \tag{2.55}$$

is solved with

$$\bar{C}(t) = \left(\int \frac{C_a(t)}{\tau} e^{\frac{t}{\lambda\tau}} dt + A\right) e^{-\frac{t}{\lambda\tau}}. \tag{2.56}$$

Further assumptions about the arterial input function $C_a(t)$ are necessary to solve the integral and to specify A. The scenario of a transient tracer bolus that arrives at $t = t_0$ with concentration C_0 can be modeled by a Dirac function

$$C_a(t) = C_0 \delta(t - t_0). \tag{2.57}$$

This, combined with the initial condition of $\bar{C}(t = 0) = 0$, yields an exponential decay of tracer particle concentration over time with the perfusion rate $1/(\tau\lambda)$,

$$\bar{C}(t) = \frac{C_0}{\tau} \Theta(t - t_0) e^{-\frac{t}{\tau\lambda}} \text{ with } \Theta \rightarrow \begin{cases} 0: & t < t_0 \\ 1: & t \geq t_0 \end{cases}, \tag{2.58}$$

Table 2.1 Diffusion—perfusion—flow—waves: mass and energy transport quantified by medical imaging modalities

Transport mechanism	Continuity equation	Flux	Equilibrium equation	Solution
Diffusion	$\dot{\rho} = -\nabla \cdot \mathbf{P}$	$\mathbf{P} = -D\nabla\rho$	$\dot{\rho} = D\Delta\rho$	$N(r,t) = \dfrac{1}{(4\pi D t)^{\frac{3}{2}}} e^{-\frac{r^2}{4Dt}}$
Elastic waves	$\dot{E} = -\nabla \cdot \mathbf{P}$	$\mathbf{P} = -\sigma \cdot \dot{\mathbf{u}}$ $\sigma = \mathbf{C}_{\text{elastic}} \cdot \boldsymbol{\varepsilon}$	$\rho\ddot{\mathbf{u}} = \left(K + \dfrac{4}{3}\mu\right)\nabla(\nabla \cdot \mathbf{u})$ $- \mu\nabla \times (\nabla \times \mathbf{u})$	$\mathbf{u} = \mathbf{u}_0 e^{i\omega\left(\frac{1}{c}\mathbf{n}\cdot\mathbf{r}-t\right)}$ $\rho c^2 = K + \dfrac{4}{3}\mu$ (compression waves) $\rho c^2 = \mu$ (shear waves)
Diffusive waves	--	--	$\rho\dfrac{\partial^{2-\alpha}\mathbf{u}}{\partial t^{2-\alpha}} = \kappa\Delta\mathbf{u}$	$\kappa = \rho D$, for $\alpha = 1$ (diffusion) $\kappa = \mu$, for $\alpha = 0$ (elastic waves) $\kappa = \mu^{1-\alpha}\mu'^\alpha$, for $0 < \alpha < 1$ (diffusive waves)
Flow[a]	$\dot{E} = -\nabla \cdot \mathbf{P}$	$\mathbf{P} = -\sigma \cdot \mathbf{v}$ $\sigma = \mathbf{C}_{\text{elastic}} \cdot \boldsymbol{\varepsilon}$ $+ \mathbf{C}_{\text{viscous}} \cdot \dot{\boldsymbol{\varepsilon}}$	$\rho(\dot{\mathbf{v}} + \mathbf{v}\cdot\nabla\mathbf{v}) =$ $-\nabla p + \eta\Delta\mathbf{v}$	–
Perfusion	$\dot{C} = -\nabla \cdot \mathbf{P}$	$\mathbf{P} = -C\,\mathbf{v}$	$\dot{C}(t) = \dfrac{1}{\tau}\left(\begin{array}{c}C_a(t) \\ -\lambda C(t)\end{array}\right)$	$C(t) = \dfrac{C_0}{\tau}\Theta(t-t_0)e^{-\frac{t}{\tau\lambda}}$

[a]Incompressible fluids

where $\Theta(t-t_0)$ denotes the Heaviside function. A more detailed analysis needs to account for specific input functions $C_a(t)$ which can simulate the experimentally observable finite increase in arterial concentrations over time.

In perfusion MRI, the blood is magnetically labeled (arterial spin labeling, ASL) right before entering the imaging volume. After a delay time, an image is acquired, in which the presence of labeled blood reduces image intensity relative to an unlabeled reference image. The delay between labeling and image acquisition corresponds to the delay time t_0 in Eq. (2.58) (see Sect. 2.7.4 and Chap. 22 for a more detailed description of perfusion MRI).

Due to the hierarchy of the vascular architecture, delay time t_0 determines the range of sensitivity of the perfusion technique. During the delay between labeling and image acquisition, the labeled blood is allowed to reach the capillaries, where it gives rise to the measured perfusion signal. Consequently, shorter waiting times shift the sensitivity of perfusion measurements toward arteriole perfusion when the blood has not yet entered the capillary bed.

A summary of the equations governing mechanical and transport-related parameters as quantified by medical imaging is given in Table 2.1.

2.7 Motion Encoding by Medical Imaging

Measurement of motion by medical imaging modalities requires encoding of tissue deflection or particle velocity into the image contrast or labeling part of the tissue (e.g., blood or water) in a series of time-resolved images. The latter approach was the initial way of motion measurement by medical imaging and is referred to as *bolus tracking*. Bolus tracking is used in almost all medical imaging methods including CT, MRI, PET, and ultrasound. For example, CT angiography uses an injection of iodine-based contrast agents to help diagnose and evaluate blood vessel disease or related conditions, such as aneurysms or steno-occlusive disease. Similarly, in MRI, either gadolinium-based contrast agents or manipulation of the proton spin magnetization are used to enhance image contrast. For instance, in arterial spin labeling, the transverse spin magnetization of instreaming blood is changed by radiofrequency pulses, allowing the measurement of signal intensity variations related to flow and perfusion without employing a contrast agent [30].

2.7.1 Motion Encoding in Medical Ultrasound

An important principle of motion encoding in ultrasound relies on the Doppler effect. If an ultrasound wave with a given frequency f scatters at moving tissue boundaries or blood particles, it undergoes a frequency shift twice: first when the moving particle acts as a receiver of an incident wave and a second time when the particle emits the wave as a moving source. The overall frequency shift is

$$\Delta f = \frac{c+v_a}{c-v_a} f \approx \left(1+\frac{2v_a}{c}\right)f.$$
(2.59)

v_a is the velocity of the signal-emitting particle in the direction of the ultrasound beam (axial direction, with positive values indicating motion toward the ultrasound probe), and c is the ultrasound wave speed in the tissue [26]. Thus, measuring the frequency shift Δf allows the calculation of v_a, which, however, is not the true particle velocity but its projection onto the ultrasound beam axis. Therefore, v_a is usually corrected for the projection angle enclosed by the vessel (flow direction) and the ultrasound beam. Δf is easy to measure from a continous wave source but extremely difficult to determine from a pulsed source. Normal ultrasound mode is pulsed since short wave packages offer higher spatial resolution. However, the lower the number of ultrasonic wave cycles in a pulse, the higher the uncertainty of Doppler frequency estimation due to decreased frequency resolution. Therefore,

Doppler methods rely on a trade-off between continuous and pulsed waves, which inherently limits both spatial resolution and the accuracy of frequency estimation. Additional limitations of measuring Δf arise from physical processes including viscous damping and frequency-dependent scattering, which can also modulate the frequency of the received signal.

Therefore, many motion-encoding modes in ultrasound rely on correlation techniques termed *color mode* [26]. In the color mode, the ultrasound signals of two lines acquired at different times are correlated with each other. Axial resolution is achieved by successively shifting the correlation window along the signal line. Typical cross-correlation methods analyze the time shift or the phase shift of the ultrasound waves. Axial displacement u can be obtained from the maximum of the correlation function between two ultrasound wave signals $y_1(x,t)$ and $y_2(x, t + \Delta t) \approx y_1(x - u, t)$, i.e.,

$$\max_u \int y_1\left(x,t\right) y_2\left(x+u,t+\Delta t\right) dx. \tag{2.60}$$

The displacement then is the value of u which corresponds to the maximum of correlation. Since time Δt has passed between the acquisition of the two ultrasound lines, the measured quantity is velocity $u/\Delta t$. An alternative approach to motion estimation by ultrasound is the phase-shift method which exploits the phase shift $\Delta\phi(u)$ of the ultrasound wave due to displacement. If the axial distance between a particle and the ultrasound probe changes by an amount u between two acquisitions, the propagation path of the ultrasonic pulse from the probe to the particle and back to the probe changes by $2u$. The second pulse is thus detected with a phase offset $\Delta\phi = 2\pi\dfrac{2u}{\lambda} = 4\pi\dfrac{uf}{c}$, where λ is the ultrasonic wavelength. The phase offset $\Delta\phi(u)$ between two complex-valued signals $y_1(x,t) \propto \exp(i[kx + \phi_1])$ and $y_2(x, t + \Delta t) \propto \exp(i[kx + \phi_2])$ is obtained by the argument function [31]:

$$\Delta\phi = \phi_1 - \phi_2 = \arg\left(\frac{y_1}{y_2}\right). \tag{2.61}$$

For velocity measurements within larger two-dimensional regions, 2D correlation techniques such as *speckle tracking* can be used [32]. Despite their random nature, speckles in the ultrasound image display characteristic patterns of the tissue of interest and can thus be used to track motion. It has to be mentioned that, in general, ultrasound is limited in measuring lateral motion as compared to its high-resolution capabilities for motion tracking along the axial direction. Full 3D motion fields are still better acquired by MRI.

2.7.2   Motion Encoding in MRI

One of the most elaborate medical imaging technologies for motion quantification is MRI.

Motion can be encoded in the phase information of the complex magnetization exploiting frequency differences due to displacements of moving spins during application of magnetic field gradients [33]. Motion-sensitive MRI methods have many applications in medical diagnosis and research. Therefore, we will briefly review the motion-encoding mechanism that is used in MRI to detect and quantify coherent and incoherent motion.

Most biological tissues contain water molecules and thus a high number of hydrogen protons. In the presence of a static magnetic field, B_0, the magnetic moment associated with the spin of protons results in a macroscopic net magnetization, which forms the basis of MRI. In particular, the signal in MRI is based on the excitation of transverse magnetization relative to B_0 by means of resonant radiofrequency (RF) pulses. Directly after excitation of such magnetization, all spin ensembles precess about the direction of the main magnetic field with precession frequency ω_0. Spin ensembles or particles with the same precession frequency are called *isochromats*. ω_0 is also known as the Larmor frequency and depends on the magnetic field strength:

$$\omega_0 = \gamma B_0, \tag{2.62}$$

where γ denotes the gyromagnetic ratio of protons ($\gamma = 2\pi$ 42.58 MHz/T). In MRI, the local precession frequency is modulated by applying linear magnetic field variations in space, which are referred to as "gradients" **G**, introducing the dependence of the resonance frequency of magnetization on the position $\mathbf{r}_0$:

$$\omega_{MRI}(\mathbf{r}_0) = \omega_0 + \gamma \, \mathbf{G} \cdot \mathbf{r}_0. \tag{2.63}$$

Quadrature detection implemented in the scanner hardware subtracts ω_0 from the resonance frequency, leaving the MRI signal in the rotating frame, i.e., $\omega'_{MRI}(\mathbf{r}_0) = \gamma \, \mathbf{G} \cdot \mathbf{r}_0$. ω'_{MRI} is the basic signal exploited by spatial frequency encoding in MRI. The phase of this signal accumulated during a time interval τ when a constant gradient **G** is deployed is

$$\varphi(\mathbf{r}_0, \tau) = \gamma \, \mathbf{G} \cdot \mathbf{r}_0 \tau. \tag{2.64}$$

This fundamental MRI phase equation is the basis for spatial encoding along the phase-encode axis (see Chap. 8), MRI motion encoding (see Chap. 9), as well as T_2^* mapping (see Chap. 15). We now assume the signal-emitting particles (e.g., blood or tissue) to move, so that their positions become time dependent. Furthermore, we account for time-varying gradients, e.g., a rectangular waveform of duration τ that changes polarity at $\tau/2$, i.e., a balanced bipolar gradient. Equation (2.64) then becomes

$$\varphi(\mathbf{r}_0, \tau) = \gamma \int_0^\tau \mathbf{G}(t) \cdot \left[\mathbf{r}_0 + \mathbf{r}(t)\right] dt. \tag{2.65}$$

$\mathbf{r}(t)$ is the trajectory of the isochromats around equilibrium position $\mathbf{r}_0$. This equation expresses the motion sensitivity of MRI conveyed by the phase of magnetization. For coherent motion phenomena, such as flow or tissue oscillations, where all isochromats move more or less in synchrony (i.e., coherent) with their neighbors, φ is a good estimate of the ensemble spin phase in a voxel (i.e., the phase of the complex sum of all spins in that voxel). Therefore, coherent motion encoding by phase-contrast MRI based on Eq. (2.65) is the method of choice for flow field or wave field detection. Expanding the trajectory $\mathbf{r}(t)$ to the first order ($\mathbf{r}(\tau) \Rightarrow \mathbf{r}_0 + \dot{\mathbf{r}}\tau$) and assuming a rectangular gradient of total duration τ that is switched on for $0 \leq t \leq \tau$ yields for the accumulated spin phase

$$\varphi(t \geq \tau) = \gamma\, \mathbf{m}_0(\tau) \cdot \mathbf{r}_0 + \gamma\, \mathbf{m}_1(\tau) \cdot \dot{\mathbf{r}}_1, \tag{2.66}$$

with the zeroth- and first-order gradient moments $\mathbf{m}_0$ and $\mathbf{m}_1$. The n-th gradient moment is defined as

$$\mathbf{m}_n(\tau) = \int_{-\infty}^{\tau} \mathbf{G}(t) \cdot t^n dt. \tag{2.67}$$

A unipolar rectangular gradient has $\mathbf{m}_0(\tau) = \mathbf{G}\tau$ and $\mathbf{m}_1(\tau) = \frac{1}{2}\mathbf{G}\tau^2$, whereas a balanced bipolar rectangular gradient has $\mathbf{m}_0(\tau) = 0$ and $\mathbf{m}_1(\tau) = \mathbf{G}\tau^2$ [33] (more details on gradient moments are provided in Chap. 9). The zeroth-order term in Eq. (2.66) vanishes when bipolar gradients are applied leading to spin phase $\varphi = \gamma\, \mathbf{G} \cdot \mathbf{r}\tau = \gamma\, g\, r\, \tau$. g and r denote gradient amplitude and effective path length that the magnetization has traveled in the direction of gradient $\mathbf{G}$ during time τ, respectively. For flow, the resulting phase is therefore a measure of how far the magnetization has traveled between the application of the two gradient lobes and is thus proportional to the velocity.

2.7.3 Diffusion-Weighted MRI

In diffusion-weighted MRI, magnetization due to water molecules is labeled by imposing a phase that depends on the position of the magnetization along an axis that is defined by the direction of a magnetic *diffusion gradient* field [34].

The technique for motion encoding in MRI was explained in the previous section. However, diffusion represents an incoherent type of motion, and isochromats with different initial positions and motion trajectories are mixed within a voxel. In particular, we can imagine that the first lobe of a bipolar gradient produces a certain concentration of phase-labeled spins in each voxel, where the phase value depends on the position of the voxel along the direction of the gradient. Then the isochromats are allowed to perform random walk motion over a given evolution time before the spin phases are rephased by the second part of the diffusion gradient with opposite polarity. Signal-emitting particles that have moved randomly between the application of the two gradients result in phase dispersion, i.e., their magnetization vectors will partly cancel. This destructive interference is proportional to the distance that the particles have moved in the gradient direction (see Eq. (2.65)). The farther the particles in one voxel have traveled randomly, the more decorrelated their phases, and thus the smaller the complex sum of all magnetization portions in the voxel, which constitutes the MRI signal. The relative strength of the signal in one voxel, compared to the signal obtained from an identical scan but without diffusion gradients, represents a measure of the *apparent diffusion coefficient* (ADC), which is the main parameter measured in diffusion-weighted MRI. The ADC indicates that the diffusion process is not free in tissues, can involve multiple compartments, and may be superposed by intra-voxel incoherent motion such as blood flow in small vessels or cerebrospinal fluid in ventricles, which also contribute to MRI signal attenuation. For brevity, we restrict our further discussion to D and calculate the loss of signal due to spin phase decorrelation by diffusion based on the Gaussian distribution for diffusive spin motion, described by Gaussian distribution $N(r,t)$, given in Eq. (2.5). As the simplest case, we analyze a bipolar pair of short diffusion gradients with amplitude g, duration τ, and temporal separation τ_Δ, as shown in Fig. 2.6a. We assume that the gradients are short enough that the spins can be assumed to be static while the gradients are applied. Under these conditions, Eq. (2.65) becomes $\varphi(r) = -\gamma g \tau r$, with r the distance by which the spin moves along the gradient axis between the application of the two gradient lobes. In a normal diffusion process, r is distributed according to a normal distribution $N(r,t)$, evaluated at $t = \tau_\Delta$. The complex-valued macroscopic MRI signal $S = S_0 e^{i\varphi}$ (see Eq. (15.5), Chap. 15) of a voxel can then be expressed by integration over all spins within that voxel:

$$S = S_0 \int_{-\infty}^{\infty} e^{i\varphi(r)} N\left(r, \tau_\Delta\right)\, dr = S_0 \int_{-\infty}^{\infty} e^{-i\gamma g \tau r} N\left(r, \tau\right)\, dr = S_0 e^{\overbrace{-\gamma^2 g^2 \tau^2 \tau_\Delta D}^{b}}. \qquad (2.68)$$

All MRI-specific parameters, such as gyromagnetic ratio γ and gradient amplitude g, encoding time τ, or other timing variables, are usually collected in the "b-value." Note that the above formula for the b-value is only valid for short gradients with $\tau_\Delta \gg \tau$. Typically, diffusion-encoding gradients comprise positive and negative gradient lobes as well as a delay between the two, so that more complicated b-values result as exemplified in Fig. 2.6 and in Eq. (17.1) of Chap. 17. The mono-exponential signal decay of Eq. (2.68) $S \sim e^{-bD}$ is considered as "regular diffusion" characterized by $\langle r_2 \rangle \propto t$ [4]. In the presence of two compartments (e.g., intra-and extracellular water) with different diffusion coefficients, one would observe a bi-exponential decay. The bi-exponential model can be generalized to a multicompart-ment model, which predicts a stretched exponential decay of the form $S \sim e^{-b^\alpha D}$ with $\alpha < 1$ for a complex static environment in which diffusion is constrained [35]. In that case of "anomalous diffusion," the mean squared displacement is a power law in time, i.e., $\langle r_2 \rangle \propto t^\alpha$. Note the similarity of this power law to the springpot given in Eq. (2.43), which is due to conceptual similarities between hierarchic structures in water diffusion and viscoelastic networks. Equation (2.68) can be further generalized by accounting for direction-dependent diffusion constants which evoke a 3×3 diffusion tensor $\mathbf{D}$, which is, similar to strain and stress, symmetric and has thus six independent entries. The eigenvalues of $\mathbf{D}$ represent three independent diffusion coefficients along the three Cartesian axes of the diffusion eigensystem

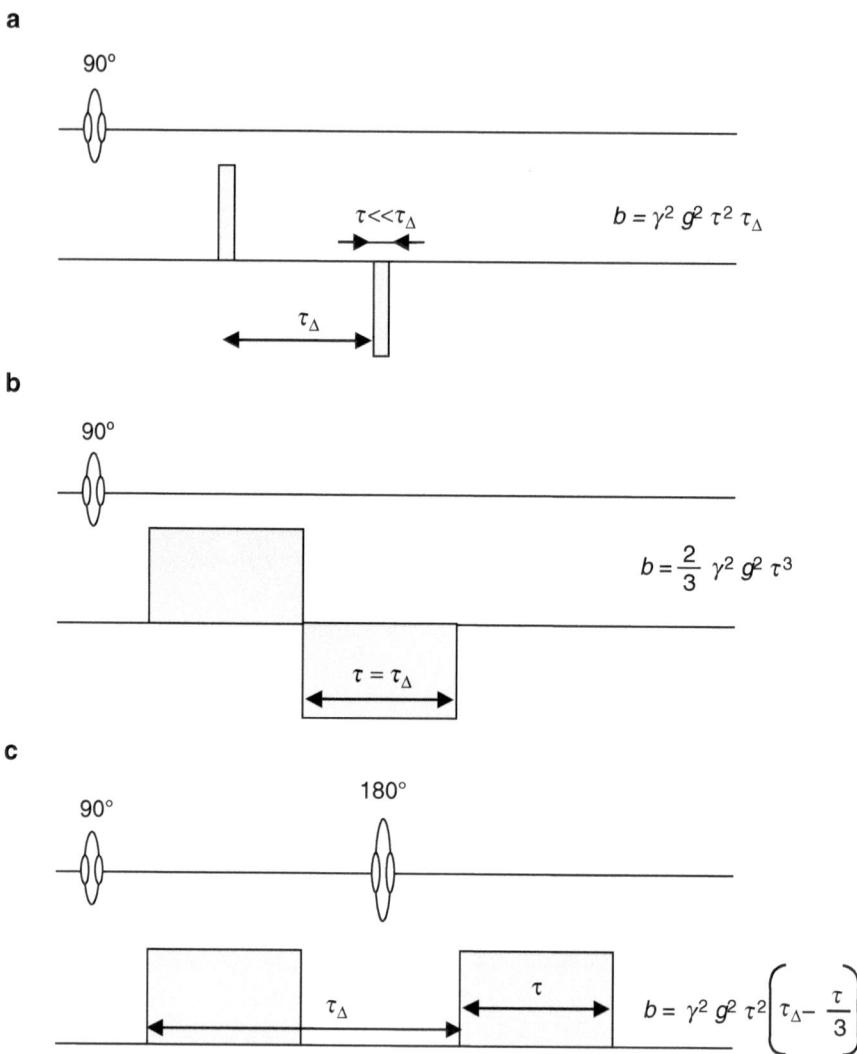

Fig. 2.6 Three basic diffusion-weighted MRI sequences and corresponding b-values. (**a,b**) Gradient echo sequence, (**c**) spin echo sequence; all with rectangular motion-encoding gradients (*gray rectangles*)

corresponding to the spread of particles in 3D and implying that diffusion is not a vector such as coherent flow. The MRI signal decay for anisotropic diffusion depends on the direction vectors **g** of the motion-encoding gradient: $S \sim e^{-b\mathbf{g}^T \cdot \mathbf{D} \cdot \mathbf{g}}$. Further details and applications of diffusion tensor imaging (DTI) are given in Chap. 17. It is worthwhile to mention that DTI normally accounts for restricted diffusion in the spatial-directional dimension. However, the restriction of water molecules moving against spatial boundaries is different at different propagation times t.

Table 2.2 Classification of diffusion as used in the literature of medical imaging

	Mean path lengths	MRI signal
Regular diffusion	$\langle \mathbf{r}^2 \rangle \propto Dt$	$S \sim e^{-bD}$
Anomalous diffusion	$\langle \mathbf{r}^2 \rangle \propto Dt^\alpha$	$S \sim e^{-b^\alpha D}$
Anisotropic diffusion	$\langle r_i r_j \rangle \propto D_{ij}t$	$S \sim e^{-b\mathbf{g}^T \cdot \mathbf{D} \cdot \mathbf{g}}$
Restricted diffusion (kurtosis imaging)	$\mathbf{r}^2 \propto Dt \cdot \left(1 + \sqrt{K}\right)_a$	$S \sim e^{-bD + \frac{1}{6}b^2 D^2 K}$

[a]K is proportional to the excess kurtosis parameter used in the literature including all model-dependent prefactors [36]

The longer the molecules travel, the higher the probability for them to hit a restricting boundary in the tissue. For that reason, the diffusion profile is in general not characterized by a Gaussian distribution but by some deviating probability function. The excess diffusion kurtosis, also called "diffusion kurtosis," measures the deviation of the probability distribution function from a Gaussian curve and can be used to quantify heterogeneity of the diffusion environment [36]. Table 2.2 summarizes the discussed cases of the diffusion MRI signal.

2.7.4 Arterial Spin Labeling

Like diffusion MRI, perfusion MRI based on arterial spin labeling (ASL, also called arterial spin tagging, AST) is a difference technique, where the change in image intensity between a motion-sensitized scan and a non-sensitized reference scan is quantified and correlated with an underlying transport process. In ASL, contrast is generated by manipulating the magnetization of inflowing blood in such a way that it reduces the MR signal [37]. The spins in the inflowing blood are inverted by an 180° RF pulse prior to flowing into the volume of interest. Within that volume, they mix with the spins of the tissue. The inverted spins cause a reduction of the total magnetization since they cancel a part of the non-inverted tissue magnetization. As a consequence, the amplitude of the MR signal is reduced compared to the non-labeled reference image, leading to slight hypointensity in well-perfused regions. However, the contrast change is very subtle (on the order of 1–2%), requiring extensive signal averaging through repeated measurements. Furthermore, additional confounders have to be excluded as far as possible, leading to more refined and complex ASL MR sequences than explained above. In general, two types of ASL techniques exist: pulsed ASL (pASL), where a labeling pulse is deployed prior to every image acquisition, and continuous ASL (cASL), which aims to establish a steady state of labeled and unlabeled magnetization, necessitating a larger number of tagging pulses. A more in-depth discussion of different ASL techniques can be found in [33], among others.

2.8   Quantification of Other Structure-Related Parameters by Medical Imaging

Quantitative mapping of biophysical parameters other than those discussed above has a long tradition in medical imaging. One of the first quantitative and system-independent parameters measured by medical imaging was volume (volumetry). Tissue volume is often related to physiological functions of organs or the progression of diseases. For example, brain volume segmentation based on 3D MRI is the current gold standard for quantification of atrophy in neurodegenerative diseases [38]. Another example is muscle volume measurement by cardiac MRI or echocardiography as a direct measure of cardiac hypertrophy, which has wide implications for heart function [39]. Furthermore, the ejection fraction of the heart is a measure of systolic function [40]. The ejection fraction is calculated by dividing the stroke volume by the end-diastolic volume of the left ventricle. Slow changes in soft tissue volumes are regularly measured in longitudinal imaging examinations for quantifying tumor growth or the response to tumor treatment [41, 42].

Related to volume and mass is density, which, however, needs assumptions when estimated by medical imaging since tissue mass is normally not directly measurable. X-ray attenuation coefficients are highly dependent on tissue density and are thus exploited for quantifying bone density. Dual-energy X-ray absorptiometry (DXA) is the current gold standard for bone (mineral) density measurement as required for diagnosing and staging osteoporosis [43]. Quantification of minerals accumulated in soft tissues, such as calcifications in breast tumors, would be highly relevant for tumor staging [44, 45]. Morphology, location, and quantity of calcifications in the breast are established markers used in the BI-RADS atlas for breast tumor classification [46]. A further example is the quantification of iron in brain matter by MRI. Iron is the most abundant trace element in the human brain and plays an important role in maturation of the central nervous system and brain metabolism. Deposits of iron in the brain could cause neurological diseases like Alzheimer and Parkinson [47]. Susceptibility mapping by MRI can be used to quantify the spatial deposition of iron, e.g., in the human brain, but cannot quantify absolute iron content [48]. For the liver, iron content measurements based on relaxometry have been demonstrated [49]. Exogenous susceptibility sources such as gadolinium-based contrast agents induce magnetic susceptibility values which are theoretically linearly proportional to the concentration of the contrast agent. Thus, susceptibility mapping has the potential for in vivo quantification of contrast agent concentrations [50]. The ability of paramagnetic metal ions like iron or gadolinium to shorten magnetic relaxation is used in MRI contrast agents for the manipulation of image contrast in regions where water protons can interact with the contrast agent. Most clinically used MRI contrast agents reduce the T_1 relaxation time of protons by interaction with the nearby contrast agent. However, quantification of the amount of contrast agent in tissue by relaxation times is difficult since magnetic relaxation depends on the MRI system. Nevertheless, relaxation times constitute the primary contrast in MRI and are useful for the quantification of tissue properties as detailed for the myocardium in Chap. 15. Measurement of different water proton relaxation mechanisms such as T_2 and effective T_2 (T_2^*) allows estimation of vessel size and blood volume in the tissue [51, 52], which are both highly relevant biophysical parameters for characterizing tissue physiology and pathology [53]. Paired with perfusion measurements, vessel size and blood volume provide the input parameters for deriving hemodynamic constants such as hydraulic conductivity of the blood through the capillary bed of the tissue based on Darcy's law (see Chap. 3) [54]. Further structure-related biophysical and medical imaging-based parameters include geometry of vessels or fibers as can be quantified by bifurcation indices, tortuosity, self-similarity, structural density, and fractal dimension [55].

Conclusion

State-of-the-art medical imaging offers a wide range of sensitivities to motion in the human body. From incoherent motion of water molecules to coherent flow, medical imaging modalities can be used to depict physiologic and dysfunctional transport phenomena and to derive quantitative imaging markers for clinical diagnosis. This chapter summarized the physics of transport from an imaging point of view and reviewed the basic concepts of motion encoding in medical imaging. It was shown that very basic assumptions such as continuity of mass and energy at the position of the measurement can lead to a rich set of equations that explain multiple phenomena observed by motion-sensitive imaging modalities including diffusion, perfusion, and flow imaging as well as elastography. Since transport-related parameters and mechanical constants are of huge importance for the quantification of constitution and function of biological tissues, this chapter presents a primer of quantitative biophysical medical imaging.

References

1. Feynman RP, Leighton RB, Sands M, Feynman P. The Feynman lectures on physics, volume II: the new millennium edition: mainly electromagnetism and matter. New York: Basic Books; 2015.
2. Feynman RP, Leighton RB, Sands M, Feynman P. The Feynman lectures on physics, volume I: the new millennium edition: mainly mechanics, radiation, and heat. New York: Basic Books; 2015.
3. Einstein A. Über die von der molekularkinetischen Theorie der Wärme geforderte Bewegung von in ruhenden Flüssigkeiten suspendierten Teilchen. Ann Phys. 1905;17(8):549–60.
4. Ben-Avraham D, Havlin S. Diffusion and reactions and disordered systems. Cambridge: Cambridge University Press; 2000.
5. Landau LD, Lifschitz EM. Theory of elasticity. Oxford: Pergamon; 1986.
6. Lai WM, Rubin D, Krempl E. Introduction to continuum mechanics. Burlington: Butterworth Heinemann; 1994. 570 p.
7. Fedorov FI. Theory of elastic waves in crystals. New York: Plenum; 1968.
8. Parker KJ, Huang SR, Musulin RA, Lerner RM. Tissue response to mechanical vibrations for "sonoelasticity imaging". Ultrasound Med Biol. 1990;16(3):241–6.
9. Muthupillai R, Ehman RL. Magnetic resonance elastography. Nat Med. 1996;2(5):601–3.
10. Manduca A, Oliphant TE, Dresner MA, Mahowald JL, Kruse SA, Amromin E, Felmlee JP, Greenleaf JF, Ehman RL. Magnetic resonance elastography: non-invasive mapping of tissue elasticity. Med Image Anal. 2001;5(4):237–54.

11. Hirsch S, Braun J, Sack I. Magnetic resonance elastography: physical background and medical applications. Weinheim: Wiley; 2017.
12. Romano AJ, Shirron JJ, Bucaro JA. On the noninvasive determination of material parameters from a knowledge of elastic displacements theory and numerical simulation. IEEE Trans Ultrason Ferroelectr Freq Control. 1998;45(3):751–9.
13. Romano AJ, Bucaro JA, Ehnan RL, Shirron JJ. Evaluation of a material parameter extraction algorithm using MRI-based displacement measurements. IEEE Trans Ultrason Ferroelectr Freq Control. 2000;47(6):1575–81.
14. Park E, Maniatty AM. Shear modulus reconstruction in dynamic elastography: time harmonic case. Phys Med Biol. 2006;51(15):3697–721.
15. Oliphant TE, Manduca A, Ehman RL, Greenleaf JF. Complex-valued stiffness reconstruction for magnetic resonance elastography by algebraic inversion of the differential equation. Magn Reson Med. 2001;45:299–310.
16. Hirsch S, Klatt D, Freimann F, Scheel M, Braun J, Sack I. In vivo measurement of volumetric strain in the human brain induced by arterial pulsation and harmonic waves. Magn Reson Med. 2012;70(3):671–83.
17. Mousavi SR, Fehlner A, Streitberger KJ, Braun J, Samani A, Sack I. Measurement of in vivo cerebral volumetric strain induced by the Valsalva maneuver. J Biomech. 2014;47(7):1652–7.
18. Aki K, Richards PG. Quantitative seismology. Sausalito: Universtiy Science Books; 2002.
19. Fink M, Montaldo G, Tanter M. Time-reversal acoustics in biomedical engineering. Annu Rev Biomed Eng. 2003;5:465–97.
20. Mainardi F. Fractional calculus and waves in linear viscoelasticity. London: Imperial College Press; 2010.
21. Magin RL. Fractional calculus in bioengineering. Crit Rev Biomed Eng. 2004;32(1):1–104.
22. Fabry B, Maksym GN, Butler JP, Glogauer M, Navajas D, Taback NA, Millet EJ, Fredberg JJ. Time scale and other invariants of integrative mechanical behavior in living cells. Phys Rev E Stat Nonlinear Soft Matter Phys. 2003;68(4 Pt 1):041914.
23. Sack I, Joehrens K, Wuerfel E, Braun J. Structure sensitive elastography: on the viscoelastic powerlaw behavior of in vivo human tissue in health and disease. Soft Matter. 2013;9(24):5672–80.
24. Tschoegl NW. The phenomenological theory of linear viscoelastic behavior. Berlin: Springer; 1989.
25. Achenbach JD. Wave propagation in elastic solids. Amsterdam: Elsevier; 1999.
26. Fenster A, Lacefield JC. In: Karellas A, editor. Ultrasound imaging and therapy. Boca Raton: CRC Press; 2015.
27. Stankovic Z, Allen BD, Garcia J, Jarvis KB, Markl M. 4D flow imaging with MRI. Cardiovasc Diagn Ther. 2014;4(2):173–92.
28. Thomas DL, Lythgoe MF, Pell GS, Calamante F, Ordidge RJ. The measurement of diffusion and perfusion in biological systems using magnetic resonance imaging. Phys Med Biol. 2000;45(8):R97–138.
29. Le Bihan D. Theoretical principles of perfusion imaging. Application to magnetic resonance imaging. Investig Radiol. 1992;27(Suppl 2):S6–11.
30. Buxton RB. Quantifying CBF with arterial spin labeling. J Magn Reson Imaging. 2005;22(6):723–6.
31. Pesavento A, Perrey C, Krueger M, Ermert A. A Time-efficient and accurate strain estimation concept for ultrasonic elastography using iterative phase zero estimation. IEEE Trans Ultrason Ferroelectr Freq Control. 1999;46(5):1057–67.
32. Bamber J, Cosgrove D, Dietrich CF, Fromageau J, Bojunga J, Calliada F, Cantisani V, Correas JM, D'Onofrio M, Drakonaki EE, Fink M, Friedrich-Rust M, Gilja OH, Havre RF, Jenssen C, Klauser AS, Ohlinger R, Saftoiu A, Schaefer F, Sporea I, Piscaglia F. EFSUMB guidelines and recommendations on the clinical use of ultrasound elastography. Part 1: basic principles and technology. Ultraschall Med. 2013;34(2):169–84.
33. Bernstein MA, King KF, Zhou XJ. Handbook of MRI pulse sequences. Burlington: Elsevier Academic; 2004.

34. Stejskal EO, Tanner JE. Spin diffusion measurements: spin echoes in the presence of a time-dependent field gradient. J Chem Phys. 1965;42(1):288–92.
35. Hall MG, Barrick TR. From diffusion-weighted MRI to anomalous diffusion imaging. Magn Reson Med. 2008;59(3):447–55.
36. Jensen JH, Helpern JA. MRI quantification of non-Gaussian water diffusion by kurtosis analysis. NMR Biomed. 2010;23(7):698–710.
37. Detre JA, Leigh JS, Williams DS, Koretsky AP. Perfusion imaging. Magn Reson Med. 1992;23(1):37–45.
38. Despotovic I, Goossens B, Philips W. MRI segmentation of the human brain: challenges, methods, and applications. Comput Math Methods Med. 2015;2015:450341.
39. Maceira AM, Mohiaddin RH. Cardiovascular magnetic resonance in systemic hypertension. J Cardiovasc Magn Reson. 2012;14:28.
40. Lang RM, Bierig M, Devereux RB, Flachskampf FA, Foster E, Pellikka PA, Picard MH, Roman MJ, Seward J, Shanewise JS, Solomon SD, Spencer KT, Sutton MS, Stewart WJ. Recommendations for chamber quantification: a report from the American Society of Echocardiography's Guidelines and Standards Committee and the Chamber Quantification Writing Group, developed in conjunction with the European Association of Echocardiography, a branch of the European Society of Cardiology. J Am Soc Echocardiogr. 2005;18(12):1440–63.
41. Andreopoulou E, Andreopoulos D, Adamidis K, Fountzila-Kalogera A, Fountzilas G, Dimopoulos MA, Aravantinos G, Zamboglou N, Baltas D, Pavlidis N. Tumor volumetry as predictive and prognostic factor in the management of ovarian cancer. Anticancer Res. 2002;22(3):1903–8.
42. Iliadis G, Kotoula V, Chatzisotiriou A, Televantou D, Eleftheraki AG, Lambaki S, Misailidou D, Selviaridis P, Fountzilas G. Volumetric and MGMT parameters in glioblastoma patients: survival analysis. BMC Cancer. 2012;12:3.
43. Blake GM, Fogelman I. Technical principles of dual energy x-ray absorptiometry. Semin Nucl Med. 1997;27(3):210–28.
44. Grigoryev M, Thomas A, Plath L, Durmus T, Slowinski T, Diekmann F, Fischer T. Detection of microcalcifications in women with dense breasts and hypoechoic focal lesions: comparison of mammography and ultrasound. Ultraschall Med. 2014;35(6):554–60.
45. Naseem M, Murray J, Hilton JF, Karamchandani J, Muradali D, Faragalla H, Polenz C, Han D, Bell DC, Brezden-Masley C. Mammographic microcalcifications and breast cancer tumorigenesis: a radiologic-pathologic analysis. BMC Cancer. 2015;15:307.
46. Bassett LW. Standardized reporting for mammography: BI-RADS. Breast J. 1997;3:207–10.
47. Ward RJ, Zucca FA, Duyn JH, Crichton RR, Zecca L. The role of iron in brain ageing and neurodegenerative disorders. Lancet. 2014;13(10):1045–60.
48. Deistung A, Schweser F, Reichenbach JR. Overview of quantitative susceptibility mapping. NMR Biomed. 2016;30:e3569.
49. Kannengiesser S. Iron quantification with LiverLab. MAGNETOM Flash. 2016;3(66):44–6.
50. de Rochefort L, Nguyen T, Brown R, Spincemaille P, Choi G, Weinsaft J, Prince MR, Wang Y. In vivo quantification of contrast agent concentration using the induced magnetic field for time-resolved arterial input function measurement with MRI. Med Phys. 2008;35(12):5328–39.
51. Tropres I, Grimault S, Vaeth A, Grillon E, Julien C, Payen JF, Lamalle L, Decorps M. Vessel size imaging. Magn Reson Med. 2001;45(3):397–408.
52. Shen Y, Pu IM, Ahearn T, Clemence M, Schwarzbauer C. Quantification of venous vessel size in human brain in response to hypercapnia and hyperoxia using magnetic resonance imaging. Magn Reson Med. 2013;69(6):1541–52.
53. Emblem KE, Mouridsen K, Bjornerud A, Farrar CT, Jennings D, Borra RJ, Wen PY, Ivy P, Batchelor TT, Rosen BR, Jain RK, Sorensen AG. Vessel architectural imaging identifies cancer patient responders to anti-angiogenic therapy. Nat Med. 2013;19(9):1178–83.
54. Hetzer S, Birr P, Fehlner A, Hirsch S, Dittmann F, Barnhill E, Braun J, Sack I. Perfusion alters stiffness of deep gray matter. J Cereb Blood Flow Metab. 2017:271678X17691530.
55. Bankman I. In: Bankman I, editor. Handbook of medical image processing and analysis. Amsterdam: Elsevier; 2009.

Mathematical Modeling of Blood Flow in the Cardiovascular System

3

Alfonso Caiazzo and Irene E. Vignon-Clementel

Abstract

This chapter gives a short overview of the mathematical modeling of blood flow at different resolutions, from the large vessel scale (three-dimensional, one-dimensional, and zero-dimensional modeling) to microcirculation and tissue perfusion. The chapter focuses first on the formulation of the mathematical modeling, discussing the underlying physical laws, the need for suitable boundary conditions, and the link to clinical data. Recent applications related to medical imaging are then discussed, in order to highlight the potential of computer simulation and of the interplay between modeling, imaging, and experiments in order to improve clinical diagnosis and treatment. The chapter ends presenting some current challenges and perspectives.

3.1 Introduction

In this chapter we provide an overview of the mathematical modeling of blood flow in living tissues and of some applications in connection with medical imaging. In particular, the first aim is to present the basic ideas behind *multi-scale* or also called *multi-domain* hemodynamics simulations, discussing the modeling of the cardiovascular system from the scale of large and medium vessels down to the scale of microcirculation within tissues. The second aim is to highlight the potential of

A. Caiazzo (✉)
Weierstrass Institute for Applied Analysis and Stochastics, Berlin, Germany
e-mail: caiazzo@wias-berlin.de

I.E. Vignon-Clementel
INRIA Paris, Paris, France
e-mail: Irene.Vignon-Clementel@inria.fr

© Springer International Publishing AG 2018
I. Sack, T. Schaeffter (eds.), *Quantification of Biophysical Parameters in Medical Imaging*, https://doi.org/10.1007/978-3-319-65924-4_3

modeling and simulation of blood flow in connection with available medical data (in particular in imaging), to improve clinical diagnosis and interventional treatment. Throughout the chapter, several references on to recent computational methods and application examples will be provided.

The chapter is organized as follows: Sect. 3.2 is dedicated to modeling at the large vessel scale, or typically where the geometry of each vessel is known, introducing the three-dimensional Navier–Stokes equations and discussing modeling approaches based on spatial dimensional reduction (one- and zero-dimensional). Section 3.3 focuses on the fluids at the tissue scale, discussing the modeling of microcirculation and tissue perfusion. Finally, Sect. 3.4 gives a few recent applications of computational fluid dynamics to improve image-based non-invasive diagnosis, while Sect. 3.5 provides a brief summary.

3.2   Modeling Fluids at the Large Vessel Scale

This section describes the modeling of blood flow within blood vessels, discussing first three-dimensional (3D) approaches and, next, the so-called reduced models, i.e., based on lower spatial dimensional (1D or 0D) descriptions.

3.2.1   Three-Dimensional Navier–Stokes Equations

The simplest three-dimensional description of blood flow considers the motion of an incompressible, Newtonian fluid within a vessel with rigid walls, i.e., without taking into account the interaction between the blood and the vessel wall.

Formally, let $\Omega \subset \mathrm{R}^3$ denote a piece of a blood vessel, and let us decompose the boundary of Ω in three subsets, an upstream cross-sectional area Γ_{up}, a downstream boundary Γ_{down}, and the vessel wall Γ_{wall} (see Fig. 3.1), upstream and downstream referring to the main direction of flow.

In order to model the hemodynamics, blood is assumed to be an incompressible, Newtonian fluid. The latter properties refer to a fluid in which viscous stresses are linearly proportional to the local strain rate, by the viscosity parameter. Although it is well known that blood is composed of particles of different nature, thus yielding complex mechanical properties, these effects become significant only when the particle size is comparable to the vessel size (i.e., for blood flow in the smallest vessels). The Newtonian approximation is valid for shear strain rate above $100 \ \mathrm{s}^{-1}$, and hence mostly used when considering large to medium-size vessels.

Using the principle of mass and momentum conservation, the above assumptions allow to describe the blood flow dynamics through the following incompressible Navier–Stokes equations:

$$\begin{cases} \rho \partial_t \mathbf{v} + \rho \mathbf{v} \nabla \mathbf{v} + \nabla p - 2\eta \nabla \cdot \dot{\varepsilon}(\mathbf{v}) = \mathbf{f} \\ \nabla \cdot \mathbf{v} = 0 \end{cases} \qquad (3.1)$$

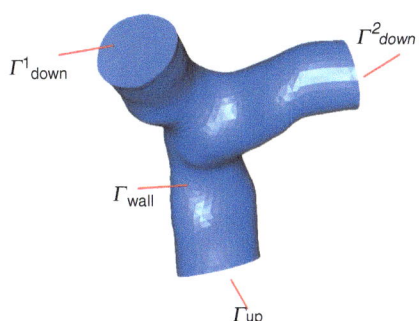

Fig. 3.1 Example of computational domain for a 3D model (here representing a pulmonary artery bifurcation), decomposing the boundary as in (3.2). In this case, the upstream cross-sectional area Γ_{up} is connected with the right ventricle, while the downstream ones Γ_{down}^{1} and Γ_{down}^{2} interface with the left and the right pulmonary arteries

which consist of a system of non-linear parabolic partial differential equations for the velocity $\mathbf{v}: \Omega \times R^{+} \to R^{3}$ and the pressure $p: \Omega \times R^{+} \to R$.

In (3.1), ρ and η stand for the blood density and viscosity, and $\dot{\varepsilon}\left(\mathbf{v}\right) = \dfrac{1}{2}\left(\nabla\mathbf{v} + \nabla\mathbf{v}^{T}\right)$ denotes the strain rate tensor. The term $\mathbf{f}$ on the right-hand side of Eq. (3.1) takes into account the external forces. In the left-hand side, the first two terms correspond to inertia, while the other two stand, respectively, for the pressure force and the viscous force.

The system of equations (3.1) has to be completed with an initial condition, which shall be given by a divergence free velocity field, and with suitable boundary conditions. A typical set of boundary conditions [1] is to prescribe the velocity at the upstream boundary and the pressure on the downstream boundary, combined with a no-slip condition on the vessel wall, i.e.,

$$\begin{cases} \mathbf{v} = \mathbf{v}_{up}, \text{on}\,\Gamma_{up} \\ \mathbf{v} = \mathbf{0}_{up}, \text{on}\,\Gamma_{wall} \\ \dot{\varepsilon}\left(\mathbf{v}\right)\mathbf{n} - p\mathbf{n} = -P_{down}\mathbf{n}, \text{on}\,\Gamma_{down} \end{cases} \qquad (3.2)$$

where $\mathbf{n}$ denotes the normal to the boundary directed outward the fluid domain and P_{down} is a prescribed downstream pressure. In practical applications, it makes more sense to in fact impose a pressure that comes from the response to blood flow of the truncated distal part of the circulation.

The concept of image-based patient-specific modeling consists in solving the Eqs. (3.1) and (3.2) constructing (1) a geometrical model segmented from the patient's clinical images (typically CT or MRA), and (2) patient-specific boundary conditions based on the available flow or pressure measurements (typically from catheterization, PC-MRI, Doppler-ultrasound—see Chap. 2). Then, the resulting system needs to be solved on the computer with appropriate numerical methods (meshing, finite elements, finite volumes, etc.). Section 3.4.1 provides some examples of applications. We refer to primers for clinicians on this topic [2, 3].

These steps require a tight communication between modelers and clinicians, as measurements can rarely be directly applied as boundary conditions. In fact, a flow rate measurement necessitates the choice of a velocity profile to lead to a velocity boundary condition on a face and, at the same time, measurements are usually not taken where boundary conditions need to be prescribed (see, e.g., [1, 4, 5]). Moreover, for predictive purposes (such as in surgical planning), it is better to impose a relationship between pressure and flow, instead of any of them. This leads to coupling of this 3D model with reduced models, and the need of their parametrization from patient data (see Sects. 3.2.3 and 3.2.4). Such so-called *multi-scale* or *multi-domain* modeling is challenging for mathematical (coupling conditions), numerical (stable and efficient algorithms), and software (handling of codes possibly from different teams) reasons. They have led to various approaches to tackle these issues [1, 6–9].

3.2.2 Fluid–Structure Interaction

Depending on the quantity of interests for the particular application, it might be necessary to take into account fluid–structure interaction (FSI), i.e., to model the interplay between the fluid flow within a vessel or a heart cavity and solid parts of the cardiovascular system. For instance, this is the case when the model is expected to provide information about fluid stresses at the moving vessel wall, or about mechanical stresses or movement of native valves and implants (artificial valves or stents).

The modeling of FSI is a rather technical subject and it is out of the scope of this chapter. For the sake of completeness, it is worth mentioning that from the mathematical point of view, FSI can be described combining the 3D Navier–Stokes equations (3.1) with structural models of varying complexity, including, among others, 2D *shell* (see, e.g., [10, 11]), tissue-support [12], or hyperelastic [8] models. These topics have been largely investigated in the last decades, in terms of mathematical properties (such as the well-posedness of coupling conditions and the stability of the resulting problem) and in terms of computational methods for the resolution of the coupled system (see, e.g., [13–17]).

3.2.3 One-Dimensional Fluid–Structure Interaction Modeling

The high complexity of three-dimensional fluid–structure interaction models makes them currently not suited for simulation of blood flow within extended vessel networks. This would require the acquisition and preprocessing of a large geometrical data set, in order to generate suitable computational domains, as well as the solution at each time step of large systems of unknowns, not to mention the parameterization aspect.

In fact, in many applications to be able to simulate flow efficiently or in extended vessel networks, while capturing wave propagation, is more important than the level

Fig. 3.2 Left: Sketch of
the one-dimensional blood
flow model for a
bifurcation (junction),
showing on a cross-
sectional area A the
associated flow-rate q and
pressure p; Right: Example
of 1D model for the human
arterial system [18]
(picture courtesy of Lucas
O. Müller, NTNU
Trondheim)

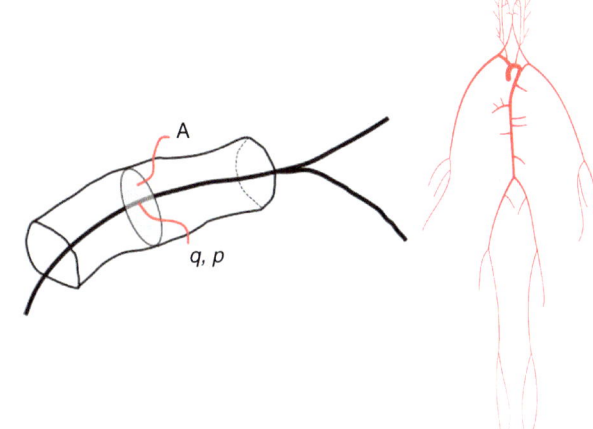

of local accuracy. In these cases, one-dimensional (1D) blood flow models represent an interesting alternative to reduce model complexity. These models can be derived by assuming that the fluid is homogeneous, in a tethered cylindrical vessel, that the non-slip condition holds at the vessel wall, and the flow is approximately uni-dimensional (i.e., mainly directed along the longitudinal direction) with a homogeneous pressure on each cross-section. Under these hypotheses, 1D models result from the integration of three-dimensional fluid equations (3.1) in a moving domain over the vessel cross-section at each location along the longitudinal direction (hence its 1D name, Fig. 3.2 left).

In order to formulate the governing equations, let us denote with x the one-dimensional coordinate parametrizing the vessel direction and with $S(x,t)$ the cross-section, introducing the (time-dependent) cross-sectional area $A(x,t) = \int S(x,t)\, dS$ and the flow rate $q(x,t) = \int S(x,t)\, \mathbf{v} dS$. With these notations, integrating the continuity equation over a cross-section, assuming there is no seepage, yields

$$\frac{\partial}{\partial t} \int_{S(x,t)} \rho + \rho \int_{S(x,t)} \nabla \cdot \mathbf{v} = 0, \tag{3.3}$$

which can be rewritten as

$$\frac{\partial A}{\partial t} + \frac{\partial q}{\partial x} = 0. \tag{3.4}$$

A further equation is obtained integrating the momentum conservation equation over the cross-section and introducing additional modeling assumptions for the convective and the viscous stress term, yielding

$$\frac{\partial q}{\partial t} + \frac{\partial}{\partial x}\left(\alpha \frac{q^2}{A} \right) + \frac{A}{\rho} \frac{\partial p}{\partial x} = -f, \tag{3.5}$$

(for a detailed derivation, see, e.g., [19], or for less restrictive assumptions [20]). In (3.5), $p(x,t)$ denotes average blood pressure over the cross-section, $f(x,t)$ stands for the friction force (other forces, such as gravity, can be included) per unit length and the parameter α (also called Coriolis coefficient) depends on the assumed velocity profile (parabolic, blunt, etc.) within the vessel segment and it is commonly taken equal to one.

Equations (3.4) and (3.5) define a first order, non-linear hyperbolic system that must be completed by a constitutive law (usually called *tube law*) that relates the internal pressure $p(x,t)$ to the area $A(x,t)$, through the strain and strain rate of the vessel wall.

To this end, the structural mechanics of the vessel are assumed to be sufficiently described using only radial displacement and by (visco-)elastic material properties. A first-order approximation, assuming a linear elastic behavior for vessel walls that are thin and incompressible, yields the relation (see, e.g., [19])

$$p(x,t) = p_e(x,t) + \frac{4Eh_0\sqrt{\pi}}{3}\left(\frac{\sqrt{A(x,t)} - \sqrt{A_0(x)}}{A_0(x)}\right) \qquad (3.6)$$

where p_e denotes the external pressure, E is the Young modulus of the vessel wall, h_0 its thickness, and A_0 denotes the cross-sectional area at a given reference pressure.

A more general approach is given by the tube-law (see, e.g., [18])

$$p(x,t) = p_e(x,t) + K(x)\phi\big(A(x,t),A_0(x)\big) + \frac{\Gamma}{A_0(x)\sqrt{A(x,t)}}\frac{\partial A}{\partial t}, \qquad (3.7)$$

where $K(x)$ is a positive function that depends on the Young modulus and on the wall thickness, while the function $\phi(A(x,t),A_0(x))$ has the form

$$\phi(A,A_0) = \left(\frac{A}{A_0}\right)^m - \left(\frac{A}{A_0}\right)^n, \qquad (3.8)$$

where the parameters m and n come from experimental measurements. Typical values for arteries are $m = 0.5$, $n = 0$ (reducing to the linear approximation (3.6)), while $m = 10$ and $n = -1.5$ are usually assumed for veins. Finally, the last term in (3.7) can be included in order to account for the viscosity Γ of the vessel wall. We refer to [18, 21] for recent detailed discussions of the one-dimensional model and the associated parameters.

Equations (3.4), (3.5), and (3.7) describe a one-dimensional model of a single vessel tract (or segment). In order to model a realistic network of vessels, these equations are combined with conservation of mass and momentum across the *junctions*, i.e., the nodes shared by multiple vessels.

In recent years, a number of approaches have been proposed in order to accurately and efficiently solve one-dimensional blood flow equations on a network of compliant vessels (see Fig. 3.2, right). The interested reader is referred, e.g., to the high order finite-volume method described in [22–24], to the one-dimensional finite element methods combined with simpler technique to handle vessel collapse [25] and to the recent numerical benchmark [26] comparing different numerical solution approaches. Similarly to the case of three-dimensional models, geometrical vessel information ideally comes from imaging data, and dynamic pressure, cross-sectional area, flow rate and velocity measurements can be integrated in the model (see, e.g., [27, 28] and the example described in Sect. 3.4.2).

3.2.4   Lumped Parameter Models

Lumped parameter models (also referred to as zero-dimensional models as they neglect spatial dimensions) are the simplest approach for describing flow through a vessel or a network of vessels, through an organ vasculature or through the heart. These models are electric analogues in the sense that they describe the relationship between flow rate and pressure analogously as the one between current and voltage in electric circuits.

In particular, each vessel tract can be characterized by three components, a resistance R, which represents the viscous resistance to flow, in series with an inductance L, which models the inertial effect and in parallel to a capacitance C, which takes into account the compliance of vessel walls. Integration of the 1D equations (3.4) and (3.5) along a vessel segment defines the relationships between incoming and outgoing flows and pressures of the vessel tract, and the component parameter values [29]. The latter are essentially a function of fluid and vessel properties (e.g., density, viscosity, radius, length, thickness, and Young modulus). Each component is thus defined by a specific flow–pressure relationship (see examples in Fig. 3.3, left) that can be linear or nonlinear, and with constant or time-varying parameters.

The usage of such 0D model is not restricted to a vessel tract. In fact, one or several complex parts of the cardiovascular system can be modeled by more general

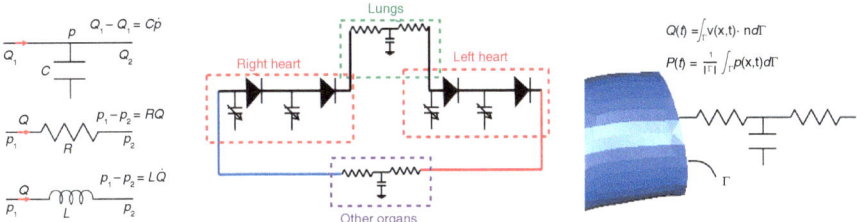

Fig. 3.3   Left: Elements of linear 0D models of a vessel tract in terms of capacity (C), resistance (R), and inductance (L), and corresponding pressure–flow relations. Center: A simple closed-loop 0D model of the whole circulation composed of four compartments: the left heart, the systemic circulation, the right heart and the pulmonary circulation. The two circulations are modeled by a Windkessel (RCR) block. The heart compartments include elements for which the pressure–flow relation is nonlinear such as diodes for the valves ( ) and effectively time-varying capacitances for the heart chambers ( ). Right: an RCR model used as boundary condition for a 3D model; in this case, average flow rate and pressure are computed from the three-dimensional *distributed* fields

lumped parameter models, connecting together different elementary components, as for an electric circuit. Each component can thus describe a single vessel or a set of vessels in an organ, or a group of organs. This methodology has been successfully applied to derive global lumped circulation models including the heart and with more or less details for the different circulations such as the hepatic or the pulmonary circulations (see, e.g., [25, 30, 31]). They include the two heart sides, with diodes representing valves and effectively time-varying capacitances representing the heart chambers (see [25, 31, 32] for different chamber models), and more components depending on the model complexity. An example of such a closed-loop modeling the circulation with four compartments is depicted in Fig. 3.3 (center). Such models result in nonlinear coupled equations, which form an algebraic-differential dynamical system. Efficient numerical schemes exist, even for stiff systems, usually involving adaptive time-stepping (e.g., [33]).

Lumped parameter models are also widely used as boundary conditions for 3D or 1D models, in order to take into account the rest of the circulation out of the domain of interest (see, e.g., [32, 34]). In these approaches, the flow rate and pressure on the three-dimensional boundary are related through simple 0D models (e.g., RCR elements, also called Windkessel model [35], see Fig. 3.3, right) or lumped models for the rest of the circulation. We refer to Sect. 3.2.1 for numerical aspects.

As standalone or coupled to higher-dimensional models, their parametrization from medical imaging and catheterization data is a research challenge, for which different strategies have been developed [4, 17, 31, 32, 36].

3.3   Modeling of Fluid at the Tissue Scale

The models presented in Sect. 3.2 describe the flow in large conduits and its effect at the organ or body scale at a *macro-scale* level. This section presents the modeling of blood flow on smaller scales, in order to better understand organ perfusion and blood flow through the vascular trees of the tissue. It focuses on two approaches. The first one models the microcirculation within a vessel's geometry, which can be either given or dynamically evolving, as it happens in angiogenesis. The second one is based on porous media flow modeling, in which the vessel geometry is not explicitly represented. The latter is extended to poroelasticity, in which the interaction between fluid and tissue is modeled considering a biphasic description of the tissue.

3.3.1   Modeling of Microcirculation

As explained in Sect. 3.2.4, the blood circulation can be modeled with electric analogues. In this context, blood flow in the microcirculation (small arteries, capillaries, small veins) is represented either by a single resistance in a top-down approach [31] or by an impedance (or resistance), computed, in a bottom-up way, recursively on structured (e.g., fractal) or morphometric vascular trees for which the impedance of individual conduits is defined based on their geometry and on the blood properties. Conservation of mass and continuity of pressure are applied at bifurcations. Microcirculation flow demand is thus embedded in 3D [1] or 1D [37, 38] arterial models, or in-between 1D arterial and venous trees [18].

When the focus is not on the effect of the microcirculation on the rest of the circulation, but rather on what is happening in the microcirculation itself, the approaches described above can be followed focusing on what is happening at the various generations [39]. However, other choices can be made. A detailed modeling of the flow in the microvasculature is out of reach, as available non-invasive image data have only a limited spatial resolution. Even if the detailed geometry can be extracted for extended networks (e.g., from micro-CT, casts), the resolution of full-scale mathematical models might result in an excessively high computational complexity.

To address the first limitation, one approach in order to obtain vasculature models beyond non-invasive image resolution is the so-called *constrained constructive optimization* (CCO): starting from a segmented vascular tree, new nodes are added finding optimal new bifurcations (see Fig. 3.4). The optimal conditions are not clear a priori, and can be defined, e.g., based on geometrical constraints (see, e.g., [40]). This approach has recently been employed for modeling the drug perfusion in the liver, considering an advection-diffusion model for the drugs driven by a constant fluid velocity in each vessel segment [41]. In other approaches, arterial and venous trees have been generated using bifurcation rules based on the prescribed boundary conditions [42]. Fractal trees, connected between the arterial and venous sides, and embedded into specific organ tissues have been the basis for hemodynamics and transport simulations [43, 44].

Less often, as it is difficult to acquire the geometry, microcirculation simulations are performed in segmented networks from invasive imaging data (e.g., [45, 46]). Flow and pressure are usually computed based on connected resistances or simplified 1D models, but 3D models have also been used on limited geometries [47]. Viscosity may vary with the vessel size and hematocrit [48].

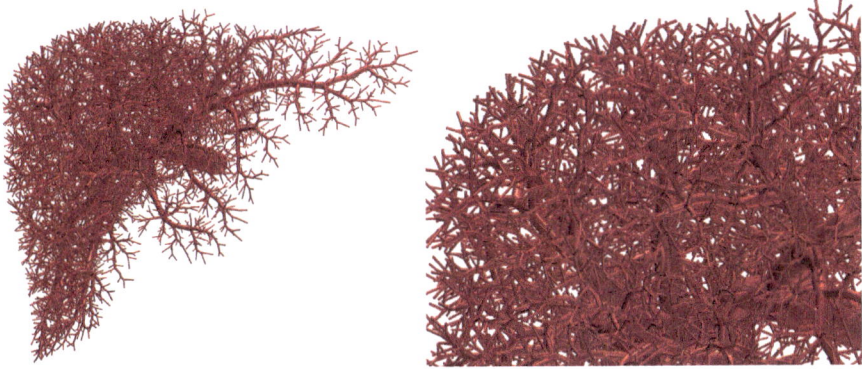

Fig. 3.4 Liver vasculature obtained computationally with a CCO algorithm [40] (Picture courtesy of Dr. Lars Ole Schwen, Fraunhofer MEVIS, Bremen)

3.3.2   Modeling of Microvascular Adaptation

The models described in Sect. 3.3.1 consider a fixed microvascular geometry. However, in several biomedical contexts it is important to take into account micro-vasculature growth and remodeling. This can be due to surgical changes in larger vessels, leading to a 3D-0D model that includes vascular distal tree adaptation [49] (see Sect. 3.2). A more complete mathematical model for vascular remodeling in response to altered vascular reactivity, as it might occur in hypertension, has been proposed in Ref. [45]. In this case, the structural and functional properties of micro-vascular networks are related to the adaptive responses of individual segments to hemodynamic and metabolic stimuli. Namely, vessels are assumed to respond to changes in wall shear stress, circumferential wall stress, and tissue metabolic status (indicated by partial pressure of oxygen), and structural changes in each segment are represented by the variation of mid-wall diameter and wall cross-section area. The flow in each vessel segment is based on Poiseuille law, taking into account the experimentally determined variation of apparent viscosity of blood with vessel diameter and hematocrit, while an advection model is used to simulate oxygen transport and consumption along the network [50].

Recently, autoregulation has been modeled in the context of retinal hemodynam-ics, considering 3D network of retinal arteries and taking into account FSI (see Sect. 3.2.2) [51]. In this case, the blood flow is modeled with the time-dependent Stokes equations, while the arterial wall model includes the endothelium and the smooth muscle fibers.

The flow within microvasculature is also relevant in the context of mathemati-cal modeling of tumor growth, in order to investigate the role of angiogenesis and vascular remodeling in the delivery of nutrients and during the development of cancer. In this case, the vasculature is not a given data, but it must be considered as a dynamically changing aspect. The potential of the mathematical model lies thus in its capability of investigating the possible evolution of vasculature captur-ing its essential features. Angiogenesis modeling by itself is thus an active field of research [52].

Besides continuum approaches that do not resolve individual vessels, individual-based (or agent-based) models treat capillaries and tissue cells as discrete objects, subject to a set of stochastic rules for cellular growth, vessel generation, and vascu-lature remodeling [53]. The dynamics of the system is described by a stochastic probability-based model defined by a set of rules on cell position, cell cycle and replication, growth, division and necrosis, vessel sprouting and remodeling, affected by local oxygen and glucose concentrations. The probabilities of different events depend on the state of the system itself.

The dynamics can be formalized by decomposing the spatial domain in lattice sites and denoting by $\mathbf{x}_k$ the state vector of the system at site k (if the site is empty, the vector state is zero), and by $X = \{x_1, x_2, \ldots\}$ the state vector of the whole system. The time evolution is then computed using the Gillespie algorithm [54] assuming that the underlying dynamics follows the master equation for the multivariate prob-ability distribution

$$\frac{\partial \wp(X,t)}{\partial t} = \sum_{X'} \left(\tau_{X' \to X}\right)^{-1} \wp(X',t) - \left(\tau_{X \to X'}\right)^{-1} \wp(X,t). \tag{3.9}$$

where $\wp(X,t)$ is the probability of the system being in configuration X at time t and the possible transitions from this X into another configuration X' are denoted by rates $(\tau_{X \to X'})^{-1}$ for each process (cell growth, division, and death). A process is then chosen with a probability that corresponds to its relative weight, calculated as the rate for this process divided by the sum of the rates for all other processes. To eliminate fluctuation effects that emerge from individual time evolution paths, observables are averaged over many realizations.

From an algorithm point of view, at each new event within the system (i.e., cell death, vascular sprouting, etc.) the blood flow model is updated (pressure, flow, shear stress—based on Poiseuille flow with Pries' viscosity model) and then the nutrients and growth factor reaction–diffusion equations are updated and solved. State transition probabilities are then computed for the vascular network and the cells, before a new event is computed to occur. An example of individual-based approach for vascularized tumor growth modeling, including an interaction between the different model components is described in Sect. 3.4.4.

3.3.3 Flow in Porous Media

This section is dedicated to Darcy's law, which is a widely used constitutive equation for describing flow through a porous medium. This is a suitable approach when the microcirculation geometry is not known in detail, but the envelope of the organ is known, and one wishes to assess the pressure gradients necessary to push the fluid within this tissue, possibly with local increase of resistance to flow.

Darcy's law can be derived starting from the Stokes equation, i.e., a simplification of the Navier–Stokes equation (3.1) in the case of small velocities (neglecting the quadratic velocity term) and of quasi-steady flow (neglecting the time derivative):

$$\nabla p - \eta \underbrace{2\nabla \cdot \mathbf{e}(\mathbf{v})}_{\text{viscous forces}} = \mathbf{f}. \tag{3.10}$$

As in (3.1), $\mathbf{v}$ denotes the velocity of the fluid, p its pressure, and $\mathbf{f}$ is related to external forces (e.g., gravity). Next, the magnitude of the viscous forces within the porous material (in other words, the resistance of the medium against the flow) is considered to be proportional to the fluid velocity, and to the porosity ϕ of the medium (i.e., the volume of fluid relative to the tissue volume), yielding

$$\nabla p + \eta \phi \mathbf{K}^{-1} \mathbf{v} = \mathbf{f}.$$

The tensor $\mathbf{K}$ is referred to as the permeability tensor, and it characterizes the resistance of a possibly anisotropic material depending on the flow direction. In particular, notice that isotropic porous media are characterized by a permeability tensor in the form $\mathbf{K} = k\mathbf{1}$ (with $k \in \mathbb{R}$). Hence, introducing the perfusion velocity $\mathbf{w} = \phi\mathbf{v}$ leads to Darcy's law

$$\mathbf{w} = -\frac{\mathbf{K}}{\eta}\left(\nabla p - \mathbf{f}\right). \tag{3.11}$$

The system is closed considering conservation of fluid mass, the fluid being considered incompressible:

$$\nabla \cdot \mathbf{w} = s, \tag{3.12}$$

where s stands for a sink or a source term, and appropriate boundary conditions (e.g., no flux or pressure boundary conditions).

Such 3D porous media approach has been used for representing microcirculatory flow in functional units of the liver [55] or a larger part of the organ [56], and has also been coupled to other models. A coupled 1D-3D approach for modeling heart perfusion has been recently described in [57]. In this case, the available vasculature is described via a 1D model (see Sect. 3.2.3) coupled, at the end-points, with a porous media model for the tissue. Simplified 1D models, but that allow for flow seepage, have been coupled with such 3D model to determine the interstitial tissue pressure, and study transport in this system [58]. A similar approach that increases the level of details of flow and tissue dynamics, coarsening the resolution at which the vasculature itself is modeled, has been described in Ref. [59]. In this model, the vessels up to image resolution are also treated as one-dimensional tracts, while the microcirculation network and the interaction with the interstitial tissue are handled using coupled compartments on different scales (3D, 1D, 0D) and a filtration model across the vessel wall. For the link of such models to perfusion imaging data (see, e.g., Chaps. 21–23), we refer, e.g., to the recent review [60].

The next section describes the extension of the static porous media to the case where the tissue deforms and interacts with the blood flowing inside it.

3.3.4   Poroelastic Modeling of Organ Perfusion

Notwithstanding the advances of anatomical description and measurements of organ trees (see, e.g., [61]) and of the corresponding physiological, physical, and numerical modeling aspects (see, e.g., [62, 63]), the simulation of blood flow inside all the vessels of an entire organ from the arteries to the veins via the capillaries is still out of reach. Thus, perfusion can be modeled within the frameworks of porous media and poroelasticity, in which the flow is described only up to a given spatial scale and phenomena on smaller scales are aggregated into macroscopic quantities (see [64] and subsequent works).

Poromechanics is a simplified mixture theory in which the complex fluid–structure interaction problem (see Sect. 3.2.2) is replaced by the superposition of fluid and solid components at every spatial point with varying fractions. In particular, linear poroelastic models for small displacement are based on Biot's theory [65, 66] (see Chap. 4) with an elastic description of the solid matrix and Darcy's law for the porous media flow (see also Sect. 3.3.3). When this theory is applied to organs, incompressibility and large strain are particularly important, yet not easy, to take into account.

This section describes a general formulation valid for finite strain poroelasticity and compatible with the incompressibility constraint [67], two important aspects in the modeling of living tissues (beating heart, deformed organ during surgery, important swelling). This is the result of a strategy—presented in Ref [68] in a linear framework—based on deriving the formulation from an appropriate free energy functional, which is crucial to guarantee that fundamental thermodynamics principles are satisfied. Based on thermodynamics considerations, we developed the constitutive laws that characterize the solid phase, the fluid phase, and their interaction.

In the following equations, we adopt a Lagrangian formulation, indicating with $\hat{x}, \hat{y}$, and $\hat{z}$ the fixed reference coordinate system. Moreover, the subscript "0" denotes the physical quantities in the fixed reference configuration. In particular, we denote with $\mathbf{u}_0$ the displacement field of the skeleton with respect to the reference configuration and with $\hat{\nabla}$ the gradient operators defined by the spatial derivatives with respect to $\hat{x}, \hat{y}$ and $\hat{z}$. Next, we introduce the deformation gradient

$$\mathbf{F} = \mathbf{1} + \hat{\nabla}_{\mathbf{u}_0}$$

and the quantities

$$J = \det \mathbf{F}, \quad \mathbf{C} = \mathbf{F}^T \mathbf{F}, \quad \mathbf{E} = \frac{1}{2}(\mathbf{C} - \mathbf{1})$$

denoting the local volume change, the right Cauchy-Green deformation tensor and the Green-Lagrange strain tensor, respectively.

The saturated porous continuum consists of a solid part—the *skeleton*—and of a fluid part that accounts for the volume fraction ϕ of the tissue volume in deformed configuration. As next, let us denote with ρ^f the fluid mass per unit of fluid volume and with ρ_0 the total mass per unit volume in the reference configuration, i.e. $\rho_0 = \phi_0 \rho_0^f + (1 - \phi_0) \rho_0^s$.

The change in fluid mass per unit volume of the reference configuration reads $m = \rho^f J \phi - \rho_0^f \phi_0$, and the general conservation of fluid mass in porous media is given by

$$\nabla \cdot (\rho^f \mathbf{w}) + \frac{1}{J} \dot{m} = \rho^f s, \tag{3.13}$$

where $\mathbf{v}^s = \mathbf{u}_0$ is the velocity of a solid particle, $\mathbf{v}^f$ the velocity of a fluid particle, $\mathbf{w} = \phi(\mathbf{v}^f - \mathbf{v}^s)$ is the perfusion velocity, and s a general sink or source term.

Moreover, the balance of momentum of the porous medium written in Lagrangian form reads

$$(\rho_0 + m)\ddot{\mathbf{u}}_0 = \hat{\nabla} \cdot (\mathbf{F} \cdot \Sigma), \tag{3.14}$$

where Σ is the second Piola-Kirchhoff stress tensor, that shall be defined by a proper constitutive law (see below). The discrepancy between the fluid and solid accelerations has been neglected in the inertia term (see [67] for details).

Next, the following constitutive equations are considered in order to derive a finite strain poromechanics formulation that is compatible with the incompressible limit of both the fluid and solid phases individually.

$$\Sigma = \frac{\partial W^{\mathrm{hyp}}}{\partial \mathbf{E}} - Hb\frac{m}{\rho_0^f}\left(f+(J-1)f'\right)J\mathbf{C}^{-1} + \frac{1}{2}H\left(\frac{m}{\rho_0^f}\right)^2 f'J\mathbf{C}^{-1}, \qquad (3.15)$$

$$p - p_0 = Hf(J)\left(b(1-J)+\frac{m}{\rho_0^f}\right) - \frac{\kappa_0}{m/\rho_0^f + \phi_0}. \qquad (3.16)$$

In (3.15), W^{hyp} denotes a hyperelastic potential to be chosen, p is the fluid pressure, H is the so-called Biot modulus (see Chap. 4), b a parameter characteristic of the skeleton, and f a function to define according to the specific hyperelastic potential considered. In particular, when using the modified Ciarlet-Geymonat expression (see [69] for the original formulation, J_1 and J_2 being the first two invariants of the left Cauchy-Green deformation tensor)

$$W^{\mathrm{hyp}} = \kappa_1(J_1 - 3) + \kappa_2(J_2 - 3) + \kappa(J-1) - \kappa \ln J, \qquad (3.17)$$

where nearly-incompressible materials are obtained by taking the bulk modulus κ large compared to the other material parameters κ_1 and κ_2, some constraints arise on parameters and on the function f.

The last term in the energy function (3.16), depending on the coefficient κ_0, is added in order to avoid the violation of the porosity constraint $0 < \phi < 1$, which may happen in the above poroelastic constitutive law. Finally, the transport of fluid mass is described by Darcy's law (see (3.11)).

Combining (3.13), (3.14), (3.16) and (3.11), neglecting body forces, the porous media dynamics is thus finally characterized by the following system. Find $\mathbf{u}_0$, $\mathbf{w}$, p, and m such that

$$\begin{cases} (\rho_0 + m)\ddot{\mathbf{u}}_0 = \hat{\nabla}\cdot(\mathbf{F}\cdot\Sigma), \\[2mm] \nabla\cdot\mathbf{w} + \frac{1}{JHf}\dot{p} - \frac{f'}{Hf^2}(p-p_0)\nabla\cdot\mathbf{v}^S = -b\nabla\cdot\mathbf{v}^s + s, \\[2mm] \mathbf{w} = -\mathbf{K}\cdot\nabla p, \\[2mm] p - p_0 = Hf(J)\left(b(1-J)+\frac{m}{\rho_0^f}\right) - \frac{\kappa_0}{m/\rho_0^f + \phi_0}. \end{cases} \qquad (3.18)$$

This system needs to be complemented by adequate boundary conditions, typical of solid and fluid equations. In order to solve the resulting equations in practice, finite element codes tailored for the solid and fluid parts can be iteratively coupled (the interested reader is referred to [67]).

Compared to modeling the solid alone or the fluid alone, this coupled model has only three parameters, two (H and b) for the fluid–solid interaction per se, and one (κ_0) to make sure that the porosity stays positive. The advantage of this poroelastic approach is that it can handle large swelling (how an increase of fluid deforms the pores and the whole skeleton) or complete drainage of tissue (how an external force is transmitted into deforming the tissue and changing fluid porosity) [67]. Section 3.4.3 provides an example of application to cardiac mechanics and perfusion.

3.4   Applications

This section presents recent selected applications in which the modeling approaches described in Sects. 3.2 and 3.3 are employed, in connection with suitable computational methods for the solution of the model equations. The aim is to enhance the information given by biomedical imaging towards non-invasive diagnosis and treatment planning.

3.4.1   Image-Based Evaluation of Indicators

Computational fluid dynamics (CFD) is a complementary tool to in vitro and in vivo experiments to gain insights on hemodynamics and to better understand the link between hemodynamics and disease development or treatment. This section shortly reviews some recent numerical studies, which, based on experimental and clinical data, focus on different aspects of the anatomical and physiological blood circulation.

In [70], the results of numerical simulations showed a decrease of the simulated infrarenal aortic wall shear stress in spinal cord injury patients compared to normal subjects, which may explain why these patients are more prone to abdominal aneurysm development. Similarly, the fact that their legs atrophy, and consequently their downstream arterial impedance changes is likely to favor aneurysm development due to the computed sustained high pressure during the cardiac cycle. These two factors, evaluated via a computational model, would not have been easily measured in patients. As observed in Sect. 3.2.1, appropriate boundary conditions are essential in order to obtain clinically relevant results [1].

Physicians have also expressed the need for a better understanding of the hemodynamics conditions in subjects with congenital heart defects, pre- and postoperatively. Integrating patient-specific clinical information (image-based geometry and hemodynamics measurements) into numerical simulations is critical to yield results which accurately represent a patient-specific condition (see, e.g., [3]). These simulations can be used to evaluate pressure loss, wall shear stress (WSS), and oscillatory shear index (OSI), which are considered to be relevant indicators of altered hemodynamics compared to healthy cases in coarctation of the aorta, even for subjects that underwent repair, under rest and simulated exercise conditions [71].

As an example, the following Figs. 3.5 and 3.6 show the result of studying the complex hemodynamics of single-ventricle patients between two stages of palliative surgeries [5]. Clinical data were acquired a few days before the patient's Fontan procedure. To build the patient-specific geometry (Fig. 3.5), magnetic resonance angiography (MRA) images were acquired with a 1.5 T scanner (Signa Twin-Speed,

Fig. 3.5 3D geometry (in full red) segmented from MRA imaging data, here shown in the background as MIP. 3D fast gradient-recalled echo sequence after injection of a gadolinium-based contrast agent. Spatial resolution: 0.6 mm × 0.6 mm × 2 mm. Reproduced from [5] with permission

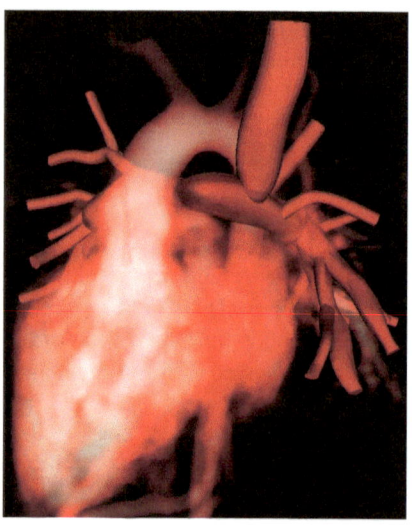

General Electric, Milwaukee, MI, USA) with a 3D fast gradient-recalled echo sequence after injection of a gadolinium-based contrast agent. The patient was intubated and ventilation was held during the acquisition. The spatial resolution was about 0.6 mm × 0.6 mm × 2 mm.

Flow rates at several locations were acquired using PC-MRI. Twenty cardiac-gated and respiratory-compensated time points per cardiac cycle were retrospectively reconstructed from the through-plane velocity acquisitions with a velocity encoding of $150 \frac{cm}{s}$. Combined with catheterization data, these imaging inputs led to the patient-specific simulations shown in Fig. 3.6. In this case, the results suggested that a lower-than-normal wall shear stress in their pulmonary arteries might explain their propensity to clotting [5].

However, although predictive modeling for surgical planning has first been demonstrated in [3] for a Fontan patient predicting the high pressure in the superior vena cava that can lead to cognitive development issues, the prediction of clinically critical hemodynamics quantities, however, is still in its infancy. The power of simulation in this context is the ability to test different 3D surgical options, which obviously is not possible on a patient [32].

3.4.2   Estimation of Model Parameters in One-Dimensional Models

As already mentioned in the modeling Sect. 3.2, in order to reproduce clinically relevant scenarios, mathematical models have to be tuned in order to represent patient-specific settings. This is the goal of *data assimilation*, i.e., the branch of mathematics aiming at improving models using available data (e.g., observations)

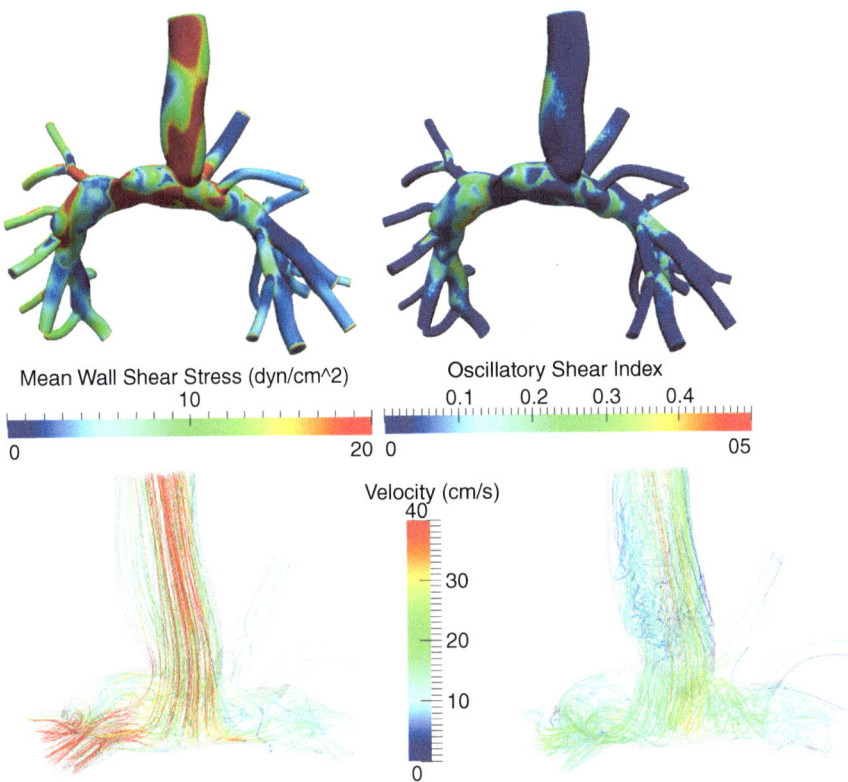

Mean Wall Shear Stress (dyn/cm^2)

Oscillatory Shear Index

Velocity (cm/s)

Fig. 3.6 Simulated hemodynamics indicators from the 3D image-based geometry (Fig. 3.5): WSS (top left), OSI (top middle), and zooms in the region of interest of streamlines at peak flow (bottom left) and deceleration (bottom middle). These quantities were obtained solving numerically the 3D Navier–Stokes equations (3.1) within the acquired patient-specific geometry. Reproduced from [5] with permission

on the system of interest. In other words, the goal is to optimize (minimize) a certain functional defined in terms of the discrepancy between the observed quantities and the result of the simulated mathematical model. The quantities to change in order to reach this optimum are usually a set of model parameters.

Addressing this inverse problem (e.g., tuning model parameters starting from a given solution) requires multiple solutions of the underlying forward or direct problem (e.g., the mathematical model), making necessary the usage of a fast fluid solver, such as a reduced order model. This is a growing field of research in hemodynamics, and we refer to recent publications and their references for different approaches involving 3D, 1D, and 0D parameter estimations [27, 31, 36, 72, 73]. This section shortly describes the approach detailed in [27], which focuses on the estimation of model parameters (e.g., wall properties) in 1D networks based on available measurements.

In recent works [74, 75], an experimental 1:1 replica of the largest conduit arteries of the systemic circulation, manufactured using crafted tubular aluminium moulds (Fig. 3.7, left) has been employed to generate a set of in vitro data, which were then used for validating different numerical methods for the solution of system (3.4), (3.5), and (3.7) with different computational methods (see, e.g., [26]). In Ref. [27] the published experimental measurements have been used to assess the parameter estimation using the reduced-order unscented Kalman filter (roUKF) [76, 77]. Namely, starting from the mathematical model of the in vitro network and from the experimental flow and pressure data, the focus of [27] was to recover the material parameters (e.g., Young modulus or vessel thicknesses).

The algorithm defining the roUKF can be summarized with the following steps:

1. Sampling step: depending on the number of parameters of the system, the parameter space is sampled, creating different instances of the mathematical model (so-called *particles*);
2. Propagation step: each particle is advanced in time according to the underlying *forward* solver, i.e., a computational method for solving Eqs. (3.4) and (3.5);

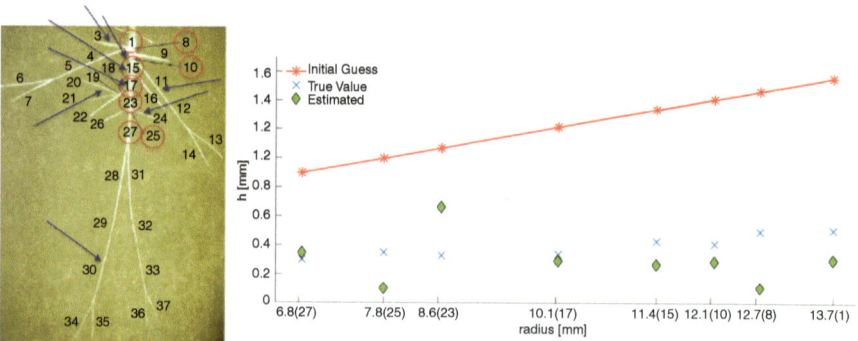

Fig. 3.7 Left: The 1:1 in vitro replica of the systemic circulation developed in Ref. [75]. The network consists of 37 crafted tubular aluminium moulds (CMBD Research Unit, Ghent University, Belgium) which were either linearly tapered or had constant cross-section (for the smallest branches). The silicone network was manually connected to a Harvard pulsatile pump, mimicking the heart, and to a set of terminal resistance tubes connected to overflow reservoirs. A 65–35% water glycerol mixture was used to mimic the blood flow. Values of length, radius, and thickness were measured at the inlet, middle point, and outlet of each arterial segment using a ruler, calipers, and micrometer. By performing tensile tests on 63 silicone sample strips, a constant Young's modulus of 1.2 MPa was measured within the range of pressures considered in the experiment. The flow rate was measured using an ultrasonic volume-flow meter (Dual Channel Flowmeter, Transonic Systems Inc., Ithaca, NY, USA) and pressure was measured simultaneously by means of two micro-tip catheter pressure transducers (Millar Instruments, Houston, TX, USA). The red circles denote the vessels belonging to the aorta, while the blue arrows indicate the location of the observed flows used for thickness estimation. Right: Results for the estimation of aorta thickness. For each vessel (the number in brackets correspond to the sketch on the left), the figure compares the initial guess, the results of the filter (estimated) and the reference values provided by [75] (referred to as *true values*). Reproduced from [75] with permission

3. Evaluation step: the discrepancies between available measurements and model results are computed (so-called *innovation*);
4. Correction/filtering step: the state variables and the parameters of the system are modified in order to minimize the expected value of the error.

The roUKF is a sequential approach for data assimilation, in the sense that it interacts with the system at each simulated time step, unlike variational methods, that typically require the solution of a whole interval (cardiac cycle, etc.) before assimilating the available data.

Moreover, the roUKF does not require the solution of a tangent problem in order to find a suitable minimization direction. In fact, the parameters are optimized choosing efficient sampling of the parameter space. A further advantage is that the algorithm described in the above steps can be applied as a black-box, having at disposal a code for performing the time integration (propagation) of the desired mathematical model.

Figure 3.7 (right) shows the results for the estimation of the thicknesses of the tubes representing the aorta, using eight measured flow rates. Even if starting from a relatively far initial guess (about three times larger value), the target values (i.e., the values reported in [78]) are approached. Numerical simulations have been carried out using the solver described in Ref. [22]. Vessel parameter estimation has the potential to help characterizing vessel diseased states [28].

3.4.3   Cardiac Perfusion

Through the poroelastic model under large deformations presented in Sect. 3.3.4 it is possible to study cardiac perfusion, which is an application where swelling and drainage occur during each cardiac beat.

Heart perfusion consists of a flow through coronary arteries, arterioles, capillaries, venules, and veins. In Ref. [67], this phenomenon has been modeled starting from a 3D analytical geometry mimicking the left ventricle (Fig. 3.8, left), with fibers of varying orientation [79], and based on a single-compartment porovisco-elastic model, which is supposed to only describe small arteries, capillaries, and small veins. The rest of the venous network is modeled as a simple sink term in the porous flow equations, while the flow is assumed to be fed by a distributed arterial source term. The specific physiologic values for boundary conditions and other model parameters are fully described in Ref. [67].

The resulting time-averaged behavior is a system in which blood flows continuously from the small arteries into the capillaries, and from the capillaries into the small veins, at a flow rate of 4 mL s^{-1}, a mean pressure of 15 mmHg (2 kPa) and a muscle volume of 260 mL. Figure 3.8 (right) shows the results of the simulated perfusion, in terms of the variation in time of pressure, muscle fluid volume (relative to a reference state), and flow rate.

Fibers contraction induces a rise of pressure. As a consequence, flow from the small arteries into the capillaries q_a is considerably reduced—which corresponds to the so-called *flow impediment* (see, e.g., [63] and the references herein)—and blood

Fig. 3.8 Simulated pressure, added blood mass, and flow rates traced over one cardiac cycle. The values of pressure, mass, and arterial and venous flow rates are traced for endocardium (endo) and epicardium (epi), at exact locations shown in the right figure (3D model of left ventricle), as well as on average over the volume. Reproduced from [67] with permission

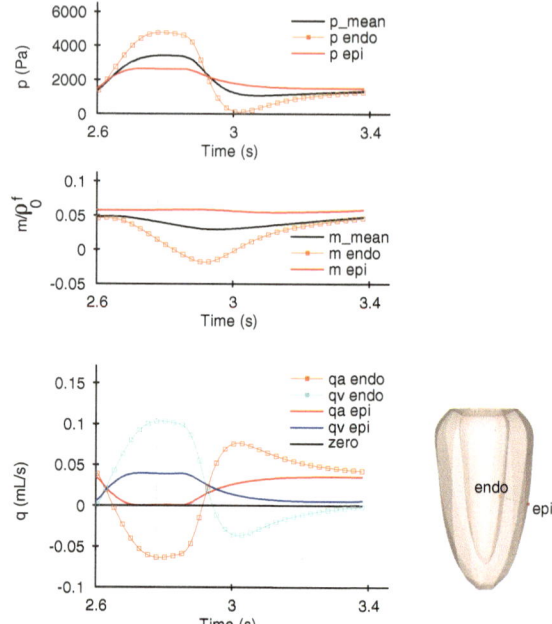

is squeezed out of the capillaries (lowering of m) into the small veins (rise of q_v). During relaxation, the opposite happens: m rises as q_a increases and q_v decreases, filling up the capacitance of the capillaries. The mainly systolic flow in the small arteries and mainly diastolic flow in the small veins are consistent with the measured velocities in small arteries and small veins in the left ventricle given in Ref. [80]. We refer to [67] for further details on this application.

To conclude, as far as cardiac perfusion is concerned, the major difficulty lies in the complexity of the physiological phenomena and the lack of sufficiently detailed measurements to validate the various modeling assumptions. Nevertheless, the results indicate that taking into account the ventricle cavity pressure is not necessary to model flow impediment during systole and to obtain a higher impediment in the endocardium than in the epicardium. At the time of [67], this 3D perfusion model represents a significant step forward of realistic simulations in a beating ventricle. Cardiac perfusion modeling and simulation have since then gained more attention [57, 78, 81].

3.4.4 Vascularized Tumor

In this section, we shortly describe the model of microvasculature proposed in Ref. [53] in the context of angiogenesis, i.e. the formation of new blood vessels, for describing the transition of tumors from a benign state to a malignant one.

The model includes the cellular (individual cells and vessels, development and death) and the molecular (oxygen, glucose and angiogenic growth factors reaction/diffusion) components, showing that experimental observations can be explained by the interplay of processes on the molecular and the cellular scale. In particular, the model of [53] combines stochastic agent-based models (cellular automaton) for cells and vessels, defined by a set of rules on agent position and processes, affected by local oxygen and glucose concentrations. The nearby vasculature releases oxygen and glucose that diffuse in the local environment and nourish the tumor. This behavior is described by a reaction–diffusion equation, where the concentration is set at the blood vessel nodes, and cells represent distributed sinks. These cancer cell consumption rates of glucose and oxygen are coupled non-linear terms. Similarly, VEGF is released by the (hypoxic) necrotic cells and diffuses into the tumor environment. In this case, the microcirculation is modeled as described in Sect. 3.3.1. Moreover, the vascular network responds to the changes of the local micro-environment by angiogenesis or remodeling.

The model can be used to explore the role of angiogenesis on the growth dynamics of small tumors by direct comparison to situations where angiogenesis would not occur. In addition to the tumor size, its structure varies significantly with the different environmental conditions. When there are neither oxygen nor nutrient limitations (Fig. 3.9, left), cells are either proliferating (yellow) or quiescent (green), but none of the cells are necrotic. In contrast, when oxygen or nutrients are lacking because their diffusions from blood vessels are not fast enough and their local concentrations are too low, necrotic cells (blue) appear in the center. This is the case before angiogenesis starts or when it is blocked by drugs (Fig. 3.9, center). As a response to hypoxia and hyponutrition, cells produce growth factors that diffuse through the tissue, reach the existing blood vessels, and finally trigger sprouting from them to create new blood vessels (Fig. 3.9, right).

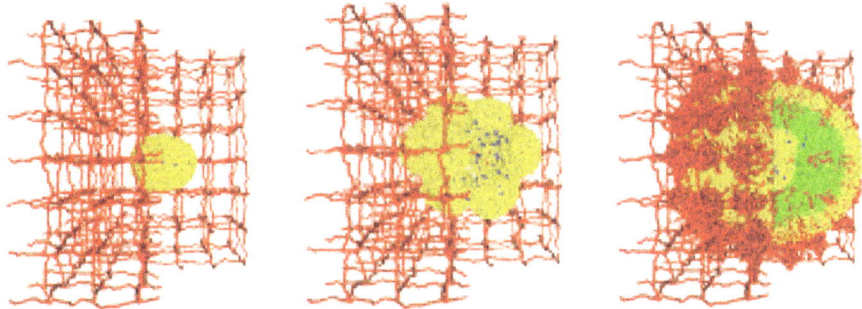

Fig. 3.9 Tumor growth from a common no-limitation state (left) into vascularized tissue without (center) and with (right) angiogenesis. Proliferating (*yellow*), quiescent (*green*), and necrotic cells (*blue*) around blood vessels (*red*). External views (left) and cropped views (right)

3.5   Summary

This chapter gave an overview on several approaches for the mathematical modeling of blood flows, discussing approaches on different scales. First, the macrocirculation level has been considered, introducing three-dimensional models based on the incompressible Navier–Stokes equations. In parallel, complexity reduction methods based on one-dimensional and zero-dimensional approximations have shortly been introduced. Second, we discussed the modeling of blood flow at the tissue scale, including mathematical modeling of microcirculation and tissue perfusion. In both cases, we considered static vessel models and extension with interactions with their environment. Finally, several examples of image-based quantification of blood flow have been presented, where morphology (e.g., from MRI, CT) as well as physiological (e.g., from MRI, Doppler or catheter) information is combined with mathematical modeling in order to acquire a deeper understanding of imaging data and to provide, through numerical simulations, quantities not easily measurable. Disease states can be understood by a new angle. Besides, disease development, virtual surgery, or novel device design can be explored by multiscale computations, which are to a certain extent predictive. In particular, computing numerically local (three-dimensional) flow disturbances sheds light on the difficulty to image blood flow in certain regions, and it could enable radiologists to re-interpret their imaging data (in PC-MRI, Doppler-US, etc.).

However, there are still a number of challenges that remain to be tackled. The access to the input medical data is still not common, and the workflow from data gathering to simulations output would benefit from having dedicated engineers in the clinics. The calibration of models based on medical data is in general difficult, and the strategy depends on the type of model, as well as on the available data. Concerning non-invasive in-vivo imaging, for which the resolution is limited, it is not completely clear how the information relates to the underlying tissue microstructure. Moreover, this limited resolution in space and time may introduce much uncertainty in the output, that remains largely to be quantified. Finally, for a model to be truly predictive, validation is a key issue, which requires a close interaction between modeling, simulation, and clinical experiments.

Acknowledgments The authors are grateful to C. Bertoglio, D. Lombardi, and L. O. Müller for the useful discussions.

References

1. Vignon-Clementel IE, Figueroa CA, Jansen KE, Taylor CA. Outflow boundary conditions for three-dimensional finite element modeling of blood flow and pressure in arteries. Comput Methods Appl Mech Eng. 2006;195(29):3776–96.
2. Arbia G, Corsini C, Moghadam M, Marsden A, Migliavacca F, Pennati G, Hsia T, Vignon-Clementel I, Allianc MCH. Numerical blood flow simulation in surgical corrections: what do we need for an accurate analysis? J Surg Res. 2014;186:44–55.

3. Vignon-Clementel IE, Marsden AL, Feinstein JA. A primer on computational simulation in congenital heart disease for the clinician. Prog Pediatr Cardiol. 2010;30:3–13.
4. Arbia G, Corsini C, Baker C, Pennati G, Hsia TY, Vignon-Clementel IE. Pulmonary hemodynamics simulations before stage 2 single ventricle surgery: patient-specific parameter identification and clinical data assessment. Cardiovasc Eng Technol. 2015;6(3):268–80.
5. Troianowski G, Taylor CA, Feinstein JA, Vignon-Clementel IE. Three-dimensional simulations in Glenn patients: clinically based boundary conditions, hemodynamic results and sensitivity to input data. J Biomech Eng Trans ASME. 2011;133(11):111006.
6. Formaggia L, Gerbeau JF, Nobile F, Quarteroni A. Numerical treatment of defective boundary conditions for the Navier-Stokes equations. SIAM J Numer Anal. 2002;40(1):376–401.
7. Malossi A, Blanco P, Deparis S, Quarteroni A. Algorithms for the partitioned solution of weakly coupled fluid models for cardiovascular flows. Int J Numer Methods Biomed Eng. 2011;27(12):2035–57.
8. Moghadam M, Vignon-Clementel I, Figliola R, Marsden A. A modular numerical method for implicit 0d/3d coupling in cardiovascular finite element simulations. J Comput Phys. 2013;244:63–79.
9. Quarteroni A, Ragni S, Veneziani A. Coupling between lumped and distributed models for blood flow problems. Comput Vis Sci. 2001;4:111–24.
10. Chapelle D, Bathe K. The finite element analysis of shells – fundamentals. Berlin Heidelberg: Springer; 2003.
11. Gerbeau JF, Vidrascu M, Frey P. Fluid-structure interaction in blood flows on geometries based on medical imaging. Comput Struct. 2005;83(2–3):155–65.
12. Moireau P, Xiao N, Astorino M, Figueroa CA, Chapelle D, Taylor C, Gerbeau JF. External tissue support and fluid–structure simulation in blood flows. Biomech Model Mechanobiol. 2012;11(1–2):1–18.
13. Astorino M, Chouly F, Fernández M. An added-mass free semi-implicit coupling scheme for fluid-structure interaction. C R Acad Sci Paris Sér I Math. 2009;347(1–2):99–104.
14. Badia S, Nobile F, Vergara C. Fluid-structure partitioned procedures based on Robin transmission conditions. J Comput Phys. 2008;227:7027–51.
15. Fernández M, Gerbeau JF. Fluid structure interaction problems in haemodynamics, Chap. 9. In: Formaggia L, Quarteroni A, Veneziani A, editors. Cardiovascular mathematics. Modeling and simulation of the circulatory system. Milano: Springer Verlag; 2009.
16. Figueroa A, Vignon-Clementel I, Jansen K, Hughes T, Taylor C. A coupled momentum method for modeling blood flow in three-dimensional deformable arteries. Comput Methods Appl Mech Eng. 2006;195(41):5685–706.
17. Quarteroni A, Veneziani A, Vergara C. Geometric multiscale modeling of the cardiovascular system, between theory and practice. Comput Methods Appl Mech Eng. 2016;302:193–252.
18. Müller LO, Toro EF. A global multiscale mathematical model for the human circulation with emphasis on the venous system. Int J Numer Methods Biomed Eng. 2014;30:681–725.
19. Formaggia L, Lamponi D, Quarteroni A. One-dimensional models for blood flow in arteries. J Eng Math. 2003;47:251–76.
20. Hughes TJ, Lubliner J. On the one-dimensional theory of blood flow in the larger vessels. Math Biosci. 1973;18(1):161–70.
21. Wang XF, Ghigo A, Nishi S, Matsukawa M, Lagrée PY, Fullana J. Fluid friction and wall viscosity of the 1D blood flow model: study with an in-vitro experimental setup. J Biomech. 2016;49:565–71.
22. Müller LO, Blanco PJ, Watanabe SM, Feijóo R. A high-order local time stepping finite volume solver for one-dimensional blood flow simulations: application to the Adan model. Int J Numer Methods Biomed Eng. 2016;32(10): n/a–n/a.
23. Müller LO, Toro EF. Well-balanced high-order solver for blood flow in networks of vessels with variable properties. Int J Numer Methods Biomed Eng. 2013;29(12):1388–411.
24. Toro EF. Riemann solvers and numerical methods for fluid dynamics: a practical introduction. 3rd ed. Berlin Heidelberg: Springer; 2009. ISBN 978-3-540-25202-3.

25. Audebert C, Bucur P, Bekheit M, Vibert E, Vignon-Clementel IE, Gerbeau JF. Kinetic scheme for arterial and venous blood flow, and application to partial hepatectomy modeling. Comput Methods Appl Mech Eng. 2017;314:102–25.
26. Boileau E, Nithiarasu P, Blanco PJ, Müller LO, Fossan FE, Hellevik LR, Donders WP, Huberts W, Willemet M, Alastruey J. A benchmark study of numerical schemes for one-dimensional arterial blood flow modelling. Int J Numer Methods Biomed Eng. 2015;31(10): n/a–n/a.
27. Caiazzo A, Caforio F, Montecinos G, Müller LO, Blanco PJ, Toro EF. Assessment of reduced order Kalman filter for parameter identification in one-dimensional blood flow models using experimental data. Tech. Rep. 2248, WIAS; 2016.
28. Dumas L, El Bouti T, Lucor D. A robust and subject-specific hemodynamic model of the lower limb based on noninvasive arterial measurements. J Biomech Eng. 2017;139(1):011002.
29. Formaggia L, Quarteroni A, Veneziani A, editors. Cardiovascular mathematics. Modeling and simulation of the circulatory system. Modeling, simulation and applications, vol. 1. Milano: Springer; 2009.
30. Liang F, Liu H. A closed-loop lumped parameter computational model for human cardiovascular system. JSME Int J. 2005;48(4):484–93.
31. Pant S, Corsini C, Baker C, Hsia TY, Pennati G, Vignon-Clementel IE, for MOCHA. Data assimilation and modelling of patient-specific single-ventricle physiology with and without valve regurgitation. J Biomech. 2016;49(11):2162–73.
32. Corsini C, Baker C, Kung E, Schievano S, Arbia G, Baretta A, Biglino G, Migliavacca F, Dubini G, Pennati G, Marsden A, Vignon-Clementel I, Taylor A, Hsia T, Dorfman A, Hearts MC. An integrated approach to patient-specific predictive modeling for single ventricle heart palliation. Comput Methods Biomech Biomed Eng. 2014;17:1572–89.
33. Serban R, Petra C, Hindmarsh A. User documentation of ida v27.0. 2015.
34. Vignon-Clementel IE, Figueroa CA, Jansen KE, Taylor CA. Outflow boundary conditions for 3d simulations of non-periodic blood flow and pressure fields in deformable arteries. Comput Methods Biomech Biomed Eng. 2010;13:625–40.
35. Frank O. Die Grundform des arteriellen Pulses. Z Biol. 1899;37:483526.
36. Pant S, Fabrèges B, Gerbeau JF, Vignon-Clementel I. A methodological paradigm for patient-specific multi-scale CFD simulations: from clinical measurements to parameter estimates for individual analysis. Int J Numer Methods Biomed Eng. 2014;30(12):1614–48.
37. Olufsen MS. Structured tree outflow condition for blood flow in larger systemic arteries. Am J Physiol Heart Circ Physiol. 1999;276(1):H257–68.
38. Spilker R, Feinstein J, Parker D, Reddy V, Taylor C. Morphometry-based impedance boundary conditions for patient-specific modeling of blood flow in pulmonary arteries. Ann Biomed Eng. 2007;35(4):546–59.
39. Debbaut C, Monbaliu D, Casteleyn C, Cornillie P, Van Loo D, Masschaele B, Pirenne J, Simoens P, Van Hoorebeke L, Segers P. From vascular corrosion cast to electrical analog model for the study of human liver hemodynamics and perfusion. IEEE Trans Biomed Eng. 2011;58(1):25–35.
40. Schwen LO, Preusser T. Analysis and algorithmic generation of hepatic vascular systems. Int J Hepatol. 2012:1–17. https://doi.org/10.1155/2012/357687.
41. Schwen LO, Krauss M, Niederalt C, Gremse F, Kiessling F, Schenk A, Preusser T, Kuepfer L. Spatio-temporal simulation of first pass drug perfusion in the liver. PLoS Comput Biol. 2014;10(3):e1003499.
42. Schreiner W, Buxbaum P. Computer-optimization of vascular trees. IEEE Trans Biomed Eng. 1993;40(5):482–91.
43. Causin P, Guidoboni G, Malgaroli F, Sacco R, Harris A. Blood flow mechanics and oxygen transport and delivery in the retinal microcirculation: multiscale mathematical modeling and numerical simulation. Biomech Model Mechanobiol. 2016;15(3):525–42.
44. Mescam M, Kretowski M, Bezy-Wendling J. Multiscale model of liver dce-mri towards a better understanding of tumor complexity. IEEE Trans Med Imaging. 2010;29(3):699–707.
45. Pries AR, Reglin B, Secomb TW. Remodeling of blood vessels. Hypertension. 2005;46(4): 725–31.

46. Stamatelos SK, Kim E, Pathak AP, Popel AS. A bioimage informatics based reconstruction of breast tumor microvasculature with computational blood flow predictions. Microvasc Res. 2014;91:8–21.
47. Debbaut C, Vierendeels J, Casteleyn C, Cornillie P, Van Loo D, Simoens P, Van Hoorebeke L, Monbaliu D, Segers P. Perfusion characteristics of the human hepatic microcirculation based on three-dimensional reconstructions and computational fluid dynamic analysis. J Biomech Eng. 2012;134(1):011003.
48. Pries AR, Secomb TW, Gessner T, Sperandio MB, Gross JF, Gaehtgens P. Resistance to blood flow in microvessels in vivo. Circ Res. 1994;75:904–15.
49. Yang W, Feinstein JA, Vignon-Clementel IE. Adaptive outflow boundary conditions improve post-operative predictions after repair of peripheral pulmonary artery stenosis. Biomech Model Mechanobiol. 2016;15(5):1345–53.
50. Pries AR, Reglin B, Secomb T. Structural adaptation of microvascular networks: functional roles of adaptive responses. Am J Phys. 2001;281:H1015–25.
51. Aletti M, Gerbeau JF, Lombardi D. Modeling autoregulation in three-dimensional simulations of retinal hemodynamics. Journal for Modeling in Ophthalmology 2016;1:88–115.
52. Secomb TW, Alberding JP, Hsu R, Dewhirst MW, Pries AR. Angiogenesis: an adaptive dynamic biological patterning problem. PLoS Comput Biol. 2013;9(3):e1002983.
53. Drasdo D, Jagiella N, Ramis-Conde I, Vignon-Clementel IE, Weens W. Modeling steps from a begnin tumor to an invasive cancer: examples of instrinsically multiscale problems. Boca Raton: Taylor & Francis; 2010. p. 379–416.
54. Gillespie DT. Exact stochastical simulations of coupled chemical reactions. J Phys Chem. 1977;81(25):2340–61.
55. Debbaut C, Vierendeels J, Siggers JH, Repetto R, Monbaliu D, Segers P. A 3d porous media liver lobule model: the importance of vascular septa and anisotropic permeability for homogeneous perfusion. Comput Methods Biomech Biomed Eng. 2014;17(12):1295–310.
56. Ricken T, Werner D, Holzhütter H, König M, Dahmen U, Dirsch O. Modeling function–perfusion behavior in liver lobules including tissue, blood, glucose, lactate and glycogen by use of a coupled two-scale pde–ode approach. Biomech Model Mechanobiol. 2015;14(3):515–36.
57. Cookson A, Lee J, Michler C, Chabiniok R, Hyde E, Nordsletten D, Sinclair M, Siebes M, Smith N. A novel porous mechanical framework for modelling the interaction between coronary perfusion and myocardial mechanics. J Biomech. 2012;45(5):850–5.
58. Cattaneo L, Zunino P. A computational model of drug delivery through microcirculation to compare different tumor treatments. Int J Numer Methods Biomed Eng. 2014;30(11):1347–71.
59. D'Angelo C, Quarteroni A. On the coupling of 1d and 3d diffusion-reaction equations: application to tissue perfusion problems. Math Models Methods Appl Sci. 2008;18(08):1481–504.
60. Nolte F, Hyde ER, Rolandi C, Lee J, van Horssen P, Asrress K, van den Wijngaard JP, Cookson AN, van de Hoef T, Chabiniok R, et al. Myocardial perfusion distribution and coronary arterial pressure and flow signals: clinical relevance in relation to multi-scale modeling, a review. Med Biol Eng Comput. 2013;51(11):1271–86.
61. Spaan J, Kolyva C, van den Wijngaard J, ter Wee R, van Horssen P, Piek J, Siebes M. Coronary structure and perfusion in health and disease. Phil Trans R Soc A. 2008;366(1878):3137–53.
62. Smith N. A computational study of the interaction between coronary blood flow and myocardial mechanics. Physiol Meas. 2004;25(4):863–77.
63. Westerhof N, Boer C, Lamberts RR, Sipkema P. Cross-talk between cardiac muscle and coronary vasculature. Physiol Rev. 2006;86(4):1263–308.
64. Huyghe J, Arts T, van Campen D. Porous medium finite element model of the beating left ventricle. Am J Phys. 1992;262:H1256–67.
65. Biot MA. Theory of propagation of elastic waves in a fluid-saturated porous solid. II higher frequency range. J Acoust Soc Am. 1956;28:179–91.
66. Biot MA. Theory of finite deformations of porous solids. Indiana Univ Math J. 1972;21:597–620.
67. Chapelle D, Gerbeau JF, Sainte-Marie J, Vignon-Clementel I. A poroelastic model valid in large strains with applications to perfusion in cardiac modeling. Comput Mech. 2010;46:91–101.
68. Coussy O. Mechanics of porous continua. New York: Wiley; 1995.

69. Ciarlet PG, Geymonat G. Sur les lois de comportement en élasticité non linéaire. C R Acad Sci Sér II. 1982;295:423–6.
70. Yeung JJ, Kim HJ, Abbruzzese TA, Vignon-Clementel IE, Draney-Blomme MT, Yeung KK, Perkash I, Herfkens RJ, Taylor CA, Dalman RL. Aortoiliac hemodynamic and morphologic adaptation to chronic spinal cord injury. J Vasc Surg. 2006;44(6):1254–65.
71. LaDisa J, John F, Dholakia RJ, Figueroa CA, Vignon-Clementel IE, Chan FP, Samyn MM, Cava JR, Taylor CA, Feinstein JA, et al. Congenit Heart Dis. 2011;6(5):432–43.
72. DeVault K, Gremaud PA, Novak V, Olufsen MS, Vernires G, Zhao P. Blood flow in the circle of Willis: modeling and calibration. Multiscale Model Simul. 2008;7(2):888–909.
73. Moireau P, Chapelle D. Reduced-order unscented Kalman filtering with application to parameter identification in large-dimensional systems. ESAIM Control Optim Calc Var. 2011;17:380–405.
74. Alastruey J, Khir AW, Matthys KS, Segers P, Sherwin SJ. Pulse wave propagation in a model human arterial network: Assessment of 1-D visco-elastic simulations against in vitro measurements. J Biomech. 2011;44:2250–8.
75. Matthys KS, Alastruey J, Peiró J, Khir AW, Segers P, Verdonck PR, Parker KH, Sherwin SJ. Pulse wave propagation in a model human arterial network: assessment of 1-D numerical simulations against in vitro measurements. J Biomech. 2007;40(15):3476–86.
76. Julier S, Uhlmann J. Reduced sigma point filters for the propagation of means and covariances through nonlinear transformations. In: Proceedings of IEEE American Control Conference; 2002. pp. 887–892.
77. Kalman R, Bucy R. New results in linear filtering and prediction theory. Trans ASME J Basic Eng. 1961;83:95–108.
78. Bradley C, Bowery A, Britten R, Budelmann V, Camara O, Christie R, Cookson A, Frangi AF, Gamage TB, Heidlauf T, et al. Opencmiss: a multi-physics & multi-scale computational infrastructure for the vph/physiome project. Prog Biophys Mol Biol. 2011;107(1):32–47.
79. Sainte-Marie J, Chapelle D, Cimrman R, Sorine M. Modeling and estimation of the cardiac electromechanical activity. Comput Struct. 2006;84:1743–59.
80. Ghista D, Ng E. Cardiac perfusion and pumping engineering. Hackensack: World Scientific; 2007.
81. Vuong AT, Yoshihara L, Wall W. A general approach for modeling interacting flow through porous media under finite deformations. Comput Methods Appl Mech Eng. 2015;283:1240–59.

A Biphasic Poroelasticity Model for Soft Tissue

Sebastian Hirsch

Abstract

Poroelastic media are characterized by biphasic structure, consisting of a solid matrix permeated by a fluid-filled pore space. Originally conceived for soil mechanics, poroelastic models have been used to study the mechanics of biological tissues, where the matrix consists of cells or the extracellular matrix, and the fluid compartment is comprised of the vascular tree, interstitial fluid, cerebro-spinal fluid, or a combination of these. Unlike simpler monophasic models, poroelasticity predicts two longitudinal (compressional) wave modes with different wave speeds. The poroelastic equations of motion depend on two interaction or coupling parameters that describe the transfer of energy or momentum between the two phases, thus leading to a system of coupled equations. In this chapter, we will explain the poroelastic tissue model, present a poroelastic version of the wave equation, and complement this theoretical part with numerical simulations.

4.1 Introduction

Poroelasticity is an extension of viscoelastic theory that aims to describe the interaction of a fluid and a solid phase in a biphasic medium. The name is derived from the assumption that the solid forms a porous, sponge-like matrix with a contiguous pore space occupied by a fluid. Poroelastic models try to capture the interactions between tissue, blood, interstitial fluid and (in the case of the brain) cerebrospinal fluid.

S. Hirsch
Institute of Medical Informatics, Charité – Universitätsmedizin Berlin, Berlin, Germany
e-mail: sebastian.hirsch@charite.de

© Springer International Publishing AG 2018   71
I. Sack, T. Schaeffter (eds.), *Quantification of Biophysical Parameters in Medical Imaging*, https://doi.org/10.1007/978-3-319-65924-4_4

Poroelastic theory was pioneered by von Terzaghi [1] for static scenarios and further extended to dynamic settings by Biot [2–6]. Originally, the theory was derived for geophysical applications, such as the analysis of soil consolidation under load, and oil prospecting. Applications in the medical field followed decades later. It was introduced into the elastography community, which tries to quantify organ stiffness by means of ultrasound or magnetic resonance imaging (MRI), by Konofagou et al. [7] (ultrasound) and Perriñez et al. [8].

The significance of poroelasticity in the context of biological tissues can be explained by tissue structure. In the simplest case, an organ can be modeled as parenchyma (functional organ tissue) with an embedded vascular tree. The vascular tree has a hierarchical structure. The main arteries and veins are usually large enough to be visible in most medical imaging modalities. However, inside the organ, they branch into smaller vessels, and this process is repeated several times, down to capillaries that only measure 5–10 microns in diameter and are thus invisible to in vivo imaging methods. For poroelastic modeling, the vascular tree is then associated with the pore space, and the surrounding tissue with the solid matrix (or skeleton). More refined poroelastic models might extend the concept to three or more compartments, accounting for cerebrospinal fluid (CSF) in the brain, or interstitial fluid. In this section we will discuss a "lumped" poroelastic model, which means that all liquids are summarized in the fluid compartment, while the solid phase includes all structural elements. With the exception of the lungs and the digestive tract there are usually no gas-filled cavities in human organs, so that the fluid compartment can be treated as a liquid (as opposed to a gas, which the theory could handle as well). If applied to the lungs, the model has to be adapted for a gas instead of a liquid, the main difference being that gases are easily compressible [9].

The poroelastic theory derived in this chapter will be based on *stress* and *strain*. A different approach to poroelasticity has been outlined in Chap. 3. Strain, designated by ϵ, quantifies deformation of a body from its equilibrium state, and stress, σ, represents the forces causing or caused by deformation. Both σ and ϵ are rank-two tensors whose meaning can be illustrated on a small cubic volume element with edges parallel to the coordinate axes (e_1, e_2, e_3), as shown in Figure 2.3 in Chap. 2.

Stress and strain were introduced in Chap. 2 (Eqs. (2.6) and (2.7), respectively), and the relationship between them is governed by *Hooke's law* with elasticity tensor C (see Eq. (2.8) in Chap. 2).

Stresses and strains can be categorized as volumetric (or axial), represented by the tensor elements on the main diagonal, and shear (or deviatoric), represented by the offdiagonal tensor elements. Volumetric deformation expands or compresses the volume of the body, whereas shear deformation preserves the volume but skews angles.

The assumption underlying poroelasticity is that the fluid and the solid compartments are deformed separately, but that there is a coupling between the deformation fields. For example, if the fluid is assumed to be incompressible (which is a typical

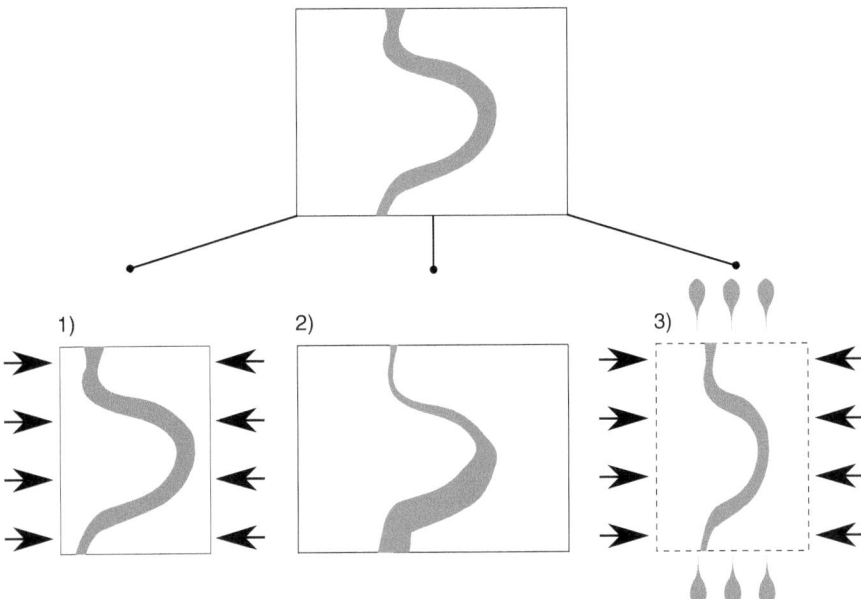

Fig. 4.1 Top: A small volume element of a poroelastic medium, consisting of the solid matrix (light gray) and a fluid-filled pore (dark gray). Bottom: Under compressive stress, three scenarios have to be considered: (1) Jacketed scenario with at least one compressible compartment: the jacket prevents material from being squeezed out of the volume element, but compressibility causes shrinking of the volume element. (2) Jacketed scenario with two incompressible constituents: the medium is incompressible; however, the spatial distribution of the two constituents can change. (3) Unjacketed scenario: the fluid can be squeezed out of a volume element, resulting in finite compressibility even if both compartments are incompressible. The poroelastic biological tissue model that will be developed in this chapter describes the two jacketed scenarios (1) and (2)

approximation made for water), then forcing more liquid into a given volume element can only be achieved if the solid contained in that volume element is either compressed or expelled. The poroelastic system can be either *jacketed* – meaning that the amount of fluid is held constant by a surrounding impermeable membrane or "jacket"–, or *unjacketed*, where fluid exchange with an external reservoir is possible. An overview of these scenarios and their response to volumetric stress is shown in Fig. 4.1.

In this chapter, we will present a poroelastic model that is based on Biot's work [4, 5], but tailored to biological tissues, making the following assumptions:

1. Each infinitesimal point of the material is occupied by either the solid or the fluid. There is no empty space, and there are no trapped gas bubbles.
2. The volume fraction, β, is defined as the amount of fluid volume per bulk material:

$$\beta = \frac{V_f}{V},$$

with V_f the volume of the fluid contained in the volume element V.

3. The pore space is contiguous, i.e., there are no isolated pores.
4. The fluid, referred to by the index f, is considered to have a very high compression modulus, K^f, on the order of 2.2 GPa (the bulk modulus of water). Furthermore, the shear modulus of the fluid is considered negligible, i.e., the fluid only resists compression and can be sheared effortlessly. Therefore, stress and strain of the fluid can be represented by a single number, σ^f and ϵ^f, respectively, rather than a tensorial representation, as will be needed for the solid.
5. The solid, referred to by the index s, resists both compression and shear deformation, as quantified by its finite moduli, K^s and μ^s. The full tensorial description is required to describe the deformation of the solid.
6. Both compartments are considered purely elastic, i.e., all moduli are real-valued and there is no dissipative energy loss.
7. There is no global fluid flow. Without any external driving force, fluid and solid are at rest relative to each other, and any motion is induced by deformation. Deformation of one compartment can lead to deformation of the other compartment through interaction.
8. Filtration (transition of liquid from the fluid compartment into the solid compartment) is negligible on the time scale that is considered relevant in the domain of dynamic MRE.

Item 7 in this list distinguishes our approach from others, such as those published by McGarry et al. [10] or Parker et al. [11, 12], who accounted for the existence of arbitrary fluid pressure gradients by incorporating an extra term in the equations of motion. In summary, the above assumptions boil down to a biphasic fluid-solid model with lossless dynamic interaction between the two compartments.

These assumptions allow us to write a poroelastic version of Hooke's law in matrix form:

$$
\begin{pmatrix}
(1-\beta)\sigma_{11}^s \\
(1-\beta)\sigma_{22}^s \\
(1-\beta)\sigma_{33}^s \\
(1-\beta)\sigma_{12}^s \\
(1-\beta)\sigma_{23}^s \\
(1-\beta)\sigma_{13}^s \\
\beta\sigma^f
\end{pmatrix}
=
\begin{pmatrix}
(1-\beta)\left(K^s+\dfrac{4}{3}\mu^s\right) & (1-\beta)\left(K^s-\dfrac{2}{3}\mu^s\right) & (1-\beta)\left(K^s-\dfrac{2}{3}\mu^s\right) & 0 & 0 & 0 & \beta(1-\beta)H \\
(1-\beta)\left(K^s-\dfrac{2}{3}\mu^s\right) & (1-\beta)\left(K^s+\dfrac{4}{3}\mu^s\right) & (1-\beta)\left(K^s-\dfrac{2}{3}\mu^s\right) & 0 & 0 & 0 & \beta(1-\beta)H \\
(1-\beta)\left(K^s-\dfrac{2}{3}\mu^s\right) & (1-\beta)\left(K^s-\dfrac{2}{3}\mu^s\right) & (1-\beta)\left(K^s+\dfrac{4}{3}\mu^s\right) & 0 & 0 & 0 & \beta(1-\beta)H \\
0 & 0 & 0 & (1-\beta)\mu^s & 0 & 0 & 0 \\
0 & 0 & 0 & 0 & (1-\beta)\mu^s & 0 & 0 \\
0 & 0 & 0 & 0 & 0 & (1-\beta)\mu^s & 0 \\
\beta(1-\beta)H & \beta(1-\beta)H & \beta(1-\beta)H & 0 & 0 & 0 & \beta K^f
\end{pmatrix}
\cdot
\begin{pmatrix}
\epsilon_{11}^s \\
\epsilon_{22}^s \\
\epsilon_{33}^s \\
\epsilon_{12}^s \\
\epsilon_{23}^s \\
\epsilon_{13}^s \\
\epsilon^f
\end{pmatrix}
$$

$$(4.1)$$

From this formula we can see that fluid strains, ϵ^f, translate into fluid stresses, σ^f, via K^f, and solid strains, ϵ^s, translate into solid stresses, σ^s, via K^s and μ^s. However, the *coupling modulus* H introduces symmetric coupling between volumetric strains in one compartment and axial stress in the other. Therefore, compression of the solid will affect the volumetric stress of the fluid, and vice versa. In the limit $\beta \to 0$, i.e., vanishing fluid fraction, Eq. (4.1) becomes identical to Hooke's law for a monophasic solid, as given in Eq. (2.10) in Chap. 2.

We want to use the poroelastic framework to model the propagation of mechanical waves through such a material. To this end, we have to find and solve the equation of motion for the above stress-strain relation. This can be achieved by using Newton's law, which, in its simplest formulation, can be written as

$$\mathbf{f} = \rho \cdot \ddot{\mathbf{u}}, \tag{4.2}$$

where $\mathbf{f}$ is the *force density* (force per unit volume), ρ is the mass density, and $\ddot{\mathbf{u}}$ is the acceleration. The acting force densities result from deformation of the solid and are given by

$$\mathbf{f} = \nabla \cdot \boldsymbol{\sigma}. \tag{4.3}$$

For a derivation of Eq. (4.3) we refer to the literature [13] or Chap. 2. For an intuitive explanation, it is important to keep in mind that a constant stress tensor would indicate the same force acting on every point of the medium. This would result in bulk motion of the body in the direction of the acting force, but it would not cause deformation. Instead, deformation is caused by variation of the acting stresses throughout the medium, which explains the use of the derivative operator ∇. The notation $\nabla \cdot \boldsymbol{\sigma}$ is shorthand for the multiplication of row vector $\nabla = \left(\dfrac{\partial}{\partial x}, \dfrac{\partial}{\partial y}, \dfrac{\partial}{\partial z} \right)$ with the 3×3 matrix $\boldsymbol{\sigma}$, yielding a column vector.

We will now evaluate Eq. (4.3) separately for the solid and fluid stress. For the solid, we obtain

$$(1-\beta)\nabla \cdot \boldsymbol{\sigma}^s = (1-\beta)\left[\left(K^s + \frac{1}{3}\mu^s \right)\nabla\left(\nabla \cdot \mathbf{u}^s\right) + \beta H \nabla \epsilon^f + \mu^s \Delta \mathbf{u}^s \right]. \tag{4.4}$$

This equation was derived by applying Eq. (4.3) to the solid part (i.e., the first six rows) of Eq. (4.1); a comprehensive derivation of this equation can be found in [14]. Equation (4.4) is structurally similar to Navier's equation of a monophasic solid (see Eq. (2.18) in Chap. 2). As a simplification, it was assumed that the spatial variation of the elastic parameters, K^s, μ^s, and H, is small. This very common assumption, known as *local homogeneity*, allows us to discard all terms that contain a spatial derivative of one of these parameters. In a similar fashion, taking the gradient of the fluid stress yields

$$\beta \nabla \boldsymbol{\sigma}^f = \beta(1-\beta)H\nabla\left(\nabla \cdot \mathbf{u}^s\right) + \beta K^f \nabla\left(\nabla \cdot \mathbf{u}^f\right). \tag{4.5}$$

We can see that Eqs. (4.4) and (4.5) constitute a coupled system of equations with coupling modulus H, translating deformation of one compartment into a stress acting on the other compartment. For the acceleration term (i.e., the right-hand side of Newton's law Eq. (4.2)), we have to introduce an analogous coupling in the form of a *coupling density*, ρ_{12}:

$$\rho_{11}\ddot{\mathbf{u}}^s + \rho_{12}\ddot{\mathbf{u}}^f = (1-\beta)\nabla\cdot\boldsymbol{\sigma}^s \tag{4.6}$$

$$\rho_{12}\ddot{\mathbf{u}}^s + \rho_{22}\ddot{\mathbf{u}}^f = \beta\nabla\boldsymbol{\sigma}^f \tag{4.7}$$

with $\rho_{11} = (1-\beta)\rho^s - \rho_{12}$ and $\rho_{22} = \beta\rho^f - \rho_{12}$. The coupling density can be parameterized as $\rho_{12} = (T-1)\beta\rho^f$, where T is the tortuosity[1] of the pores. The coupling density is thus always negative. Intuitively, the physical meaning of the coupling density can be understood as follows: Assume that the fluid is accelerated into one direction, whereas the solid initially is static. Through interaction, the fluid will exert a force on the solid, causing an acceleration of the solid in the same direction. From the fluid's perspective, the solid is slowing the acceleration down, which is the reason for the coupling density being negative. While its effect is similar to that of friction, it does not cause dissipative loss of kinetic energy but rather a transfer of kinetic energy from one compartment to the other. In summary, interactions in a biphasic poroelastic medium are characterized by two interaction parameters, one that couples strain energies and a second one that couples kinetic energies.

The solid stress term (Eq. (4.4)) contains compression and shear deformation. Since shear waves and compression waves propagate at different velocities and are governed by different elastic moduli, it is customary in MRE to separate these two deformation modes. This can be achieved by using the Helmholtz theorem, which states that any well-behaved three-dimensional vector field $\boldsymbol{\chi}$ is the sum of a longitudinal (curl-free) and a transverse (divergence-free) field:

$$\boldsymbol{\chi} = \boldsymbol{\chi}_L + \boldsymbol{\chi}_T \tag{4.8}$$

with

$$\nabla\times\boldsymbol{\chi}_L = \mathbf{0} \tag{4.9}$$

and

$$\nabla\cdot\boldsymbol{\chi}_T = 0. \tag{4.10}$$

---

[1] Tortuosity is the ratio of the length of a curve (in this case, a pore) to the Euclidean distance of its end points. Therefore, $T \geq 1$, with larger values indicating a more "curled" shape of the pore space.

4.2   Compression Waves

Thus, for analyzing compression wave propagation, we apply the divergence opera-
tor to Eqs. (4.6) and (4.7). We also introduce the shorthand notation $\theta \equiv \nabla \cdot \mathbf{u}^f = \epsilon^f$
and $\zeta \equiv \nabla \cdot \mathbf{u}^s$. Furthermore, we invoke the vector calculus identity $\Delta \mathbf{u}^s = \nabla^2 \mathbf{u}^s$
$= \nabla(\nabla \cdot \mathbf{u}^s) - \nabla \times \nabla \times \mathbf{u}^s$ and discard its rotational part, since it vanishes for longitudinal waves:

$$\rho_{11}\,\ddot{\zeta} + \rho_{12}\ddot{\theta} = \left(1-\beta\right)\left(K^s + \frac{4}{3}\mu^s \right)\nabla \cdot \nabla\zeta + \beta\left(1-\beta\right)H\nabla \cdot \nabla\theta$$

$$\rho_{22}\ddot{\theta} + \rho_{12}\ddot{\zeta} = \beta\left(1-\beta\right)H\nabla \cdot \nabla\zeta + \beta K^f\nabla \cdot \nabla\theta$$

As the final step, we define the P-wave (pressure wave) modulus, $M^s = K^s + \frac{4}{3}\mu^s$,
and use the scalar Laplacian, $\Delta\psi \equiv \nabla \cdot \nabla\psi$:

$$\rho_{11}\ddot{\zeta} + \rho_{12}\ddot{\theta} = \left(1-\beta\right)M^s\Delta\zeta + \beta\left(1-\beta\right)H\Delta\theta \tag{4.11}$$

$$\rho_{22}\ddot{\theta} + \rho_{12}\ddot{\zeta} = \beta\left(1-\beta\right)H\Delta\zeta + \beta K^f\Delta\theta. \tag{4.12}$$

Thus, the dynamics of fluid and solid compression are coupled via the previously
introduced coupling parameters H and ρ_{12}. Equations (4.11) and (4.12) have the
form of a wave equation, with a second-order temporal derivative on the left-hand
side and a second-order spatial derivative on the right-hand side. These wave equations are solved by plane waves:

$$\zeta\left(\mathbf{r},t\right) = \zeta_0 \cdot e^{i(\mathbf{k}\cdot\mathbf{r}-\omega t)} \tag{4.13}$$

$$\theta\left(\mathbf{r},t\right) = \theta_0 \cdot e^{i(\mathbf{k}\cdot\mathbf{r}-\omega t)} \tag{4.14}$$

Both waves have the same $\mathbf{k}$-vector (and thus the same wavelength $\lambda = \dfrac{2\pi}{|\mathbf{k}|}$)
because of the coupling between them. However, their amplitudes and phases, as
parameterized by the complex amplitudes, θ_0 and ζ_0, are not yet determined. We will
show now that two independent pressure wave modes exists, one in which the solid
and the fluid oscillate in phase and one in which they oscillate with opposite phases
(see Fig. 4.2).

Substituting the plane waves (Eqs. (4.13) and (4.14)) into the equations of motion
(4.11) and (4.12) and setting $q \equiv k^2$ results in

$$\left(\underbrace{\left(1-\beta\right)M^s\,q - \rho_{11}\omega^2}_{\equiv\, a} \right)\zeta_0 = \left(\underbrace{\rho_{12}\omega^2 - \beta\left(1-\beta\right)H\,q}_{\equiv\, b} \right)\theta_0 \tag{4.15}$$

Fig. 4.2 Illustration of the two longitudinal wave modes in a 1D poroelastic medium. The arrows indicate displacements from the equilibrium positions (dots)

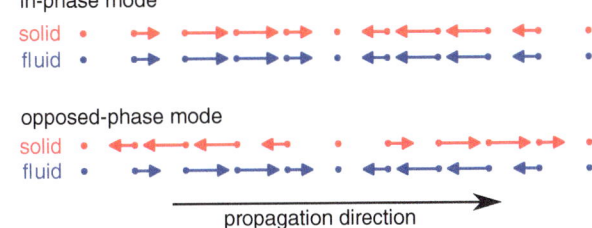

$$\left(\underbrace{\beta K^{\mathrm{f}}}_{\equiv c} q - \rho_{22}\omega^2\right)\theta_0 = \left(\rho_{12}\omega^2 - \underbrace{\beta(1-\beta)H}_{\equiv b} q\right)\zeta_0 \tag{4.16}$$

We introduce the parameters a, b, $c \geq 0$ to simplify the following calculations. We can now reorder Eqs. (4.15) and (4.16) and obtain

$$\begin{pmatrix} aq - \rho_{11}\omega^2 & bq - \rho_{12}\omega^2 \\ bq - \rho_{12}\omega^2 & cq - \rho_{22}\omega^2 \end{pmatrix} \cdot \begin{pmatrix} \zeta_0 \\ \theta_0 \end{pmatrix} \equiv \mathbf{R} \cdot \begin{pmatrix} \zeta_0 \\ \theta_0 \end{pmatrix} = \mathbf{0}, \tag{4.17}$$

which is a linear system of two equations that is solved by the two complex oscillation amplitudes, ζ_0 and θ_0. However, being a homogeneous system, Eq. (4.17) only has nontrivial solutions if $\det(\mathbf{R}) = 0$. This imposes a condition on the values of q which we previously defined as the square of the wave vector:

$$q_{\pm} = \frac{c\rho_{11} - 2b\rho_{12} + a\rho_{22} \pm \sqrt{(c\rho_{11} - a\rho_{22})^2 + 4(a\rho_{12} - b\rho_{11})(c\rho_{12} - b\rho_{22})}}{2(ac - b^2)}\omega^2 \tag{4.18}$$

These two solutions correspond to two wave modes with two different wave numbers, $k_{\pm}^2 = q_{\pm}$. By inserting the two solutions, q_+ and q_-, into Eq. (4.17) and solving for the complex wave amplitudes, ζ_0 and θ_0, we obtain the relationships

$$\left[\begin{array}{l} \zeta_0 = \alpha, \\ \theta_0 = \alpha \dfrac{c\rho_{11} - a\rho_{22} + \sqrt{(c\rho_{11} - a\rho_{22})^2 + 4(a\rho_{12} - b\rho_{11})(c\rho_{12} - b\rho_{22})}}{2(b\rho_{22} - c\rho_{12})} \end{array}\right] \quad \text{for } q_- \tag{4.19}$$

and

$$\left[\begin{array}{l} \zeta_0 = \alpha, \\ \theta_0 = \alpha \dfrac{c\rho_{11} - a\rho_{22} - \sqrt{(c\rho_{11} - a\rho_{22})^2 + 4(a\rho_{12} - b\rho_{11})(c\rho_{12} - b\rho_{22})}}{2(b\rho_{22} - c\rho_{12})} \end{array}\right] \quad \text{for } q_+ \tag{4.20}$$

with $\alpha \in \mathbb{C}$. The fraction in the solution for θ_0 determines the phase relation between the waves in the solid and the fluid. Since all parameters in that expression, except ρ_{12}, are positive, the value of the square root is greater than $c\rho_{11} - a\rho_{22}$. Thus, the nominator is positive in the first solution and negative in the second one, whereas the denominator is always positive. This means that in the first solution, ζ_0 and θ_0 have the same complex phase (and are thus in phase), whereas in the second solution they have opposite signs and are opposed-phase.

Since we defined $q = k^2$, each of the solutions in Eq. (4.18) represents two k-values with opposite signs: $q_+ \rightarrow (k_+, -k_+)$ and $q_- \rightarrow (k_-, -k_-)$. The propagation speed can be calculated according to

$$v_\pm = \frac{\omega}{|k_\pm|} = \frac{\omega}{\sqrt{q_\pm}}. \tag{4.21}$$

The dependence of the two compression wave speeds on the poroelastic parameters is illustrated in Fig. 4.4.

In summary, in a poroelastic medium there are two longitudinal wave modes, one where the two compartments oscillate in-phase (designated by $-$), and one where they oscillate opposed-phase (designated by $+$). Their wavelengths, $\lambda_\pm = \frac{2\pi}{|k_\pm|}$, and propagation speeds, $v_\pm$, differ. Actual values of these quantities can be calculated for a given material by substituting the expressions for a, b and c from Eqs. (4.15) and (4.16) into Eq. (4.18) and Eq. (4.21). The in-phase mode propagates at high speed, comparable to the compression wave speed in a monophasic medium with similar mechanical properties. The opposed-phase mode, on the other hand, is slower. The fast mode is similar to a compression wave in a monophasic medium, imposing a periodic modulation of bulk density. If both compartments are nearly incompressible, the achievable density changes are minuscule. However, the slow mode represents motion of the two compartments relative to each other, so that in a small volume element, the concentration of one compartment increases whereas the concentration of the other one decreases. This does not have to lead to a modulation of bulk density, so that larger wave amplitudes are possible for this mode.

Both modes were detected simultaneously in a poroelastic medium made of sintered glass spheres and water using ultrasound as described in [15]. The propagation speeds of both wave modes in porcine lung tissue were quantified in [9].

4.3 Shear Waves

Most applications of elastography focus on the detection of shear waves and derive values for the shear wave speed or shear modulus as the quantity of interest. In this section, we will therefore look at the dependence of the "effective" shear properties of a biphasic poroelastic medium.

In analogy to the derivation of the compression wave by applying the divergence operator to Eqs. (4.6) and (4.7) we can also obtain the shear wave equation by using the curl operator instead. Using the shorthand notation $\mathbf{c}^{s/f} = \nabla \times \mathbf{u}^{s/f}$ we can write the equations of motion as

$$(1-\beta)\rho^s \ddot{\mathbf{c}}^s + \rho_{12}(\ddot{\mathbf{c}}^f - \ddot{\mathbf{c}}^s) = (1-\beta)\nabla \times \nabla \cdot \boldsymbol{\sigma}^s \tag{4.22}$$

$$\beta\rho^f \ddot{\mathbf{c}}^f + \rho_{12}(\ddot{\mathbf{c}}^s - \ddot{\mathbf{c}}^f) = \beta\nabla \times \nabla\boldsymbol{\sigma}^f = \mathbf{0} \tag{4.23}$$

On the right-hand side of the second equation we used the fact that the curl of any gradient vanishes. This allows us to establish a relationship between $\mathbf{c}^s$ and $\mathbf{c}^f$:

$$\ddot{\mathbf{c}}^f = -\frac{\rho_{12}\ddot{\mathbf{c}}^s}{\beta\rho^f - \rho_{12}}, \tag{4.24}$$

which we can insert into Eq. (4.22):

$$\ddot{\mathbf{c}}^s \left((1-\beta)\rho^s - \rho_{12}\left(1 + \frac{\rho_{12}}{\beta\rho^f - \rho_{12}}\right)\right) = (1-\beta)\nabla \times \nabla \cdot \boldsymbol{\sigma}^s. \tag{4.25}$$

To calculate the right-hand side, we apply the curl operator to Eq. (4.4), exploiting the fact that $\nabla \times \nabla\xi = \mathbf{0}$ for any scalar function ξ:

$$
\begin{aligned}
(1-\beta)\nabla \times \nabla \cdot \boldsymbol{\sigma}^s &= (1-\beta)\left(\begin{array}{c} \left(K^s + \frac{1}{3}\mu^s\right)\underbrace{\nabla \times \nabla\left(\nabla \cdot \mathbf{u}^s\right)}_{=0} \\ + fH\underbrace{\nabla \times \nabla \mathbf{u}^f}_{=0} + \mu^s \nabla \times \Delta \mathbf{u}^s \end{array}\right) \\
&= (1-\beta)\mu^s \nabla \times \Delta \mathbf{u}^s \\
&= (1-\beta)\mu^s \Delta \mathbf{c}^s
\end{aligned}
\tag{4.26}
$$

By combining the left-hand side of Eq. (4.25) with the right-hand side of Eq. (4.26), we can compose the wave equation for the shear wave in a poroelastic medium:

$$\underbrace{\left((1-\beta)\rho^s - \rho_{12}\left(1 + \frac{\rho_{12}}{\beta\rho^f - \rho_{12}}\right)\right)}_{\equiv \rho_{\text{effective}}}\ddot{\mathbf{c}}^s = \underbrace{(1-\beta)\mu^s}_{\equiv \mu_{\text{effective}}} \Delta \mathbf{c}^s. \tag{4.27}$$

Hence, for shear waves, the poroelastic medium behaves like a monophasic medium with density $\rho_{\text{effective}}$ and shear modulus $\mu_{\text{effective}}$. In contrast to compression waves, there is only a single shear wave mode, since the fluid does not support shear waves itself ($\mu^f = 0$) and only acts as a "parasitic mass" through interaction with the solid, quantified by coupling density ρ_{12}. The effective shear modulus is always less

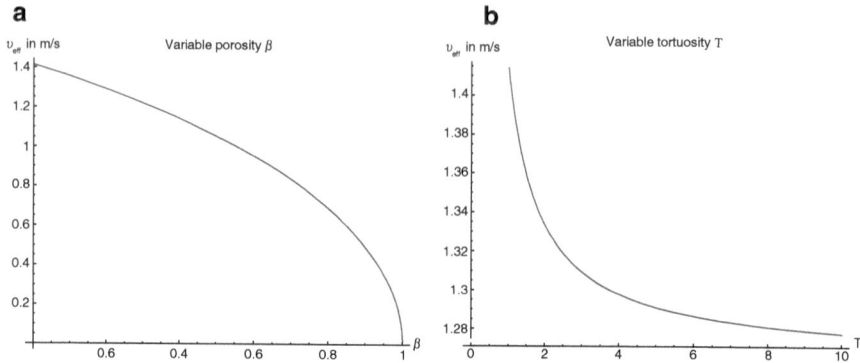

Fig. 4.3 Effective shear wave velocities according to Eq. (4.28). The following parameters were used (unless indicated otherwise): $\beta = 0.2$, $\mu^s = 2000$ Pa, $T = 5$. The shear wave velocity in a monophasic solid material with the same shear modulus as the solid compartment is equal to the values for $\beta = 0$ (in (**a**)) and $T = 1$ (in (**b**)). Note that in (**b**), $T = 1$ is the lower bound for the domain of T, and v_{eff} is well-behaved at this point; there is no singularity

than the shear modulus of the solid, since $0 \leq (1 - \beta) \leq 1$. The effective density, on the other hand, is larger than the porosity-weighted density $(1 - \beta)\rho^s$, since $\rho_{12} < 0$. We can represent the effective density as $\rho_{\text{effective}} = (1 - \beta)\rho^s + X$, $X > 0$. The effective shear wave speed is thus

$$v_{\text{effective}} = \sqrt{\frac{\mu_{\text{effective}}}{\rho_{\text{effective}}}} = \sqrt{\frac{(1-\beta)\mu^s}{(1-\beta)\rho^s + X}} < \sqrt{\frac{\mu^s}{\rho^s}} = v^s. \tag{4.28}$$

The poroelastic shear wave is therefore always slower than a shear wave in a monophasic medium (v^s) with density ρ^s and shear modulus μ^s, which is also an effect of the fluid "clinging" to the solid, thus increasing the mass to be accelerated without affecting the accelerating stress. A plot of $v_{\text{effective}}$ as a function of β and T is shown in Fig. 4.3. In this model, fluid pressure has no influence on effective shear wave speed and shear modulus. However, the shear modulus is sensitive to changes in the fluid volume fraction, β. As discussed in Chap. 20, β values above the physiological normal state lead to higher shear moduli in the liver and lower values in the spleen and pancreas.

4.4 Poroelastic Magnetic Resonance Imaging

In this section we will review the assessment of poroelastic tissue properties by means of magnetic resonance elastography (MRE). MRE is based on inducing low-frequency mechanical vibrations in the tissue of interest, and then acquiring MR images at different time points of the oscillation cycle. The motion-encoding mechanism explained in Chap. 2 is utilized to store information about the instantaneous tissue displacement in the phase of the complex MR signal. This allows one to restore the induced time-harmonic displacement field through post-processing. In a subsequent step, this wave field can be used to invert the wave equation, resulting in maps of elastic moduli, such as the shear modulus.

In its conventional form, MRI detects the signal from excited hydrogen nuclei in a strong magnetic field. Since we assume that in biological tissue both the matrix and the fluid are comprised mainly of water, the MR signal is a superposition of the solid and the fluid signals, and there is no straightforward way to separate them. However, for analysis of the opposed-phase wave mode, the MRI signal constitutes the superposition of motion in opposite directions. Luckily, the MRI signal equation allows us to draw conclusions even from this lumped signal, as we will show in the following paragraphs.

The MRI signal from a voxel $\mathbf{r}$ is typically represented in polar form as

$$S(\mathbf{r}) = \overline{S}(\mathbf{r}) \cdot e^{i\phi(\mathbf{r})}$$

with signal magnitude $\overline{S} \geq 0$ and signal phase ϕ. If the voxel contains fluid and solid components, the resulting signal is a superposition of the two:

$$S = \overline{S}^f \cdot e^{i\phi^f} + \overline{S}^s \cdot e^{i\phi^s}. \tag{4.29}$$

Note that we dropped the dependence on $\mathbf{r}$ for the sake of legibility. Furthermore, we will only consider the motion-induced part of the phase that was generated by the motion-encoding gradients, and we will ignore all other contributions to the MR signal phase, such as the susceptibility background. If the two compartments move in-phase, with the same amplitude, then

$$\phi^s = \phi^f \equiv \phi$$

and thus

$$S = \left(\overline{S}^s + \overline{S}^f\right) \cdot e^{i\phi}.$$

On the other hand, if the two compartments move in opposite directions with the same amplitude in the opposed-phase wave mode, their signal phases have opposite sign:

$$\phi^s = \phi^f + \pi,$$

so that we can simplify Eq. (4.29) to

$$S = \left(\overline{S}^f - \overline{S}^s\right) \cdot e^{i\phi^s}.$$

The implication is that, if the two signals do not have the same magnitude, the compound signal will be representative of the stronger of the two signals. The absolute signal strengths from the two compartments, $\overline{S}^f$ and $\overline{S}^s$, cannot be quantified in a straightforward way, since they depend on too many parameters with complex interactions (such as hydrogen content, iron content causing faster signal decay, relaxation time constants, echo time of the MR sequence, proton mobility to name just a few). In first-order approximation one can assume a linear relationship between the volume fractions and signal amplitudes:

$$\frac{\overline{S}^{\mathrm{f}}}{\overline{S}^{\mathrm{s}}} \approx \frac{\beta}{1-\beta}.$$

Hence, for typical volume fractions on the order of $\beta \approx 0.2$ [16], the solid accounts for the larger portion of the MR signal, so that the compound signal can be considered to represent the motion of the solid in a semiquantitative way. However, this statement only applies if it is assumed that the amplitudes of the solid and fluid oscillations are equal, which is not required by the theory. If this assumption is to be dropped, then more elaborate strategies have to be deployed to solve Eqn. (4.29).

4.5   Discussion

In this section we will discuss some of the implications of this version of poroelasticity theory. All calculations were performed for biological tissues, assuming equal densities for the two compartments, $\rho^{\mathrm{s}} = \rho^{\mathrm{f}} = 1000$ kg/m^3, and equal moduli[2] $M^{\mathrm{s}} \approx K^{\mathrm{f}} = 2.2$ GPa. The results might be different for poroelastic media for which these assumptions do not hold.

Figure 4.4 shows the dependence of the two pressure wave velocities on the poroelastic parameters β, H and T. Notably, the fast wave mode (v_-) is less strongly affected by these parameters than the slow mode (v_+). With a bulk modulus $M^{\mathrm{s}} \approx K^{\mathrm{f}} = 2.2$ GPa (equal to that of water), the fast wave propagates at approximately 1500 m/s, in agreement with the value found as the speed of sound in medical ultrasound applications.

Next we analyze the relative volumetric strain amplitudes that the wave induces in the solid and the fluid for each of the two wave modes. From Eqs. (4.19) and (4.20) we can see that the solid volumetric strain amplitude ζ_0 and the fluid volumetric strain amplitude θ_0 are both proportional to a complex number α, which allows us to scale the amplitudes to arbitrary values. We set $\alpha = 1$, so that $\zeta_0 = 1$, and analyze the dependence of the fluid strain on porosity β. The result for a medium with tissue-like properties is shown in Fig. 4.5.

The bulk volumetric strain resulting from either wave mode is given by

$$A = \left(1-\beta\right)\zeta + \beta\theta,$$

or

$$|A| = \left|\left(1-\beta\right)\zeta_0 + \beta\theta_0\right|. \tag{4.30}$$

As stated above, in-phase motion requires compressibility of at least one of the two compartments, whereas opposed-phase motion can also occur in biphasic media

---

[2] The value of 2.2 GPa is the bulk modulus of pure water, which is the main constituent of biological tissues. We neglect the influence of the solid shear modulus in the formula $M^{\mathrm{s}} = K^{\mathrm{s}} + \frac{4}{3}\mu^{\mathrm{s}}$, since, in biological tissues, μ^{s} is three orders of magnitude smaller than compression modulus K^{s}. We thus approximate $M^{\mathrm{s}} \approx K^{\mathrm{s}}$.

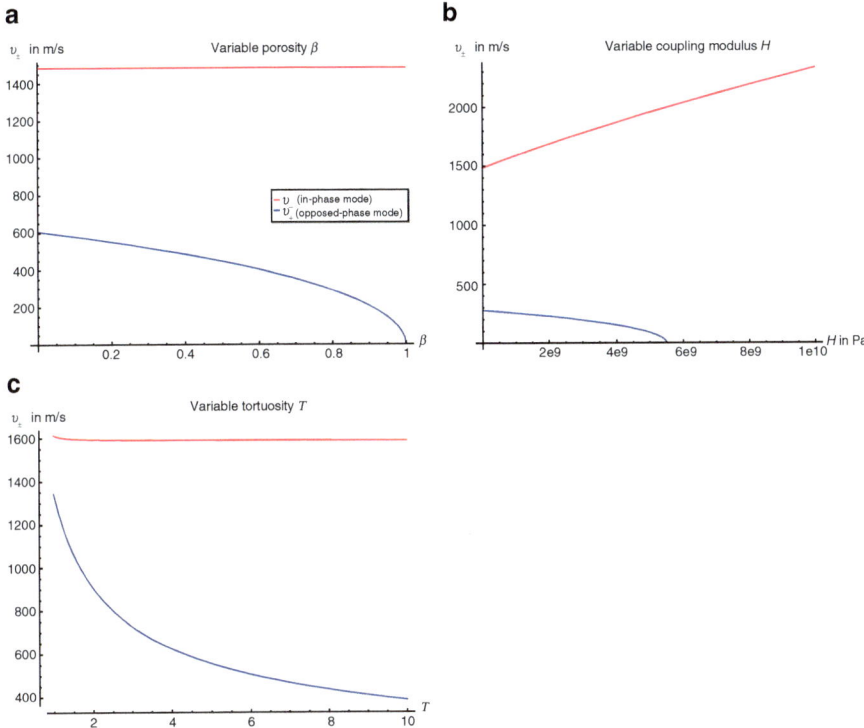

Fig. 4.4 Pressure wave velocities according to Eqs. (4.18) and (4.21) with the following parameters (unless indicated otherwise): $M^s = K^T = 2.2$ GPa, $\beta = 0.2$, $T = 5$, $\rho^s = \rho^f = 1000$ kg/m^3, $H = 1$ MPa

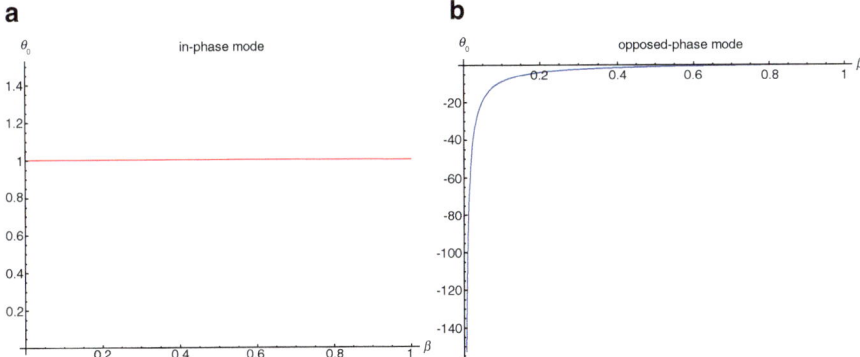

Fig. 4.5 Fluid displacement amplitude (θ_0) according to Eq. (4.19) (in-phase mode, **a**) and Eq. (4.20) (opposed-phase mode, **b**) if we set the solid displacement amplitude to $\zeta_0 = \alpha = 1$. For the in-phase mode, the fluid displacement amplitude is mostly independent of porosity, whereas, for the opposed-phase mode, it follows approximately a $-\dfrac{1}{\beta}$ law. The reason for this and its implications are discussed in the text. The same parameters were used as in Fig. 4.4

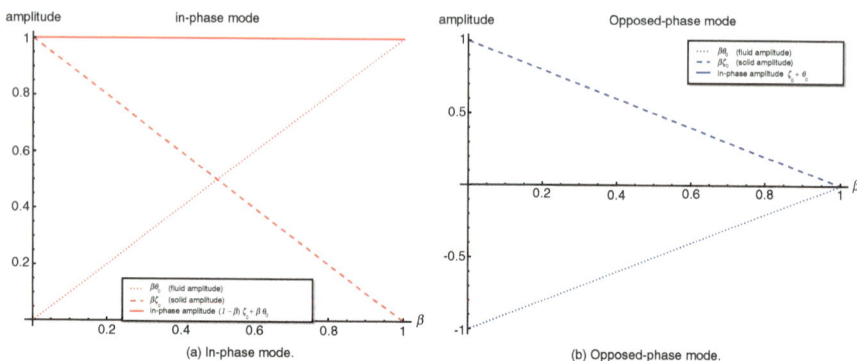

Fig. 4.6 The relative porosity-weighted volumetric strain amplitudes of the wave in the solid (ζ_0) and the fluid (θ_0) and the bulk volumetric strain if we set $\alpha = 1$ in Eq. (4.19) (**a**) and Eq. (4.20) (**b**)

where both compartments are incompressible, since the opposed-phase wave mode is based on the exchange of mass rather than a change in density. In other words, in the opposed-phase mode, influx of one compartment into a volume element requires expulsion of the other compartment in such a way that the two effects on the bulk volumetric strain cancel. Mathematically, this implies that $|A| = 0$, i.e., that there is no net volumetric strain. This explains why θ_0 diverges for $\beta \to 0$ in Fig. 4.5b: if the solid strain amplitude is fixed at $\zeta_0 = 1$, as β decreases, θ_0 has to tend toward $-\infty$ to ensure that $|A| = 0$.

The in-phase wave mode, on the other hand, requires that at least one of the two compartments be compressible. In that case, total volumetric strain does not have to vanish. The resulting volumetric strain is therefore independent of β (in the simple case of $\rho^s = \rho^f$ and $M^s = K^f$), as shown in Fig. 4.5a.

The resulting bulk volumetric strain according to Eq. (4.30) and its two constituents, $(1 - \beta)\zeta_0$ and $\beta\theta_0$, are shown in Fig. 4.6. It is clearly visible that the bulk strain amplitudes do not depend on porosity. However, this only applies for the special case where the densities and the bulk moduli of the compartments are assumed to be equal; otherwise, more complex relationships can emerge.

For practical applications, the number of viscoelastic parameters, H, ρ_{12} (or T), β, K^f and K^s, precludes inversion of the poroelastic wave equations, (Eqs. (4.11), (4.12) and (4.27)), which is routinely performed for monophasic media in MRE, as the corresponding equations are underdetermined if only the compound MRI signal (Eq. (4.29)) is measured. Instead, an *effective medium* approach, as outlined above for shear waves, is often used, resulting in an effective shear or compression modulus. Effective moduli do not include any information on the poroelastic properties of the medium; however, their change between different physiological states, such as high pressure/low pressure or inspiration/expiration, can be explained by the underlying poroelastic model. Such pressure-dependent changes were observed by MRE, e.g., in the lungs [17, 18], liver [19] and brain [20], even though not all of these findings were discussed in a poroelastic context. More examples and further discussion can be found in Chap. 20.

Conclusion

Poroelasticity provides a more apt description of biological tissues than the more common monophasic linear (visco)elasticity theory. However, the number of poroelastic parameters and the difficulties associated with their in vivo measurement render its application in the context of ultrasound or magnetic resonance elastography challenging. One of the main benefits of poroelastic theory is that it can be formulated to incorporate hydrostatic pore pressure and its gradient, which would allow noninvasive pressure measurement as an alternative to catheter-based invasive methods. Furthermore, recent research has shown a correlation between tissue perfusion and shear stiffness (see Chap. 20). While the exact origin of that correlation remains to be further analyzed, poroelastic effects clearly play a role, as outlined in this chapter.

References

1. von Terzaghi K. Theoretical soil mechanics. Hoboken: Wiley; 1943.
2. Biot MA. General theory of three-dimensional consolidation. J Appl Phys. 1941;12(2):155–64.
3. Biot MA. Theory of elasticity and consolidation for a porous anisotropic solid. J Appl Phys. 1955;26(2):182–5.
4. Biot MA. Theory of propagation of elastic waves in a fluid-saturated porous solid. I. Low-frequency range. J Acoust Soc Am. 1956;28(2):168–78.
5. Biot MA. Theory of propagation of elastic waves in a fluid-saturated porous solid. II. Higher frequency range. J Acoust Soc Am. 1956;28(2):179–91.
6. Biot MA, Willis DG. The elastic coefficients of the theory of consolidation. J Appl Mech. 1957;24:594–601.
7. Konofagou EE, Harrigan TP, Ophir J, Krouskop TA. Poroelastography: imaging the poroelastic properties of tissues. Ultrasound Med Biol. 2001;27(10):1387–97.
8. Perriñez PR, Kennedy FE, Van Houten EEW, Weaver JB, Paulsen KD. Modeling of soft Poroelastic tissue in time-harmonic MR Elastography. IEEE Trans Biomed Eng. 2009;56(3):598–608.
9. Dai Z, Peng Y, Mansy HA, Sandler RH, Royston TJ. Comparison of poroviscoelastic models for sound and vibration in the lungs. J Vib Acoust. 2014;136(5):0510121–5101211.
10. McGarry MDJ, Johnson CL, Sutton BP, Georgiadis JG, Van Houten EEW, Pattison AJ, et al. Suitability of poroelastic and viscoelastic mechanical models for high and low frequency MR elastography. Med Phys. 2015;42(2):947–57.
11. Parker KJ. A microchannel flow model for soft tissue elasticity. Phys Med Biol. 2014;59(15):4443–57.
12. Parker KJ. Experimental evaluations of the microchannel flow model. Phys Med Biol. 2015;60(11):4227–42.
13. Landau LD, Lifschitz EM. Theory of elasticity. 3rd ed. Oxford: Butterworth Heinemann; 1986.
14. Hirsch S, Braun J, Sack I. Magnetic resonance elastography - physical background and medical applications. 1st ed. Weinheim: Wiley-VCH; 2017.
15. Plona TJ. Observation of a second bulk compressional wave in a porous medium at ultrasonic frequencies. Appl Phys Lett. 1980;36(4):259–61.
16. Sykova E, Nicholson C. Diffusion in brain extracellular space. Physiol Rev. 2008;88(4):1277–340.
17. Mariappan YK, Kolipaka A, Manduca A, Hubmayr RD, Ehman RL, Araoz PA, et al. Magnetic resonance elastography of the lung parenchyma in an in situ porcine model with a noninvasive mechanical driver: correlation of shear stiffness with trans-respiratory system pressures. Magn Reson Med. 2012;67(1):210–7.

18. Hirsch S, Posnansky O, Papazoglou S, Elgeti T, Braun J, Sack I. Measurement of vibration-induced volumetric strain in the human lung. Magn Reson Med. 2013;69(3):667–74.
19. Hirsch S, Guo J, Reiter R, Schott E, Somasundaram R, Braun J, et al. Towards compression-sensitive magnetic resonance elastography of the liver: sensitivity of harmonic volumetric strain to portal hypertension. J Magn Reson Imaging. 2014;39(2):298–306.
20. Hirsch S, Klatt D, Freimann FB, Scheel M, Braun J, Sack I. In vivo measurement of volumetric strain in the human brain induced by arterial pulsation and harmonic waves. Magn Reson Med. 2013;70(3):671–83.

Physical Properties of Single Cells and Collective Behavior

5

Hans Kubitschke, Erik W. Morawetz, Josef A. Käs, and Jörg Schnauß

Abstract

Cells display a high degree of functional organization, largely attributed to the intracellular biopolymer scaffold known as the cytoskeleton. This inherently complex structure drives the system out of equilibrium by constantly consuming energy to conserve or reorganize its structure. Thus, the active, structurally organized cytoskeleton is the key player for the emergent mechanical properties of cells, which further determine properties of cell clusters and even multicellular organisms. In this spirit, this chapter introduces the physical principles on the different levels of biological complexity ranging from single biopolymers to tissues. The emergent mechanical properties and their respective effects on each level will be highlighted with a strong emphasis on their intertwined nature.

5.1 Introduction

The tremendous complexity of biological matter emerges from the interplay between intertwined levels or scales, with each level contributing a rich repertoire of physical principles. To uncover these principles and their interplay has proven to be a nontrivial task since processes, which we consider the fundamentals of life, exist far from thermodynamic equilibrium. Thus, traditional, purely reductionist approaches are unsuitable to fully elucidate and describe biological soft matter [1–3].

H. Kubitschke • E.W. Morawetz • J.A. Käs
Peter Debye Institute for Soft Matter Physics, Universität Leipzig, Leipzig, Germany
e-mail: hans.kubitschke@uni-leipzig.de

J. Schnauß (✉)
Peter Debye Institute for Soft Matter Physics, Universität Leipzig, Leipzig, Germany

Fraunhofer Institute for Cell Therapy and Immunology, Leipzig, Germany
e-mail: joerg.schnauss@uni-leipzig.de

© Springer International Publishing AG 2018   89
I. Sack, T. Schaeffter (eds.), *Quantification of Biophysical Parameters in Medical Imaging*, https://doi.org/10.1007/978-3-319-65924-4_5

Generally, complex systems are difficult to capture by intuitive understanding, which rather impedes an abstraction of the system in form of a model. When dealing with living matter, we can refer to different levels of complexity, which can be assigned to physical scales [4]. In this framework, a higher level contains the lower level, and complexity necessarily increases, which can even lead to entirely new properties [1]. If principles of the macrostate are absent in the underlying microstate, they are considered emergent (e.g., a single fish cannot exhibit swarm behavior). The term emergence describes the process leading to novel emergent properties, with the prefixes "micro" and "macro" not referring to definite length scales but to different levels of biological complexity [4]. For biological matter, a possible hierarchy might be described as

protein → filament → network → cell → tissue.

Using this hierarchy as a point of departure, we can aim at describing a given system only on the basis of the next underlying level of complexity, an approach termed hierarchical reductionism or coarse-graining [5, 6]. A biological tissue, for example, can be described as an accumulation of cells and extracellular matrix without considering subcellular structures. Further, cell mechanics can be described in terms of principles of networks of filaments. In this process, the problem is reduced, losing the detail of lower levels, similar to the use of computers, where not every single transistor has to be considered to operate the machine.

This chapter describes the physical principles emerging at the different levels of complexity and how they can be scaled up. In this context, it is important to clearly distinguish the concepts of self-organization (processes driven by energy dissipation) and self-assembly (processes driven by minimization of free energy, i.e., no energy is dissipated) [7–10]. With these terms at hand, we begin by introducing the lowest level of complexity, i.e., monomers and filaments, and proceed to the higher levels, successively describing the physical principles of cells, cell clusters, and tissues.

5.2   The Cytoskeleton

The cytoskeleton is a scaffold lending cells mechanical integrity and stability. It consists of three main constituents: actin, intermediate filaments (IFs), and microtubules (MTs). These components form fibers in the micrometer range by polymerizing their monomers into specific arrangements, resulting in a different intrinsic filament stiffness for each class [11] (Fig. 5.1). The stiffness is commonly characterized via the so-called persistence length (l_p) [12, 13]. This material-specific parameter is a measure of the fluctuation correlation along the filament backbone, quantifying over which distance an oscillation at a specific point (S_0) at the backbone becomes uncorrelated to the movement of another point (S_2) at the filament (Fig. 5.1). The persistence length can be directly observed, for instance, by analyzing the average transverse fluctuations of filaments observed over time or by

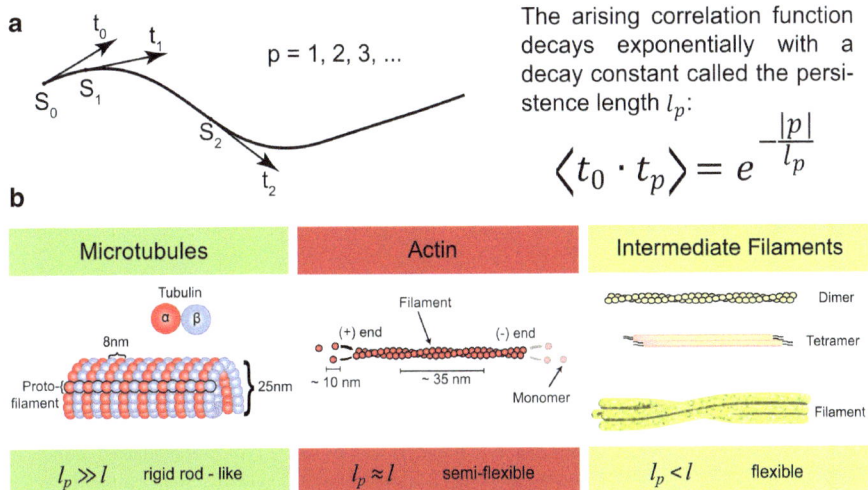

Fig. 5.1 (a) Points (*S*) along the contour of a semiflexible polymer have different tangent vectors (*t*). If points are close to each other (*S*$_0$ and *S*$_1$), their tangent vectors are correlated and roughly point in the same direction. When points are further apart (*S*$_0$ and *S*$_2$), their tangent vectors are uncorrelated and point in different directions. (**b**) illustrates the stiffness regimes of the three major cytoskeletal components—microtubules (MTs), actin, and intermediate filaments (IFs). Different mechanical properties are a direct result of the differing filament architectures. *l* denotes the length of the filament and *l*$_p$ the persistence length [7]

evaluating their tangent–cosine correlation function [14]. Based on these methods, actin filaments have an l_p of ~10 μm [15, 16], while MTs have a much longer l_p in the range of millimeters [17]. Note that l_p of natural biopolymers cannot be freely tuned and new model systems have to be used to derive the respective scaling laws [13, 18–20]. Since l_p is derived via thermal fluctuations (imagine a fluctuating cooked spaghetti), it is a temperature-dependent parameter and cannot be considered a material-defining constant. However, by multiplying thermal energy, k_BT, with the temperature, T, and the Boltzmann constant, k_B, a new temperature-independent parameter called bending stiffness, $\kappa = k_BTl_p$, can be derived.

Besides their mechanical properties, which have a static function and serve as the cellular skeleton, the cytoskeletal filaments are also very dynamic structures, enabling rapid adaptive organization of the entire cytoskeleton to fulfill functions such as cell migration or division. Fascinatingly, the cell can use the same components for these somehow contradictory tasks, which is only possible because of permanent energy dissipation, permitting rapid transition between different states. Furthermore, although the cytoskeletal building blocks are preserved in almost every eukaryotic cell, induced morphologies vary substantially among different cell types. Even within a single cell, the cytoskeleton spatially organizes into various different structures responsible for differing sets of functions—a strategy known as multifunctionality.

The different filament architectures not only result in a wide range of different bending rigidities but also determine their role in dynamic processes. MTs, for instance, are very rigid and thus typically appear as individual fibers, extending

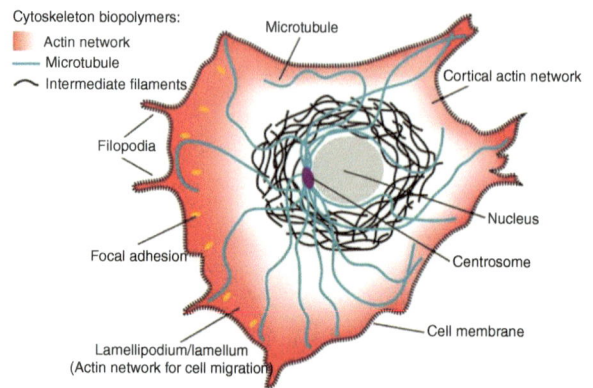

Fig. 5.2 Schematic drawing of a crawling cell on a 2D substrate showing the most prominent locations for the three types of cytoskeleton biopolymers. MTs are typically nucleated at the centrosome and span the largest portion of the cell. IFs are most commonly found around the cell nucleus, whereas actin filaments form dense networks close to the cell membrane. Particularly, dense and dynamic actin networks are found at the leading edge of migrating cells (forming lamellipodia and filopodia) [7]

from the cell interior to the membrane (Fig. 5.2). Due to their outreach and rather straight structures, they are especially well suited for intracellular transport and for providing directed forces during mitosis and for organelle positioning [7]. Actin filaments, on the other hand, are semiflexible polymers and are typically arranged into networks and bundles driving processes such as cell migration. Actin filaments polymerize near the membrane (leading to high local concentrations as illustrated in Fig. 5.2) and effectively push the boundary of the cell forward. In this dynamic process, actin monomers depolymerize at the filament ends pointing toward the cell center. These monomers subsequently travel to the front to reenter the polymerization cycle, a process called treadmilling [7, 21]. Due to their highly dynamic nature, actin filaments can trigger rapid cellular changes. Additional components such as cross-linking proteins or active myosin motors substantially enrich both their mechanical and dynamic phase spaces. It should be noted here that biological force generation is commonly attributed to the activity of molecular motors [11]. However, recent studies have demonstrated that actin as well as MT-based force generation can be driven solely by entropic arguments without requiring any energy dissipation [22–28].

In general, actin turnover and interactions with molecular motors are persistent processes in biological matter, resulting in substantial energy consumption. In eukaryotic cells, for instance, actin turnover alone can reach up to ~50% of the total energy consumption [29, 30], which in turn indicates that minimizing energy consumption has not been the most dominant evolutionary factor. Apart from molecular motors, all other actin accessory proteins influence the filament or network properties without consuming energy in a form of ATP or GTP. Accordingly, regulatory functions can be roughly classified as modifications of either polymerization dynamics, cross-linking, or filament nucleation [7, 21].

IFs are much less studied than the other two main components of the cytoskeleton. Additionally, IFs describe not only one specific polymer but a rather heterogeneous class of biopolymers, which form extended networks and thus substantially contribute to cell mechanics [31, 32]. Different types of IFs performing specific cellular tasks have been identified [33]. However, a general feature of all these cytoskeletal biopolymers is that they undergo growth and shrinkage by addition or subtraction of monomers. Therefore, their length is adjustable in a dynamic fashion and highly depends on stochastic fluctuations [32, 34, 35]. Further, their dynamic organization is largely determined by a complex interplay with a multitude of molecular accessory proteins, which nucleate, sever, cross-link, weaken, strengthen, or transport individual filaments [11]. The dynamic, self-organizing cytoskeleton is powered by energy-dissipating ATP or GTP consumption, mainly fueling two key processes: hydrolysis-powered depolymerization/polymerization of filaments and molecular motor-driven filament/motor transport [7]. Unlike IFs, MTs and actin are polar structures due to their asymmetrical polymerization and depolymerization dynamics (treadmilling) caused by their differing critical concentrations at the two ends. These two different critical concentrations are a direct result of ATP or GTP consumption, thus reflecting the intrinsic non-equilibrium process, which exerts substantial pushing forces [36]. The arising polarity is crucial for molecular motors to be able to move in a specific direction, enabling controlled cargo transport as well as directed pulling forces [37].

5.3   Rheology

Rheology is the study of deformation responses of materials to applied forces. The deformation response to constantly acting forces depends on whether the material is categorized as a solid or a fluidlike material. In solid materials, the magnitude of the deformation, typically elongation, scales with the applied force, e.g., an elastic spring under tension. Solid responses may also include plastic deformations such as overstretching a spring beyond its elastic limit, which permanently deforms it. The so-called viscous deformation response of a fluidlike material describes how the deformation *rate* scales with the applied force, e.g., ketchup flowing out of a bottle or squeezing glue out of a tube.

For biological samples—in this chapter single cells and soft tissues—viscoelastic responses to small forces have two distinct time scales: on short time scales, from split seconds to minutes, tissue deformation is proportional to applied forces and will recover to return to its initial form after stress release. This is easily confirmed by pressing against muscular or fatty tissue, where responses are nearly elastic. On long time scales (days to months), tissues tend to behave like highly viscous fluids, enabling body modification such as stretching lips by inserting lip discs, as, for example, practiced by the tribes of Mursi and Surma residing in Ethiopia [38] and the south American peoples of the Kayapo and Botocudo [39], or earspools ("flesh tunnel") in western subcultures and various African and American tribes [38, 40]. Materials which are governed by both elastic and viscous behaviors, such as most biological tissues, are considered viscoelastic materials. Examples of deformation responses are presented in Fig. 2.5 of Chap. 2.

Besides these passive responses, cells and tissues can actively react to environmental cues. Well-known active force generators are myosin motors, which exert pulling forces on actin filaments. This interaction is crucial on the cellular level for processes such as single-cell motility as well as on the tissue level, e.g., for muscle contraction. Many active processes are complexly intertwined with cellular pathways or immune responses and can highly influence the material properties of cells and tissues.

While the qualitative description of biological materials is straightforward, quantitative descriptions involve a profound theoretical background and mathematical models. The main goal of quantitative description is to gather material parameters from biological tissues. Since material parameters should describe intrinsic properties, they should be—in the best possible case—independent of features of the experimental setup such as applied forces, size of the tissue, or type of experiment conducted. This chapter will outline the *theoretical minimum* needed to adequately describe single cells and biological tissues later on and will introduce the concepts and terminology needed to understand the following chapters. Special cases such as nonlinearity and temperature dependencies are deliberately neglected here and are partially discussed later.

5.3.1   Step Experiment

As a starting point, consider a cuboid of tissue. The deformation response will depend on the strength of the force and how it is applied, i.e., on which side and in which direction. Vice versa, if a given deformation is forced upon the material, an internal force will arise accordingly. For the sake of simplicity—mathematical and explanatory—we restrict the possible types of force and deformation application to the types illustrated in Fig. 5.3: a longitudinal sudden force experiment and a transverse shear experiment.

In the stretching experiment, the force is applied equally on two counter-facing sides (red-dashed lines), resulting in an applied stress σ (force per area). The material will expand by Δx in the stretching direction and retract by $-\Delta y$ perpendicular to it. The resulting elongation is measured as strain, i.e., relative extension $\gamma = \Delta x/x$ (a tensor notation of linear strain is given in Eq. (2.6) of Chap. 2). Contraction occurs due to internal forces of the material, usually since many materials, such as water, are nearly incompressible. The relation between axial strain and transverse strain is captured in the Poisson ratio ν given by

$$\nu = -\frac{\Delta y / y}{\Delta x / x} = \frac{1}{2}\left(1 - \frac{\Delta V}{V}\frac{x}{\Delta x}\right),\tag{5.1}$$

with volume V of the cuboid and all further parameters as sketched in Fig. 5.3a. The Poisson ratio is a dimensionless unit and typically ranges from 0 to 0.5. A value

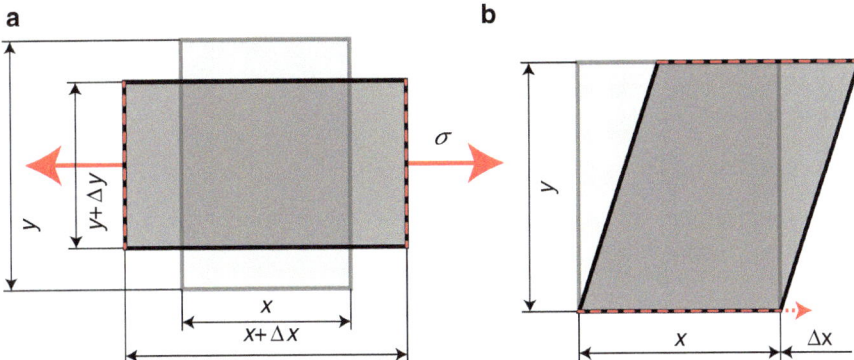

Fig. 5.3 The two archetypes of deformation response experiments in rheology. (**a**) shows the stretching mode, where two counter-facing sides are pulled apart perpendicular to the surface. The material elongates in the direction of the applied force and contracts perpendicular to it. (**b**) shows the shear mode. A strain is applied on the upper side, and the strain response is measured on the lower side

close to 0 means nearly no lateral contraction upon pulling, while a value close to 0.5 means that the material is nearly incompressible. The Poisson ratio for biological microtissue samples was found to be on the order of $\nu = 0.45$ [41]. Lower values can be found in multiphasic tissues, in which a fluid phase is allowed to freely move (see Chaps. 3 and 4).

For simplicity, only constant step stresses, as illustrated in Fig. 2.5 of Chap. 2, are considered in the following. When applying such a constant stress profile, σ_0, starting at $t = 0$, strain γ can be expressed as

$$\gamma(t) = D(t)\sigma_0 \quad \text{or} \quad D(t) = \frac{\gamma(t)}{\sigma_0}, \tag{5.2}$$

where t denotes time and $D(t)$ denotes *tensile creep compliance*. This parameter is usually unique to the type of material measured. For a perfect elastic, springlike material, tensile creep compliance reduces to a constant, $D(t) = 1/E$, where Young's modulus E describes the stiffness of a solid material under forces very similar to the spring constant of an elastic spring. The higher it is, the stiffer the material. Muscle tissue, for example, has an average Young's modulus of 2.12 ± 0.91 kPa [42], while cancellous/trabecular bone has 14.8 ± 1.4 GPa [43, 44]. An overview of elastic properties of tissues has been presented in a review by Akhtar et al. [45].

For a perfect viscous, fluidlike material, the tensile creep compliance will follow $D(t) = t/\eta$, where η is the viscosity of the fluid. The higher the viscosity of a material, the slower the flow speed for given forces will be. Honey, for instance, has a viscosity between 2.54 and 23.4 Pa·s (at 25°C, depending on moisture and sugar composition) [46], while blood has 4 mPa·s [47].

For viscoelastic materials, the detailed time-dependent response will be more complicated. Examples for illustration are presented in Fig. 2.5 and Fig. 5.4.

Fig. 5.4 Graph of the strain response of cells (blue, including confidence interval) under a step stress in an optical stretcher. The applied stress is proportional to the laser power (green). After 2 s, the stress is released and the cell relaxes again. A detailed description of the optical stretcher can be found in the next chapter

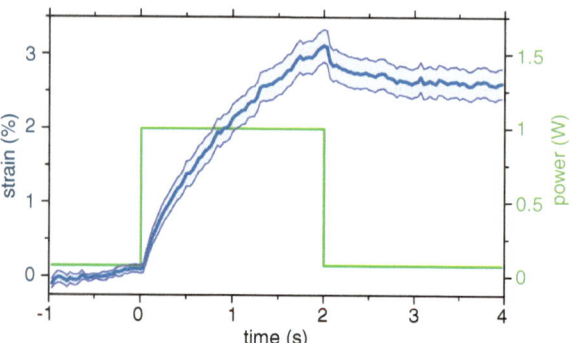

If a step strain γ_0 is applied, instead of a step stress, the corresponding stress response $\sigma(t)$ will be given by

$$\sigma(t) = E(t)\gamma_0 \quad \text{or} \quad E(t) = \frac{\sigma(t)}{\gamma_0}, \tag{5.3}$$

with the *elastic modulus* $E(t)$, also known as time-dependent Young's modulus, which is in principle the inverted tensile creep compliance. The two material parameters—tensile creep compliance and elastic modulus—are not independent. In fact, one can translate tensile creep compliance to elastic modulus and vice versa. The conversion can be done using Laplace transform $\mathcal{L}$:

$$\mathcal{L}\big[E(t)\big]\mathcal{L}\big[D(t)\big] = \frac{1}{s^2}, \tag{5.4}$$

with the complex frequency parameter $s = \sigma + i\omega$. Because the interpretation of elastic modulus $E(t)$ is more straightforward—the higher the modulus, the stiffer the material—it is more commonly used in the scientific community than creep compliance.

5.3.2 Oscillatory Experiment

The second measurement archetype is the shear experiment (Fig. 5.3b), in which a cuboid of material is fixed between two plates. On one plate, a shear stress is applied, and, in the opposite plate, the strain response is measured. Vice versa, applying a strain and measuring a stress response would give the same qualitative result. Commonly, forced strain γ_{in} is a sinusoidal alternation at frequency ω with a chosen maximum strain amplitude γ_0:

$$\gamma_{in}(t) = \gamma_0 \cos(\omega t). \tag{5.5}$$

The response to stress on the second plate will depend on the material, either elastic, viscous, or viscoelastic. In Fig. 5.5 the three types of responses are illustrated for applying a shear strain and measuring the stress response and vice versa.

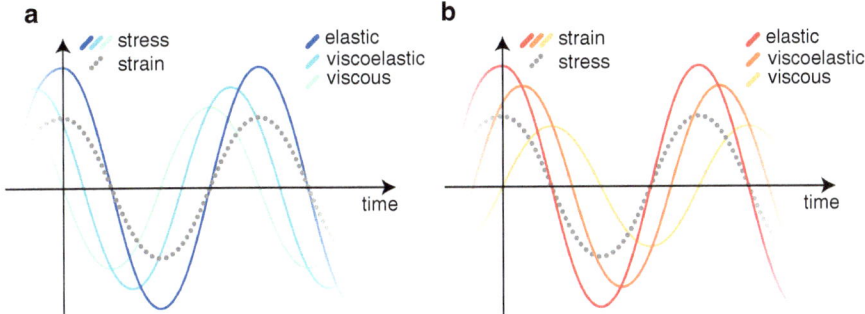

Fig. 5.5 Sketches of the shear–strain relation in the shear experiment. In (**a**), a shear strain is applied (gray), and the stress response is measured, while in (**b**) a shear stress is applied, and strain is measured. An elastic response (blue and red, respectively) is in phase, while a viscous response is phase shifted by 90° (positive direction in (**a**), negative direction in (**b**)). The phase shift will be between 0° and 90° for a viscoelastic response

In either case, the response is also sinusoidal but phase shifted, if the material is not purely elastic. Since cases (a) and (b) in Fig. 5.5 give principally the same results, but with inverted cause and effect, we henceforth focus on case (a), applying strain and measuring stress, as this is the more common experimental method.

As the viscoelastic stress response will be sinusoidal with an added phase shift—the strain lags behind the stress—it is possible to use some basic trigonometric identities for dealing with sines and cosines in an elegant way. Nonetheless, the solution for viscoelastic materials can already be explained qualitatively since the response of elastic and viscous materials is already known.

For elastic materials, the resulting stress, σ_{ela} , is proportional to the applied strain, γ_{in}, giving

$$\sigma_{\mathrm{ela}}\left(t\right)=\sigma_{0,\mathrm{ela}}\cos\left(\omega t\right),\tag{5.6}$$

since γ_{in} is also a cosine function. In contrast, for a viscous material, the strain *rate* is proportional to stress, meaning that stress will be highest when strain changes the fastest and stress will be zero when strain is constant. Therefore, the viscous stress response will be out of phase by 90°:

$$\sigma_{\mathrm{vis}}\left(t\right)=-\sigma_{0,\mathrm{vis}}\sin\left(\omega t\right).\tag{5.7}$$

Intuitively, the stress response of a viscoelastic material can be found as the sum of the elastic and the viscous response:

$$\sigma_{\mathrm{VE}}\left(t\right)=\sigma_{0,\mathrm{ela}}\cos\left(\omega t\right)-\sigma_{0,\mathrm{vis}}\sin\left(\omega t\right).\tag{5.8}$$

Analogous to the tensile creep compliance presented in the previous chapter, we can define the *complex shear modulus* given by

$$G^{*}\left(\omega,t\right)=\frac{\sigma_{\mathrm{VE}}\left(\omega,t\right)}{\gamma_{0}}=G'\left(\omega\right)\cos\left(\omega t\right)-G''\left(\omega\right)\sin\left(\omega t\right),\tag{5.9}$$

with storage modulus G' and loss modulus G''.

A viscoelastic material can therefore be characterized by the ratio of the viscous and the elastic stress response amplitude for a given frequency [13]. The higher the viscous amplitude relative to the elastic amplitude, the more viscous than elastic a material is at that given frequency. The elastic and viscous response can be different for varying shear frequencies (see Fig. 5.6. The ratio of both amplitudes also defines phase shift angle δ, like in Fig. 5.5, given by the following equation (see Eq. (2.44) in Chap. 2):

$$\tan \delta (\omega) = \frac{\sigma_{0,\text{vis}}}{\sigma_{0,\text{ela}}} = \frac{G''(\omega)}{G'(\omega)}. \tag{5.10}$$

For the quantitative part, the basic trigonometric identities mentioned above are needed. As the elastic and viscous amplitudes cannot be measured independently in shear experiments, we have to convert G^* from a sum of a sine and a cosine to a single cosine (or sine) including phase shift. After conversion, we obtain

$$G^* (\omega,t) = \left| G^* (\omega) \right| \cos (\omega t + \delta), \tag{5.11}$$

where $|G^*(\omega)|$ denotes the measured absolute amplitude given by

$$\left| G^* (\omega) \right| = \sqrt{G'(\omega)^2 + G''(\omega)^2}. \tag{5.12}$$

The storage and loss modulus can be recovered via $G'(\omega) = |G^*| \cos \delta$ and $G''(\omega) = |G^*| \sin \delta$.

This representation of complex shear modulus G^* is most suited for experimental data since it includes absolute amplitude $|G^*(\omega)|$ and phase shift δ, both of which are quantities that are easily measurable in any oscillatory shear experiment for any frequency. Complex shear modulus $G^*(\omega)$ is the favored material parameter in the scientific community since it has a more convenient interpretation, basically the

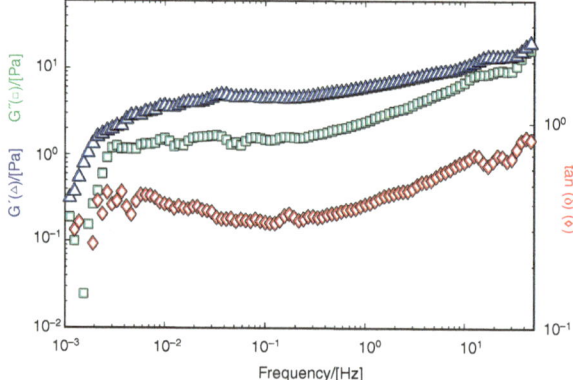

Fig. 5.6 Graph of the elastic and viscous shear modulus (extracted from the complex shear modulus) and phase angle of an actin polymer network measured with a rotation shear rheometer. For low frequencies (10^{-2} Hz and lower), actin is easier to deform on long time scales of 100 s, which cause a significantly lower storage (blue) and modulus (green). Deformability, both elastic and viscous, increases with frequency. As the phase angle increases with frequency, actin becomes more and more viscous in its response

same as elastic modulus E. Also, rotation shear rheometers are predominantly used for oscillatory shear experiments, which can directly measure the storage and loss modulus for a broad frequency range. If instead a shear stress is applied and a strain response is measured (Fig. 5.5a), the calculations above can be done analogously and will result in the complex inverse of $G^*(\omega)$, i.e., complex compliance $J^*(\omega) = 1/G^*(\omega)$. A typical complex shear modulus graph is depicted in Fig. 5.6, showing the frequency-dependent response of an actin network under strain load.

In addition, the previously introduced elastic modulus $E(t)$ and tensile creep compliance $D(t)$ can also be converted to complex shear modulus $G^*(\omega)$. Details of the conversion will be omitted since it involves advanced calculus and will not give more insight into the material properties of biological tissues.

5.3.3 Modeling Viscoelasticity

Modeling the complex time dependence of tensile creep compliance or the elastic modulus of viscoelastic materials is often done via constitutive equations and/or models. In a simplified, coarse-grained vision, cells can be considered as polymer scaffolds (cytoskeleton) filled with a viscous fluid (cytosol) and functional entities (organelles, which are obstacles in the polymer meshwork), or, put simply, the cytoplasm responds like a *water-filled sponge* [48]. The cytoskeleton is the main contributor to elastic behavior, while the flow and friction of the cytosol and organelles contribute to the viscous response. The combination of both results in an overall viscoelastic response.

An ideal spring with its elastic response, simulating the cytoskeleton, and an ideal dashpot, simulating the viscosity of the cytosol, can be interconnected to set up toy models simulating viscoelastic responses. We will give here only a short and shallow overview on how viscoelastic properties can arise from the combination of perfectly elastic and viscous subunits. When combining a spring and a dashpot in parallel in the so-called Kelvin–Voigt model, the applied stress is distributed between the spring and the dashpot as captured in the following simple equation:

$$\sigma_{total} = \sigma_{spring} + \sigma_{dashpot}. \tag{5.13}$$

In addition, the strain of the spring and dashpot will be the same as the total strain:

$$\gamma_{total} = \gamma_{spring} = \gamma_{dashpot}. \tag{5.14}$$

With this set of equations, the time-dependent strain response $\gamma(t)$ of the Kelvin–Voigt model (and analogously for the Maxwell model) for *any* given time-dependent stress $\sigma(t)$ can be calculated. With the recipe given above, more complex models, possibly featuring more material details for different time scales, can be set up using more than one spring and dashpot. Introduction of another dashpot in series, for instance, accounts for permanent plastic deformation. Furthermore, these models can be applied to any type of experiment—shear and pulling/pushing mode, stress or strain application, and oscillatory and stepping mode—rendering them universally applicable. These models are therefore widely used in the scientific community as a first top–down approach.

Another modeling approach originates from polymer physics allowing to derive many scaling laws from basic principles [12]. Scaling laws are powerful, predictive tools and are generally found in biological systems [49–51]. For instance, the basal metabolic rate P of mammals is approximately proportional to their mass M to the power of three fourths ($P \propto M^{3/4}$). For single cells and tissues, scaling laws can be found for the strain response under stress load ($\gamma \propto \sigma_0 \cdot t^\alpha$) [52–54]. For a scaling exponent α between 0 and 1, a viscoelastic response can be modeled, including the limit cases of $\alpha = 0$ and $\alpha = 1$, corresponding to a purely elastic and purely viscous response, respectively. Indeed, scaling behaviors of material parameters, such as G^* and E, can be found in various biological systems [55]. They can be also derived from more fundamental principles of polymer interactions [12] and hold even for advanced theories, i.e., the glassy wormlike chain [56, 57] including nonlinearities like strain hardening and softening [58]. More modeling approaches for cells and tissues can be found in [59, 60].

However, many approaches assume a passive material, which might not be the case for biological matter on time scales of minutes or longer [60]. Introducing active responses, and therefore active force generation, in models is a challenging task since many active processes cannot be described with ease in a coarse-grained manner. Modeling force generation of myosin motors, however, has made significant advances, and appropriate models have been introduced [61–65].

Besides active responses, when probing the mechanical properties of biological matter, the effect of temperature should not be neglected. The temperature should always be in a physiological range since many processes in organisms are highly temperature dependent, e.g., polymerization and depolymerization rates of actin and microtubules [66, 67] as well as motor activity of myosin, dynein, and kinesin [68]. Temperature also affects passive material properties as many materials become less viscous at higher temperatures. Honey, for instance, is much more viscous at lower temperatures [46]. In detail, the viscosity of honey follows an Arrhenius law [69]:

$$\eta(T) = \eta(T_{\text{ref}}) e^{-\left(\frac{E_A}{k_B T}\right)}, \tag{5.15}$$

where $\eta(T_{\text{ref}})$ is the viscosity of the material at a given reference temperature, k_B the Boltzmann constant, and E_A the activation energy of a transition of states, usually energy barriers of chemical reactions or binding energies. This effect, commonly known as time–temperature superposition, can be observed for single cells [70]. However, many cells do not show this clear relation, and the temperature dependency of their responses is more complex [71].

5.4 Mechanics on the Cellular Level

Modeling is often limited by strong interactions across multiple levels of complexity. Already on a cellular level, the many cell organelles and functional groups make it difficult to grasp the cell "as a whole" in terms of a coarse-grained system. As the internal structures of cells are already highly anisotropically distributed, it might appear counterintuitive to describe the behavior of a whole cell as a coarse-grained system. It turns out, however, that many of the introduced concepts are no oversimplifications *per se* and that biological systems can often be described in such a simplified manner. As long as cells (and/or tissues) are not actively reacting to applied forces and deformations, cell mechanics can be understood as an emergent consequence of the cytoskeletal network level. To push this conceptual approach even further, key aspects of cell migration [72] or cell shape [73] can be described without considering details of the filamentous or the molecular level by solely using very fundamental hydrodynamics-based descriptions [74].

Biological cells can be structurally regarded as polymer-filled entities enclosed by a nearly impenetrable membrane. Due to the considerable backbone stiffness of actin filaments and microtubules, mechanical integrity is already given for relatively large mesh sizes and low-volume fraction. An analogy for this concept is a tent, which mechanically stabilizes a certain volume with enough space for passive and active molecular transport. As described in the cytoskeleton chapter, the three main components, actin, microtubules, and intermediate filaments, can form emergent structures such as networks and bundles [7, 75]. Although cells come in a broad phenotypic diversity, including keratinocytes, fibroblasts, and neurons, the cytoskeletal composition does not necessarily need a thorough overhaul. Usually, slight compositional variation or introduction of active processes, i.e., molecular motors, is often sufficient to generate this rich pool of structural appearances. Additionally, many other cellular components contribute to the mechanical behavior.

The main actors in this scenario are the nucleus, the cytoskeleton, and the cell membrane. While these structures are functionally and mechanically intertwined, effects can be specifically attributed to certain subcellular structures. Whole-cell deformations like squeezing or stretching of the cell body are mostly affected by the cytoskeleton as it is the most extensive structure in the cell. Small deformations in the linear regime (up to 5% [58]) are dominated by actin and intermediate filaments [76]. For larger deformations between 5 and 25% strains, nonlinear effects of the actin cytoskeleton can result in nonlinear responses, e.g., strain stiffening [77–79] and—paradoxically—also strain softening [80, 81] (the paradox is resolved in [58]). Even larger elongations are intercepted by microtubules and ultimately limited by the integrity of the membrane. Lamellipodia and other protrusions also rely on the stabilization by cytoskeletal filaments. Nuclear mechanics come into play when cells are moving in narrow spaces or are heavily compressed [82, 83]. Passing of narrow channels is dependent on the viscous response of the nucleus to such confinement [83–85], and the nuclear lamina will actively respond to environmental stiffness [86, 87]. The mechanical response of cells is also influenced by their membrane, and diseases such as cancer can often cause membrane alterations [88, 89]. Small indentations (tenth of cell diameter) and pulling forces (sub-nano-Newton) are a matter of membrane bending rigidity, while global (and quasi-global) indentations are influenced by effective membrane tension [89, 90]. Furthermore, membrane rigidity influences the extent of mechanosensing signaling pathways. This also holds for self-induced invaginations and blebbing, i.e., exo- and endocytosis, receptor binding, and treadmilling. When discussing processes beyond these circumstances and on the scale of a whole cell, the entanglement of these differently behaving structures has to be taken into account. Still, the cytoskeleton can be considered the most influential part of the cell regarding mechanical behavior.

The strong influence of the cytoskeleton on the overall mechanical properties is striking when looking at the elastic modulus E of most cells. It ranges from unexpected low values [52] of some hundred Pa in glial or neuronal cells [91] to tens of kPa in human thrombocytes [92], illustrating the high variability and adaptivity of cells compared to classical and synthetic materials. As parts of the cytoskeleton are in constant treadmilling, appearance and mechanical structure are not persistent and allow the cell to adapt to its environment, rendering the cytoskeleton self-regulatory.

From the broad range of elastic moduli of different cell types and completely different functions, it is apparent that cell mechanics is a vital component of cellular functioning [93–97] including mitosis, where the cytoskeleton undergoes a significant overhaul enabling controlled cell division [98].

5.4.1 Probing Techniques

Cell rheology probes the response of cells to applied disturbances. As already explained in detail in the rheology chapter, responses and disturbances are usually forces and deformation or vice versa. Since material parameters like the complex shear modulus are in principle independent of the probing technique, a variety of techniques have been established based on different physical concepts to generate stresses or strains. Table 5.1 summarizes the most common probing techniques for single cells. Another reason for this variety is that every technique has its own working range of stresses and strains and temporal resolution and probes a cell either locally or globally. Nonetheless, due to the high structural heterogeneity of single cells including local mechanical alterations, it has turned out that directly comparable, consistent results are difficult to obtain. One eminent question is *what* exactly is probed since the main components of the cytoskeleton already differ in their mechanical properties. Also, adherent cells, which form prominent stress fibers on substrates [99], differ in their responses from suspended cells, in which actin conglomerates to a shell-like cortex below the membrane [100, 101]. It remains an open question whether (and how) results obtained for adhered and suspended cells compare.

The comparably simple and inexpensive micropipettes were one of the first tools for characterizing cell mechanics via micropipette aspiration [117], albeit it is limited by inherently lower throughput due to long preparation and measurement times. In the earliest application of this method, blood cells with different diameters were used and analyzed with regard to their response to higher or lower suction pressure (=stress). In general, any suspended cell can be probed including isolated cells from tissues [103]. If a very small pipette diameter is chosen, the local mechanical properties can be probed, whereas larger pipettes can be used to suck in cells for global probing on time scales from seconds to hours [102], and deformations far from the linear regime can be obtained.

Table 5.1   Cell mechanics probing techniques

Technique	Range of application
Micropipette aspiration	Local or global probing of strain and stress depending on diameter Time range 1–1000 s [102, 103]
AFM/SFM indentation	Low and high strains possible Local or averaged probing Frequency range 1–300 Hz Force range pN–nN [93, 104, 105]
Active and passive optical trap rheology	Probing of overall force and fraction of force transmitted to the environment Local or global probing of stresses (Pa) Force range pN–nN [106]
Passive bead rheology	Passive method Local properties of viscosity via diffusion Energies of order $k_B T$ [107–109]
Magnetic bead rheology	Local probing of elastic response Frequency range 0.01–1000 Hz Torques up to 130 Pa, linear [54, 110]
Optical stretcher	Local or global probing of stresses (Pa) Forces 0.1 nN Small strains of 1–10% [97, 101, 111–113]
Real-time deformation cytometer (RT-DC)	Global probing of deformation under high pressures Stresses up to 500 Pa [114–116]
Micro-constriction array	Global probing of deformation of cells Cell nuclei deformation probing Stresses up to 400 Pa [83]

Another well-established method to determine mechanical properties of cells, which can be also extended to small tissue samples, is atomic (scanning) force microscope (AFM). Depending on the geometry of the cantilever, cells can be probed very locally by using a pointy tip [93], broadly locally by using a beaded tip [91, 104, 118], or globally by using a flat tip [119]. In terms of stress and strain application, the AFM is versatile, allowing indention times from milliseconds to minutes, only limited by drifting of experimental stages (which can be stabilized with slight sophistication [120]). The force application ranges from pN to nN, including forces beyond the linear limit [121, 122]. Furthermore, using an oscillating cantilever to induce oscillatory stresses allows complex shear modulus measurements [41, 105]. One drawback of the AFM is the comparably low throughput due to possible long preparation times of the experiments, and since adherent cells are very flat, substrate stiffness and roughness affect the results and have to be considered [104]. Stretching of cells can also be done by letting the cell adhere to the cantilever first and subsequently pulling it away [123].

To circumvent the bottleneck of low throughput, further techniques make use of parallel preparation of cells by incorporating tracing beads and using passive Brownian motion (passive microrheology) or probing beads and actively displacing them (active microrheology) [124, 125]. The established method of analysis is based on cross-correlation of the motion of different beads and correction for local heterogeneities [109, 126]. Since artificial beads may be invasive, naturally present "tracers" such as storage granules, mitochondria, and other submicron particles were used with success [108, 127].

With the small size of the beads relative to cell size, both active and passive microrheologies are best suited to probe local rather than global mechanical properties. As microrheology is a contact method, it is highly dependent on the type and strength of linkage between the beads and the surrounding cellular structures [124, 128] and, due to the heterogeneity of cells, a controlled binding affinity is yet to be achieved [124].

The most common active microrheological method is magnetic twisting cytometry, which involves manipulation (usually twisting) of magnetic beads with an external magnetic field [54, 110, 129]. Since oscillating magnetic fields can be easily generated, many cells can be probed in parallel at once across a range of over four decades of frequencies. Limiting factors degrading measurement accuracy, however, include the exact determination of the bead's magnetic moment along with bead-to-bead variation and the applied external magnetic field.

All probing techniques discussed so far are contact based and are therefore prone to be invasive and might measure cell mechanics in an altered state. Furthermore, for all techniques, cells have to be at least weakly adherent, introducing additional problems due to substrate influences even for techniques based on optically trapped probing beads [106]. These limitations and difficulties can be overcome by using optical manipulation and microfluidic techniques to measure single suspended cells.

For optical manipulation, the optical stretcher has been established. This technique is based on the momentum transfer of photons on interfaces with changing refractive indices. Two antiparallel laser fibers with divergent beam profiles can generate an optical pressure force [130] enabling optical trapping at lower laser powers and deformation of cells at higher powers [101]. The optical stretcher can apply step stresses over a time-range from 0.1 s to tens of seconds, enabling creep compliance measurements [111, 112]. Applied stresses are in the Pa range, corresponding to sub-nano-Newton forces, which depend on fiber-to-fiber distance and cell size. Cell mechanics can be probed locally for large cells and small fiber distances or globally for small cells and increased fiber distances. Since cells can be optically trapped for a prolonged period of time, active responses of single cells can be observed, for instance, active contractions of cells under force load [131]. Oscillatory stress application is also possible, however, with the restriction that only positive stresses (stretching of cells) and no negative stresses (squeezing of cells) can be applied [132, 133]. Therefore, the stress pattern will include an offset stress: $\sigma(t) = \sigma_0 + \sigma_0 \sin(\omega t)$. Embedded in a microfluidic setup, the optical stretcher allows serial measurements of up to 300 cells per hour and subsequent sorting, which can be challenged by the global heterogeneity of cell stiffness [97, 101].

Related techniques are able to deform cells in a similar contact-free manner but are based on hydrodynamic instead of optical forces [116]. Here, cells are pushed through capillaries in a continuous flow. Sudden changes in capillary geometry alter the flow profile locally. Shear flow velocities are applied that are sufficient to generate force differences large enough to deform whole-cell bodies. With these techniques, immense throughputs can be achieved [114–116]. However, the very limited observation time for one cell (millisecond range) impedes long-term deformation measurement, reducing measurable cell mechanics to the relative deformation of the cells after entering the measurement channel, i.e., the (time-independent) elastic modulus E.

Further details and comparisons of the different commonly applied probing techniques can be found in [52, 55, 60, 121, 134, 135].

5.4.2 Comparability and Interpretation

The broad range of experimental techniques and their intrinsic advantages and disadvantages make it challenging to compare results obtained with different techniques, especially quantitative results. Responses of suspended and adherent cells (as well as resting and migrating) will inherently differ from each other due to their altered geometries. Furthermore, probing a cell locally might not yield the same results as probing it globally. Even focusing on a single technique, defining the mechanical properties of a certain cell type is already nontrivial as cell-type stiffness follows a broad, non-Gaussian distribution (which can be tackled by averaging over many cells in large cell monolayer shear measurements [136]).

Despite the given quantitative challenges, the broad range of experimental techniques has yielded a comprehensive qualitative picture of cell mechanics covering various orders of stress and strain regimes [52, 124]. For instance, measurements with different techniques show a common power-law behavior of the complex shear modulus with only a slightly varying power-law exponent [107, 124, 134, 137, 138].

Nonetheless, in order to compare experimental techniques, one has to overcome some drawbacks of these techniques, which often involve poor statistics and the lack of standardized measurement protocols. For some of the methods presented here, it is difficult to obtain enough data during the short time in which the mechanical properties are altered actively in response to environmental changes. On the other hand, the diversity of probing techniques is also an advantage. As different cell types occur in different environments, cell mechanics differ to suit their environment, e.g., red blood cells are more elastic since they have to squeeze through capillaries [11]. Thus, the most suitable technique for a given cell type can be chosen. Suspended cells like RBCs, for instance, can be measured more easily and rapidly in an optical stretcher or real-time deformation cytometer [83, 112, 115, 139].

5.5 Tissues

The mechanics of systems consisting of multiple cells are changed drastically by two elements that are not present at the single-cell level: adhesive contacts between cells and between cells and the extracellular matrix (ECM), enabling active responses of cells to their neighbors. Cell–cell communication via very basic interactions, such as chemical feedback loops, can already lead to collective behavior of cells, e.g., swarm-like collective migration of keratinocytes [140] and collective cell migration during invasion and metastatic spread of malignant tumors [141]. Furthermore, strength and type of adhesive contacts direct force transmission and modulate cell migration and motility. As a consequence, cells can show collective structural behavior on different length scales ranging from organization of smaller subdomains in tissues with individual properties up to global quasi-frozen, glass-like states with no (relative) cell migration as in epithelial layers [142], which commonly occurs as soon as increased adhesion force and cell density are introduced. Since active processes of tissues are emergent phenomena, they can strongly influence mechanical responses. Adhesive cell–cell contacts and ECM can highly contribute to the stiffness and fluidity of tissues and span a phase space ranging from quasi-solid responses, e.g., epithelial tissue, to quasi-fluid materials such as migratory cells in mesenchymal tissues.

5.5.1 Cell–Cell and Cell–Matrix Adhesion

Most types of cell adhesion are mediated by proteins of the family of cell adhesion molecules (CAMs). CAMs are typically transmembrane proteins with binding sites for the cytoskeleton as well as sites for *trans-* and *cis*-interactions with other CAMs or the ECM in the extracellular domain. In addition to adhesion, they function as cytoskeletal anchors and play significant roles in mechanosignaling [11], which elegantly illustrates the intertwined nature of the different levels of complexity.

CAMs are often calcium dependent, and one of the most prominent CAM families is cadherins (a portmanteau word combining *ca*lcium and *adhering*). Cadherins come in three flavors and are usually associated with certain tissues (but are not restricted to these): E-cad, epithelial cells; N-cad, neuronal cells, and P-cad, pancreatic cells. They can appear as single free molecules but are usually ordered as nanoclusters linked to actin or as more complex desmosomes linked to the keratin cytoskeleton [143]. Differences in the function of these proteins are still under investigation, but changes in their expression rate can be correlated to changes of phenotype and behavior. In epithelial-to-mesenchymal transition (EMT), for example, cells lose their well-structured belt-like distribution of E-cadherins in favor of P- and N-cadherins [144]. At the same time, cells will regain their ability to move through tissue, a hallmark step in tumorigenesis and metastasis [145, 146]. In malignant tumors, this switch can promote directed invasion into the surrounding tissue of small cell clusters, often regulated by messenger RNA (mRNA). For instance, upregulation of miR-9, a short, noncoding RNA gene involved in gene regulation, leads to increased cell motility and invasiveness [147]. The increased expression of P-cad triggers polarization and directed movement [148]. This results in cells moving in single file (Indian filing) out of the tumor, with a tumor cell or fibroblast as the leading cell [149]. The polarization of the fibroblast that moves away from the cancer cells is mediated by N-cad adhesive sites [150].

Contact to the ECM is mainly established by integrins, which consist of two subunits, the α- and β-units, and can bind collagen, glycoproteins of the ECM (e.g., fibronectin), or both. There are several types of α- and β-units and consequentially a broad range of different integrins. Like cadherins, they cluster into functional domains by focal adhesions. Binding of integrins to extracellular structures induces signaling cascades that intervene with basic cellular functions such as cell growth and apoptosis. Depending on the range of expression of different integrins, and subsequently the composition of focal adhesions, signaling pathways can be promoted or suppressed. Integrins such as $\alpha_v\beta_3$ or $\alpha_5\beta_1$, for instance, are often found in cancer cells and seem play a role in cancer development. They influence the mechanical behavior [151], invasiveness in ECM-rich surroundings [152], and cellular survival [153]. In addition, some integrins are known to form complexes with growth factors that induce EMT [154] and increase proliferation [155].

The mechanical feedback of cell–cell and cell–ECM interactions is a fundamental parameter for cellular regulation and shows its drastic influence in diseases such as cancer [156]. Cells can modify their morphology and mechanical properties in response to a changing microenvironment [99]. Especially since cells can and do modulate their ECM and CAMs, they can be a strong promoter of metastasis formation [157]. This active reaction of cells to their microenvironment becomes stronger with increasing malignancy, e.g., metastatic cells can mimic mechanical properties of neuronal cells [158] by reactivating (epigenetically) silenced genes [159–161].

5.5.2 Tissue Dynamics and Collective Cell Behavior

It is a great challenge to define simple mechanical parameters for tissues. At small time scales and low strains, tissues show frequency-dependent complex shear moduli [162]. At larger time scales, tissues can lose their viscoelastic behavior and show fluidlike mechanics—usually coupled to movement of single cells in the tissue and in a regime far from equilibrium dynamics. The two established main models describe tissues as glass-like, amorphous materials or as yield stress fluids. With both models, the active contribution of cells in the system is of interest. Softness of the cell body, adhesion strength to neighboring cells, and matrix as well as active forces determine whether a cell is able to migrate or not [163].

Analogies to glassy materials can be found in liquid- to solid-like glass transition. When the temperature of a glassy material drops below the glass transition point, molecules are strongly confined in their motion and fixed in a chaotic lattice. A cell system that reaches a certain density due to proliferation will exhibit similar behavior. Additionally, density-independent transitions can be observed when adhesion and stiffness of cells are modulated [164]. Inhibited migration and proliferation mark the point at which the system goes into a static (glassy) state [165, 166], a concept which can be transferred to tissues under high stresses. Epithelial layers, for example, have been shown to exert constant pulling forces that can influence their behavior [167].

Yield stress fluids show viscoelastic behavior to the point where a certain energy barrier has been overcome, when the material will start to flow. In this model, single-cell mechanics and stresses acting on the tissue are the main contributors [168]. Stress on the tissue, intrinsic or extrinsic, interacts with local adhesive mechanics, giving rise to either fluidlike or solid-like behavior. Especially homeostatic stresses determine the flow of the tissue, and a tissue with higher homeostatic stress will invade surrounding tissues either as small, separate islands or as a front [169] (Fig. 5.7A1, A2).

In both cases, transitions can occur for the whole tissue at once, cell clusters in confinement, or for single cells within a tissue. Even when a tissue is above the transition point and remains static, some cells might be in a different state and still able to pass through it (Fig. 5.7).

A rheological approach to access the different states of tissues is measuring phase angle δ. Although this approach does not account for active cell migration, it allows estimating the fluidity of the tissue. δ near $0°$ (purely elastic) indicates a jammed state, while viscous, flowing tissue approaches $90°$. To probe local properties of tissues, scanning force microscopy can be employed to create a map of viscoelasticity of a tissue slice that can be attributed to processes in the sample [170]. With standard bulk shear rheometers, global measurements can be performed to determine the overall viscoelastic properties [171]. Methods such as magnetic resonance elastography enable direct measurement of complex shear moduli of a whole tissue or with spatial resolution [162, 172].

When parameters of single cells, such as adhesive and viscoelastic properties, are known, these can be used to draw conclusions regarding tissue dynamics. The differential adhesion hypothesis (DAH) states that a mixture of cells

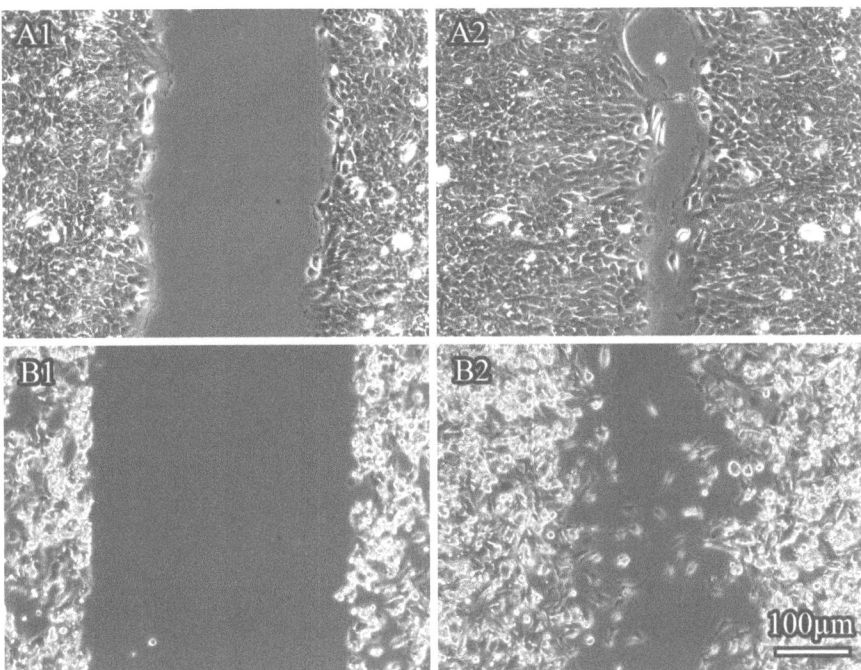

Fig. 5.7 Wound-healing assay of two different cell types. In the upper panel, A1 and A2, an epithelial cell layer starts at a certain time (A1) and closes the wound after 30 h (A2). The epithelial cell layer maintains its cell front and shows a coordinated, collective motion. A mesenchymal cell layer, B1 and B2, loses its front (B1) in this process and shows randomly walking single cells and no coordinated, collective motion after 30 h (B2)

with different adhesiveness develops into an ordered state, separating subpopulations accordingly [173]. The DAH holds true in morphogenesis, where cells demix like fluids with different surface tensions [174], but fails for cancer development. When cells undergo EMT, demixing can be observed, which seems to follow a more complex behavior [175]. Local jamming, unjamming, and tumor cell invasion are strongly related to these observations. When a malignant neoplasm forms, it constitutes a clearly separated bulk of tissue within the healthy stroma. Cells in the tumor exhibit altered mechanical behavior and heavily remodel the ECM, but there is no obvious reason why a distinct boundary exists. As soon as the cells lose their epithelial phenotype, they show increased invasion [176]. Still, the primary tumor grows to a certain size before cells escape, which might have its origins in cellular jamming. Fibrosis, for instance, leads to a very stiff ECM [177], and the growth of the tumor creates pressure on the surrounding tissue and the tumor itself [178]. In other words, the tumor embeds itself in a strong matrix. This goes hand in hand with a tumor being a rigid mass, although single malignant cells tend to be softer when becoming more invasive [179]. While the self-driven confinement creates a jammed state within the tumor [180], the mechanical feedback leads to further transitioning toward more malignant phenotypes [181]. Over time, the distribution of cellular softness becomes broader [97], and expression of CAMs is altered [144] up to the point where some cells undergo an unjamming transition and start to move out of the tumor [182]. It has to be noted that this cellular escape does not occur in a random pattern. Similar to embryogenesis, where cells follow the DAH, self-organization within the tissue is the first step. Collective migration in 2D assays can be observed, when the confluency of the layer confines cells, before the layer becomes jammed [183] (Fig. 5.7 A2). In 3D, this cellular streaming has also been observed: small conglomerates of cells in an unjammed state form and migrate collectively, often following a leader cell [184]. This marks the start of metastatic spread, and the moment at which invasion begins heavily depends on the individual neoplasm. Some tumors will grow to immense sizes over months or even years before cells pass the boundaries, while others metastasize within weeks of the original tumor formation.

Conclusion

The eukaryotic cell is well studied with decades of research dedicated to various branches and aspects ranging from classification of whole-cell types down to molecular details of protein folding processes [11, 185]. With the advancement of techniques and detailed insights into biological matter, the cause and effect of many diseases could be attributed to certain functional or structural units and levels of complexity within the cell; for instance, sickle cell disease is often caused by only a single-nucleotide polymorphism (SNP) of the beta-globin gene, which results in strand-like clustering of defective hemoglobin and consequently stiffening of red blood cells [186, 187]. Especially cancer—as one of the most prominent maladies—is well studied on many levels of complexity [188], and crucial developmental steps and many biological changes in tumorigenesis were found and described [145, 146]. The development of cancer is accompanied by major changes of the cytoskeleton, which are necessary for cancer cells to migrate and invade other tissues [97, 189]. The cytoskeleton usually gets softer with increased malignancy, as shown with different cellular probing techniques [115, 116, 179, 188–194]. These important insights render rheological material characterization of a viable tumor marker. At the tissue level, however, tumors are found to be stiffer than healthy surrounding tissue due to a stiffer stroma and elevated cytoskeletal tension [177] although they are constituted of softer cells. While many emergent phenomena on the microtissue level will influence mechanical properties, many of them are physically characterized and quantified. While the biophysics of tumorigenesis and tissue mechanics is qualitatively well studied. Still, the quantitative description of demixing, jamming, and surface tension is under current investigation and remains promising with ongoing research on this frontier of science [163, 164, 183, 195–201].

Acknowledgments We thank Till Möhn, Jürgen Lippoldt, Martin Glaser and Benjamin Wolf for their very helpful comments, discussions, advices for illustrations, and language editing.

Glossary

Self-organization: Self-organization is an active, non-equilibrium process of an open system where energy is constantly dissipated and needs to be resupplied, for instance, to generate forces (such as the actin–myosin power stroke) or to organize dynamic structures (such as the lamellipodium for cell migration) far from the thermodynamic equilibrium [10].

Self-assembly: Self-assembly processes are solely based on equilibrium dynamics and are independent of energy dissipation. They occur spontaneously and tend to minimize the free energy of the system driving it toward its thermodynamic equilibrium without an additional energy source such as ATP or GTP. Self-assembly can occur in closed systems [10].

Persistence Length: Mechanical property quantifying the stiffness of a polymer relative to its length. The length scale on which the direction vectors of both ends of a filament lose their correlation.

Bending Stiffness: The resistance of a beam with unit diameter and length while undergoing bending. The higher the bending stiffness, the harder to flex the unit beam.

Molecular Motors: Molecular machines which consume energy (e.g., ATP, GTP) and convert it into motion or mechanical work.

Treadmilling: Steady-state phenomenon of cytoskeletal filaments, mostly actin, where one filament end depolymerizes and the other end polymerizes, leading to shrinkage and growth at the ends with no net length change of the filament.

Tensile Creep Compliance: The magnitude of the creep response of a unit bulk material for a given unit force load. The higher the tensile creep compliance, the easier it deforms under force load.

Elastic Modulus (Young's Modulus) or Shear Modulus: The resistance of a unit bulk material under axial load or under shearing load, respectively. The higher the elastic modulus or shear modulus, the harder to deform the material. In incompressible materials, the elastic modulus is three times the shear modulus.

Exocytosis: Active transport of molecules out of the cell via a secretory vesicle as transport carrier.

Endocytosis: Active transport of molecules into the cell via encapsulation of the molecules with the cell membrane and formation of a vesicle as transport carrier.

Receptor Binding: Binding of signaling molecules to transmembrane proteins used for cellular and tissue response.

Mechanosignaling: Sensing and signaling of cells induce a response to mechanical, environmental cues.

Glass-like Material: Solid-phase state of a material, where the strong, noncrystalline entanglement of the molecules, usually polymer chains, prevents an unhindered liquid-like flow and movement of the molecules for low thermal energy. Above the glass transition temperature, the material can flow again.

Amorphous Material: Noncrystalline solid with no long-range order, usually consisting of many clustered domains with different (crystalline) orientations and substructures.

Yield Stress Fluid: A fluid which only starts to flow above a critical stress, the yield stress. For stresses below the yield stress, it behaves like a solid.

Homeostatic Stress: Internal stress of a tissue generated by adhesion forces, which is actively regulated to remain close to constant.

Differential Adhesion Hypothesis: Hypothesis for cellular movement in tissues of different cell types based on thermodynamic principles. Cells with different adhesion forces will minimize their free energy by moving to other cells with similar adhesion forces in order to maximize bonding strength.

Jamming: A quasi-phase transition of a material, where rigidity suddenly increases and fluidity suddenly decreases when the density of cells (or molecules) increases above a critical level.

References

1. Anderson PW. More is different. Science. 1972;177(4047):393–6. https://doi.org/10.1126/science.177.4047.393.
2. Laughlin RB, Pines D. The theory of everything. Proc Natl Acad Sci U S A. 2000;97(1):28–31. https://doi.org/10.1073/pnas.97.1.28.
3. Schrödinger E. What is life?: the physical aspect of the living cell, Canto. Cambridge: Cambridge University Press; 2010.
4. Ryan AJ. Emergence is coupled to scope, not level. Complexity. 2007;13(2):67–77. https://doi.org/10.1002/cplx.20203.
5. Dawkins R The blind watchmaker: why the evidence of evolution reveals a universe without design. New edition [reissue] ed. New York: W. W. Norton; 1996.
6. Schuster P. A beginning of the end of the holism versus reductionism debate?: molecular biology goes cellular and organismic. Complexity. 2007;13(1):10–3. https://doi.org/10.1002/cplx.20193.
7. Huber F, Schnauss J, Ronicke S, et al. Emergent complexity of the cytoskeleton: from single filaments to tissue. Adv Phys. 2013;62(1):1–112. https://doi.org/10.1080/00018732.2013.771509.
8. Huber F, Kas J. Self-regulative organization of the cytoskeleton. Cytoskeleton. 2011;68(5):259–65. https://doi.org/10.1002/cm.20509.
9. Halley JD, Winkler DA. Classification of emergence and its relation to self-organization. Complexity. 2008;13(5):10–5. https://doi.org/10.1002/cplx.20216.
10. Halley JD, Winkler DA. Consistent concepts of self-organization and self-assembly. Complexity. 2008;14(2):10–7. https://doi.org/10.1002/cplx.20235.
11. Alberts B. Molecular biology of the cell: [MBOC]. 6th ed. New York: GS Garland Science; 2015.
12. Doi M, Edwards SF. The theory of polymer dynamics, The international series of monographs on physics, vol. 73. Oxford: Clarendon Press; 2003.
13. Schuldt C, Schnauß J, Händler T, et al. Tuning synthetic semiflexible networks by bending stiffness. Phys Rev Lett. 2016;117(19):197801. https://doi.org/10.1103/PhysRevLett.117.197801.
14. Isambert H, Venier P, Maggs A, et al. Flexibility of actin filaments derived from thermal fluctuations. Effect of bound nucleotide, phalloidin, and muscle regulatory proteins. J Biol Chem. 1995;270(19):11437–44. https://doi.org/10.1074/jbc.270.19.11437.

15. Isambert H, Maggs AC. Dynamics and rheology of actin solutions. Macromolecules. 1996;29(3):1036–40. https://doi.org/10.1021/ma946418x.
16. Greenberg MJ, Wang C-LA, Lehman W, et al. Modulation of actin mechanics by caldesmon and tropomyosin. Cell Motil Cytoskeleton. 2008;65(2):156–64. https://doi.org/10.1002/cm.20251.
17. Janson ME, Dogterom M. A bending mode analysis for growing microtubules: evidence for a velocity-dependent rigidity. Biophys J. 2004;87(4):2723–36. https://doi.org/10.1529/biophysj.103.038877.
18. Yin P, Hariadi RF, Sahu S, et al. Programming DNA tube circumferences. Science. 2008;321(5890):824–6. https://doi.org/10.1126/science.1157312.
19. Schiffels D, Liedl T, Fygenson DK. Nanoscale structure and microscale stiffness of DNA nanotubes. ACS Nano. 2013;7(8):6700–10. https://doi.org/10.1021/nn401362p.
20. Glaser M, Schnauß J, Tschirner T, et al. Self-assembly of hierarchically ordered structures in DNA nanotube systems. New J Phys. 2016;18(5):55001. https://doi.org/10.1088/1367-2630/18/5/055001.
21. Pollard TD, Blanchoin L, Mullins RD. Molecular mechanisms controlling actin filament dynamics in nonmuscle cells. Annu Rev Biophys Biomol Struct. 2000;29:545–76. https://doi.org/10.1146/annurev.biophys.29.1.545.
22. Schnauß J, Händler T, Käs J. Semiflexible biopolymers in bundled arrangements. Polymers. 2016;8(8):274. https://doi.org/10.3390/polym8080274.
23. Schnauss J, Golde T, Schuldt C, et al. Transition from a linear to a harmonic potential in collective dynamics of a multifilament actin bundle. Phys Rev Lett. 2016;116(10):108102. https://doi.org/10.1103/PhysRevLett.116.108102.
24. Lansky Z, Braun M, Ludecke A, et al. Diffusible crosslinkers generate directed forces in microtubule networks. Cell. 2015;160(6):1159–68. https://doi.org/10.1016/j.cell.2015.01.051.
25. Braun M, Lansky Z, Hilitski F, et al. Entropic forces drive contraction of cytoskeletal networks. BioEssays. 2016;38(5):474–81. https://doi.org/10.1002/bies.201500183.
26. Ward A, Hilitski F, Schwenger W, et al. Solid friction between soft filaments. Nat Mater. 2015;14(6):583–8. https://doi.org/10.1038/nmat4222.
27. Hilitski F, Ward AR, Cajamarca L, et al. Measuring cohesion between macromolecular filaments one pair at a time: depletion-induced microtubule bundling. Phys Rev Lett. 2015;114(13):138102. https://doi.org/10.1103/PhysRevLett.114.138102.
28. Huber F, Strehle D, Schnauß J, et al. Formation of regularly spaced networks as a general feature of actin bundle condensation by entropic forces. New J Phys. 2015;17(4):43029. https://doi.org/10.1088/1367-2630/17/4/043029.
29. Daniel JL, Molish IR, Robkin L, et al. Nucleotide exchange between cytosolic ATP and F-actin-bound ADP may be a major energy-utilizing process in unstimulated platelets. Eur J Biochem. 1986;156(3):677–83. https://doi.org/10.1111/j.1432-1033.1986.tb09631.x.
30. Bernstein BW, Bamburg JR. Actin-ATP hydrolysis is a major energy drain for neurons. J Neurosci. 2003;23(1):1–6.
31. Block J, Schroeder V, Pawelzyk P, et al. Physical properties of cytoplasmic intermediate filaments. Biochim Biophys Acta. 2015;1853(11 Pt B):3053–64. https://doi.org/10.1016/j.bbamcr.2015.05.009.
32. Herrmann H, Bar H, Kreplak L, et al. Intermediate filaments: from cell architecture to nanomechanics. Nat Rev Mol Cell Biol. 2007;8(7):562–73. https://doi.org/10.1038/nrm2197.
33. Herrmann H, Aebi U. Intermediate filaments: molecular structure, assembly mechanism, and integration into functionally distinct intracellular scaffolds. Annu Rev Biochem. 2004;73:749–89. https://doi.org/10.1146/annurev.biochem.73.011303.073823.
34. Howard J, Hyman AA. Dynamics and mechanics of the microtubule plus end. Nature. 2003;422(6933):753–8. https://doi.org/10.1038/nature01600.
35. Vavylonis D, Yang Q, O'Shaughnessy B. Actin polymerization kinetics, cap structure, and fluctuations. Proc Natl Acad Sci U S A. 2005;102(24):8543–8. https://doi.org/10.1073/pnas.0501435102.

36. Footer MJ, Kerssemakers JWJ, Theriot JA, et al. Direct measurement of force generation by actin filament polymerization using an optical trap. Proc Natl Acad Sci U S A. 2007;104(7):2181–6. https://doi.org/10.1073/pnas.0607052104.
37. Howard J. Mechanics of motor proteins and the cytoskeleton. Sunderland: Sinauer Associates; 2006.
38. Lockot HW, Uhlig S. Bibliographia aethiopica, Aethiopistische forschungen, vol. 41. Wiesbaden: Steiner; 1998.
39. Bell A, Macfarquhar C. Encyclopaedia britannica: or, a dictionary of arts and sciences, three volumes. Scotland: Edinburgh; 1771.
40. Borel F, Taylor JB, Paris IM. The splendor of ethnic jewelry: from the Colette and Jean-Pierre Ghysels collection, Pbk. ed. New York: H. N. Abrams; 2001
41. Mahaffy RE, Shih CK, MacKintosh FC, et al. Scanning probe-based frequency-dependent microrheology of polymer gels and biological cells. Phys Rev Lett. 2000;85(4):880–3. https://doi.org/10.1103/PhysRevLett.85.880.
42. Chen EJ, Novakofski J, Jenkins WK, et al. Young's modulus measurements of soft tissues with application to elasticity imaging. IEEE Trans Ultrason Ferroelect Freq Contr. 1996;43(1):191–4. https://doi.org/10.1109/58.484478.
43. Rho J-Y, Kuhn-Spearing L, Zioupos P. Mechanical properties and the hierarchical structure of bone. Med Eng Phys. 1998;20(2):92–102. https://doi.org/10.1016/S1350-4533(98)00007-1.
44. Rho JY, Ashman RB, Turner CH. Young's modulus of trabecular and cortical bone material: ultrasonic and microtensile measurements. J Biomech. 1993;26(2):111–9. https://doi.org/10.1016/0021-9290(93)90042-D.
45. Akhtar R, Sherratt MJ, Cruickshank JK, et al. Characterizing the elastic properties of tissues. Mater Today. 2011;14(3):96–105. https://doi.org/10.1016/S1369-7021(11)70059-1.
46. Yanniotis S, Skaltsi S, Karaburnioti S. Effect of moisture content on the viscosity of honey at different temperatures. J Food Eng. 2006;72(4):372–7. https://doi.org/10.1016/j.jfoodeng.2004.12.017.
47. Kesmarky G, Kenyeres P, Rabai M, et al. Plasma viscosity: a forgotten variable. Clin Hemorheol Microcirc. 2008;39(1-4):243–6.
48. Zhou EH, Martinez FD, Fredberg JJ. Cell rheology: mush rather than machine. Nat Mater. 2013;12(3):184. https://doi.org/10.1038/nmat3574.
49. Spence AJ. Scaling in biology. Curr Biol. 2009;19(2):R57–61. https://doi.org/10.1016/j.cub.2008.10.042.
50. Brown JH, West GB, editors. Scaling in biology. Santa Fe Institute studies in the sciences of complexity. Oxford: Oxford University Press; 2000.
51. West GB. A general model for the origin of allometric scaling laws in biology. Science. 1997;276(5309):122–6. https://doi.org/10.1126/science.276.5309.122.
52. Kollmannsberger P, Fabry B. Linear and nonlinear rheology of living cells. Annu Rev Mater Res. 2011;41(1):75–97. https://doi.org/10.1146/annurev-matsci-062910-100351.
53. Sandersius SA, Newman TJ. Modeling cell rheology with the subcellular element model. Phys Biol. 2008;5(1):15002. https://doi.org/10.1088/1478-3975/5/1/015002.
54. Fabry B, Maksym GN, Butler JP, et al. Scaling the microrheology of living cells. Phys Rev Lett. 2001;87(14):148102. https://doi.org/10.1103/PhysRevLett.87.148102.
55. Chen DT, Wen Q, Janmey PA, et al. Rheology of Soft materials. Annu Rev Condens Matter Phys. 2010;1(1):301–22. https://doi.org/10.1146/annurev-conmatphys-070909-104120.
56. Kroy K. Dynamics of wormlike and glassy wormlike chains. Soft Matter. 2008;4(12):2323. https://doi.org/10.1039/B807018K.
57. Kroy K, Glaser J. The glassy wormlike chain. New J Phys. 2007;9(11):416. https://doi.org/10.1088/1367-2630/9/11/416.
58. Wolff L, Fernandez P, Kroy K. Resolving the stiffening-softening paradox in cell mechanics. PLoS One. 2012;7(7):e40063. https://doi.org/10.1371/journal.pone.0040063.
59. Rodriguez ML, McGarry PJ, Sniadecki NJ. Review on cell mechanics: experimental and modeling approaches. Appl Mech Rev. 2013;65(6):60801. https://doi.org/10.1115/1.4025355.

60. Lim CT, Zhou EH, Quek ST. Mechanical models for living cells—a review. J Biomech. 2006;39(2):195–216. https://doi.org/10.1016/j.jbiomech.2004.12.008.

61. Herant M, Marganski WA, Dembo M. The mechanics of neutrophils: synthetic modeling of three experiments. Biophys J. 2003;84(5):3389–413. https://doi.org/10.1016/S0006-3495(03)70062-9.

62. Dai J, Ting-Beall HP, Hochmuth RM, et al. Myosin I contributes to the generation of resting cortical tension. Biophys J. 1999;77(2):1168–76.

63. Peskin CS, Odell GM, Oster GF. Cellular motions and thermal fluctuations: the Brownian ratchet. Biophys J. 1993;65(1):316–24. https://doi.org/10.1016/S0006-3495(93)81035-X.

64. Mogilner A, Oster G. Cell motility driven by actin polymerization. Biophys J. 1996;71(6):3030–45. https://doi.org/10.1016/S0006-3495(96)79496-1.

65. Kuusela E, Alt W. Continuum model of cell adhesion and migration. J Math Biol. 2009;58(1-2):135–61. https://doi.org/10.1007/s00285-008-0179-x.

66. Zimmerle CT, Frieden C. Effect of temperature on the mechanism of actin polymerization. Biochemistry. 1986;25(21):6432–8.

67. Kis A, Kasas S, Kulik AJ, et al. Temperature-dependent elasticity of microtubules. Langmuir. 2008;24(12):6176–81. https://doi.org/10.1021/la800438q.

68. Yengo CM, Takagi Y, Sellers JR. Temperature dependent measurements reveal similarities between muscle and non-muscle myosin motility. J Muscle Res Cell Motil. 2012;33(6):385–94. https://doi.org/10.1007/s10974-012-9316-7.

69. Oroian M, Amariei S, Escriche I, et al. A viscoelastic model for honeys using the time–temperature superposition principle (TTSP). Food Bioprocess Technol. 2013;6(9):2251–60. https://doi.org/10.1007/s11947-012-0893-7.

70. Kießling TR, Stange R, Käs JA, et al. Thermorheology of living cells—impact of temperature variations on cell mechanics. New J Phys. 2013;15(4):45026. https://doi.org/10.1088/1367-2630/15/4/045026.

71. Schmidt BUS, Kießling TR, Warmt E, et al. Complex thermorheology of living cells. New J Phys. 2015;17(7):73010. https://doi.org/10.1088/1367-2630/17/7/073010.

72. Joanny J, Prost J. Active gels as a description of the actin-myosin cytoskeleton. HFSP J. 2009;3(2):94–104. https://doi.org/10.2976/1.3054712.

73. Joanny J-F, Ramaswamy S. A drop of active matter. J Fluid Mech. 2012;705:46–57. https://doi.org/10.1017/jfm.2012.131.

74. Pearson JE. Complex patterns in a simple system. Science. 1993;261(5118):189–92. https://doi.org/10.1126/science.261.5118.189.

75. Strehle D, Schnauss J, Heussinger C, et al. Transiently crosslinked F-actin bundles. Eur Biophys J. 2011;40(1):93–101. https://doi.org/10.1007/s00249-010-0621-z.

76. Goldman RD, Khuon S, Chou YH, et al. The function of intermediate filaments in cell shape and cytoskeletal integrity. J Cell Biol. 1996;134(4):971–83.

77. Pourati J, Maniotis A, Spiegel D, et al. Is cytoskeletal tension a major determinant of cell deformability in adherent endothelial cells? Am J Physiol. 1998;274(5 Pt 1):C1283–9.

78. Wang N, Im T-N, Chen J, et al. Cell prestress. I. Stiffness and prestress are closely associated in adherent contractile cells. Am J Physiol Cell Physiol. 2002;282(3):C606–16. https://doi.org/10.1152/ajpcell.00269.2001.

79. Fernandez P, Pullarkat PA, Ott A. A master relation defines the nonlinear viscoelasticity of single fibroblasts. Biophys J. 2006;90(10):3796–805. https://doi.org/10.1529/biophysj.105.072215.

80. Trepat X, Deng L, An SS, et al. Universal physical responses to stretch in the living cell. Nature. 2007;447(7144):592–5. https://doi.org/10.1038/nature05824.

81. Krishnan R, Park CY, Lin YC, et al. Reinforcement versus fluidization in cytoskeletal mechanoresponsiveness. PLoS One. 2009;4(5):e5486. https://doi.org/10.1371/journal.pone.0005486.

82. Wolf K, Te Lindert M, Krause M, et al. Physical limits of cell migration: control by ECM space and nuclear deformation and tuning by proteolysis and traction force. J Cell Biol. 2013;201(7):1069–84. https://doi.org/10.1083/jcb.201210152.

83. Lange JR, Steinwachs J, Kolb T, et al. Microconstriction arrays for high-throughput quantitative measurements of cell mechanical properties. Biophys J. 2015;109(1):26–34. https://doi.org/10.1016/j.bpj.2015.05.029.
84. Friedl P, Wolf K, Lammerding J. Nuclear mechanics during cell migration. Curr Opin Cell Biol. 2011;23(1):55–64. https://doi.org/10.1016/j.ceb.2010.10.015.
85. Dahl KN, Ribeiro AJ, Lammerding J. Nuclear shape, mechanics, and mechanotransduction. Circ Res. 2008;102(11):1307–18. https://doi.org/10.1161/CIRCRESAHA.108.173989.
86. Swift J, Discher DE. The nuclear lamina is mechano-responsive to ECM elasticity in mature tissue. J Cell Sci. 2014;127(Pt 14):3005–15. https://doi.org/10.1242/jcs.149203.
87. Harada T, Swift J, Irianto J, et al. Nuclear lamin stiffness is a barrier to 3D migration, but softness can limit survival. J Cell Biol. 2014;204(5):669–82. https://doi.org/10.1083/jcb.201308029.
88. Händel C, Schmidt BUS, Schiller J, et al. Cell membrane softening in human breast and cervical cancer cells. New J Phys. 2015;17(8):83008. https://doi.org/10.1088/1367-2630/17/8/083008.
89. Braig S, Schmidt BUS, Stoiber K, et al. Pharmacological targeting of membrane rigidity: implications on cancer cell migration and invasion. New J Phys. 2015;17(8):83007. https://doi.org/10.1088/1367-2630/17/8/083007.
90. Gracià RS, Bezlyepkina N, Knorr RL, et al. Effect of cholesterol on the rigidity of saturated and unsaturated membranes: fluctuation and electrodeformation analysis of giant vesicles. Soft Matter. 2010;6(7):1472. https://doi.org/10.1039/b920629a.
91. Lu Y-B, Franze K, Seifert G, et al. Viscoelastic properties of individual glial cells and neurons in the CNS. Proc Natl Acad Sci U S A. 2006;103(47):17759–64. https://doi.org/10.1073/pnas.0606150103.
92. Radmacher M, Fritz M, Kacher CM, et al. Measuring the viscoelastic properties of human platelets with the atomic force microscope. Biophys J. 1996;70(1):556–67. https://doi.org/10.1016/S0006-3495(96)79602-9.
93. Radmacher M. Studying the mechanics of cellular processes by atomic force microscopy. In: Cell mechanics, vol. 83. London: Elsevier; 2007. p. 347–72.
94. Janmey PA, Winer JP, Murray ME, et al. The hard life of soft cells. Cell Motil Cytoskeleton. 2009;66(8):597–605. https://doi.org/10.1002/cm.20382.
95. Mierke CT, Rosel D, Fabry B, et al. Contractile forces in tumor cell migration. Eur J Cell Biol. 2008;87(8-9):669–76. https://doi.org/10.1016/j.ejcb.2008.01.002.
96. Engler AJ, Sen S, Sweeney HL, et al. Matrix elasticity directs stem cell lineage specification. Cell. 2006;126(4):677–89. https://doi.org/10.1016/j.cell.2006.06.044.
97. Fritsch A, Höckel M, Kiessling T, et al. Are biomechanical changes necessary for tumour progression? Nat Phys. 2010;6(10):730–2. https://doi.org/10.1038/nphys1800.
98. Thery M, Bornens M. Cell shape and cell division. Curr Opin Cell Biol. 2006;18(6):648–57. https://doi.org/10.1016/j.ceb.2006.10.001.
99. Yeung T, Georges PC, Flanagan LA, et al. Effects of substrate stiffness on cell morphology, cytoskeletal structure, and adhesion. Cell Motil Cytoskeleton. 2005;60(1):24–34. https://doi.org/10.1002/cm.20041.
100. Thoumine O, Cardoso O, Meister J-J. Changes in the mechanical properties of fibroblasts during spreading: a micromanipulation study. Eur Biophys J. 1999;28(3):222–34. https://doi.org/10.1007/s002490050203.
101. Wottawah F, Schinkinger S, Lincoln B, et al. Optical rheology of biological cells. Phys Rev Lett. 2005;94(9):98103. https://doi.org/10.1103/PhysRevLett.94.098103.
102. Schmid-Schönbein GW, Sung KL, Tözeren H, et al. Passive mechanical properties of human leukocytes. Biophys J. 1981;36(1):243–56. https://doi.org/10.1016/S0006-3495(81)84726-1.
103. Thoumine O, Ott A. Comparison of the mechanical properties of normal and transformed fibroblasts. Biorheology. 1997;34(4-5):309–26. https://doi.org/10.1016/S0006-355X(98)00007-9.
104. Mahaffy RE, Park S, Gerde E, et al. Quantitative analysis of the viscoelastic properties of thin regions of fibroblasts using atomic force microscopy. Biophys J. 2004;86(3):1777–93. https://doi.org/10.1016/S0006-3495(04)74245-9.

105. Alcaraz J, Buscemi L, Grabulosa M, et al. Microrheology of human lung epithelial cells measured by atomic force microscopy. Biophys J. 2003;84(3):2071–9. https://doi.org/10.1016/S0006-3495(03)75014-0.
106. Mizuno D, Bacabac R, Tardin C, et al. High-resolution probing of cellular force transmission. Phys Rev Lett. 2009;102(16):168102. https://doi.org/10.1103/PhysRevLett.102.168102.
107. Hoffman BD, Massiera G, van Citters KM, et al. The consensus mechanics of cultured mammalian cells. Proc Natl Acad Sci U S A. 2006;103(27):10259–64. https://doi.org/10.1073/pnas.0510348103.
108. Yamada S, Wirtz D, Kuo SC. Mechanics of living cells measured by laser tracking microrheology. Biophys J. 2000;78(4):1736–47. https://doi.org/10.1016/S0006-3495(00)76725-7.
109. Crocker JC, Valentine MT, Weeks ER, et al. Two-point microrheology of inhomogeneous soft materials. Phys Rev Lett. 2000;85(4):888–91. https://doi.org/10.1103/PhysRevLett.85.888.
110. Fabry B, Maksym GN, Butler JP, et al. Time scale and other invariants of integrative mechanical behavior in living cells. Phys Rev E Stat Nonlin Soft Matter Phys. 2003;68(4 Pt 1):41914. https://doi.org/10.1103/PhysRevE.68.041914.
111. Guck J, Ananthakrishnan R, Moon TJ, et al. Optical deformability of soft biological dielectrics. Phys Rev Lett. 2000;84(23):5451–4. https://doi.org/10.1103/PhysRevLett.84.5451.
112. Guck J, Ananthakrishnan R, Mahmood H, et al. The optical stretcher: a novel laser tool to micromanipulate cells. Biophys J. 2001;81(2):767–84. https://doi.org/10.1016/S0006-3495(01)75740-2.
113. Brunner C, Niendorf A, Käs JA. Passive and active single-cell biomechanics: a new perspective in cancer diagnosis. Soft Matter. 2009;5(11):2171. https://doi.org/10.1039/b807545j.
114. Mietke A, Otto O, Girardo S, et al. Extracting cell stiffness from real-time deformability cytometry: theory and experiment. Biophys J. 2015;109(10):2023–36. https://doi.org/10.1016/j.bpj.2015.09.006.
115. Otto O, Rosendahl P, Mietke A, et al. Real-time deformability cytometry: on-the-fly cell mechanical phenotyping. Nat Methods. 2015;12(3):199–202. https://doi.org/10.1038/nmeth.3281.
116. Gossett DR, Tse HTK, Lee SA, et al. Hydrodynamic stretching of single cells for large population mechanical phenotyping. Proc Natl Acad Sci U S A. 2012;109(20):7630–5. https://doi.org/10.1073/pnas.1200107109.
117. Evans EA. Bending elastic modulus of red blood cell membrane derived from buckling instability in micropipet aspiration tests. Biophys J. 1983;43(1):27–30. https://doi.org/10.1016/S0006-3495(83)84319-7.
118. Schulze C, Muller K, Kas JA, et al. Compaction of cell shape occurs before decrease of elasticity in CHO-K1 cells treated with actin cytoskeleton disrupting drug cytochalasin D. Cell Motil Cytoskeleton. 2009;66(4):193–201. https://doi.org/10.1002/cm.20341.
119. Jonas O, Duschl C. Force propagation and force generation in cells. Cytoskeleton. 2010;67(9):555–63. https://doi.org/10.1002/cm.20466.
120. Fuhs T, Reuter L, Vonderhaid I, et al. Inherently slow and weak forward forces of neuronal growth cones measured by a drift-stabilized atomic force microscope. Cytoskeleton. 2013;70(1):44–53. https://doi.org/10.1002/cm.21080.
121. Thoumine O, Ott A, Cardoso O, et al. Microplates: a new tool for manipulation and mechanical perturbation of individual cells. J Biochem Biophys Methods. 1999;39(1-2):47–62. https://doi.org/10.1016/S0165-022X(98)00052-9.
122. Fernandez P, Ott A. Single cell mechanics: stress stiffening and kinematic hardening. Phys Rev Lett. 2008;100(23):238102. https://doi.org/10.1103/PhysRevLett.100.238102.
123. Benoit M, Gabriel D, Gerisch G, et al. Discrete interactions in cell adhesion measured by single-molecule force spectroscopy. Nat Cell Biol. 2000;2(6):313–7. https://doi.org/10.1038/35014000.
124. Hoffman BD, Crocker JC. Cell mechanics: dissecting the physical responses of cells to force. Annu Rev Biomed Eng. 2009;11:259–88. https://doi.org/10.1146/annurev.bioeng.10.061807.160511.

125. Golde T, Schuldt C, Schnauss J, et al. Fluorescent beads disintegrate actin networks. Phys Rev E Stat Nonlinear Soft Matter Phys. 2013;88(4):44601. https://doi.org/10.1103/PhysRevE.88.044601.

126. Levine L. One- and two-particle microrheology. Phys Rev Lett. 2000;85(8):1774–7. https://doi.org/10.1103/PhysRevLett.85.1774.

127. Lau AWC, Hoffman BD, Davies A, et al. Microrheology, stress fluctuations, and active behavior of living cells. Phys Rev Lett. 2003;91(19):198101. https://doi.org/10.1103/PhysRevLett.91.198101.

128. Mijailovich SM, Kojic M, Zivkovic M, et al. A finite element model of cell deformation during magnetic bead twisting. J Appl Physiol. 2002;93(4):1429–36. https://doi.org/10.1152/japplphysiol.00255.2002.

129. Massiera G, van Citters KM, Biancaniello PL, et al. Mechanics of single cells: rheology, time dependence, and fluctuations. Biophys J. 2007;93(10):3703–13. https://doi.org/10.1529/biophysj.107.111641.

130. Kreysing MK, Kießling T, Fritsch A, et al. The optical cell rotator. Opt Express. 2008;16(21):16984. https://doi.org/10.1364/OE.16.016984.

131. Gyger M, Stange R, Kiessling TR, et al. Active contractions in single suspended epithelial cells. Eur Biophys J. 2014;43(1):11–23. https://doi.org/10.1007/s00249-013-0935-8.

132. Maloney JM, Lehnhardt E, Long AF, et al. Mechanical fluidity of fully suspended biological cells. Biophys J. 2013;105(8):1767–77. https://doi.org/10.1016/j.bpj.2013.08.040.

133. Maloney JM, van Vliet KJ. Chemoenvironmental modulators of fluidity in the suspended biological cell. Soft Matter. 2014;10(40):8031–42. https://doi.org/10.1039/C4SM00743C.

134. van Vliet K, Bao G, Suresh S. The biomechanics toolbox: Experimental approaches for living cells and biomolecules. Acta Mater. 2003;51(19):5881–905. https://doi.org/10.1016/j.actamat.2003.09.001.

135. Pullarkat P, Fernandez P, Ott A. Rheological properties of the eukaryotic cell cytoskeleton. Phys Rep. 2007;449(1-3):29–53. https://doi.org/10.1016/j.physrep.2007.03.002.

136. Fernández P, Heymann L, Ott A, et al. Shear rheology of a cell monolayer. New J Phys. 2007;9(11):419. https://doi.org/10.1088/1367-2630/9/11/419.

137. Deng L, Trepat X, Butler JP, et al. Fast and slow dynamics of the cytoskeleton. Nat Mater. 2006;5(8):636–40. https://doi.org/10.1038/nmat1685.

138. Weihs D, Mason TG, Teitell MA. Bio-microrheology: a frontier in microrheology. Biophys J. 2006;91(11):4296–305. https://doi.org/10.1529/biophysj.106.081109.

139. Roth KB, Neeves KB, Squier J, et al. High-throughput linear optical stretcher for mechanical characterization of blood cells. Cytometry A. 2016;89(4):391–7. https://doi.org/10.1002/cyto.a.22794.

140. Szabó B, Szöllösi GJ, Gönci B, et al. Phase transition in the collective migration of tissue cells: experiment and model. Phys Rev E. 2006;74(6):61908. https://doi.org/10.1103/PhysRevE.74.061908.

141. Deisboeck TS, Couzin ID. Collective behavior in cancer cell populations. BioEssays. 2009;31(2):190–7. https://doi.org/10.1002/bies.200800084.

142. Sander EE, van Delft S, ten KJP, et al. Matrix-dependent Tiam1/Rac signaling in epithelial cells promotes either cell-cell adhesion or cell migration and is regulated by phosphatidylinositol 3-kinase. J Cell Biol. 1998;143(5):1385–98.

143. Wu Y, Kanchanawong P, Zaidel-Bar R. Actin-delimited adhesion-independent clustering of E-cadherin forms the nanoscale building blocks of adherens junctions. Dev Cell. 2015;32(2):139–54. https://doi.org/10.1016/j.devcel.2014.12.003.

144. Kalluri R, Weinberg RA. The basics of epithelial-mesenchymal transition. J Clin Invest. 2009;119(6):1420–8. https://doi.org/10.1172/JCI39104.

145. Hanahan D, Weinberg RA. The hallmarks of cancer. Cell. 2000;100(1):57–70. https://doi.org/10.1016/S0092-8674(00)81683-9.

146. Hanahan D, Weinberg RA. Hallmarks of cancer: the next generation. Cell. 2011;144(5):646–74. https://doi.org/10.1016/j.cell.2011.02.013.

147. Ma L, Young J, Prabhala H, et al. miR-9, a MYC/MYCN-activated microRNA, regulates E-cadherin and cancer metastasis. Nat Cell Biol. 2010;12(3):247–56. https://doi.org/10.1038/ncb2024.
148. Plutoni C, Bazellieres E, Le Borgne-Rochet M, et al. P-cadherin promotes collective cell migration via a Cdc42-mediated increase in mechanical forces. J Cell Biol. 2016;212(2):199–217. https://doi.org/10.1083/jcb.201505105.
149. Gaggioli C, Hooper S, Hidalgo-Carcedo C, et al. Fibroblast-led collective invasion of carcinoma cells with differing roles for RhoGTPases in leading and following cells. Nat Cell Biol. 2007;9(12):1392–400. https://doi.org/10.1038/ncb1658.
150. Ladoux B, Mege R-M, Trepat X. Front-rear polarization by mechanical cues: from single cells to tissues. Trends Cell Biol. 2016;26(6):420–33. https://doi.org/10.1016/j.tcb.2016.02.002.
151. Mierke CT. The integrin alphav beta3 increases cellular stiffness and cytoskeletal remodeling dynamics to facilitate cancer cell invasion. New J Phys. 2013;15(1):15003. https://doi.org/10.1088/1367-2630/15/1/015003.
152. Seftor RE, Seftor EA, Gehlsen KR, et al. Role of the alpha v beta 3 integrin in human melanoma cell invasion. Proc Natl Acad Sci U S A. 1992;89(5):1557–61. https://doi.org/10.1073/pnas.89.5.1557.
153. Aoudjit F, Vuori K. Integrin signaling inhibits paclitaxel-induced apoptosis in breast cancer cells. Oncogene. 2001;20(36):4995–5004. https://doi.org/10.1038/sj.onc.1204554.
154. Munger JS, Huang X, Kawakatsu H, et al. A mechanism for regulating pulmonary inflammation and fibrosis: the integrin αvβ6 binds and activates latent TGF β1. Cell. 1999;96(3):319–28. https://doi.org/10.1016/S0092-8674(00)80545-0.
155. Soldi R, Mitola S, Strasly M, et al. Role of alphavbeta3 integrin in the activation of vascular endothelial growth factor receptor-2. EMBO J. 1999;18(4):882–92. https://doi.org/10.1093/emboj/18.4.882.
156. Bissell MJ, Radisky DC, Rizki A, et al. The organizing principle: microenvironmental influences in the normal and malignant breast. Differentiation. 2002;70(9-10):537–46. https://doi.org/10.1046/j.1432-0436.2002.700907.x.
157. Dvorak HF, Weaver VM, Tlsty TD, et al. Tumor microenvironment and progression. J Surg Oncol. 2011;103(6):468–74. https://doi.org/10.1002/jso.21709.
158. Heine P, Ehrlicher A, Käs J. Neuronal and metastatic cancer cells: unlike brothers. Biochim Biophys Acta Mol Cell Res. 2015;1853(11):3126–31. https://doi.org/10.1016/j.bbamcr.2015.06.011.
159. Ibragimova I, de Cáceres II, Hoffman AM, et al. Global Reactivation of Epigenetically Silenced Genes in Prostate Cancer. Cancer Prev Res (Phila). 2010;3(9):1084–92. https://doi.org/10.1158/1940-6207.CAPR-10-0039.
160. Karpf AR, Jones DA. Reactivating the expression of methylation silenced genes in human cancer. Oncogene. 2002;21(35):5496–503. https://doi.org/10.1038/sj.onc.1205602.
161. Cameron EE, Bachman KE, Myohanen S, et al. Synergy of demethylation and histone deacetylase inhibition in the re-expression of genes silenced in cancer. Nat Genet. 1999;21(1):103–7. https://doi.org/10.1038/5047.
162. Riek K, Klatt D, Nuzha H, et al. Wide-range dynamic magnetic resonance elastography. J Biomech. 2011;44(7):1380–6. https://doi.org/10.1016/j.jbiomech.2010.12.031.
163. Bi D, Yang X, Marchetti MC, et al. Motility-driven glass and jamming transitions in biological tissues. Phys Rev X. 2016;6(2):021011. https://doi.org/10.1103/PhysRevX.6.021011.
164. Bi D, Lopez JH, Schwarz JM, et al. A density-independent rigidity transition in biological tissues. Nat Phys. 2015;11(12):1074–9. https://doi.org/10.1038/NPHYS3471.
165. Angelini TE, Hannezo E, Trepat X, et al. Glass-like dynamics of collective cell migration. Proc Natl Acad Sci U S A. 2011;108(12):4714–9. https://doi.org/10.1073/pnas.1010059108.
166. Bi D, Lopez JH, Schwarz JM, et al. Energy barriers and cell migration in densely packed tissues. Soft Matter. 2014;10(12):1885–90. https://doi.org/10.1039/c3sm52893f.

167. Farhadifar R, Roper J-C, Aigouy B, et al. The influence of cell mechanics, cell-cell interactions, and proliferation on epithelial packing. Curr Biol. 2007;17(24):2095–104. https://doi.org/10.1016/j.cub.2007.11.049.
168. Basan M, Prost J, Joanny J-F, et al. Dissipative particle dynamics simulations for biological tissues: rheology and competition. Phys Biol. 2011;8(2):26014. https://doi.org/10.1088/1478-3975/8/2/026014.
169. Podewitz N, Jülicher F, Gompper G, et al. Interface dynamics of competing tissues. New J Phys. 2016;18(8):83020. https://doi.org/10.1088/1367-2630/18/8/083020.
170. Zhu Y, Dong Z, Wejinya UC, et al. Determination of mechanical properties of soft tissue scaffolds by atomic force microscopy nanoindentation. J Biomech. 2011;44(13):2356–61. https://doi.org/10.1016/j.jbiomech.2011.07.010.
171. Schuldt C, Karl A, Korber N, et al. Dose-dependent collagen cross-linking of rabbit scleral tissue by blue light and riboflavin treatment probed by dynamic shear rheology. Acta Ophthalmol. 2015;93(5):e328–36. https://doi.org/10.1111/aos.12621.
172. Reiss-Zimmermann M, Streitberger K-J, Sack I, et al. High resolution imaging of viscoelastic properties of intracranial tumours by multi-frequency magnetic resonance elastography. Clin Neuroradiol. 2015;25(4):371–8. https://doi.org/10.1007/s00062-014-0311-9.
173. Steinberg MS. On the mechanism of tissue reconstruction by dissociated cells. I. Population kinetics, differential adhesiveness. and the absence of directed migration. Proc Natl Acad Sci. 1962;48(9):1577–82. https://doi.org/10.1073/pnas.48.9.1577.
174. Foty RA, Steinberg MS. The differential adhesion hypothesis: a direct evaluation. Dev Biol. 2005;278(1):255–63. https://doi.org/10.1016/j.ydbio.2004.11.012.
175. Pawlizak S, Fritsch AW, Grosser S, et al. Testing the differential adhesion hypothesis across the epithelial–mesenchymal transition. New J Phys. 2015;17(8):83049. https://doi.org/10.1088/1367-2630/17/8/083049.
176. Albini A, Iwamoto Y, Kleinman HK, et al. A rapid in vitro assay for quantitating the invasive potential of tumor cells. Cancer Res. 1987;47(12):3239–45.
177. Paszek MJ, Zahir N, Johnson KR, et al. Tensional homeostasis and the malignant phenotype. Cancer Cell. 2005;8(3):241–54. https://doi.org/10.1016/j.ccr.2005.08.010.
178. Young JS, Llumsden CE, Stalker AL. The significance of the "tissue pressure" of normal testicular and of neoplastic (Brown-Pearce carcinoma) tissue in the rabbit. J Pathol. 1950;62(3):313–33. https://doi.org/10.1002/path.1700620303.
179. Plodinec M, Loparic M, Monnier CA, et al. The nanomechanical signature of breast cancer. Nat Nanotechnol. 2012;7(11):757–65. https://doi.org/10.1038/nnano.2012.167.
180. Nnetu KD, Knorr M, Käs J, et al. The impact of jamming on boundaries of collectively moving weak-interacting cells. New J Phys. 2012;14(11):115012. https://doi.org/10.1088/1367-2630/14/11/115012.
181. Mouw JK, Yui Y, Damiano L, et al. Tissue mechanics modulate microRNA-dependent PTEN expression to regulate malignant progression. Nat Med. 2014;20(4):360–7. https://doi.org/10.1038/nm.3497.
182. Haeger A, Krause M, Wolf K, et al. Cell jamming: collective invasion of mesenchymal tumor cells imposed by tissue confinement. Biochim Biophys Acta. 2014;1840(8):2386–95. https://doi.org/10.1016/j.bbagen.2014.03.020.
183. Park J-A, Kim JH, Bi D, et al. Unjamming and cell shape in the asthmatic airway epithelium. Nat Mater. 2015;14(10):1040–8. https://doi.org/10.1038/nmat4357.
184. Farina KL, Wyckoff JB, Rivera J, et al. Cell motility of tumor cells visualized in living intact primary tumors using green fluorescent protein. Cancer Res. 1998;58(12):2528–32.
185. Wang Y-L, Discher DE, editors. Cell mechanics, Methods in cell biology. London: Elsevier; 2007.
186. Diggs LW. Pathology of sickle cell disease. JAMA. 1971;218(4):600. https://doi.org/10.1001/jama.1971.03190170078040.
187. Platt OS. Sickle cell anemia as an inflammatory disease. J Clin Invest. 2000;106(3):337–8. https://doi.org/10.1172/JCI10726.
188. Weinberg RA. The biology of cancer. 2nd ed. New York: Garland Science; 2014.

189. Seltmann K, Fritsch AW, Käs JA, et al. Keratins significantly contribute to cell stiffness and impact invasive behavior. Proc Natl Acad Sci U S A. 2013;110(46):18507–12. https://doi.org/10.1073/pnas.1310493110.

190. Cross SE, Jin Y-S, Tondre J, et al. AFM-based analysis of human metastatic cancer cells. Nanotechnology. 2008;19(38):384003. https://doi.org/10.1088/0957-4484/19/38/384003.

191. Lichtman MA. Rheology of leukocytes, leukocyte suspensions, and blood in leukemia possible relationship to clinical manifestations. J Clin Invest. 1973;52(2):350–8.

192. Baker EL, Bonnecaze RT, Zaman MH. Extracellular matrix stiffness and architecture govern intracellular rheology in cancer. Biophys J. 2009;97(4):1013–21. https://doi.org/10.1016/j.bpj.2009.05.054.

193. Mofrad MR. Rheology of the cytoskeleton. Annu Rev Fluid Mech. 2009;41(1):433–53. https://doi.org/10.1146/annurev.fluid.010908.165236.

194. Guck J, Schinkinger S, Lincoln B, et al. Optical deformability as an inherent cell marker for testing malignant transformation and metastatic competence. Biophys J. 2005;88(5):3689–98. https://doi.org/10.1529/biophysj.104.045476.

195. Merkel M, Manning ML. Using cell deformation and motion to predict forces and collective behavior in morphogenesis. Semin Cell Dev Biol. 2016;67:167. https://doi.org/10.1016/j.semcdb.2016.07.029.

196. Manning ML, Collins E-MS. Focus on physical models in biology: multicellularity and active matter. New J Phys. 2015;17(4):40201. https://doi.org/10.1088/1367-2630/17/4/040201.

197. Pegoraro AF, Fredberg JJ, Park J-A. Problems in biology with many scales of length: cell–cell adhesion and cell jamming in collective cellular migration. Exp Cell Res. 2016;343(1):54–9. https://doi.org/10.1016/j.yexcr.2015.10.036.

198. Weigelin B, Friedl P. Cancer cells: stemness shaped by curvature. Nat Mater. 2016;15(8):827–8. https://doi.org/10.1038/nmat4711.

199. Te Boekhorst V, Preziosi L, Friedl P. Plasticity of cell migration in vivo and in silico. Annu Rev Cell Dev Biol. 2016;32:491–526. https://doi.org/10.1146/annurev-cellbio-111315-125201.

200. Collins C, Nelson WJ. Running with neighbors: coordinating cell migration and cell–cell adhesion. Cell Adhes Migr. 2015;36:62–70. https://doi.org/10.1016/j.ceb.2015.07.004.

201. Kashef J, Franz CM. Quantitative methods for analyzing cell–cell adhesion in development. Dev Biol. 2015;401(1):165–74. https://doi.org/10.1016/j.ydbio.2014.11.002.

The Extracellular Matrix as a Target for Biophysical and Molecular Magnetic Resonance Imaging

Angela Ariza de Schellenberger, Judith Bergs, Ingolf Sack, and Matthias Taupitz

Abstract

All tissues and organs are composed of cells and extracellular matrix (ECM). The components of the ECM have important functional and structural roles in tissues. On the one hand, the ECM often dominates the biomechanical properties of soft tissues and provides mechanical support to the tissue. On the other hand, ECM components maintain tissue homeostasis, pH, and hydration of the micromilieu and, via signal transduction, also play a key role in ECM-cell interactions which in turn control cell migration, differentiation, growth, and death. Inflammation, fibrosis, tumor invasion, and injury are associated with the transition of the ECM from homeostasis to remodeling which can dramatically alter the biochemical and biomechanical properties of ECM components. Hence, it is possible to detect and characterize disease by sensing biochemical and biomechanical changes of the ECM when appropriate imaging methods are used. This chapter discusses ECM-specific magnetic resonance imaging (MRI) based on contrast agents and elastography from a clinical radiological perspective in a variety of diseases including atherosclerosis, cardiomyopathy, inflammation, and liver fibrosis.

6.1 Introduction

All tissues and organs of vertebrates are composed of cellular and noncellular components, and the latter are often collectively referred to as *extracellular matrix* (ECM). Connective tissues of skin, cartilage, vessel walls, and intervertebral discs are rich in ECM and have the best-characterized ECM. As early as 1929, the

A. Ariza de Schellenberger • J. Bergs • I. Sack • M. Taupitz (✉)
Department of Radiology, Charité – Universitätsmedizin Berlin, Berlin, Germany
e-mail: Matthias.Taupitz@charite.de

© Springer International Publishing AG 2018
I. Sack, T. Schaeffter (eds.), *Quantification of Biophysical Parameters in Medical Imaging*, https://doi.org/10.1007/978-3-319-65924-4_6

adhesive and sticky properties of cells were linked to their self-produced ECM [1]. Initially, the ECM was thought to be a structural element of tissues only, but its active role in cell differentiation, development, cell migration and tissue homeostasis is increasingly recognized [2, 3].

While there is an overwhelming diversity of structures and forms of biological tissues, the variety in the chemical ECM composition is fairly small. Essentially, fibrous proteins, glycoproteins (GP), proteoglycans (PG), glycosaminoglycans (GAGs), salts, and water (approx. 80% of the ECM wet weight) make up the ECM and thereby determine the biophysical and chemical environment of cells [3].

The biophysical properties of many tissues are largely governed by their ECM components and their interactions. The stroma of the eye's cornea, for example, consists of regularly arranged collagen fibers that withstand large tensile forces giving rise to a large elastic modulus greater than 50 MPa [4]. In contrast, the vitreous body of the eye is mainly made up of hydrated polysaccharide gels immersed by sparse collagen fibers, resulting in an elastic modulus below 2 Pa [4]. This huge difference in elasticity on the order of 10^6 illustrates the crucial role of ECM structures for the mechanical properties of tissues. The components of the ECM are produced by tissue-resident cells and secreted into the ECM. Once the molecules have been secreted, they aggregate within the existing matrix to support cellular viability [3]. This insight led to the once predominant view of cells as the key components of tissues. Many discoveries in recent years have led to a reappraisal, pointing to the importance of the ECM for many vital processes [5–8]. The ECM provides structural support, separates cells at tissue interfaces, and provides cellular signaling through cell-matrix connections such as focal adhesions. The latter influence cell morphology, movement, functions, and even cell fate. Cells and the ECM form a symbiotic unit that defines both the physiological functions and macroscopic properties of a biological tissue. The ECM is highly dynamic: chemical and structural changes occur during embryonic development and tissue homeostasis and can also be induced by pathophysiological stimuli and injury. These insights motivate research in the field of biomedical imaging aimed at identifying ECM-specific targets. The prospect of developing biophysical and molecular probes for ECM imaging offers a promising direction of molecular medical imaging in parallel to the development of cell-specific imaging probes. In this chapter, we will review the foundations of ECM-specific medical imaging with a focus on clinical magnetic resonance imaging (MRI). Before continuing we give a brief overview of the composition and constituents of the ECM in soft tissues of vertebrates.

6.2 Composition of the ECM

The ECM is a mesh-like structure composed of the *basement membrane* (BM) and the *interstitial matrix* (Fig. 6.1). The BM is a highly cross-linked constituent of the ECM that is composed of sheetlike depositions secreted by epithelial cells and localized between the epithelial cells and the underlying connective tissue. The BM separates the epithelium from the stroma of any given tissue. The BM is always in

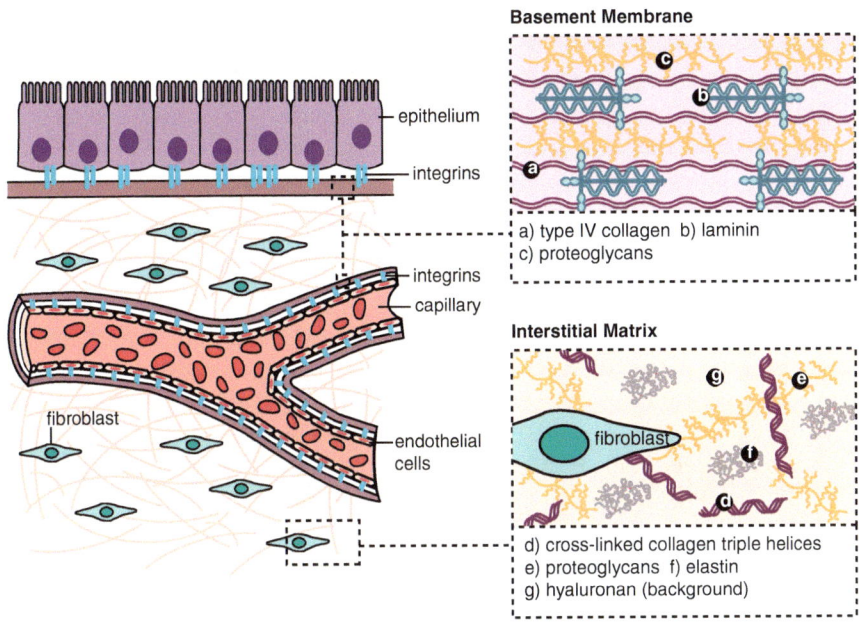

Fig. 6.1 Diagram of the histologic structure of the ECM. The basic subdivision of the ECM into basement membrane and interstitial matrix is shown along with major structural components (collagen and elastin) as well as the background matrix made up of proteoglycans and HA

contact with the cells, provides structural support, divides tissues into compartments, and regulates cell behavior [9, 10]. Overall, about 50 proteins are known to make up the BM of which approx. 50% are of the collagen type (especially type IV collagen). The BM is formed by three layers: the lamina densa, the lamina lucida, and a network of reticular collagen type III [1]. The lamina densa (*collagen IV*, perlecan, heparan sulfate proteoglycans (*HSPG*)) and the lamina lucida (*laminin*, integrins entactin and dystroglycans) form networks connected by nidogen (entactin), which is also known to bind proteoglycans and fibulins [11]. The variability of the BM in different tissues is mainly due to the ratio of type IV collagen, laminin, and HSPG [9, 12]. Collagen IV fibers aggregate into a network and provide the scaffold of a matrix with laminin as the central part. This matrix is essential for cells and other BM components that interact with each other such as perlecan, nidogen (entactin), fibulin, and collagen XVIII [13–18]. The laminin family of ECM glycoproteins is the major noncollagenous constituent of the BM and is also involved in cellular processes such as differentiation, migration, signaling, and tumor invasion. Laminins are composed of three nonidentical chains that combined form different heterotrimeric laminin isoforms, e.g., laminin 1 is a $\alpha1\beta1\gamma1$ heterotrimer [19]. Laminin incorporation into the ECM is dependent on its interactions with other ECM molecules such as collagen IV, nidogen, fibulins, and other laminin molecules. The major cell surface receptors for laminins are integrins, which have a high binding specificity for different cell types. Laminin function is altered by posttranslational modifications, i.e., covalent or enzymatic modifications during or after protein synthesis. Therefore, the detection of both laminin and integrin receptors is required in order to study cellular function [20–23]. The components of the BM can be modified by structural or functional changes, e.g., as required for blood vessel formation (neoangiogenesis) and cancer progression [9, 23–25]. Furthermore, fibronectin attaches cells to different ECM components and forms part of the connective tissue that is the ECM at interfaces and organ boundaries.

The *interstitial* or *intercellular space* is the other major ECM compartment next to the BM. Structural proteins such as collagens are mainly immersed in the interstitial space. Collagens are the most abundant proteins of the ECM and account for roughly 30% of the total mass of proteins in the body. There are over 28 different types of collagen fibrils. Collagen types I, II, and III are the most abundant, and collagen I constitutes nearly 90% of all the collagen in the human body. Collagens are mostly synthesized by fibroblasts, but epithelial cells also contribute to the synthesis of some collagens in the ECM. Skin, cartilage, tendons, and ligaments are particularly rich in collagens, which form mechanical networks and supportive structures that determine the shear modulus and limit the extensibility of these tissues. In contrast, tissues with high stretching or bending capacity are rich in elastic fibers characterized by cross-linked elastin interspersed with fibrillins. Elastin is particularly abundant in the walls of large arteries, skin, and lungs, influencing the linear tensile elastic properties of these tissues.

The major volume-occupying components of interstitial spaces are glycosaminoglycan chains (GAGs) attached to the core proteins, forming proteoglycans (PGs). PGs are macromolecules consisting of a core protein heavily glycosylated by one or more covalently bound carbohydrate chains. The biophysical function of PGs depends on their bound GAGs such as keratan sulfate (KS) and dermatan sulfate (DS) chains, whose strong negative charge allows them to bind large amounts of water and to fill most of the interstitial space as hydrated gels [2] (see Fig. 6.2). Interstitial PGs can interact with collagens and thereby also influence the structural organization of tissues. Because of the incompressibility of water, hydrated proteoglycan gels constitute the main mechanical support of tissues against compressive forces, whereas fibrillar proteins such as collagen and elastin resist stretching forces [3]. PGs, such as aggrecan, provide lubricating function to joints and structural integrity to cells and allow cell migration and diffusion of cell factors essential for cell communication [26, 27]. PGs are also involved in the organization of BM structures and thereby influence epithelial cell migration, proliferation, and differentiation. Thus, many biological functions involved in tissue development and homeostasis are mediated by specific binding of GAGs to macromolecules, mostly proteins. For example, PGs can accumulate in secretory vesicles, maintain proteases in their active state, and regulate biological activities after secretion, such as coagulation, host defense, and wound repair. Specifically, blood coagulation is controlled by GAGs such as heparin and, to a lesser extent, by heparan sulfate (HS) binding antithrombin III. When no binding proteases (thrombin, factors IVa and XIa) are present, thrombus formation is inactivated or slowed down, and blood coagulation is distorted [28]. PGs can also bind cytokines, chemokines, growth factors, and

Fig. 6.2 Illustration of the relative volumes occupied by collagen, globular proteins, and a single hydrated HA molecule. MW denotes the relative molecular weight

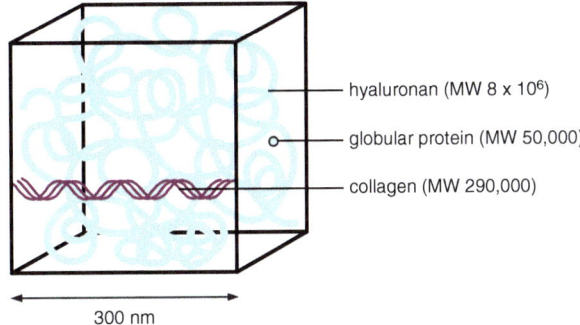

hyaluronan (MW 8 x 10^6)

globular protein (MW 50,000)

collagen (MW 290,000)

300 nm

morphogens, protecting them against proteolysis. These interactions involve a reservoir of regulatory factors that can be released by selective ECM degradation. For example, PGs can bind growth factors from the fibroblast growth factor (FGF) family, which also bind GAGs such as heparin or HS trapped inside the ECM. These GAGs can alter the conformation of this complex and distort normal release of FGF, thereby influencing cell proliferation or activation [29, 30].

In general, biological PG function depends on the interaction of GAG chains with different protein ligands, e.g., integrins and other cell receptors assisting in cell attachment, cell interaction, and cell migration [31, 32].

6.2.1   Glycosaminoglycans (GAGs) in the ECM

GAGs consist of linear polysaccharide chains composed of repeating disaccharide units, which consist of a hexose or hexuronic acid (or a galactose for keratan sulfate) and a hexosamine [28, 33]. They are highly hydrophilic and can adopt extended conformations essential for hydrogel formation. Therefore, GAGs in connective tissue, while constituting less than 10% of the structural proteins, occupy most of the extracellular space (Fig. 6.2) [3]. GAGs can be subdivided into sulfated GAGs (chondroitin sulfate (CS), dermatan sulfate (DS), keratan sulfate (KS), heparan sulfate (HS), and heparin)) and nonsulfated GAGs (hyaluronan, HA). Only sulfated GAGs are covalently bound to proteins, forming a total of 43 different types of proteoglycans [34]. GAGs are primarily located on cell surfaces and in the ECM but are also found in secretory vesicles in some types of cells. The distribution of GAGs differs with the type of tissue: while HA is a major component of the ECM of cartilage, CS is bound to the cell surface, e.g., in the brain and cartilage [35–37]. DS is present in skin and the nervous system, often bound to CS, while KS is more abundant in cartilage and cornea [37–40]. The function of GAGs depends on the protein with which they interact and the orientation and nature of GAG-bound sulfate groups. In general, the properties of GAGs tend to dominate the chemical properties of PG. GAGs can influence physiological processes, such as development, angiogenesis, and innate immunity, but also cancer and neurodegenerative diseases [38–43]. HS is the most common GAG and can sequester growth factors in the ECM.

It is involved in inflammatory processes and malignant processes such as invasive growth and metastatic spread [29] as well as other pathophysiological and physiological processes [29, 44–46]. HA is a unique GAG because it does not contain any sulfate group and has not been found to be covalently bound to a core protein to form PG [34, 47]. Instead, it is often noncovalently linked to PG and can immobilize large amounts of water. Due to the large amount of bound water, hydrogels formed by GAGs are nearly incompressible, which makes them ideal for lubrication in the joints. For instance, the unique biophysical properties of HA are fundamental for the mechanical functioning of synovial fluid or the resistance of connective tissues to compressive forces [48]. HA is predominantly necessary to maintain indirect interaction with other components of the ECM such as fibronectin, laminin, collagen, and PG. Together with its cellular receptor (CD44), HA plays an essential role in cell migration, tissue development, and inflammation and is often used in regenerative medicine [49–53]. One special feature of GAGs is their structural heterogeneity, mostly due to chemical modifications during and after synthesis. This makes the development of analytical techniques for GAGs (glycosaminoglycomics) challenging [33].

Different physiological activities arise from the variability of GAGs in the amount of negative charges and the presence of carboxylate and sulfate groups, which can bind cationic ions. Since GAGs are virtually polyanionic polymers, Ando et al. postulated that GAGs, especially in their sulfated forms, are similar to ion-exchange resins [54]. One special property of GAGs is that, with increasing disease severity, they are modified by corresponding enzymes, e.g., resulting in enhanced sulfation [55]. The degree of GAG sulfation in several inherited disorders and pathologies, as well as inflammation-involved sulfated GAGs, have been described [28, 45, 46, 56].

In normal tissue, GAGs are predominantly organized with the complexing groups forming hydrogen bonds [57], but in pathological processes, ranging from inflammation to tumor invasion, an increase in one or more GAG components takes place, and potentially complex-forming groups become exposed [55, 58–62]. It has been demonstrated in models of chemically induced tumors that characteristic changes in the ECM, especially alterations in GAG components, already take place at the preneoplastic stage [63]. In patients with malignancy, the patterns of GAG composition have prognostic significance for the subsequent disease course, in particular with regard to the risk of recurrence and metastatic spread [64]. GAGs are also a major component of the glycocalyx-a paracellular matrix located on the surface of endothelial cells. In fact, the endothelium is negatively charged due to the high proportion of highly sulfated GAGs included in the glycocalyx [65]. The glycocalyx contributes to the barrier function of the epithelium between the vascular and interstitial compartments and regulates the transport by diffusion of low-molecular-weight substances, the controlled transport of macromolecules and particles, and the passage of cells. It has been shown that an increase in the shear stress on vascular endothelium leads to a change in the GAG composition of the glycocalyx with an increase in HA [66], which provides further evidence for the central role of mechanical signal transduction between cells and their ECM. Higher levels of shear stress have been reported to be associated with an increase in GAG sulfation [67]. It is known that the glycocalyx has a function in mechanotransduction—a mechanism by which cells sense mechanical stimuli such as hemodynamic changes, e.g., blood flow velocity and pressure, and translate them into cellular responses, e.g., enhanced NO synthesis [68]. Pathological tissue changes, especially inflammation, are associated with an alteration in the glycocalyx of the local vascular endothelium, which affects the adhesion and transendothelial transport of endogenous and exogenous low-molecular-weight and macromolecular substances [65, 69].

Another observation worth mentioning in the context of GAG function is that carboxyl and sulfate groups not only bind to physiologically occurring ions such as K^+, Na^+, Ca^{2+}, Zn^{2+}, and Cu^{2+} but also have high affinity to lanthanide ions like Gd^{3+} and La^{3+} [70–72]. This property offers a target for ECM-directed imaging that can be exploited by targeting GAG components of the ECM using, for example, cationic ions or molecules that function as signal-generating moieties in the imaging modality used. On the other hand, GAGs may contribute to the retention of Gd after intravenous injection of Gd-based contrast media, as frequently used in clinical routine. The exact mechanisms for tissue retention of Gd after intravenous injection of Gd-based contrast media are still not fully understood [73–81].

Glossary

Glycosaminoglycans (GAGs): Long, negatively charged, linear chains of disaccharide repeats. Major GAGs include HS (component of basement membrane), CS (cartilage and neural ECM), DS (skin, blood vessels, tendons, lungs), HA, and KS (cornea, cartilage, bone).

Proteoglycans (PG): Heavily glycosylated proteins formed by GAG chains covalently linked to a core protein. Proteoglycans provide hydration and compressive resistance to ECM and hold numerous other biological functions including support of cell signaling, proliferation and migration, wound repair by binding of growth factors, cytokines, and ECM proteins.

Fibrous proteins: Also called scleroproteins, these proteins are characterized by their insolubility and their fibrous structure and, among others, include collagen, elastin, fibronectin, and laminin. These proteins are produced by fibroblasts and are released in a precursor form; their subsequent incorporation into the ECM is guided by fibroblasts according to the functional needs of the resident tissue.

Fibronectin: High-molecular-weight multi-domain protein with binding capacities to cell surfaces through integrins and other biologically important molecules such as collagen and HSPGs. Fibronectin plays a major role in cell adhesion, growth, migration, and differentiation and is involved in wound healing.

Laminins: Glycoproteins that constitute the structural scaffolding of all basement membranes. Laminins bind integrins, dystroglycans, and numerous other receptors. These interactions are critical for cell differentiation, movement, cell shape, and survival.

Integrins: Transmembrane linkers that mediate interactions between the ECM and intracellular cytoskeleton or intercellular interactions. Therefore, integrins can interact with fibronectin (in the ECM) and actin (in the cytoskeleton of cells) and trigger intracellular signal transduction pathways via protein kinases (focal adhesion kinase and integrin-linked kinase), which are essential for cell migration, growth, and survival.

Glycocalyx: A glycoprotein-polysaccharide that forms a filamentous coating on the apical surface of some bacteria and interspersed among cells of the digestive tract microvilli, as well as in vascular endothelial cells. The glycocalyx also consists of a wide range of enzymes and proteins. The glycocalyx enables cells to recognize each other, helping the body to identify healthy cells and transplanted tissues or invading microorganisms. In addition, it also guides cellular movement during embryogenesis. In endothelial vascular tissue, the glycocalyx plays a major role in maintenance of plasma and vessel wall homeostasis.

Collagen: Main structural protein in mammals constituting approximately 30% of all proteins in the body. It has the capacity to bind to cell surface receptors, proteins, GAGs, and nucleic acids. The most abundant collagens are type I (component of, e.g., organs, bone, and skin) and type III (component of reticular fibers). The fibrillar collagen types I and III self-assemble into hierarchical structures are capable of withstanding tensile forces and provide mechanical integrity to the interstitial matrix. The nonfibrillar collagen type IV is a major component of the basement membrane that forms loose, sheetlike structures, which influence cell differentiation, migration, and adhesion and angiogenesis.

Elastin: Fibrous protein which is, to a major portion, responsible for the elastic (mechanical energy restoring) properties of soft biological tissues. Elastin is composed of cross-linked tropoelastin, a 60–70 kDa monomeric protein with hydrophobic and lysine-containing crosslinking domains [82]. The elastic properties of elastin networks are thought to be due to their random coil structures of cross-linked elastin molecules which allow the network to stretch and recoil like a rubber band [3].

6.3   Alteration of the ECM in Disease

The potential role of ECM-targeted medical imaging can be discussed for a variety of diseases with high clinical relevance. Therefore, the focus will be on the following conditions:

- Atherosclerosis and aortic aneurysm
- Cardiomyopathy
- Neuroinflammation
- Inflammatory bowel disease
- Liver fibrosis

The following sections present a brief overview of the current state of knowledge on the most important components of the ECM involved in these disease entities.

6.3.1 Atherosclerosis and Aortic Aneurysm

The development and progression of atherosclerotic plaques are accompanied by significant changes in the composition of the ECM of the vascular wall, eventually leading to plaque rupture. The chondroitin sulfate proteoglycan (CS-PG) versican appears to play a key role in this process. Versican is increased already at an early stage in the development of atherosclerotic plaques [83, 84]. Versican interacts with HA, the content of which also increases in the atherosclerotic vascular wall. Versican plays an important role in the regulation of plaque progression. It binds lipoproteins and leads to their accumulation in the affected vascular wall dependent on the length of the CS chains [53, 83]. In addition to versican, other proteoglycans are involved in atherosclerosis including biglycan, decorin, and perlecan. The content of elastin and collagen in the vessel wall decreases as atherosclerotic wall lesions become destabilized, while an increase in elastin and collagen indicates stabilization of the diseased vessel wall [85]. Type I, II, IV, V, and VI collagens are involved in the process of plaque formation, depending on the stage of the plaque and its location in the vascular wall. Atherosclerotic plaques can be classified using the methods of Stary [86] and Virmani [87], which are based on staining the components of the plaque ECM by Movat's pentachrome stain. The vascular wall of aortic aneurysms shows an inflammatory component and is characterized by medial degeneration [88], marked mainly by a degradation of elastin and by unorganized deposition of collagen. Loss of elastin leads to dilatation of the vascular wall, and collagen depletion increases the risk of rupture. GAG accumulation in the vascular wall and the associated pathological changes in the biomechanical properties of the aortic wall are considered a major factor underlying dissection in aortic aneurysm [89].

6.3.2 Cardiomyopathy

Myocardial damage leads to remodeling, which includes increased formation of glycoproteins, proteoglycans, and GAGs for regulation of inflammation, fibrosis, and angiogenesis [90]. Furthermore, the pathological cascade in myocardial damage involves increased accumulation of collagen, eventually leading to dilated cardiomyopathy. With progressing fibrosis and scarring, the damaged myocardium reacts by local accumulation of chondroitin sulfate (CS) and dermatan sulfate (DS) [91, 92]. This process is accompanied by an increase in sulfation of these GAGs immediately after the damage has occurred [93].

6.3.3 Neuroinflammation

The ECM also plays a key role in the growth and structural development of the central nervous system, regulating interactions between cells including neurons, glia, and inflammatory cells [7]. Normal aging of the brain is associated with a reduction of the CS-PG content of the ECM [94]. This change in the ECM alters the diffusion of small molecules and water, thereby altering the biochemical properties of CNS tissue and eventually the global biomechanical properties of the brain [95]. Sulfation of GAGs in inflammatory CNS pathologies and after mechanical tissue damage regulates cell migration as well as anti- and proinflammatory processes. For example, perivascular accumulation of HS alone or as a PG-component was found in MS lesions [96]. Dense networks of ECM components, i.e., CS- and DS-PGs and HA, have been observed in active MS lesions but not in chronic inactive MS lesions [97]. Local HA accumulation was already identified in early inflammatory demyelinating CNS lesions [98].

6.3.4   Inflammatory Bowel Disease

Marked ECM alterations also occur during the development of chronic inflammatory bowel diseases. In Crohn's disease, inflamed sections of the bowel wall have increased levels of HSPG, CS, and DS. As the disease progresses, the formation of collagen also increases, leading to fibrosis and changes in the biomechanical properties of the bowel wall [99]. Intestinal structures are more likely to occur as a clinical complication in Crohn's disease than in ulcerative colitis [100]. Roughly 30% of patients with Crohn's disease develop intestinal fibrosis with stenosis, but only 5% develop ulcerative colitis [101].

6.3.5   Liver Fibrosis and Cirrhosis

In the development of liver fibrosis and cirrhosis, the composition of the ECM becomes impaired at an early stage due to alterations in the synthesis and degradation of ECM components. The collagen content rises with the degree of fibrosis, resulting in an increase in tissue rigidity [8, 102–104]. A distinct increase in HA, CS, and DS has also been found in various models of liver damage and liver fibrosis [63, 105, 106].

Overall, inflammatory processes lead to a marked increase in the amount of ECM components as well as significant structural changes within the ECM—all of which affect the biochemical and biophysical properties of the tissue. Elastography, which is sensitive to mechanical tissue structures, has been validated as the most precise noninvasive biomarker for staging hepatic fibrosis [107]. However, it remains open which ECM components and cells contribute to the mechanical scaffold of liver tissue and how these structures change with different etiologies [108]. Since early structural changes in the liver do not impact overall hepatic function, blood biomarkers or functional breath tests are less accurate for the assessment of early fibrotic processes in the liver [104].

6.4   ECM-Specific Medical Imaging

In current clinical radiology, the majority of substances used for the enhancement of image contrast are unspecific. The contrast-enhancing agents used in clinical X-ray radiography, CT, or MRI are all based on iodinated or gadolinium-containing low-molecular-weight substances. Following intravenous injection, these agents are distributed in the intravascular and interstitial space, and a large proportion is rapidly eliminated via the kidneys. These clinical contrast agents enhance signal in regions where pathological processes are present in an unspecific manner. Unspecific pathological processes include inflammation, neovascularization, changes in local blood volume, altered permeability of the vessel wall, and increased volume of extracellular spaces. Noteworthy, the limited specificity of most contrast agents in clinical routine is the reason for their commercial success since the number of potential applications is high. Nevertheless, research groups worldwide attempt to target molecular structures at cell surfaces or metabolic processes by developing highly specific imaging probes [109]. Some very interesting disease-specific molecular imaging approaches using MRI, sonography, radionuclide imaging, and optical imaging were demonstrated to be experimentally feasible for detection of inflammatory [110–112], malignant [113–115], and degenerative processes [116–118]. However, except for radionuclide imaging, which is sensitive already at very low radionuclide concentrations (see Chap. 21), the translation of molecular probes into clinical use has been successful only in a very limited number of cases. The MRI contrast agent Primovist has been shown to enable hepatocyte-specific imaging [119, 120]. Another MRI contrast agent, Resovist, allows targeting of phagocytosing cells in liver neoplasms [121, 122]. In general, the development of molecular imaging probes is challenged by the high complexity of both molecular targets and probes, the limited accessibility of extravascular targets, unspecific sequestration of macromolecules used to deliver contrast agents like liposomes in the liver and spleen, and the relatively low number of cells suited as imaging targets within a diseased tissue.

For these reasons, alternative molecular imaging approaches are needed. One strategy to overcome the limitations of currently available contrast agents in clinical radiology and molecular imaging research is to develop ECM-specific contrast agents such as GagCEST as explained in Chap. 10. The ECM as a target for medical imaging is appealing for many reasons—three of which will be briefly discussed in the following paragraphs.

6.4.1   The Amount of ECM

The amount of ECM varies considerably between tissue types and pathophysiological states. In the *liver* (a cell-rich organ), the ECM volume fraction is less than 3% under normal conditions and can increase up to 30% in cirrhotic livers due to excessive connective tissue accumulation. In *muscle*, the total amount of ECM is approximately 5% and has a very strong mechanical influence on muscle function (see Fig. 6.3). Therefore, pathological ECM accumulation in muscle is of high clinical relevance and directly alters its function [124]. Myocardial fibrosis is a major cause of cardiac dysfunction, leading to myocardial hypertrophy with systolic and diastolic dysfunction and infarction. In the *brain*, ECM constitutes about 10–20% of total brain volume [125] with substantial changes in quantity and structure during brain development [126]. Brain ECM undergoes constant reorganization in response to learning, neuronal activity, and plasticity. The elastic properties of each tissue depend on the mechanical characteristics of its cells and their surrounding ECM. The cellular responses to disease or physiological changes relate to small fractions of the total tissue mass, while the ECM responses to disease, such as abnormal ECM deposition in fibrosis, cancer, or atherosclerosis, can have devastating effects on tissues [127].

In the evolution of atherosclerosis, growth of plaques and ECM deposition in the wall of arteries are a major pathological hallmark of the disease and are often related to severe clinical complications, as previously described [6, 53, 128–131].

Overall, the changing amount of ECM in the progression of diseases could be translated into sensitive biomarkers. Moreover, the ECM is a tissue component that contributes to effective medium properties as explained in the next section.

6.4.2   Multiscale Physical Properties of the ECM

The size of cells usually defines the microscopic scale in imaging sciences which is the voxel size of the acquired image. Imaging signals are normally averaged within the voxel volume and therefore represent effective medium properties over all scales of tissue structures from the micro to the macro level. Details of tissue interactions on the microscopic level have to be enhanced by contrast agents or molecular imaging probes to be made visible by the macroscopic image contrast. Alternatively, biophysical imaging probes rely on multiscale networks in biological tissues which translate microstructural information from the cellular level into clinical image contrast [132, 133]. Proteins and polysaccharides of the ECM often form hierarchic and highly structured networks extending up to the macroscale size. The effect of large-scale networks on the global response of a tissue is well illustrated by adding a minor amount (less than 1 mass percent) of polysaccharides to water [133]. The established sugar chains cause a phase transition from liquid water into a solid gel by bridging long-range distances up to the percolation limit (i.e., bridging over the full sample size). It should be mentioned here that cellular networks such as the neuronal network or hierarchic vascular trees can upscale microscopic interactions in a similar way as ECM networks. This is exploited, e.g., by diffusion MRI based on restricted water mobility (see Chap. 17) or by MR elastography based on the mechanical integrity of neuronal tissue [108, 134, 135]. Also, collective cell behavior, such as jamming of tumor cells (see Chap. 5), induces macroscopic properties that reflect adhesion of cells at the microscopic level. However, in general, cellular networks are embedded in a mesh of ECM components that dominate how the tissue is organized across multiple scales. The organization of skeletal muscle tissue by the collagen endomysial network is an example of the dominating structural properties of a minor fraction of tissue components and is shown for illustration in Fig. 6.4. Given that muscle tissue comprises only 5% ECM, the dominating collagen properties in mechanical muscle tests, especially under prestretched conditions, indicate that collagen is organized in extended and highly cross-linked networks, providing permanent support to the entire muscle (Figs. 6.3 and 6.4). Similarly, liver collagen infiltration in the course of fibrogenesis establishes a hierarchic network that turns liver tissue from a very soft and viscous material into a stiff and elastic

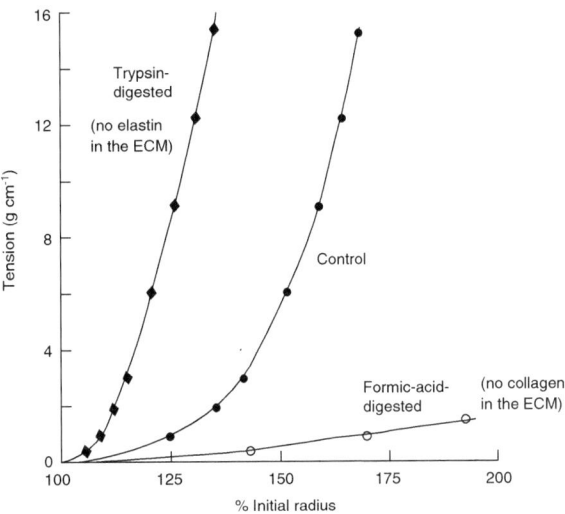

Fig. 6.3 Tension–length response of human iliac arteries with different amounts of collagen and elastin. When elastin is removed by trypsin digestion, the curve (diamonds) represents the properties of the remaining collagen. Alternatively, when collagen is removed by formic acid digestion, the curve (open circles) represents the properties of the elastin fibers. The broken curve (filled circles) is for an untreated artery (from [123] with permission)

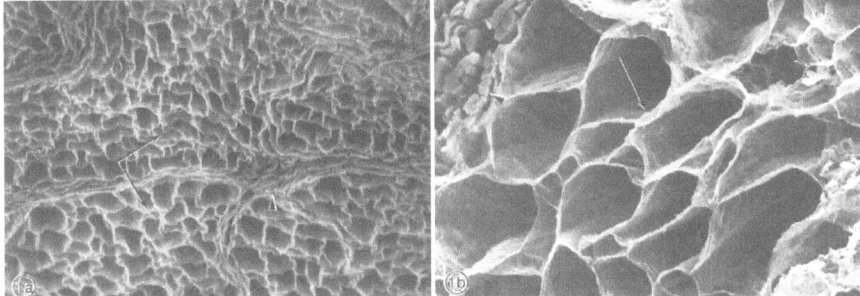

Fig. 6.4 Scanning electron microscopy of the collagen network around muscle fibers observed after digesting muscle fibers with NaOH. (**a**) Low magnification overview reveals an array of tubes into which muscle fibers insert (endomysium, arrows) as well as a thickened area surrounding the fibers (perimysium, arrowhead). (**b**) Higher-power view reveals the fine structure of the endomysial surfaces (arrow) as well as some undigested muscle fibers (arrowhead). This image demonstrates that muscle fibers are embedded in a tight matrix of connective tissue and are intimately associated with the ECM (from [136] with permission)

rubber-like solid [108, 137]. This change in global biophysical properties is the result not only of the increasing amount of collagen but also of the hierarchy of ECM architecture, which supports the entire organ on the macroscopic scale [104]. These observations raise the possibility of exploiting the multiscale nature of many ECM structures for translating microscopic biophysical and biochemical signals into macroscopic medical image contrast by using ECM-specific imaging probes.

6.4.3   ECM Function

The functions of the ECM can be divided into structural support and biochemical activity. Structural support is the mechanical function of the ECM and mainly relies on the structural ECM components such as collagen, elastin, and fibrillins. Under normal conditions, these components serve a scaffolding function, maintaining tissue architecture and homeostasis while providing tensile and compressive strength, thereby attenuating shear forces. The viscoelastic properties of mammalian tissues are tailored to their specific functions: pulmonary tissue, for example, is soft and elastic to enable large volume changes during breathing. Figure 6.3 demonstrates how the macroscopic behavior of muscle tissue changes with different amounts of collagen and elastin in the muscle ECM.

A hallmark of several diseases, including lung emphysema and arthritis or fibrosis and sclerosis, is a chronic change in the tissue's mechanical properties [127, 138]. During inflammatory processes, in particular, the activity of proteases can lead to a reduction in the content of elastin and thus to a decrease in the macroscopic shear modulus of the involved tissue. An increase in the amount of collagen has also been observed in the course of fibrosis and scar formation due to wound healing after tissue damage, regardless of the tissue or organ type.

An additional factor influencing the mechanical function of the ECM includes the balance between ECM-degrading and ECM synthesis processes. Therefore, it is not surprising that a distorted equilibrium in these processes is regarded as a hallmark of cancer [2]. ECM degradation can be induced enzymatically, for example, by matrix metalloproteases (MMP), while ECM deposition can be accomplished by crosslinking, as well as by the expression of ECM proteins and inhibitors of ECM-degrading enzymes. Furthermore, chemical signals such as cellular growth factors and cytokines are sequestered in the ECM and can later be released to diffuse through the tissue, supporting both cell and ECM homeostasis. This dynamic mechanical and biochemical interaction between cells and their ECM can influence gene transcription and therefore protein expression [139, 140]. This has implications for cell- and tissue-specific processes. For example, the ECM plays a role in promotion or inhibition of cell division [141], neuronal reorganization, and axonal outgrowth in the adult brain [142]. ECM-cell interactions are able to restrict neuronal reorganization and axonal outgrowth in the adult brain [142], making the ECM an essential determining factor of neural plasticity [126, 143, 144]. In case of disease, the ECM has been shown to contribute to cancer cell stiffness [60]. Cancer cells have been described to soften when exposed to rigid fibrotic stroma, thereby transforming to a more aggressive state with higher metastatic potential [145]. These examples demonstrate the essential role of the ECM in normal tissue function and health preservation and stress the importance for development of ECM-specific imaging probes.

6.5   The State of the Art in Research on Molecular and Biophysical Imaging of the ECM

6.5.1   ECM Targeting with Imaging Probes

6.5.1.1 Collagen and Elastin

From a biochemical perspective, collagen and elastin are relatively well defined, which facilitates their detection by imaging probes. For the specific visualization of collagen components of the ECM, Caravan et al. [146] have developed a low-molecular-weight, gadolinium-based MR imaging probe, named EP-3533, which binds to type I collagen [147, 148]. In this probe, a Gd-containing complex is coupled to a peptide with high affinity to type I collagen and identified by phage display. This probe allowed assessment of the extent of fibrotic changes in the liver parenchyma in an animal model of liver fibrosis by determining the amount of collagen using noninvasive MRI [147, 148]. In an experimental model of myocardial infarction, the probe identified the area of infarction based on its collagen content [149]. An earlier in vitro approach proposed by Sanders et al. was based on MRI contrast agent containing liposomes linked to an adhesion protein for collagen [150]. Alternatively, the ECM protein elastin was targeted using a small-molecular, elastin-specific, gadolinium-based imaging probe developed for MRI and tested on various models. The suitability of this probe to analyze the elastin content of tissue by means of MRI has been demonstrated in an atherosclerosis model [151], a murine aneurysm model [152], and a myocardial infarction model [153, 154]. The results obtained in the atherosclerosis model provided strong evidence that the elastin-specific probe is suitable for noninvasive quantification of the clinically validated parameter plaque burden at different stages of plaque development [155, 156]. It has also been demonstrated that the elastin content of a plaque matrix can be noninvasively quantified by MRI, which has the potential to be used for differentiating between vulnerable and stable atherosclerotic plaques [157]. In a rabbit model of atherosclerosis, it was demonstrated that gadofluorin M accumulates in areas of atherosclerotic plaques with high amounts of ECM deposits [158]. Noteworthy, the accumulation of gadofluorin M was not associated with the lipid content of the vessel wall.

6.5.1.2 GAGs

In contrast to the ECM components collagen and elastin, GAGs are characterized by relatively high biochemical variability. Hence, GAG-specific peptides or antibodies for immunofluorescence in histological specimens are not yet available. However, already in the early 1980s, Ando et al. demonstrated that the characteristic complex-forming properties of GAGs with regard to lanthanides can be exploited to label the pathological increase in GAG components in tumors by [67]Ga (a trivalent metal ion) or [69]Tm (a trivalent lanthanide), both administered in the form of citrate complexes [159, 160]. Using this approach, Ando et al. identified HS, KS, and heparin—particularly heavily sulfated GAGs—as the binding GAGs for [67]Ga [159, 160]. In this context, characterization of the pharmacokinetic properties of nonspecific Gd-based contrast agents as routinely applied in clinical MRI examinations remains a highly relevant research question. Since the advent of nonspecific Gd-based contrast agents in clinical routine, it has been known that a small amount of the gadolinium is not rapidly eliminated from the body, suggesting a two-compartment pharmacokinetic process. Early pharmacokinetic studies with [153]Gd-DTPA and Gd-[^{14}C] DTPA in healthy rats showed that, for a small amount (<1%) of the administered intravenous dose, free Gd^{3+} ions were released from the contrast medium complex (dechelation process) and partially accumulated in the bones, liver, and spleen of the animals [161]. Further pharmacokinetic studies with [153]Gd-DTPA in healthy rats by Wedeking et al. postulated a three-compartment model with slow exchange of the third compartment, characteristic of a drug-binding process [162]. Experimental and clinical pharmacokinetic studies using dynamic contrast-enhanced MRI with Gd-DTPA and similar substances also showed that, in various disease entities such as myocardial infarction, myocardial fibrosis and malignancy, the course of signal intensity fitted a three-compartment model better than a two-compartment model [163–168]. The third compartment is generally characterized by slow exchange constants. Current research shows that such a third compartment exists for both linear Gd contrast agents and, to a lesser extent, for macrocyclic Gd contrast agents [169]. The substrate of this third compartment is not yet known. However, it has been demonstrated that interaction of the Gd-based contrast agent with proteins does not play a role [162]. Studies conducted to determine the volume fractions of the different tissue components of the myocardium show a high binding level of La^{3+} to polyanionic sugar polymers, i.e., to the GAG components of the ECM [170], which in healthy myocardium account for 23% of the ECM volume [170]. After the discovery of nephrogenic systemic fibrosis (NSF), a severe illness observed after intravenous injection of linear Gd-based contrast agents in patients with end-stage renal disease, it was confirmed that Gd can leave the contrast agent complex by partial dechelation in vivo [171, 172]. The substrate that binds the free Gd^{3+} ions in a transchelation process has not yet been identified. However, GAGs are likely to play a role in the binding of free Gd^{3+}ions, thereby contributing to prolonged retention of these ions in tissue. Currently, the discovery of Gd deposition in brain tissue, particularly the nucleus dentatus and globus pallidus, after repeated administration of Gd-based contrast agents in patients with normal renal function gives rise to the question as to which tissue component binds the contrast agent molecule as a whole or as dechelated Gd [173, 174].

Some interesting evidence regarding GAG labeling can be derived from histochemical procedures used to identify these ECM components in histological specimens. For histochemical identification of GAGs, the sections are incubated with colloidal iron oxide and then stained with Perls' Prussian blue iron stain as a secondary stain [175]. The staining of GAGs in histological sections with the cationic alcian blue stain is also based on their complex-forming properties [176]. Thus, pathologically altered GAGs can also be targeted in vivo by intravenous injection of appropriate iron oxide nanoparticles, which serve as contrast-enhancing agents for detection by MRI. Very small superparamagnetic iron oxide nanoparticles (VSOPs) with citrate as the coating material [177, 178] have the property to target GAGs and cells in vivo. In a rabbit atherosclerosis model, it was demonstrated that GAG-binding VSOPs can be used to detect pathologically elevated levels of GAGs and can discriminate between stable atherosclerotic plaques and those at risk of rupture [179]. In a mouse model of atherosclerosis, endothelial cells and macrophages of atherosclerotic plaques were identified as main target for VSOP by electron microscopy [180]. A subsequent study found that the VSOPs were endocytosed by endothelial cells already 10 min after intravenous injection [181]. In a mouse model of neuroinflammation, VSOPs also improved the capacity to detect an altered brain-blood barrier (BBB) [182] and the different localizations of VSOPs in brain with disrupted BBB suggested multiple entry mechanisms of VSOPs into the central nervous system [182, 183]. It was also shown for atherosclerosis and neuroinflammation, by using animal models, that adhesion of VSOPs to GAG-based molecules on the endothelial glycocalyx is a major factor in mediating their cellular uptake and transendothelial transport [181, 184]. These results were found for VSOP variants which do not carry any other target-specific molecules apart from their stabilizing coating. Current studies suggest that binding of VSOPs to GAGs takes place via transchelation [179], i.e., they lose their weakly bound citrate coating in the presence of GAGs upon the formation of a strong complex. The VSOP enrichment in different animal models shows the potential of these iron oxide nanoparticles for MRI and stimulates further studies to characterize their targets in vivo.

6.5.2 Biophysical Imaging of the ECM

For decades, dynamic mechanical tests collectively known as "rheometry" have been used for the mechanical-based analysis of micro-architectural properties of polymer samples [185]. More recently, mechanical test methods have been used for studies of biological soft matter or model networks prevalent in cells and tissues [186–188]. The rheological behavior of biological tissues is linked to the hierarchy of underlying structures [189]. Scale-invariant properties of rheological constants of cells or tissues can give insight into the organization of structure elements below the resolution limit of the measurement system [190, 191]. On the microscopic scale, rheological constants can be measured by various methods such as cell-deformation-based experiments [192, 193] or scanning force microscopy [187]. Macrorheological methods include oscillatory shear stress rheology [194], dynamic shear tests [195], stress-relaxation measurements [196], tensile tests [197], and macro-indentation [198]. Most rheological methods are surface based, i.e., mechanical stimulation is locally applied, and the resulting tissue response is measured at the surface. Elastography (see Chaps. 12 and 20) induces and measures shear waves inside the bulky tissue—in vivo for diagnostic applications [199] or ex vivo for basic studies of soft tissue's rheological behavior [200]. In general, the measurement of intrinsic material properties inside a tissue volume by MRE is less susceptible to the geometry, texture, and composition of the sample surface [191]. In biological tissues, the relationship between stiffness and number of crosslinks can often be modeled by a powerlaw [132, 190]. MR elastography findings are consistent with observations made by oscillatory rheometry on macromolecular elastomers, which reveal that network elasticity originates from the cross-linked backbone of the network, while dissipativity originates from the unlinked parts of the network [201]. In general, motility of tissue elements results in enhanced lossy properties of the tissue. For example, water molecules demobilized by GAGs represent a gel-like tissue component with highly elastic and low viscous properties. For this reason, GAG-depleted tissue has higher lossy properties than tissue with high GAG content in the ECM [202]. The literature on tensile testing of tissues suggests that GAGs have a major influence on viscosity rather than on elasticity [203]. Figure 6.5 compares the MRE-measured viscoelastic dispersion function of collagen with that of HA, both with similar concentrations in water at room

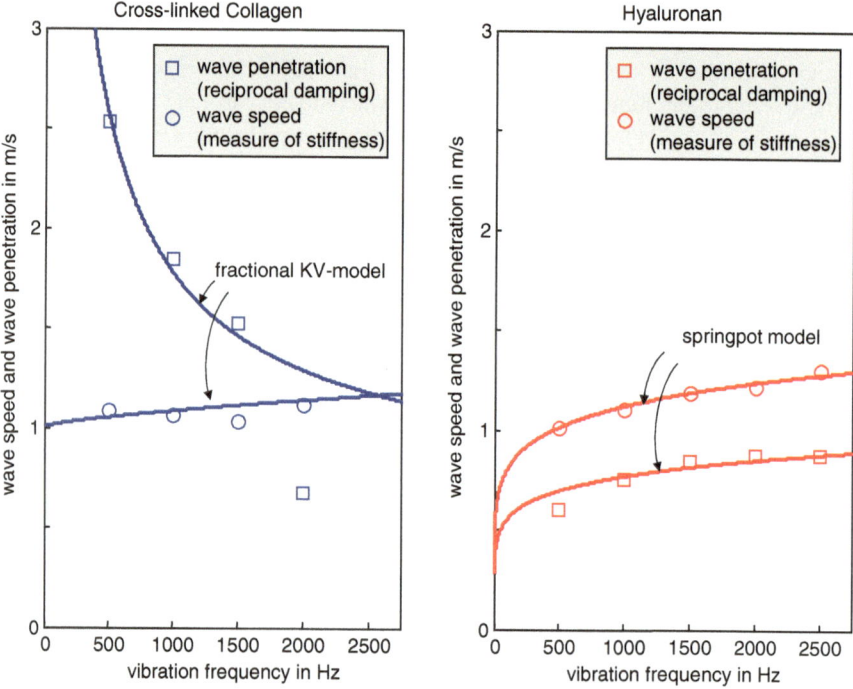

Fig. 6.5 Viscoelastic dispersion functions for single ECM components measured by tabletop MRE in small sample volumes [204]. Different viscoelastic properties of collagen and GAG are clearly seen by wave speed and wave penetration, which quantifies elasticity and reciprocal viscosity, respectively [205]. Overall, collagen properties are well predicted by a solid body model, in particular at lower frequencies, as described by the three-parameter fractional Kelvin-Voigt (KV) model [206]. In contrast, HA displays viscoelastic scaling properties as predicted by the two-parameter springpot model. Note that wave speed values in both materials are similar despite their distinct viscous behaviors

temperature. Comparison between 4% porcine collagen and HA hydrogels (Fig. 6.5) demonstrates that shear wave speed related to elasticity is similar in both materials, while wave penetration, which is reciprocally related to viscosity (inverse damping), differs markedly between collagen and HA. The fractional Kelvin-Voigt model predicts a decrease in viscosity with lowering the driving frequency, whereas in the springpot, both elasticity and viscosity have a constant ratio over all frequencies (see Chap. 2, Fig. 2.5, and Eq. (2.44) therein). This example demonstrates the importance of measuring both elasticity and viscosity to characterize the composition of the ECM in terms of networks made up of both fibrous proteins such as collagen and water-binding GAGs. So far, only indirect correlation analyses between ECM structures and tissue viscoelasticity have been performed. In Reiter et al. [104], the amount of connective tissue in fibrotic livers was quantified and compared with springpot-based wideband MRE, showing that the amount of collagen is less correlated with the degree of fibrosis than MRE.

Conclusion

The ECM of biological soft tissues consists of structural components which determine the macroscopic biophysical properties of the tissue such as elasticity in the linear and nonlinear regime. Disease can significantly alter the content of these components, leading to marked changes in mechanical tissue properties. Furthermore, GAGs, which fill most of the extracellular space due to strong water-binding capacity, provide mechanical support against compression and influence the lossy properties of the tissue as quantified by viscosity. GAGs and other ECM components might change quantitatively and qualitatively in many disease processes such as inflammation, affecting both the biomechanical and the biochemical properties of the ECM. Innovative imaging probes may thus target GAG components of the ECM in both the interstitial space and the basement membrane.

References

1. Ascoli M. Bulletin of the National Research Council, 69, Bulletin of the National Research Council, National Research Council, U.S. Washington: National Academies; 1929.
2. Frantz C, Stewart KM, Weaver VM. The extracellular matrix at a glance. J Cell Sci. 2010;123(Pt 24):4195–200.
3. Alberts B, et al. Molecular biology of the cell. 4th ed. New York: Garland Science; 2002.
4. Murphy W, Black J, Hastings G. Handbook of biomaterial properties. 2nd ed. New York: Springer; 2016.
5. Spinale FG, Zile MR. Integrating the myocardial matrix into heart failure recognition and management. Circ Res. 2013;113(6):725–38.
6. Katsuda S, Kaji T. Atherosclerosis and extracellular matrix. J Atheroscler Thromb. 2003;10(5):267–74.
7. Smith PD, et al. "GAG-ing with the neuron": the role of glycosaminoglycan patterning in the central nervous system. Exp Neurol. 2015;274(Pt B):100–14.
8. Baiocchini A, et al. Extracellular matrix molecular remodeling in human liver fibrosis evolution. PLoS One. 2016;11(3):e0151736.

9. Kalluri R. Basement membranes: structure, assembly and role in tumour angiogenesis. Nat Rev Cancer. 2003;3(6):422–33.

10. Pozzi A, Yurchenco PD, Iozzo RV. The nature and biology of basement membranes. Matrix Biol. 2017;57-58:1–11.

11. Paulsson M. Basement membrane proteins: structure, assembly, and cellular interactions. Crit Rev Biochem Mol Biol. 1992;27(1-2):93–127.

12. Timpl R, Brown JC. The laminins. Matrix Biol. 1994;14(4):275–81.

13. Timpl R. Macromolecular organization of basement membranes. Curr Opin Cell Biol. 1996;8(5):618–24.

14. Timpl R, et al. Laminin, proteoglycan, nidogen and collagen IV: structural models and molecular interactions. Ciba Found Symp. 1984;108:25–43.

15. Aumailley M, et al. Binding of nidogen and the laminin-nidogen complex to basement membrane collagen type IV. Eur J Biochem. 1989;184(1):241–8.

16. Tsiper MV, Yurchenco PD. Laminin assembles into separate basement membrane and fibrillar matrices in Schwann cells. J Cell Sci. 2002;115(Pt 5):1005–15.

17. Charonis AS, et al. Binding of laminin to type IV collagen: a morphological study. J Cell Biol. 1985;100(6):1848–53.

18. Sasaki T, et al. Deficiency of β1 integrins in teratoma interferes with basement membrane assembly and laminin-1 expression. Exp Cell Res. 1998;238(1):70–81.

19. Aumailley M, et al. A simplified laminin nomenclature. Matrix Biol. 2005;24(5):326–32.

20. Veitch DP, et al. Mammalian tolloid metalloproteinase, and not matrix metalloprotease 2 or membrane type 1 metalloprotease, processes laminin-5 in keratinocytes and skin. J Biol Chem. 2003;278(18):15661–8.

21. Koshikawa N, et al. Membrane-type matrix metalloproteinase-1 (MT1-MMP) is a processing enzyme for human laminin gamma 2 chain. J Biol Chem. 2005;280(1):88–93.

22. Koshikawa N, et al. Proteolytic processing of laminin-5 by MT1-MMP in tissues and its effects on epithelial cell morphology. FASEB J. 2004;18(2):364–6.

23. Qin Y, et al. Laminins and cancer stem cells: partners in crime? Semin Cancer Biol. 2016;45:3.

24. Feller W. An introduction to probability theory and its applications. New York: Willey; 1968.

25. Hynes RO. Integrins: bidirectional, allosteric signaling machines. Cell. 2002;110(6):673–87.

26. Voet D, Voet J, Pratt CW. Fundamentals of biochemistry. New York: Wiley; 1999.

27. Lieleg O, Ribbeck K. Biological hydrogels as selective diffusion barriers. Trends Cell Biol. 2011;21(9):543–51.

28. Esko JD, Kimata K, Lindahl U. Proteoglycans and sulfated glycosaminoglycans. In: Varki A, et al., editors. Essentials of glycobiology. New York: Cold Spring Harbor; 2009.

29. Sasaki N, et al. Cell surface localization of heparanase on macrophages regulates degradation of extracellular matrix heparan sulfate. J Immunol. 2004;172(6):3830–5.

30. Hacker U, Nybakken K, Perrimon N. Heparan sulphate proteoglycans: the sweet side of development. Nat Rev Mol Cell Biol. 2005;6(7):530–41.

31. Bishop JR, Schuksz M, Esko JD. Heparan sulphate proteoglycans fine-tune mammalian physiology. Nature. 2007;446(7139):1030–7.

32. Gubbiotti MA, Neill T, Iozzo RV. A current view of perlecan in physiology and pathology: a mosaic of functions. Matrix Biol. 2017;57-58:285–98.

33. Ricard-Blum S, Lisacek F. Glycosaminoglycanomics: where we are. Glycoconj J. 2017;34:339–49.

34. Iozzo RV, Schaefer L. Proteoglycan form and function: a comprehensive nomenclature of proteoglycans. Matrix Biol. 2015;42:11–55.

35. Bishnoi M, et al. Chondroitin sulphate: a focus on osteoarthritis. Glycoconj J. 2016;33(5):693–705.

36. Zamfir AD, et al. Brain chondroitin/dermatan sulfate, from cerebral tissue to fine structure: extraction, preparation, and fully automated chip-electrospray mass spectrometric analysis. In: Rédini F, editor. Proteoglycans: methods and protocols. Totowa: Humana Press; 2012. p. 145–59.

37. Malmström A, et al. Iduronic acid in chondroitin/dermatan sulfate: biosynthesis and biological function. J Histochem Cytochem. 2012;60(12):916–25.
38. Mongiat M, et al. Extracellular matrix, a hard player in angiogenesis. Int J Mol Sci. 2016;17(11):E1822.
39. Mercier F. Fractones: extracellular matrix niche controlling stem cell fate and growth factor activity in the brain in health and disease. Cell Mol Life Sci. 2016;73(24):4661–74.
40. Sethi MK, Zaia J. Extracellular matrix proteomics in schizophrenia and Alzheimer's disease. Anal Bioanal Chem. 2017;409(2):379–94.
41. Xu D, Esko JD. Demystifying heparan sulfate-protein interactions. Annu Rev Biochem. 2014;83:129–57.
42. Mizumoto S, Yamada S, Sugahara K. Molecular interactions between chondroitin-dermatan sulfate and growth factors/receptors/matrix proteins. Curr Opin Struct Biol. 2015;34:35–42.
43. Meneghetti MC, et al. Heparan sulfate and heparin interactions with proteins. J R Soc Interface. 2015;12(110):0589.
44. Parish CR. The role of heparan sulphate in inflammation. Nat Rev Immunol. 2006;6(9):633–43.
45. Pomin VH. Sulfated glycans in inflammation. Eur J Med Chem. 2015;92:353–69.
46. Jin-Ping L. Heparin, heparan sulfate and heparanase in cancer: remedy for metastasis? Anti Cancer Agents Med Chem. 2008;8(1):64–76.
47. Maytin EV. Hyaluronan: more than just a wrinkle filler. Glycobiology. 2016;26(6):553–9.
48. Laurent TC. The chemistry, biology and medical applications of hyaluronan and its derivatives. London: Portland Press; 1998.
49. Balazs EA, Denlinger JL. Viscosupplementation: a new concept in the treatment of osteoarthritis. J Rheumatol Suppl. 1993;39:3–9.
50. Brandt KD, Smith GN Jr, Simon LS. Intraarticular injection of hyaluronan as treatment for knee osteoarthritis: what is the evidence? Arthritis Rheum. 2000;43(6):1192–203.
51. Cohen MD. Hyaluronic acid treatment (viscosupplementation) for OA of the knee. Bull Rheum Dis. 1998;47(7):4–7.
52. George E. Intra-articular hyaluronan treatment for osteoarthritis. Ann Rheum Dis. 1998;57(11):637–40.
53. Viola M, et al. Extracellular matrix in atherosclerosis: hyaluronan and proteoglycans insights. Curr Med Chem. 2016;23(26):2958–71.
54. Ando A, et al. Mechanism of tumor and liver concentration of 111In and 169Yb: 111In and 169Yb binding substances in tumor tissues and liver. Eur J Nucl Med. 1982;7(7):298–303.
55. Taylor KR, Gallo RL. Glycosaminoglycans and their proteoglycans: host-associated molecular patterns for initiation and modulation of inflammation. FASEB J. 2006;20(1):9–22.
56. Varki A, Freeze HH. Glycans in acquired human diseases. In: Varki A, Cummings RD, Esko JD, et al., editors. Essentials of glycobiology. Cold Spring Harbor: Cold Spring Harbor Laboratory Press; 2009.
57. Mitra AK, et al. Dermatan sulfate: molecular conformations and interactions in the condensed state. J Mol Biol. 1983;169(4):873–901.
58. Corte MD, et al. Analysis of the expression of hyaluronan in intraductal and invasive carcinomas of the breast. J Cancer Res Clin Oncol. 2010;136(5):745–50.
59. Takeuchi J, et al. Variation in glycosaminoglycan components of breast tumors. Cancer Res. 1976;36(7 PT 1):2133–9.
60. Pickup MW, Mouw JK, Weaver VM. The extracellular matrix modulates the hallmarks of cancer. EMBO Rep. 2014;15(12):1243–53.
61. Rowlands D, Sugahara K, Kwok J. Glycosaminoglycans and glycomimetics in the central nervous system. Molecules. 2015;20(3):3527.
62. Moretto P, et al. Regulation of hyaluronan synthesis in vascular diseases and diabetes. J Diabetes Res. 2015;2015:167283.
63. Abdel-Hamid NM. Premalignant variations in extracellular matrix composition in chemically induced hepatocellular carcinoma in rats. J Membr Biol. 2009;230(3):155–62.
64. Schwertfeger KL, et al. Hyaluronan, inflammation, and breast cancer progression. Front Immunol. 2015;6:236.

65. Kolarova H, et al. Modulation of endothelial glycocalyx structure under inflammatory conditions. Mediat Inflamm. 2014;2014:694312.
66. Gouverneur M, et al. Fluid shear stress stimulates incorporation of hyaluronan into endothelial cell glycocalyx. Am J Physiol Heart Circ Physiol. 2006;290(1):H458–2.
67. Elhadj S, Akers RM, Forsten-Williams K. Chronic pulsatile shear stress alters insulin-like growth factor-I (IGF-I) binding protein release in vitro. Ann Biomed Eng. 2003;31(2):163–70.
68. Pahakis MY, et al. The role of endothelial glycocalyx components in mechanotransduction of fluid shear stress. Biochem Biophys Res Commun. 2007;355(1):228–33.
69. Tarbell JM, Cancel LM. The glycocalyx and its significance in human medicine. J Intern Med. 2016;280(1):97–113.
70. Rabenstein DL, Robert JM, Peng J. Multinuclear magnetic resonance studies of the interaction of inorganic cations with heparin. Carbohydr Res. 1995;278(2):239–56.
71. Casu B, et al. Stereoselective effects of gadolinium ions on the relaxation properties of 13C and 1H nuclei of aldohexuronic acids and poly(glycosiduronic acids). Carbohydr Res. 1975;41(1):C6–8.
72. Rej RN, Holme KR, Perlin AS. Marked stereoselectivity in the binding of copper ions by heparin. Contrasts with the binding of gadolinium and calcium ions. Carbohydr Res. 1990;207(2):143–52.
73. Joffe P, Thomsen HS, Meusel M. Pharmacokinetics of gadodiamide injection in patients with severe renal insufficiency and patients undergoing hemodialysis or continuous ambulatory peritoneal dialysis. Acad Radiol. 1998;5(7):491–502.
74. Gibby WA, Gibby KA, Gibby WA. Comparison of Gd DTPA-BMA (Omniscan) versus Gd HP-DO3A (ProHance) retention in human bone tissue by inductively coupled plasma atomic emission spectroscopy. Investig Radiol. 2004;39(3):138–42.
75. Marckmann P, et al. Nephrogenic systemic fibrosis: suspected causative role of gadodiamide used for contrast-enhanced magnetic resonance imaging. J Am Soc Nephrol. 2006;17(9):2359–62.
76. McDonald RJ, et al. Intracranial gadolinium deposition after contrast-enhanced MR imaging. Radiology. 2015;275(3):772–82.
77. Radbruch A, et al. Gadolinium retention in the dentate nucleus and globus pallidus is dependent on the class of contrast agent. Radiology. 2015;275(3):783–91.
78. Kanda T, et al. Gadolinium-based contrast agent accumulates in the brain even in subjects without severe renal dysfunction: evaluation of autopsy brain specimens with inductively coupled plasma mass spectroscopy. Radiology. 2015;276(1):228–32.
79. Runge VM. Safety of the gadolinium-based contrast agents for magnetic resonance imaging, focusing in part on their accumulation in the brain and especially the dentate nucleus. Investig Radiol. 2016;51(5):273–9.
80. Taupitz M, et al. Gadolinium-containing magnetic resonance contrast media: investigation on the possible transchelation of Gd(3)(+) to the glycosaminoglycan heparin. Contrast Media Mol Imaging. 2013;8(2):108–16.
81. Schlemm L, et al. Gadopentetate but not gadobutrol accumulates in the dentate nucleus of multiple sclerosis patients. Mult Scler. 2017;23:963.
82. Theocharis AD, et al. Extracellular matrix structure. Adv Drug Deliv Rev. 2016;97:4–27.
83. Kolodgie FD, et al. Differential accumulation of proteoglycans and hyaluronan in culprit lesions: insights into plaque erosion. Arterioscler Thromb Vasc Biol. 2002;22(10):1642–8.
84. Wight TN, Merrilees MJ. Proteoglycans in atherosclerosis and restenosis: key roles for versican. Circ Res. 2004;94(9):1158–67.
85. Chen W, et al. Collagen-specific peptide conjugated HDL nanoparticles as MRI contrast agent to evaluate compositional changes in atherosclerotic plaque regression. JACC Cardiovasc Imaging. 2013;6(3):373–84.
86. Stary HC, et al. A definition of advanced types of atherosclerotic lesions and a histological classification of atherosclerosis. A report from the Committee on Vascular Lesions of the Council on Arteriosclerosis, American Heart Association. Circulation. 1995;92(5):1355–74.

87. Virmani R, et al. Lessons from sudden coronary death: a comprehensive morphological classification scheme for atherosclerotic lesions. Arterioscler Thromb Vasc Biol. 2000;20(5):1262–75.
88. Daugherty A, Cassis LA. Mechanisms of abdominal aortic aneurysm formation. Curr Atheroscler Rep. 2002;4(3):222–7.
89. Humphrey JD. Possible mechanical roles of glycosaminoglycans in thoracic aortic dissection and associations with dysregulated transforming growth factor-beta. J Vasc Res. 2013;50(1):1–10.
90. Rienks M, et al. Myocardial extracellular matrix. An ever-changing and diverse entity. Circ Res. 2014;114(5):872–88.
91. Shetlar MR, Shetlar CL, Kischer CW. Healing of myocardial infarction in animal models. Tex Rep Biol Med. 1979;39:339–55.
92. Judd JT, et al. Myocardial connective tissue metabolism in response to injury. II. Investigation of the mucopolysaccharides involved in isoproterenol-induced necrosis and repair in rat hearts. Circ Res. 1970;26(1):101–9.
93. Judd JT, Wexler BC. Sulfur 35 uptake in acid mucopolysaccharides of the rat heart following injury. Am J Phys. 1973;224(2):312–7.
94. Sykova E, et al. Learning deficits in aged rats related to decrease in extracellular volume and loss of diffusion anisotropy in hippocampus. Hippocampus. 2002;12(2):269–79.
95. Sack I, et al. The impact of aging and gender on brain viscoelasticity. NeuroImage. 2009;46(3):652–7.
96. van Horssen J, et al. Basement membrane proteins in multiple sclerosis-associated inflammatory cuffs: potential role in influx and transport of leukocytes. J Neuropathol Exp Neurol. 2005;64(8):722–9.
97. van Horssen J, et al. Extensive extracellular matrix depositions in active multiple sclerosis lesions. Neurobiol Dis. 2006;24(3):484–91.
98. Back SA, et al. Hyaluronan accumulates in demyelinated lesions and inhibits oligodendrocyte progenitor maturation. Nat Med. 2005;11(9):966–72.
99. Baumgart DC, et al. US-based real-time elastography for the detection of fibrotic gut tissue in patients with stricturing Crohn disease. Radiology. 2015;275(3):889–99.
100. Burke JP, et al. Fibrogenesis in Crohn's disease. Am J Gastroenterol. 2007;102(2):439–48.
101. Latella G, et al. Mechanisms of initiation and progression of intestinal fibrosis in IBD. Scand J Gastroenterol. 2015;50(1):53–65.
102. Iredale JP. Models of liver fibrosis: exploring the dynamic nature of inflammation and repair in a solid organ. J Clin Investig. 2007;117(3):539–48.
103. Mallat A, Lotersztajn S. Cellular mechanisms of tissue fibrosis. 5. Novel insights into liver fibrosis. Am J Physiol Cell Physiol. 2013;305(8):C789–99.
104. Reiter R, et al. Wideband MRE and static mechanical indentation of human liver specimen: sensitivity of viscoelastic constants to the alteration of tissue structure in hepatic fibrosis. J Biomech. 2014;47(7):1665–74.
105. Guedes PLR, et al. Increase of glycosaminoglycans and metalloproteinases 2 and 9 in liver extracellular matrix on early stages of extrahepatic cholestasis. Arq Gastroenterol. 2014;51:309–15.
106. Scott JE, et al. The chemical morphology of extracellular matrix in experimental rat liver fibrosis resembles that of normal developing connective tissue. Virchows Arch. 1994;424(1):89–98.
107. Bonekamp S, et al. Can imaging modalities diagnose and stage hepatic fibrosis and cirrhosis accurately? J Hepatol. 2009;50(1):17–35.
108. Sack I, et al. Structure sensitive elastography: on the viscoelastic powerlaw behavior of in vivo human tissue in health and disease. Soft Matter. 2013;9(24):5672–80.
109. Weissleder R, Mahmood U. Molecular imaging. Radiology. 2001;219(2):316–33.
110. Geven EJW, et al. S100A8/A9, a potent serum and molecular imaging biomarker for synovial inflammation and joint destruction in seronegative experimental arthritis. Arthritis Res Ther. 2016;18(1):247.

111. Withana NP, et al. Dual-modality activity-based probes as molecular imaging agents for vascular inflammation. J Nucl Med. 2016;57(10):1583–90.
112. Jorgensen NP, et al. Cholinergic PET imaging in infections and inflammation using 11C-donepezil and 18F-FEOBV. Eur J Nucl Med Mol Imaging. 2017;44(3):449–58.
113. Bwatanglang IB, et al. Folic acid targeted Mn:ZnS quantum dots for theranostic applications of cancer cell imaging and therapy. Int J Nanomedicine. 2016;11:413–28.
114. Chatterjee S, et al. A humanized antibody for imaging immune checkpoint ligand PD-L1 expression in tumors. Oncotarget. 2016;7(9):10215–27.
115. Chen C, et al. Molecular imaging with MRI: potential application in pancreatic cancer. Biomed Res Int. 2015;2015:10.
116. Eisenmenger LB, et al. Advances in PET imaging of degenerative, cerebrovascular, and traumatic causes of dementia. Semin Nucl Med. 2016;46(1):57–87.
117. Gomperts SN, et al. Tau positron emission tomographic imaging in the Lewy body diseases. JAMA Neurol. 2016;73(11):1334–41.
118. Farrar CT, et al. RNA aptamer probes as optical imaging agents for the detection of amyloid plaques. PLoS One. 2014;9(2):e89901.
119. Huppertz A, et al. Improved detection of focal liver lesions at MR imaging: multicenter comparison of gadoxetic acid-enhanced MR images with intraoperative findings. Radiology. 2004;230(1):266–75.
120. Hamm B, et al. Phase I clinical evaluation of Gd-EOB-DTPA as a hepatobiliary MR contrast agent: safety, pharmacokinetics, and MR imaging. Radiology. 1995;195(3):785–92.
121. Hamm B, et al. Contrast-enhanced MR imaging of liver and spleen: first experience in humans with a new superparamagnetic iron oxide. J Magn Reson Imaging. 1994;4(5):659–68.
122. Reimer P, Balzer T. Ferucarbotran (Resovist): a new clinically approved RES-specific contrast agent for contrast-enhanced MRI of the liver: properties, clinical development, and applications. Eur Radiol. 2003;13(6):1266–76.
123. Roach MR, Burton AC. The reason for the shape of the distensibility curves of arteries. Can J Biochem Physiol. 1957;35(8):681–90.
124. Gillies AR, Lieber RL. Structure and function of the skeletal muscle extracellular matrix. Muscle Nerve. 2011;44(3):318–31.
125. Hrabětová S, Nicholson C. Biophysical properties of brain extracellular space explored with ion-selective microelectrodes, integrative optical imaging and related techniques. In: Michael A, Borland L, editors. Electrochemical methods for neuroscience. Boca Raton: CRC Press/Taylor & Francis; 2007.
126. Carulli D, et al. Chondroitin sulfate proteoglycans in neural development and regeneration. Curr Opin Neurobiol. 2005;15(1):116–20.
127. Butcher DT, Alliston T, Weaver VM. A tense situation: forcing tumour progression. Nat Rev Cancer. 2009;9(2):108–22.
128. Asplund A, et al. Macrophages exposed to hypoxia secrete proteoglycans for which LDL has higher affinity. Atherosclerosis. 2011;215(1):77–81.
129. Karangelis DE, et al. Glycosaminoglycans as key molecules in atherosclerosis: the role of versican and hyaluronan. Curr Med Chem. 2010;17(33):4018–26.
130. Tran-Lundmark K, et al. Heparan sulfate in perlecan promotes mouse atherosclerosis: roles in lipid permeability, lipid retention, and smooth muscle cell proliferation. Circ Res. 2008;103(1):43–52.
131. Theocharis AD, et al. Chondroitin sulfate as a key molecule in the development of atherosclerosis and cancer progression. Adv Pharmacol. 2006;53:281–95.
132. Posnansky O, et al. Fractal network dimension and viscoelastic powerlaw behavior: I. A modeling approach based on a coarse-graining procedure combined with shear oscillatory rheometry. Phys Med Biol. 2012;57(12):4023–40.
133. Guo J, et al. Fractal network dimension and viscoelastic powerlaw behavior: II. An experimental study of structure-mimicking phantoms by magnetic resonance elastography. Phys Med Biol. 2012;57(12):4041–53.

134. Freimann FB, et al. MR elastography in a murine stroke model reveals correlation of macroscopic viscoelastic properties of the brain with neuronal density. NMR Biomed. 2013;26(11):1534–9.
135. Klein C, et al. Enhanced adult neurogenesis increases brain stiffness: in vivo magnetic resonance elastography in a mouse model of dopamine depletion. PLoS One. 2014;9(3):e92582.
136. Trotter JA, Purslow PP. Functional morphology of the endomysium in series fibered muscles. J Morphol. 1992;212(2):109–22.
137. Asbach P, et al. Viscoelasticity-based staging of hepatic fibrosis with multifrequency MR elastography. Radiology. 2010;257(1):80–6.
138. Ingber DE. Mechanobiology and diseases of mechanotransduction. Ann Med. 2003;35(8):564–77.
139. DuFort CC, Paszek MJ, Weaver VM. Balancing forces: architectural control of mechanotransduction. Nat Rev Mol Cell Biol. 2011;12(5):308–19.
140. Nelson CM, Bissell MJ. Modeling dynamic reciprocity: engineering three-dimensional culture models of breast architecture, function, and neoplastic transformation. Semin Cancer Biol. 2005;15(5):342–52.
141. Mizuguchi S, et al. Chondroitin proteoglycans are involved in cell division of Caenorhabditis elegans. Nature. 2003;423(6938):443–8.
142. Soleman S, et al. Targeting the neural extracellular matrix in neurological disorders. Neuroscience. 2013;253:194–213.
143. Happel M, Frischknecht R. Neuronal plasticity in the juvenile and adult brain regulated by the extracellular matrix. In: Travascio F, editor. Composition and function of the extracellular matrix in the human body. Rijeka: Intech; 2016.
144. Pizzorusso T, et al. Reactivation of ocular dominance plasticity in the adult visual cortex. Science. 2002;298(5596):1248–51.
145. Levental KR, et al. Matrix crosslinking forces tumor progression by enhancing integrin signaling. Cell. 2009;139(5):891–906.
146. Caravan P, et al. Collagen-targeted MRI contrast agent for molecular imaging of fibrosis. Angew Chem Int Ed Engl. 2007;46(43):8171–3.
147. Fuchs BC, et al. Molecular MRI of collagen to diagnose and stage liver fibrosis. J Hepatol. 2013;59(5):992–8.
148. Polasek M, et al. Molecular MR imaging of liver fibrosis: a feasibility study using rat and mouse models. J Hepatol. 2012;57(3):549–55.
149. Spuentrup E, et al. Molecular magnetic resonance imaging of myocardial perfusion with EP-3600, a collagen-specific contrast agent: initial feasibility study in a swine model. Circulation. 2009;119(13):1768–75.
150. Sanders HM, et al. Morphology, binding behavior and MR-properties of paramagnetic collagen-binding liposomes. Contrast Media Mol Imaging. 2009;4(2):81–8.
151. Phinikaridou A, et al. Vascular remodeling and plaque vulnerability in a rabbit model of atherosclerosis: comparison of delayed-enhancement MR imaging with an elastin-specific contrast agent and unenhanced black-blood MR imaging. Radiology. 2014;271(2):390–9.
152. Okamura H, et al. Assessment of elastin deficit in a Marfan mouse aneurysm model using an elastin-specific magnetic resonance imaging contrast agent. Circ Cardiovasc Imaging. 2014;7(4):690–6.
153. Protti A, et al. Assessment of myocardial remodeling using an elastin/tropoelastin specific agent with high field magnetic resonance imaging (MRI). J Am Heart Assoc. 2015;4(8):e001851.
154. Wildgruber M, et al. Assessment of myocardial infarction and postinfarction scar remodeling with an elastin-specific magnetic resonance agent. Circ Cardiovasc Imaging. 2014;7(2):321–9.
155. Makowski MR, et al. Noninvasive assessment of atherosclerotic plaque progression in ApoE-/- mice using susceptibility gradient mapping. Circ Cardiovasc Imaging. 2011;4(3):295–303.
156. Stone GW, et al. A prospective natural-history study of coronary atherosclerosis. N Engl J Med. 2011;364(3):226–35.

157. Makowski MR, et al. Assessment of atherosclerotic plaque burden with an elastin-specific magnetic resonance contrast agent. Nat Med. 2011;17(3):383–8.
158. Meding J, et al. Magnetic resonance imaging of atherosclerosis by targeting extracellular matrix deposition with Gadofluorine M. Contrast Media Mol Imaging. 2007;2(3):120–9.
159. Ando A, et al. Mechanism of tumor and liver concentration of 67Ga: 67Ga binding substances in tumor tissues and liver. Int J Nucl Med Biol. 1983;10(1):1–9.
160. Ando A, et al. Affinity of 167Tm-citrate for tumor and liver tissue. Eur J Nucl Med. 1983;8(10):440–6.
161. Kasokat T, Urich K. Quantification of dechelation of gadopentetate dimeglumine in rats. Arzneimittelforschung. 1992;42(6):869–76.
162. Wedeking P, et al. Pharmacokinetic analysis of blood distribution of intravenously administered 153Gd-labeled Gd(DTPA)2- and 99mTc(DTPA) in rats. Magn Reson Imaging. 1990;8(5):567–75.
163. Knowles BR, et al. Pharmacokinetic modeling of delayed gadolinium enhancement in the myocardium. Magn Reson Med. 2008;60(6):1524–30.
164. Goldfarb JW, Zhao W, Han J. Three-compartment (3C) pharmacokinetic modeling is more accurate than two-compartment (2C) modeling of myocardial fibrosis gadolinium kinetics. J Cardiovasc Magn Reson. 2012;14(1):P248.
165. Port RE, et al. Multicompartment analysis of gadolinium chelate kinetics: blood-tissue exchange in mammary tumors as monitored by dynamic MR imaging. J Magn Reson Imaging. 1999;10(3):233–41.
166. Port RE, et al. Noncompartmental kinetic analysis of DCE-MRI data from malignant tumors: application to glioblastoma treated with bevacizumab. Magn Reson Med. 2010;64(2):408–17.
167. Franiel T, et al. Differentiation of prostate cancer from normal prostate tissue: role of hotspots in pharmacokinetic MRI and histologic evaluation. AJR Am J Roentgenol. 2010;194(3):675–81.
168. Lüdemann L, et al. Comparison of dynamic contrast-enhanced MRI with WHO tumor grading for gliomas. Eur Radiol. 2001;11(7):1231–41.
169. Lancelot E. Revisiting the pharmacokinetic profiles of gadolinium-based contrast agents: differences in long-term biodistribution and excretion. Investig Radiol. 2016;51(11):691–700.
170. Frank JS, Langer GA. The myocardial interstitium: its structure and its role in ionic exchange. J Cell Biol. 1974;60(3):586–601.
171. Robic C, et al. The role of phosphate on Omniscan(®) dechelation: an in vitro relaxivity study at pH 7. Biometals. 2011;24(4):759–68.
172. Idee JM, et al. Involvement of gadolinium chelates in the mechanism of nephrogenic systemic fibrosis: an update. Radiol Clin N Am. 2009;47(5):855–69. vii
173. Murata N, et al. Macrocyclic and other non-group 1 gadolinium contrast agents deposit low levels of gadolinium in brain and bone tissue: preliminary results from 9 patients with normal renal function. Investig Radiol. 2016;51(7):447–53.
174. Kanda T, et al. High signal intensity in dentate nucleus on unenhanced T1-weighted MR images: association with linear versus macrocyclic gadolinium chelate administration. Radiology. 2015;275(3):803–9.
175. Hale CW. Histochemical demonstration of acid polysaccharides in animal tissues. Nature. 1946;157:802.
176. Scott JE, Dorling J. Differential staining of acid glycosaminoglycans (mucopolysaccharides) by alcian blue in salt solutions. Histochemie. 1965;5(3):221–33.
177. Taupitz M, et al. New generation of monomer-stabilized very small superparamagnetic iron oxide particles (VSOP) as contrast medium for MR angiography: preclinical results in rats and rabbits. J Magn Reson Imaging. 2000;12(6):905–11.
178. Wagner S, et al. Monomer-coated very small superparamagnetic iron oxide particles as contrast medium for magnetic resonance imaging: preclinical in vivo characterization. Investig Radiol. 2002;37(4):167–77.

179. Wagner S, et al. Contrast-enhanced MR imaging of atherosclerosis using citrate-coated superparamagnetic iron oxide nanoparticles: calcifying microvesicles as imaging target for plaque characterization. Int J Nanomedicine. 2013;8:767–79.

180. Scharlach C, et al. Synthesis of acid-stabilized iron oxide nanoparticles and comparison for targeting atherosclerotic plaques: evaluation by MRI, quantitative MPS, and TEM alternative to ambiguous Prussian blue iron staining. Nanomedicine. 2015;11(5):1085–95.

181. Poller WC, et al. Uptake of citrate-coated iron oxide nanoparticles into atherosclerotic lesions in mice occurs via accelerated transcytosis through plaque endothelial cells. Nano Res. 2016;9(11):3437–52.

182. Tysiak E, et al. Beyond blood brain barrier breakdown - in vivo detection of occult neuroinflammatory foci by magnetic nanoparticles in high field MRI. J Neuroinflammation. 2009;6:20.

183. Millward JM, et al. Iron oxide magnetic nanoparticles highlight early involvement of the choroid plexus in central nervous system inflammation. ASN Neuro. 2013;5(1):e00110.

184. Ludwig A, et al. Rapid binding of electrostatically stabilized iron oxide nanoparticles to THP-1 monocytic cells via interaction with glycosaminoglycans. Basic Res Cardiol. 2013;108(2):328.

185. Tschoegl NW. The phenomenological theory of linear viscoelastic behavior. Berlin: Springer; 1989.

186. Fletcher DA, Mullins RD. Cell mechanics and the cytoskeleton. Nature. 2010;463(7280):485–92.

187. Plodinec M, et al. The nanomechanical signature of breast cancer. Nat Nanotechnol. 2012;7(11):757–65.

188. Jonietz E. Mechanics: the forces of cancer. Nature. 2012;491(7425):S56–7.

189. Fung Y. Biomechanics: mechanical properties of living tissue. New York: Springer-Verlag; 1993.

190. Fabry B, et al. Time scale and other invariants of integrative mechanical behavior in living cells. Phys Rev E Stat Nonlinear Soft Matter Phys. 2003;68(4 Pt 1):041914.

191. Lambert SA, et al. Bridging three orders of magnitude: multiple scattered waves sense fractal microscopic structures via dispersion. Phys Rev Lett. 2015;115(9):094301.

192. Ozawa H, et al. Comparison of spinal cord gray matter and white matter softness: measurement by pipette aspiration method. J Neurosurg. 2001;95(2 Suppl):221–4.

193. Guck J, et al. The optical stretcher: a novel laser tool to micromanipulate cells. Biophys J. 2001;81(2):767–84.

194. Tan K, et al. Characterising soft tissues under large amplitude oscillatory shear and combined loading. J Biomech. 2013;46(6):1060–6.

195. Kiss MZ, Varghese T, Hall TJ. Viscoelastic characterization of in vitro canine tissue. Phys Med Biol. 2004;49(18):4207–18.

196. Parker KJ. Experimental evaluations of the microchannel flow model. Phys Med Biol. 2015;60(11):4227–42.

197. Bilston LE, Thibault LE. The mechanical properties of the human cervical spinal cord in vitro. Ann Biomed Eng. 1996;24(1):67–74.

198. Samani A, Zubovits J, Plewes D. Elastic moduli of normal and pathological human breast tissues: an inversion-technique-based investigation of 169 samples. Phys Med Biol. 2007;52(6):1565–76.

199. Venkatesh SK, Yin M, Ehman RL. Magnetic resonance elastography of liver: technique, analysis, and clinical applications. J Magn Reson Imaging. 2013;37(3):544–55.

200. Othman SF, et al. Microscopic magnetic resonance elastography (microMRE). Magn Reson Med. 2005;54(3):605–15.

201. Urayama K, Kawamura T, Kohjiya S. Structure-mechanical property correlations of model siloxane elastomers with controlled network topology. Polymer. 2009;50(2):347–56.

202. Mendoza-Novelo B, et al. Decellularization of pericardial tissue and its impact on tensile viscoelasticity and glycosaminoglycan content. Acta Biomater. 2011;7(3):1241–8.

203. Al Jamal R, Roughley PJ, Ludwig MS. Effect of glycosaminoglycan degradation on lung tissue viscoelasticity. Am J Physiol Lung Cell Mol Physiol. 2001;280(2):L306–15.
204. Ipek-Ugay S, et al. Tabletop magnetic resonance elastography for the measurement of viscoelastic parameters of small tissue samples. J Magn Reson. 2015;251:13–8.
205. Tzschatzsch H, et al. Tomoelastography by multifrequency wave number recovery from time-harmonic propagating shear waves. Med Image Anal. 2016;30:1–10.
206. Hirsch S, Braun J, Sack I. Magnetic resonance elastography: physical background and medical applications. Weinheim: Wiley-VCH; 2017.

Part II

Medical Imaging Technologies

Mathematical Methods in Medical Image Processing

7

Gitta Kutyniok, Jackie Ma, and Maximilian März

Abstract

Medical imaging problems, such as magnetic resonance imaging, can typically be modeled as inverse problems. A novel methodological approach which was already proven to be highly effective and widely applicable is based on the assumption that most real-life images are intrinsically of low-dimensional nature. This sparsity property can be revealed by representation systems from the area of applied harmonic analysis such as wavelets or shearlets. The inverse problem itself is then solved by sparse regularization, which in certain situations is referred to as compressed sensing. This chapter shall serve as an introduction to and a survey of mathematical methods for medical imaging problems with a specific focus on sparsity-based methods. The effectiveness of the presented methods is demonstrated with a small case study from sparse parallel magnetic resonance imaging.

7.1 Imaging as a Linear Inverse Problem

Imaging methodologies such as computed tomography (CT) or magnetic resonance imaging (MRI) play a continuously increasing role in medical diagnosis [1, 2]. This calls for highly effective algorithms to acquire and analyze medical images. A

G. Kutyniok (✉) • M. März
Department of Mathematics, Technische Universität Berlin, Berlin, Germany
e-mail: kutyniok@math.tu-berlin.de; maerz@math.tu-berlin.de

J. Ma
Image and Video Coding Group, Fraunhofer Institute for Telecommunications–Heinrich Hertz Institute, Berlin, Germany
e-mail: jackie.ma@hhi.fraunhofer.de

© Springer International Publishing AG 2018   153
I. Sack, T. Schaeffter (eds.), *Quantification of Biophysical Parameters in Medical Imaging*, https://doi.org/10.1007/978-3-319-65924-4_7

common feature of many imaging problems is the fact that they can be modeled as a linear inverse problem.

7.1.1 Modeling of Data Acquisition as Linear Measurements

An imaging problem essentially consists of two parts: The sampling part, i.e., the physical acquisition of data, and the reconstruction part, a computational process from the acquired data back to the object of interest. Such sampling and reconstruction problems arise, for instance, in CT and MRI.

For many imaging modalities the measuring process can be modeled as an operator equation of the form

$$y = Ax + \eta, \qquad (7.1)$$

where A is a linear operator, e.g., a matrix, that represents the acquisition device in our modeling, and η represents measuring errors such as noise. Equation (7.1) is often referred to as the *forward model*. However, for the reconstruction of images we are interested in the *inverse problem*, i.e., determining x from the knowledge of y and A. We thereby say that the inverse problem is *ill-posed* if (7.1) does not have a solution, or its solution is not unique, or if it does not depend on the data y in a continuous fashion [3]. It is thus strongly desired for A to have "good" mathematical properties in order to allow for efficient recovery as well as theoretical guarantees for an object x to be reconstructed from the data y. At the same time A should model the actual sampling procedure as accurately as possible and should not be too oversimplified.

The *Fourier transform* is an example of an operator that has many interesting mathematical properties but is also important for practical problems in medical imaging. In fact, the Fourier transform is essentially the measurement operator A that is used in MRI. Another important example is the *Radon transform* which is at the heart of X-ray CT, see [1] for further information. Due to its importance for MRI, we proceed with a short discussion of the Fourier transform and refer to [4] for a mathematically detailed presentation on this topic.

7.1.2 Example: The Fourier Transform

The Fourier transform has had significant influence in mathematics and physics, in particular, it is the underlying principle behind the measurement process of MRI [1]. The Fourier transform is defined as

$$F : L^1(\mathbb{R}) \to C_0(\mathbb{R}), \ f \mapsto \int_{\mathbb{R}} f(x) e^{-2\pi i x(\cdot)} dx,$$

where $L^1(\mathbb{R})$ is the space of (Lebesgue) integrable functions and $C_0(\mathbb{R})$ denotes the space of continuous functions vanishing at infinity [4]. Further, we restrict our presentation to the one dimensional case; but generalizations to higher dimensions are straightforward, see, for instance, [4]. It is also sometimes more convenient to work with the Fourier transform defined on the Hilbert space $L^2(\mathbb{R})$—the space of square Lebesgue integrable functions defined on $\mathbb{R}$—which is mathematically justified by an extension process starting from a suitable dense subspace. For such an extension procedure, we refer the interested reader to [4].

The Fourier transform F can be used to model the acquisition of frequencies represented by a so-called k-space signal m_f of the form

$$m_f(\xi) = \int_{\mathbb{R}} f(x)e^{-2\pi i x \xi}\,dx.$$

The sampling positions represented by ξ can be chosen in a uniform or a nonuniform way. For simplicity we will continue with a discussion that assumes a Cartesian-type sampled k-space.

If we assume that the function f is defined on the one dimensional torus [4], then one can prove

$$f(x) = \sum_{\xi \in \mathbb{Z}} m_f(\xi)e^{2\pi i x \xi} \tag{7.2}$$

holds almost everywhere [4]. Equation (7.2) shows that if we were able to collect all Fourier measurements $(m_f(\xi))_{\xi \in \mathbb{Z}}$, then they would allow for a perfect recovery of f. However, since we are not able to measure and process an infinite amount of information, one can at best—if one follows this approach—work with an approximation of the form

$$f(x) \approx \sum_{\xi \in \Xi} m_f(\xi)e^{2\pi i \xi x} \tag{7.3}$$

with $\Xi \subset \mathbb{Z}$ being of finite cardinality. Depending on how many Fourier measurements are given in (7.3) the reconstruction suffers from ringing effects near discontinuities, cf. Fig. 7.1. These ringing artifacts are referred to as *Gibbs phenomenon* and are very well documented in the literature, see, for example, [5].

For practical tasks the continuous model that is computing Fourier integrals of a suitable function is not applicable and the infinite dimensional operators and functions have to be discretized. In case of the Fourier transform and the function spaces we obtain a solely discrete setting by replacing the functions with vectors and operators by matrices. However, such an approximation of the infinite dimensional object has to be done in a faithful manner. The most prominent approach is to use a (partial) *discrete Fourier transform* (DFT) $A \in \mathbb{R}^{m \times n}$ to replace the Fourier transform F. A fast computational method to compute the finite dimensional Fourier data $y = Ax \in \mathbb{R}^m$ is given by the *Fast Fourier Transform*.

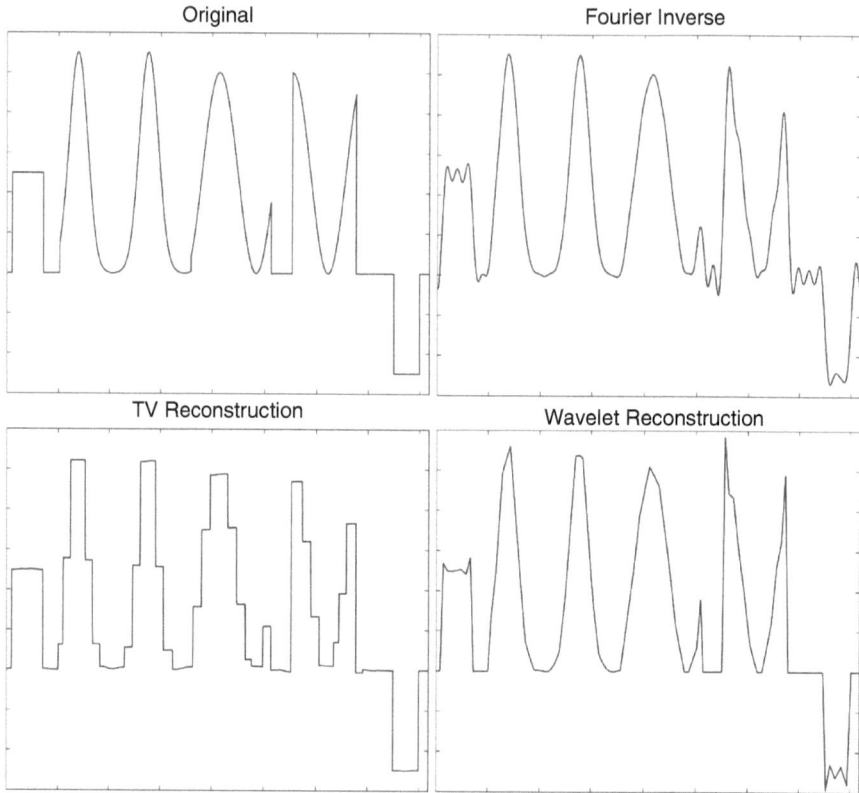

Fig. 7.1 Reconstructions from Fourier measurements based on different priors

7.2 Treating Ill-Posed Inverse Problems

For the purpose of this short introduction, we limit our scope to a treatment of Eq.
(7.1) in the finite dimensional case. In particular, we will focus on the ill-posed situation, where $A \in \mathbb{R}^{m \times n}$ with $m < n$, i.e., when less measurements are available so that the system equation in (7.1) becomes underdetermined. Such problems typically appear, for instance, in the data acquisition process of MRI.

In the following, we will discuss some of the most important mathematical concepts for treating such problems in a fairly chronological order. Due to limited space, this discussion is far from complete and we refer the interested reader to the referenced literature.

7.2.1 Tikhonov Regularization

Since we are considering an ill-posed problem, we usually cannot hope for an exact recovery of our signal of interest, but we can at least aim for a sufficiently accurate approximation of it. An intuitive condition for such an approximation is that it should match our measured data *reasonably well*. Such a consistency condition can be expressed in mathematical terms and is in fact obtained by solving the problem

$$\min_{x \in \mathbb{R}^n} \|Ax - y\|_2^2 \qquad (7.4)$$

There is an abundance of mathematical methods concerned with an analysis of the problem (7.4), for instance, considering the associated *normal equations* or using *iterative regularization methods*, such as *Landweber iterations* or *LSQR*. We refer the interested reader to [6, 7] for more details on this subject and continue our discussion by introducing the *Tikhonov regularization* method.

A general idea is to incorporate further a-priori knowledge about the signal of interest which can be done, for instance, by adding a further *regularization term* to the minimization problem (7.4), i.e.,

$$\min_{x \in \mathbb{R}^n} \|Ax - y\|_2^2 + \alpha \cdot R(x), \qquad (7.5)$$

for a *regularization function* $R: \mathbb{R}^n \to \mathbb{R}$ and a carefully chosen *regularization parameter* $\alpha > 0$. Solving this optimization program gives a preference to solutions that yield small values of R and therefore enforcing certain structural properties in the set of solutions.

One of the most successful and widespread choices is due to Andrey Tikhonov, who considered the regularizer $R(x) = \|Lx\|_2^2$ [8], where L is the so-called *Tikhonov matrix* and has to be chosen carefully depending on the application at hand. A frequent choice of the Tikhonov matrix is $L = \mathrm{Id}$. This choice results in a preference for more regular solutions with a small energy, i.e., with a small ℓ_2-norm. The design of suitable matrices L and the choice of the parameter α is an active branch of research in mathematical imaging and the study of inverse problems in general. The differentiability of the objective function in case of a Tikhonov regularizer makes the problem (7.5) numerically feasible and often very effective for various large scale imaging problems. However, a well-known drawback is that such an approach tends to find "oversmoothed" solutions that have lost their edge information. We refer to the standard literature such as [1, 6, 7] for further information.

7.2.2 Sparse Regularization and Compressed Sensing

Since images, which are our main objects in this chapter, are typically governed by edge information, a natural demand for treating inverse problems is that sharp features should be preserved in the reconstruction process. A change of paradigm occurred in the last three decades that consists in replacing the ℓ_2-norm in Tikhonov regularization by the ℓ_1-norm, i.e., considering the optimization problem

$$\min_{x \in \mathbb{R}^n} \|Ax - y\|_2^2 + \alpha \cdot \|Lx\|_1. \qquad (7.6)$$

While this appears to be a minor change at first sight, it is actually a fundamental change from a mathematical point of view. Its success can probably be described

best by an excursion to the recently established field of *compressed sensing*, introduced to allow for highly efficient data acquisition following the philosophy to
merely acquire the—typically much smaller—information content.

The mathematical framework of compressed sensing was introduced in [9, 10]
and is inspired by multiple disciplines such as linear algebra, probability theory,
harmonic analysis, and convex optimization. A fundamental observation that has
led to the development of compressed sensing is that many high-dimensional signals of interest are in fact mainly governed by much lower-dimensional structures.
A mathematical tool that allows to capture this observation is the concept of spar-
sity: A signal $x \in \mathbb{R}^n$ is called *s-sparse* if $\|x\|_0 := |\{i : x_i \neq 0\}| = s$, i.e., if at most s of its
components are nonzero. Most real-world signals are not sparse by themselves, but
for many signal classes there exist transformations $\Psi \in \mathbb{R}^{n \times n}$, such that the coefficients $\Psi x \in \mathbb{R}^n$ are sparse (see Sect. 7.2.3).

Given the measurements $y = Ax \in \mathbb{R}^m$, a natural approach to find the sparsest solution that is still consistent with the measured data is to consider the program

$$\min_{x \in \mathbb{R}^n} \|\Psi x\|_0 \quad \text{s.t.} \ Ax = y. \tag{$\ell 0$}$$

However, it turns out that ($\ell 0$) is in general *NP-hard* [11], in particular, it is in
general numerically intractable. The key idea of compressed sensing is to replace
($\ell 0$) by a *convex relaxation*, usually called *basis pursuit*, that reads as

$$\min_{x \in \mathbb{R}^n} \|\Psi x\|_1 \quad \text{s.t.} \quad Ax = y. \tag{$\ell 1$}$$

The latter minimization program is a non-differentiable, convex optimization
problem for which many well-studied algorithms exist. We refer, for instance, to
[12] as a good starting point for a systematical study of solving such optimization
programs.

Of course, the main question is now, whether the program ($\ell 1$) is a reasonable
alternative for ($\ell 0$), i.e., whether the solutions of the tractable ℓ_1-problem coincide
with those of ($\ell 0$). Finding conditions under which this property is indeed true lies
in the heart of compressed sensing and many mathematical concepts such as the *null
space property, restricted isometry property*, or other concepts from convex geometry have been proposed to tackle it [11].

We confine ourselves to presenting an important result that might be particularly
interesting for applications in medical imaging and which is quite effective in pointing out the important quantities for the success of solving ($\ell 1$). In order to do so we
need to introduce a notion of *incoherence* between the sampling and the reconstruction system. For simplicity, let us assume that the m measurement vectors (the rows
of A) are taken from a basis $\Phi = (\phi_1, \ldots, \phi_n)$ (for instance, the Fourier atoms) and that
also the sparsifying transform Ψ originates from a system $(\psi_1, \ldots, \psi_n) \subseteq \mathbb{R}^n$. We can
then define the *coherence* between Φ and Ψ as $\mu(\Phi, \Psi) := \sqrt{n} \cdot \max_{1 \leq i, j \leq n} |\langle \phi_i, \psi_j \rangle|$.
If the elements of Φ and Ψ are correlated, then μ will be large and if the two bases are
somewhat perpendicular then μ will be small.

The following (simplified) theorem shows an interaction of *probability theory, incoherence,* and *sparsity* that is responsible for the success of ($\ell 1$) by providing an insightful lower bound on the number m of measurements sufficient for recovery:

Theorem 1 ([13]) *Let Φ, Ψ be ONBs and $x \in \mathbb{R}^n$. Further assume that the coefficients Ψx are s-sparse. Let A be defined by selecting its rows uniformly at random from the elements of the sampling basis Φ. Then, if*

$$m \geq C \cdot s \cdot \mu(\Phi, \Psi)^2 \cdot \log(n),$$

for a fixed constant $C > 0$, then x is the unique solution of ($\ell 1$) with high probability.

This theorem shows that the minimization program ($\ell 1$) is indeed a good choice for recovering a signal of interest from a limited number of randomly selected measurements. A sufficient number of such measurements are thereby essentially determined by the effective complexity of the signal (i.e., its sparsity s) and the interaction of the sampling process with the underlying atoms of the signal (i.e., the coherence μ).

For practical applications, the formulation of the above theorem seems to be too idealistic, since real-world signals are usually not exactly s-sparse and measurements are typically corrupted by noise. However, by replacing the equality constraint in ($\ell 1$) by the condition $\|Ax - y\|_2 \leq \sigma$, where $\sigma \geq 0$ is a bound for the noise level, Theorem 1 can be extended to noisy measurements. Furthermore improved results that account for sparsity defects are available. Many more details and a highly non-trivial proof of this theorem can be found in [13].

Let us now return from this excursion to compressed sensing to our originally stated optimization problem (7.6). We have seen that the (geometrical) properties of the ℓ_1-norm allow for incorporating the prior information of sparsity with respect to certain representation systems. While the constrained formulation as in ($\ell 1$) is usually considered in compressed sensing, the unconstrained version as stated in (7.6) is mostly used for practical implementations. In fact, it can be shown that for certain choices of the involved parameters both problems are equivalent [11]. Let us further remark that there are many more alternative formulations of ℓ_1-minimization such as the *Dantzig selector* [14] or the *LASSO* [15].

Finally, a particularly important model for mathematical imaging shall be mentioned, namely the *Rudin-Osher-Fatemi-model (ROF)* or *total variation (TV) regularization* [16]. Therein L of (7.6) is chosen as a discrete gradient operator, i.e., one considers the program

$$\min_{x \in \mathbb{R}^n} \|Ax - y\|_2^2 + \alpha \cdot \|\nabla x\|_1,$$

where ∇ denotes a finite differences operator. In contrast to an ℓ_2-norm penalty of the gradient, as in Tikhonov regularization, the sparsity promoting ℓ_1-norm of the gradient allows for sharp edges in the reconstruction. Since its introduction over 25 years ago, the mathematical properties of TV have been analyzed carefully and it has proven itself to be very effective in various applications, cf. [17].

7.2.3 Wavelets and Shearlets as Sparsifying Systems

In the 1980s Morlet and Grossmann first invented the name *wavelets* and since then there has been an avalanche of research going on to further study and develop the so-called *wavelet systems*. These are affine systems with many mathematical properties that underline their efficiency for solving certain practical problems such as image compression [18], image restoration [18], and MRI reconstruction [2]. However, these systems are not only heavily used in applications, but they also form a very active theoretical field of mathematical research. A good overview of the basics of wavelet theory can be found in [19]. Many systems similar to wavelets have been developed over the past years, which have shown to outperform classical wavelets in terms of their approximation properties, subsequently initiating new trends in the design of sparsifying systems. An example for classes of such systems are so-called *shearlet systems*. The main purpose of this section is to introduce both wavelet and shearlet systems due to their significance for medical applications.

7.2.3.1 Wavelet Systems
We begin with *wavelet systems* in the 1D setting. A wavelet system is an affine system of the form

$$\Psi := \left\{ \psi_{j,m} := 2^{j/2} \psi \left(2^j \cdot -m \right) : j, m \in \mathbb{Z} \right\}$$

for a generating function $\psi \in L^2(\mathbb{R})$—the so-called *mother wavelet* (Fig. 7.2). Intuitively, the dyadic scaling obtained by the factor 2^j and the translation by m allows for efficient microlocal analyses of data.

Such wavelet systems can be constructed to form an orthonormal basis for $L^2(\mathbb{R})$ by the help of a *multiresolution analysis* (MRA) [19]. The MRA allows for fast and efficient computations of all wavelet coefficients which then give rise to a full representation of every element f of $L^2(\mathbb{R})$. For our purposes it suffices to discuss the approximation of piecewise-regular functions by using wavelets. Moreover, with a view towards the approximation of images, it is sufficient to

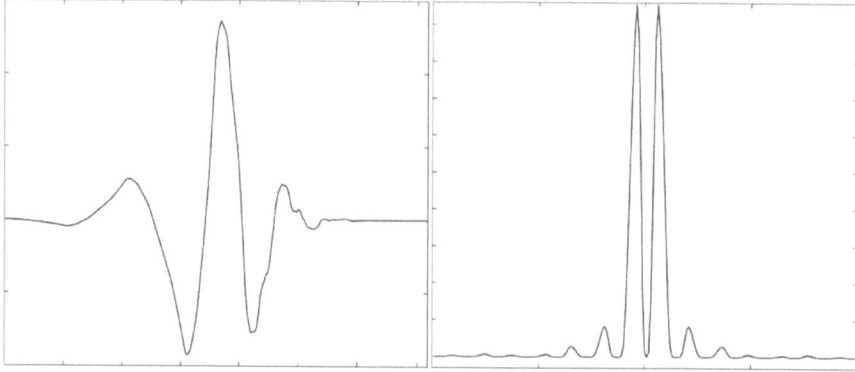

Fig. 7.2 Left: A wavelet in time domain. Right: Modulus of a wavelet in Fourier domain

continue the discussion for the Hilbert space $L^2([0, 1])$ as images are only finitely supported. A fast approximation rate of such functions explains the great success of using wavelets as a sparsifying system. In fact, since these systems form an orthonormal basis the decay of the error rate is given by the decay of the largest wavelet coefficients.

We are then, in particular, interested in the best approximation error using $N \in \mathbb{N}$ coefficients, i.e., how fast $\|f - f_N\|^2$ decays to zeros as N goes to infinity, where f_N is a representation using N wavelets, that is $f_N = \sum_{(j,m) \in \Lambda_N} \langle f, \psi_{j,m} \rangle \psi_{j,m}$ where $\Lambda_N \subset \mathbb{Z}^2$ with $\# \Lambda_N = N$. Clearly, the best choice of Λ_N is the one that minimizes the error $\|f - f_N\|$ and that is the index set containing the N largest wavelet coefficients $\left(\langle f, \psi_{j,m} \rangle \right)_{j,m \in \mathbb{Z}^2}$ in modulus.

Theorem 2 ([18]) *Let Ψ be a wavelet ONB for $L^2([0, 1])$ with a mother wavelet that is q-times differentiable and has q vanishing moments. If $f \in L^2([0, 1])$ has finitely many discontinuities and is p-times continuously differentiable with $p < q$, then*

$$\|f - f_N\|_{L^2} \le CN^{-p},$$

holds for some $C > 0$, where f_N is the wavelet expansion using the N largest wavelet coefficients in modulus.

Theorem 2 concerns the approximation of 1D functions that are piecewise regular. Wavelets for higher dimensions can be constructed by using tensor products with many structures that are known in 1D carrying over to the multivariate case, see [18, 19]. When generalizing the approximation result to the 2D scenario one has to be somewhat more subtle with the choice of singularity-types. In particular, the model has to be chosen with great care in order to be close to natural images on the one hand, but on the other hand still have enough structure for providing mathematical analysis. We will present here the class of so-called *cartoon-like functions* that was first introduced by Donoho in [20], which is a mathematical simplified model for natural images.

Definition 1 The class of *cartoon-like functions* $E(\mathbb{R}^2)$ is the set of functions $f: \mathbb{R}^2 \to \mathbb{C}$ of the form

$$f = g + h \cdot \chi_B,$$

where $B \subset [0, 1]^2$ is a set whose boundary is a closed C^2-curve with bounded curvature and $g, h \in C^2(\mathbb{R}^2)$ are functions supported in $[0, 1]^2$ with $\|g\|_{C^2}, \|h\|_{C^2} \le 1$.

As natural images are typically governed by edges which are modeled by the discontinuities of a function, it is natural to ask what the approximation rate for such functions is, when classical 2D wavelets are used. Indeed, the approximation rate is N^{-1} [18, 21]. However, the rate N^{-1} is not optimal as the following theorem shows.

Theorem 3 ([21]) *Let f be a cartoon-like function. For $N \in \mathbb{N}$ there exists a triangulation of $[0, 1]^2$ with N triangles such that the piecewise linear interpolation f_N of these triangles satisfies*

$$\|f - f_N\|_{L^2}^2 \leq C \cdot N^{-2}, \text{ as } N \to \infty,$$

where C is some positive constant.

Theorem 3 shows that the rate of wavelets can be improved, however, it is based on an adaptive triangulation. It is therefore of utmost desire to have a fixed system, similar to the wavelet system, that can reach N^{-2}. One of such systems is a *shearlet system* [21] which we will now introduce.

7.2.3.2 Shearlet Systems

Shearlets were first introduced in 2005 [22, 23]. We will now recap the basic definitions and properties of such systems and refer the interested reader to [21] for a more detailed introduction. The structural properties of shearlets are significantly different from wavelets. Recall that a wavelet system was defined using a dyadic scaling and translations. A two dimensional wavelet system obtained by tensor products thus involves an isotropic scaling and a translation along both directions. For a shearlet system one replaces the isotropic scaling with an anisotropic scaling and adds a directional component to the system by including a shearing action, cf. Fig. 7.3. Thus, we have the following definition of a *cone-adapted shearlet system*.

Definition 2 Consider the *parabolic scaling matrices* A_{2^j}, $\tilde{A}_{2^j}$ with *scaling parameter* $j \in \mathbb{N}_0$ and the *shear matrix* S_k with *shearing parameter* $k \in \mathbb{Z}$, defined as follows:

$$A_{2^j} = \begin{pmatrix} 2^j & 0 \\ 0 & 2^{j/2} \end{pmatrix}, \quad \tilde{A}_{2^j} = \begin{pmatrix} 2^{j/2} & 0 \\ 0 & 2^j \end{pmatrix}, \quad S_k = \begin{pmatrix} 1 & k \\ 0 & 1 \end{pmatrix}.$$

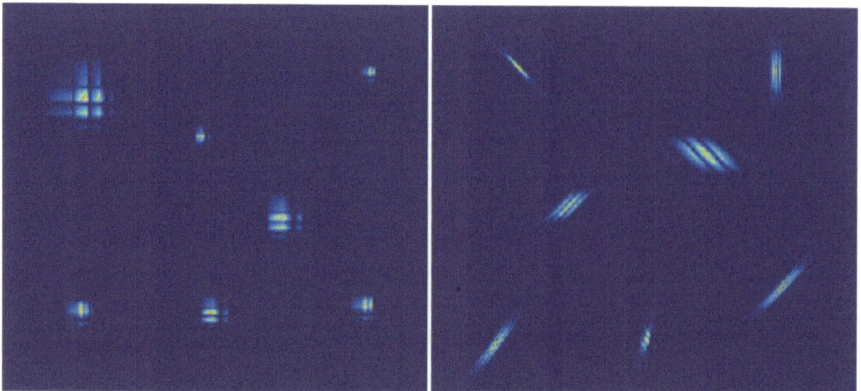

Fig. 7.3 Heat maps of 2D wavelets and shearlets in spatial domain. Left: Wavelets at different scales. Right: Shearlets at different scales and different shearing directions

Further, let $\phi, \psi, \tilde{\psi} \in L^2\left(\mathbb{R}^2\right)$ be the *generating functions* and $c = (c_1, c_2) \in \mathbb{R}^+ \times \mathbb{R}^+$. Then the *(cone-adapted discrete) shearlet system* is defined as

$$S\left(\phi, \psi, \tilde{\psi}, c\right) = \Phi\left(\phi, c_1\right) \cup \Psi\left(\psi, c\right) \cup \tilde{\Psi}\left(\tilde{\psi}, c\right),$$

where

$$\Phi\left(\phi, c_1\right) = \left\{\phi\left(\cdot - c_1 m\right) : m \in \mathbb{Z}^2\right\},$$

$$\Psi\left(\psi, c\right) = \left\{\psi_{j,k,m} = 2^{3j/4} \psi\left(\left(S_k A_{2^j}\right) \cdot - cm\right) : j \ge 0, |k| \le 2^{j/2}, m \in \mathbb{Z}^2\right\},$$

$$\tilde{\Psi}\left(\tilde{\psi}, c\right) = \left\{\tilde{\psi}_{j,k,m} = 2^{3j/4} \tilde{\psi}\left(\left(S_k^T \tilde{A}_{2^j}\right) \cdot - \tilde{c}m\right) : j \ge 0, |k| \le 2^{j/2}, m \in \mathbb{Z}^2\right\},$$

and the multiplication of c and $\tilde{c} = (c_2, c_1)$ with the translation parameter m should be understood entry wise.

Note that due to the shearing and the anisotropy of the scaling strongly localized elements can follow the structure of a curve significantly better than an isotropically scaled object. It is this geometric flexibility that allows the shearlet system to improve upon the previous approximation results of wavelets.

There are several different constructions of shearlet systems known in the literature [21, 24], in particular there are constructions based on band-limited or compactly supported generators. Furthermore, they can be constructed in such a way that the approximation rate is almost optimal. Indeed, the following result holds.

Theorem 4 ([21]) *There exist shearlet systems $S\left(\phi, \psi, \tilde{\psi}, c\right)$ such that for $f \in E\left(\mathbb{R}^2\right)$ there exists a positive constant $C > 0$ with*

$$\left\|f - f_N\right\|_{L^2}^2 \le C \cdot \log(N)^3 \cdot N^{-2}, \text{ as } N \to \infty.$$

In contrast to wavelets, no shearlet orthonormal bases are known, and shearlets are generally redundant systems. However, they can be constructed to form *frames* [25], which is a generalization of orthonormal bases to stable redundant systems.

The construction of shearlets is not limited to the 2D setting. They can also be generalized, for instance, to the 3D case. However, notice that similar as the extension from 2D to 3D the choice of the model has to be again carefully selected, see [26].

7.3   Recovery from Incomplete Fourier Measurements with 3D Shearlets

In this final section, we want to demonstrate how the use of shearlets as a sparsifying transform may improve the reconstruction quality of 3D parallel MRI imaging (pMRI). Putting most underlying physical principles aside for the sake of clarity, we are following an image-domain based reconstruction method using *sensitivity encoding* (SENSE) [27], see also Chap. 7. This means that L different coils acquire k-space data $y_i = MFS_i x \in \mathbb{R}^m$ in a parallel fashion, where M describes a radial subsampling pattern, F denotes the discrete Fourier transform and S_i is the sensitivity map belonging to the i-th coil. In conclusion, writing the resulting measurement matrix as $A := \left(MFS_1,,\ldots,,MFS_L \right)$, we consider the convex optimization problem

$$\min_{x \in \mathbb{C}^n} \|Sx\|_1 \quad \text{s.t.} \quad \|Ax - y\|_2^2 \le \sigma^2,$$

where S denotes the 3D shearlet transform as implemented in the ShearLab package [28]. For solving this high-dimensional convex optimization problem we are using an implementation based on *Alternating Direction Method of Multipliers (ADMM)*, see, for instance, [29], and to further promote the sparsity of the coefficients Sx we utilize an *iterative reweighting* strategy adapted to the multilevel structure of the shearlet coefficients, as recently proposed in [30]. In Fig. 7.4, we compare our results with a non-Cartesian iterative SENSE reconstruction [27] and a

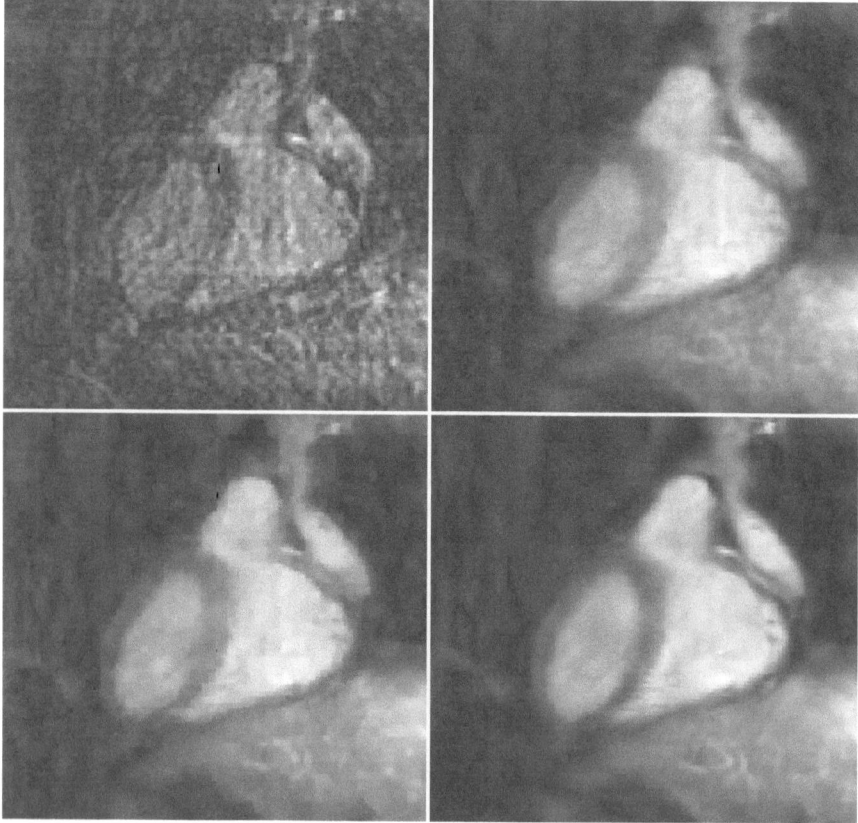

Fig. 7.4 A comparison of a 3D pMRI scan of the whole heart obtained on a 1.5T machine. Top left: Fourier inverse, top right: iterative SENSE, bottom left: total variation, bottom right: shearlet based reconstruction

non-Cartesian iterative SENSE reconstruction with a spatial total variation constraint. A relatively small fraction, i.e., $\frac{1}{24}$-th of the k-space data is acquired using $L = 8$ different coils. For didactic reasons we also display the non-regularized Fourier inverse of the zero-filled k-space data. A more detailed comparison and a systematic image quality assessment can be found in [31].

Conclusion

We presented several mathematical methods used for treating inverse problems as they arise in medical imaging tasks. Among them is compressed sensing, which is a modern reconstruction technique based on the concept of sparsity, that allows to obtain improved reconstructions compared to conventional methods. However, in its analysis formulation the success depends on the acquisition scheme, i.e., the sampling operator such as the subsampled Fourier transform, the sparsity prior, i.e., the analysis operator such as the shearlet transform and in particular, the interaction between these operators measured by their incoherence. In conclusion, there is a variety of mathematical concepts from different areas such as harmonic analysis, linear algebra, and optimization that, if applied correctly, can be used to enhance solutions to medical imaging problems.

Acknowledgements G. Kutyniok acknowledges partial support by the Einstein Foundation Berlin, the Einstein Center for Mathematics Berlin (ECMath), the European Commission-Project DEDALE (contract no. 665044) within the H2020 Framework Program, DFG Grant KU 1446/18, DFG-SPP 1798 Grants KU 1446/21 and KU 1446/23, the DFG Collaborative Research Center TRR 109 Discretization in Geometry and Dynamics, and by the DFG Research Center Matheon "Mathematics for Key Technologies" in Berlin. J. Ma acknowledges partial support by the DFG Collaborative Research Center TRR 109 Discretization in Geometry and Dynamics. M. März acknowledges partial support by the DFG SPP 1798 Compressed Sensing in Information Processing.

References

1. Epstein CL. Introduction to the mathematics of medical imaging. Philadelphia: SIAM; 2008.
2. Lustig M, Donoho D, Pauly JM. Sparse MRI: the application of compressed sensing for rapid MR imaging. Magn Reson Med. 2007;58:1182–95.
3. Hadamard J. Lectures on Cauchy's problem in linear differential equations. New Haven: Yale University Press; 1923.
4. Grafakos L. Classical Fourier analysis. New York: Springer; 2008.
5. Foster J, Richards FB. The Gibbs phenomenon for piecewise-linear approximation. Am Math Mon. 1991;98:47–9.
6. Bertero M, Boccacci P. Introduction to inverse problems in imaging. London: Institute of Physics Publishing; 1998.
7. Mueller JL, Siltanen S. Linear and nonlinear inverse problems with practical applications. Philadelphia: SIAM; 2012.
8. Tikhonov AN, Arsenin VY. Solutions of ill-posed problems. New York: Wiley; 1977.
9. Candès EJ, Tao T. Decoding by linear programming. IEEE Trans Inf Theory. 2005;51:4203–15.
10. Donoho DL. Compressed sensing. IEEE Trans Inf Theory. 2006;52:1289–306.

11. Boche H, Calderbank R, Kutyniok G, Vybiral J. A survey of compressed sensing. In: Boche H, et al., editors. Compressed sensing an its applications. Berlin: Springer; 2015.

12. Tropp JA, Wright SJ. Computational methods for sparse solution of linear inverse problems. Proc IEEE. 2010;98:948–58.

13. Candès EJ, Plan Y. A probabilistic and RIPless theory of compressed sensing. IEEE Trans Inf Theory. 2010;57:7235–54.

14. Candès EJ, Tao T. The Dantzig selector: statistical estimation when p is much larger than n. Ann Statist. 2007;35:2313–51.

15. Tibshirani R. Regression shrinkage and selection via the Lasso. J R Stat Soc Ser B Methodol. 1996;58:267–88.

16. Rudin L, Osher S, Fatemi E. Nonlinear total variation based noise removal algorithms. Physica D. 1992;60:256–86.

17. Chambolle A, Lions P. Image recovery via total variation minimization and related problems. Numer Math. 1997;76:167–88.

18. Mallat S. A wavelet tour of signal processing: the sparse way. 3rd ed. Amsterdam: Elsevier/ Academic; 2009. With contributions from Gabriel Peyré

19. Daubechies I. Ten lectures on wavelets. Philadelphia: SIAM; 1992.

20. Donoho DL. Sparse components of images and optimal atomic decomposition. Constr Approx. 2001;17:353–82.

21. Kutyniok G, Labate D. Shearlets: multiscale analysis for multivariate data. Boston: Birkhäuser Basel; 2012.

22. Guo K, Kutyniok G, Labate D. Sparse multidimensional representations using anisotropic dilation and shear operators. In: Wavelets and Splines (Athens, GA, 2005). Nashville: Nashboro Press; 2006. p. 189–201.

23. Labate D, Lim W-Q, Kutyniok G, Weiss G. Sparse multidimensional representation using shearlets. In: Papadakis M, Laine MF, Unser MA, editors. Wavelets XI, SPIE proc. 5914. Bellingham, WA: SPIE; 1974. p. 254–62.

24. Kittipoom P, Kutyniok G, Lim W-Q. Construction of compactly supported shearlet frames. Constr Approx. 2012;35:21–72.

25. Christensen O. An introduction to frames and Riesz bases, Applied and numerical harmonic analysis. Boston: Birkhäuser Boston; 2003.

26. Kutyniok G, Lemvig J, Lim W-Q. Optimally sparse approximations of 3D functions by compactly supported shearlet frames. SIAM J Math Anal. 2012;44:2962–3017.

27. Pruessmann KP, Weiger M, Scheidegger MB, Boesiger P. SENSE: sensitivity encoding for fast MRI. Magn Reson Med. 1999;42:952–62.

28. Kutyniok G, Lim W-Q, Reisenhofer R. ShearLab 3D: faithful digital shearlet transforms based on compactly supported shearlets. ACM Trans Math Softw. 2016;42:1–42.

29. Combettes PL, Pesquet J-C. Proximal splitting methods in signal processing. In: Fixed-point algorithms for inverse problems in science and engineering. New York: Springer; 2011. p. 185–212.

30. Ma J, März M. A multilevel based reweighting algorithm with joint regularizers for sparse recovery (submitted).

31. Ma J, März M, Funk S, Schulz-Menger J, Kutyniok G, Schäffter T, Kolbitsch C. Shearlet-based compressed sensing for fast 3D MR imaging using iterative reweighting (submitted).

Acceleration Strategies for Data Sampling in MRI

Christoph Kolbitsch and Tobias Schaeffter

Abstract

Magnetic resonance imaging (MRI) is a highly versatile imaging technique widely used in clinical practice. It can provide anatomical images with excellent soft tissue contrast and quantitative measurements of motion and flow. In addition, microscopic tissue structures such as neurological pathways and heart muscle fiber orientation can be assessed. One of the main challenges of MRI is long acquisition times to ensure high spatial and/or temporal resolution and full 3D coverage of the region of interest.

This chapter will give an overview of how the MR signal is created and spatially encoded and how image information can be reconstructed from this raw data. Furthermore, approaches which reduce scan times by making data acquisition faster or by reducing the amount of required raw data will be discussed.

8.1   MR Signal Equation

MRI uses magnetic fields and field gradients to create, encode, and record the signal of nuclear magnetic moments. The most commonly used nucleus in medical MRI is hydrogen due to its abundance in the human body and high gyromagnetic ration γ which ensures high signal strength for a wide range of applications.

C. Kolbitsch (✉) • T. Schaeffter
Physikalisch-Technische Bundesanstalt (PTB), Braunschweig and Berlin, Germany

Division of Imaging Sciences and Biomedical Engineering, King's College London, London, UK
e-mail: Christoph.Kolbitsch@ptb.de

© Springer International Publishing AG 2018   167
I. Sack, T. Schaeffter (eds.), *Quantification of Biophysical Parameters in Medical Imaging*, https://doi.org/10.1007/978-3-319-65924-4_8

As already outlined in Chap. 2, the magnetic moments of hydrogen nuclei precess with a Larmor frequency

$$\omega_0 = \gamma B_0 \tag{8.1}$$

with γ of $2\pi \cdot 42.58$ MHz/T in a static magnetic field B_0. The sum of all microscopic magnetic moments creates a magnetization $\mathbf{M}$, which is oriented along the main magnetic field B_0. A so-called radio-frequency (RF) pulse is used to perturb this equilibrium state, which results in a change of the orientation of $\mathbf{M}$ away from the main magnetic field. The amplitude and duration determine the angle between $\mathbf{M}$ and B_0 which is thus called the flip angle. As a consequence of this angle $\mathbf{M}$ starts to precess around the main magnetic field, and a signal can be measured with receiver coils. The spin system returns to equilibrium by independent processes. First, there is a loss of coherence by spin interactions (transversal or T_2 relaxation) and internal main magnetic field perturbations (T_2^* relaxation). This leads to a decay of the signal, which is also called free induction decay (FID). In addition, the spin system returns to its equilibrium state by exchanging energy with the surrounding lattice described by the longitudinal or T_1 relaxation.

Instead of using the FID signal directly, data acquisition in MRI is commonly carried out during the formation of signal echoes created either by additional RF pulses (spin echo (SE)) or gradients with opposite polarity (gradient echo (GE)) (Fig. 8.1) [1–4]. The time interval of the two RF pulses determines the time point

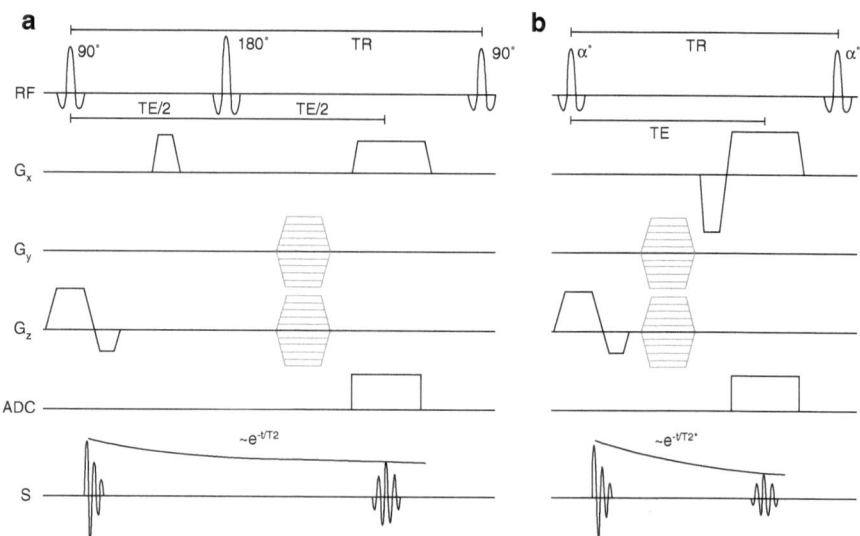

Fig. 8.1 (a) Spin echo sequence using a 90° RF pulse for excitation and a 180° RF pulse to create a spin echo. (b) For gradient echo sequences RF pulses with different flip angles (α) can be used for excitation. A bipolar readout gradient (G_x) is used to create a gradient echo. Spatial encoding in y and z direction is achieved with phase encoding prior to the data acquisition (ADC). *TR* repetition time, *TE* echo time

of the echo signal (T_E), whereas the time interval between two successive excitations is called repetition time T_R. The obtained signal depends on the acquisition protocol (e.g., flip angle, repetition time, echo time) and on tissue-specific parameters (e.g., density of nuclear magnetic moments, T_1 or T_2 relaxation times). Images with varying contrasts highlighting different pathologies can be acquired making MRI a highly flexible imaging technique.

8.2   MR Signal Encoding

The recorded MRI signal is the sum of the signals from all magnetic moments $m(\boldsymbol{r})$ in a predefined region of interest (ROI). In order to spatially localize the signal to individual voxels and obtain a 2D or 3D image, magnetic gradients $\boldsymbol{G}$ are used to modify the Larmor frequency of the nuclear magnetic moments as a function of space. If $\boldsymbol{G}$ are applied during the data acquisition, the frequency of the recorded signal is modulated as a function of space:

$$\omega(\boldsymbol{r}) = \gamma B_0 + \gamma \boldsymbol{G}(t)\boldsymbol{r} \tag{8.2}$$

and if $\boldsymbol{G}$ are switched on for a duration T prior to the actually data acquisition, the nuclear magnetic moments acquire a spatially dependent phase:

$$\phi(\boldsymbol{r}) = \int_0^T \omega(\boldsymbol{r})\,\mathrm{d}t = \phi_0 + \gamma \boldsymbol{r} \int_0^T \boldsymbol{G}(t)\,\mathrm{d}t. \tag{8.3}$$

The obtained signal S can therefore be described as

$$S(\boldsymbol{k}(t)) = e^{i\omega_0 t} \int_V m(\boldsymbol{r}) e^{-2\pi i \boldsymbol{r}\boldsymbol{k}(t)}\,\mathrm{d}\boldsymbol{r} \tag{8.4}$$

where $\boldsymbol{k}$ describes the spatial encoding achieved with the gradients

$$\boldsymbol{k}(t) = \frac{\gamma}{2\pi} \int_0^t \boldsymbol{G}(\tau)\,\mathrm{d}\tau \tag{8.5}$$

The recorded MR signal $S(\boldsymbol{k}(t))$ is therefore the Fourier transform of the magnetization $m(\boldsymbol{r})$. The diagnostic MR image information is obtained from the acquired k-space signal by applying an inverse Fourier transform to S. In order to ensure an artefact-free image, the acquired k-space has to fulfill certain sampling requirements which will be discussed in the next section.

The locations in k-space $\boldsymbol{k}(t)$ where data is acquired (k-space trajectories) only depend on the temporal behavior of the magnetic field gradients. This makes MR data acquisition very flexible and easily adaptable. The most widely used k-space trajectory is a Cartesian sampling pattern, which obtains k-space information in a set of parallel readout lines. It provides an excellent signal to noise ratio (SNR), and image data can be reconstructed with a fast Fourier transform in a fast and efficient way [5]. Popular 2D non-Cartesian trajectories are radial or spiral acquisition patterns.

There are hardware and safety constraints regarding the gradient coils, such as maximal strength of field gradients and how fast these gradients can be changed. Rather than restricting the possible k-space locations, these specifications determine the time required to go from one k-space location to the next.

After one RF excitation pulse, usually one frequency-encoded k-space line (e.g., 256 k-space points) at a certain phase-encoded k-space position is acquired (MR sequences which acquire more data after one RF excitation is discussed in "Fast sequences"). However, for image reconstruction, a fully sampled 2D or 3D k-space needs to be obtained. For this, RF excitation, echo formation, signal encoding, and data acquisition have to be repeated multiple times to obtain the required number of phase-encoded k-space lines. For fast sequences, the repetition time T_R between two RF pulses is in the range of 3–5 ms. Therefore, a 2D k-space with 256 × 256 k-space points leads to a total acquisition time of approximately 1 s, and a 3D k-space acquisition then takes a few minutes to acquire.

8.3   Nyquist Sampling Theorem

The acquired MRI signal $\hat{S}$ is not continuous but obtained on a finite number of discrete k-space locations k_i. In order to ensure that an accurate image can be reconstructed, the obtained k-space has to fulfill certain sampling requirements.

The following derivation of these sampling requirements is done for a simple 1D case of N_k regularly sampled k-space points k_i but can easily be extended to higher dimensions and non-Cartesian trajectories. The acquired signals can be described by the following sampling function:

$$\hat{S}(k_i) = S(k) \sum_{n=-\infty}^{\infty} \delta(k - n\Delta k) r\left(\frac{k + \dfrac{\Delta k}{2}}{F_k} \right) \tag{8.6}$$

where $S(k)$ is the continuous k-space signal and $F_k = \Delta k N_k$ is the extend of the sampled k-space. The rect function $r(k)$ is defined as

$$r(k) = 1 \text{ if } |k| \leq 0.5 \text{ otherwise } r(k) = 0 \tag{8.7}$$

The Fourier transform of the discrete k-space signal leads to

$$\hat{m}(r) = m(r) * \sum_{n=-\infty}^{\infty} \delta\left(r - \frac{n}{\Delta k} \right) * F_k sinc(\pi F_k r) e^{-i\pi r \Delta k} \tag{8.8}$$

The reconstructed image signal $\hat{m}(r)$ is therefore the original image signal $m(r)$ obtained from the continuous k-space $S(k)$ convolved with a comb and sinc function. The comb function leads to copies of the imaged object shifted by multiples of $1/\Delta k$ in image space. In order to avoid any overlap (so-called aliasing, fold-over, or backfolding artefacts) of these multiple copies of the object, the extend F_r of the imaged object (i.e., field of view, FOV) has to be smaller than $\dfrac{1}{\Delta k}$. A visual overview is given in Fig. 8.2.

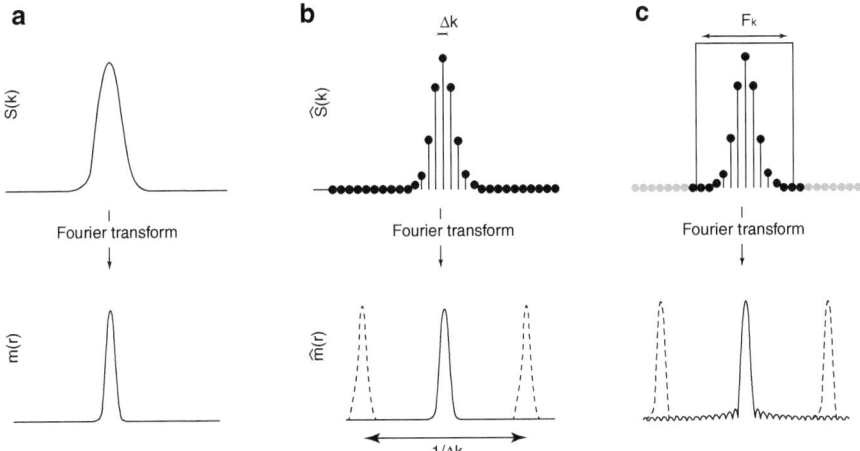

Fig. 8.2 Discrete data acquisition. (**a**) Ideal case of continuous Gaussian function $S(k)$ and its Fourier transform $m(r)$. (**b**) Discrete sampling in k-space leads to multiple copies of the images in image space. (**c**) Finite sampling in k-space causes Gibbs ringing artefacts in image space

These acquisition requirements can also be derived directly from the Nyquist-Shannon sampling criterion which states that the sample spacing Δt of a signal f bandlimited to f_{max} needs to be smaller than $\dfrac{1}{2f_{max}}$. In the above case, the sample spacing is Δk, and the bandlimit is half of the extend of the object in image space F_r.

The sinc functions lead to Gibbs ringing artefacts which can occur close to strong signal intensity changes (i.e., edges) [6]. Their frequency is proportional to the number of samples N_k, but their amplitude of $\approx 9\%$ of the original signal is constant. Low-pass filtering can be applied to the obtained k-space to reduce Gibbs ringing, but this also leads to a reduced image resolution.

The reconstructed signal $\dot{m}(r)$ is also discrete and truncated with a pixel spacing Δr and a signal extend F_r. Applying the above considerations to $m(r)$ and assuming that both k-space and image signal have the same number of samples $N_k=N_r=N$ lead to

$$\Delta k \Delta r = N \qquad (8.9)$$

For an image with a desired resolution of Δr and a field of view of F_r, the k-space has to be sampled from $-k_{max}+1$ to k_{max} on discrete sampling points separated by Δk with

$$\Delta k \leq \frac{1}{F_r} \text{ and } 2k_{max} \geq \frac{1}{\Delta r} \qquad (8.10)$$

If the Nyquist-Shannon sampling criterion is not met, so-called aliasing or undersampling artefacts appear in the reconstructed images. The appearance of these artefacts depends on which points in k-space have been acquired. One approach to study the effect of different sampling schemes on the final image quality is by assessing the point spread function (PSF). The PSF describes the response of a system to a delta peak. The system is described by the sampling of k-space along different trajectories and the subsequent Fourier transform. Consequently, the effect of sampling strategies is represented by convoluting each pixel in the MR image by the PSF. Figure 8.3 shows Cartesian, radial, and spiral undersampling patterns and the corresponding PSFs. Regular Cartesian undersampling leads to multiple copies of the PSF separated by $\frac{1}{\Delta k}$. The convolution of the image with such PSF would result in well-known "ghost images" at the position of the PSF peaks with the intensity given by the magnitude of the peak. For higher undersampling factors, Δk increases, and thus there are more peaks closer together. For the radial sampling scheme shown in Fig. 8.3, Δk increases for k-space points further away from the k-space center. The corresponding PSF is more complex with many small peaks leading to incoherent streaking artefacts [7]. In order to ensure the same undersampling artefact-free FOV as a Cartesian sampling scheme with $N \times N$ k-space points along k_x and k_y, N points need to be acquired along the radial direction at $\frac{\pi}{2} N$ different equally spaced angles [8]. The increased number of required k-space points is due to the fact that Δk increases with the distance from the k-space center, and in

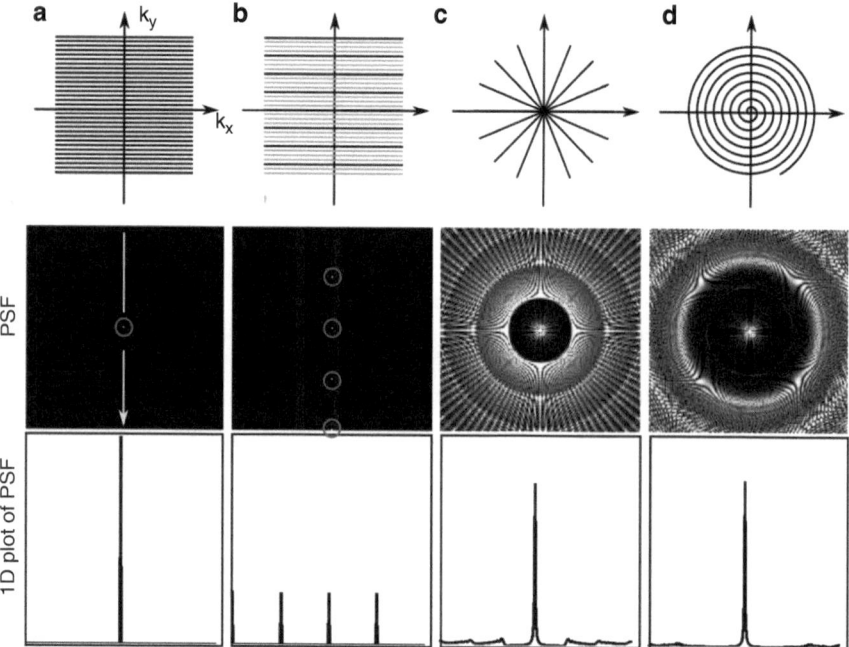

Fig. 8.3 Sampling pattern and PSF for (**a**) fully sampled Cartesian, (**b**) undersampled Cartesian, (**c**) undersampled radial, and (**d**) undersample spiral k-space trajectories. The trajectories in (**b–d**) are undersampled by a factor of four compared to the fully sampled k-space in (**a**). The PSF of the fully sampled Cartesian trajectory shows one single peak. For a regular undersampling factor of four, four peaks are visible in the PSF shifted by FOV/4 along the undersampling direction. The undersampling artefacts for radial and spiral trajectories are much more incoherent

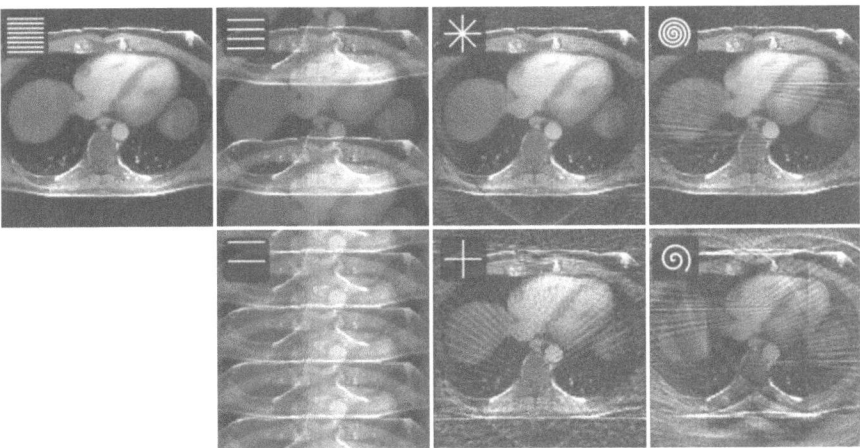

Fig. 8.4 Undersampling artefacts for Cartesian, radial, and spiral MR trajectories (symbolized in the corner of each image) with an undersampling factor of two (first row) and four (second row). The first image is the fully sampled reference image

order to ensure there are no aliasing artefacts, the Nyquist-Shannon sampling criterion also has to be fulfilled at the end of each radial spoke, where Δk between two neighboring spokes is largest. The total energy of the undersampling artefacts (i.e., the integral over the PSF without the main peak) is similar for radial and Cartesian undersampling, but for radial sampling schemes, artefacts are incoherently spread out, which is usually more acceptable for visual inspection. For the spiral sampling scheme with constant density, similar artefacts appear in the PSF. Figure 8.4 shows transversal images of a human heart obtained with a Cartesian, radial, and spiral trajectory with varying degree of undersampling.

A straightforward measure to quantify undersampling artefacts is the peak to sidelobe ratio (PSR). For a regularly undersampled Cartesian trajectory, the ratio between the main peak of the PSF and the sidelobes is 1 and therefore indicates highly regular undersampling artefacts. For radial or spiral undersampling, the PSR is smaller than 1 which indicates much more incoherent aliasing artefacts.

After discretizing the acquired k-space and the reconstructed image, the MR signal equation can also be written in matrix notation

$$s = Em \text{ with } E = LF \tag{8.11}$$

where s is the acquired k-space, m is the image, E is the encoding operator, F is the inverse fast Fourier operator, and L is the sampling operator which determines which k-space locations have been acquired. E describes the process from image information to acquired MR data. It can be extended to include further steps of the imaging process, such as the use of receiver coils with spatially varying sensitivities (which will be discussed in the Sect. 8.6) or even changes of the anatomy due to physiological motion such as breathing or heartbeat.

The inversion of the above linear equation provides the solution for the reconstruction of image *m*. For regular Cartesian sampling schemes, E can be inverted in a straightforward way and corresponds to the discrete fast Fourier transform. Especially for non-Cartesian trajectories, the encoding matrix is too large to invert directly and therefore techniques which minimize the L_2-norm between acquired and reconstructed data can be used:

$$m = \min_{\tilde{m}} \left\| E\tilde{m} - s \right\|_2^2 . \tag{8.12}$$

As already mentioned before, the frequency-encoded k-space lines are sampled one by one in a serial acquisition. This can lead to long acquisition times which are not always feasible in clinical practice. Therefore, a wide range of fast imaging sequences and acceleration techniques has been developed to speed up MR scans.

There are basically two approaches to reduce acquisition times. One approach increases the ratio between time used for data sampling relative to T_R by acquiring more than one readout line in each T_R (fast sequences). The other approach is to acquire less data, than the Nyquist-Shannon sampling criterion would actually require. In order to avoid undersampling artefacts, different approaches have been proposed that utilize additional information, such as the Hermitian properties of k-space (partial k-space acquisitions) [9], additional information about the spatial sensitivity of receiver coils (parallel imaging) [10–12], or the temporal coherence of images (temporal acceleration) [13]. Recently, techniques have also been developed which utilize information about the image (e.g., that the image is piecewise constant or sparse) to remove undersampling artefacts without impairing the actual image content [14].

8.4   Fast Sequences

Acquisition times can be strongly reduced by obtaining more than one frequency-encoded k-space line within each T_R. For SE imaging this can be achieved by applying multiple 180° RF-refocusing pulses after one excitation pulse (Fig. 8.5a). Each refocusing pulse creates one spin echo which can then be spatially encoded with standard phase encoding to sample multiple lines of k-space within each T_R.

The main challenge of this approach is that the MR signal strength decreases due to T_2 relaxation processes after the RF excitation pulse. The varying signal strength of the spin echoes leads to a varying intensity over the obtained k-space which acts like a filter and can lead to blurring and a reduced resolution of the reconstructed images. This effect becomes more severe for increasing number of echoes.

One possibility to overcome this limitation is to adapt the strength of the refocusing RF pulses during data acquisition. Rather than using only 180° RF pulses, varying flip angle patterns are used to achieve an optimal image contrast.

The most efficient GE approach is echo planar imaging (EPI) which obtains a fully sampled k-space within one T_R (Fig. 8.5b). A repeated bipolar gradient is used for frequency encoding producing multiple echoes after the RF excitation pulse. Phase encoding is achieved with so-called blip gradients, which lead to a small increase or decrease of the current phase encoding position.

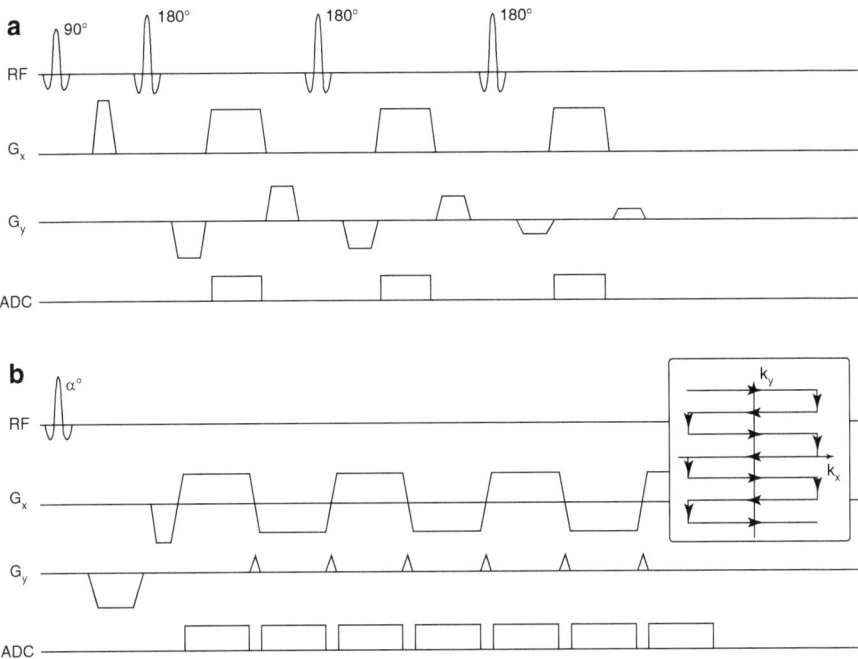

Fig. 8.5 Fast MR sequences. (**a**) For a fast spin echo sequence multiple 180° RF-refocusing pulses are applied after each 90° excitation pulse. Standard phase encoding is used to encode the multiple echoes. (**b**) Echo planar imaging (EPI) uses a bipolar readout gradient which creates multiple gradient echoes after one excitation pulse. Small phase encoding blips are used to spatially encode each echo. The small insert shows the k-space sampling pattern of EPI sequences

Similar to the fast spin echo sequence, EPI is affected by T_2^* relaxation during the echo acquisition. EPI also requires very high gradient performance and fast gradient switching times. Any hardware-related inaccuracies in the gradients lead to strong image artefacts. In addition, EPI sequences are very susceptible to magnetic field inhomogeneities and off-resonance effects which can also lead to image distortions.

Both of these approaches are mainly carried out in 2D or combined with a traditional 3D phase encoding. 3D EPI sequences are available, but they can only achieve moderate image resolution due to the limitations discussed above.

8.5   Partial k-Space Acquisitions

The objects imaged with MRI are real objects, and therefore the reconstructed images should also be real-valued. The Fourier conjugate of a real signal is Hermitian symmetric such that

$$S\big(k(t)\big) = S\big(-k(t)\big)^*$$ (8.13)

Therefore, k-space would only need to be acquired between $[0 \; k_{max}]$, and the missing data could be substituted by the complex conjugate of the obtained samples or by simple zero filling the missing k-space data and applying a Fourier transformation. Commonly partial k-space acquisitions are carried out which sample from $-(PF-0.5)^*k_{max}$ to k_{max} with a partial Fourier factor $PF \geq 0.5$. If the partial Fourier acquisition is carried out along the frequency encoding direction, it is sometimes also referred to as partial echo acquisition and can lead to reduced scan times.

A more accurate image can be obtained with a homodyne reconstruction approach which uses $PF > 0.5$ to acquire a symmetrically sampled central k-space region. MR images can have a nonconstant phase due to field inhomogeneities, eddy currents, or motion occurring during data acquisition, which violates the Hermitian symmetry. Nevertheless, this phase is assumed to be slow varying over the image and can be captured by the symmetrically sampled central k-space region allowing for the reconstruction of low-resolution images from the central part. Commonly a PF between 0.6 and 0.75 is used to provide enough information to estimate and correct this image phase. As a first step, a weighting function is applied in k-space to ensure the symmetrically and asymmetrically sampled k-space regions contribute equally to the image reconstruction. In particular, k-space values in the center are weighted by one and outside by a factor of two. For the second step, a low-resolution phase map obtained from the symmetrically sampled central k-space region is reconstructed and then applied to the partial Fourier image data to compensate for any image phase. The homodyne method can also be carried out iteratively to further improve image quality. Figure 8.6 compares the image quality achieved with simple zero filling and with the homodyne reconstruction.

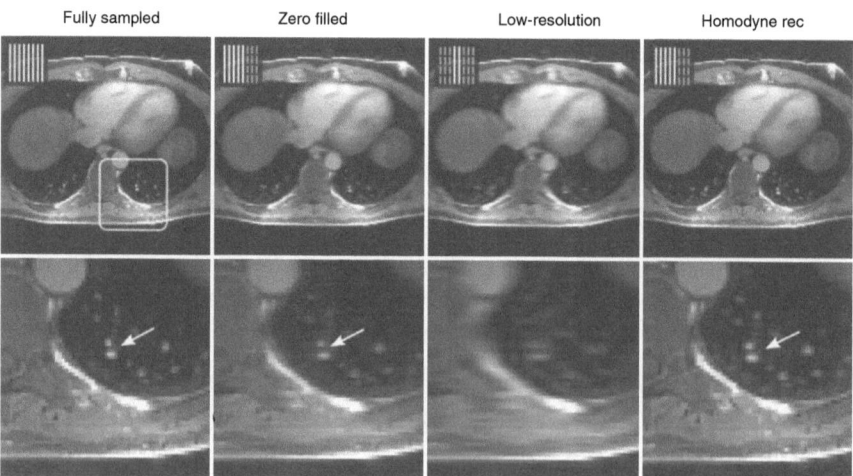

Fig. 8.6 Partial Fourier acquisition. Simple zero filling of the missing k-space data leads to strong blurring in the image. For the homodyne reconstruction, phase information is estimated from a low-resolution image reconstructed from the symmetrically sampled k-space center. The k-space is then weighted, and the phase information is used to reduce image blurring. Dashed lines: k-space data not used for image reconstruction

There are of course exceptions where the obtained MR image contains diagnostic phase information which needs to be preserved. In phase-contrast flow imaging, quantitative information about blood flow velocities can be encoded into the phase information. Water-fat separation techniques require information about the phase difference between fat and water due to their different resonance frequencies in order to be able to separate the two tissue types. Partial k-space acquisitions are still possible in these cases but require dedicated image reconstruction to preserve the correct phase information.

8.6   Parallel Imaging

The MRI signal is the sum over all the signals from the different magnetic moments. It is a time-dependent signal, and information about the spatial distribution of magnetic moments has to be obtained using encoding gradients as discussed in the first part of this chapter. Another option for spatial encoding is to exploit the spatially depending receiver sensitivity of multiple receiver coils [15]. Therefore, each receiver coil only records signals from magnetic moments in its vicinity, and signal is spatially weighted by the coil sensitivity. Using multiple receiver coils n with sensitivities $c_n(r)$ changes the MR signal equation to

$$S_n\left(k\left(t\right)\right) = e^{i\omega_0 t}\int_V m\left(r\right)c_n\left(r\right)e^{-2\pi i r k(t)}\mathrm{d}r \qquad (8.14)$$

leading to n different sets of k-space information. Reconstruction of $S_n(k(t))$ yields multiple images $m_n(r)$, each weighted differently by $c_n(r)$.

The additional information provided by the multiple receiver coils can be utilized during image reconstruction to remove undersampling artefacts [10–12, 16]. Therefore, less k-space data can be acquired without impairing image quality making these parallel imaging techniques one of the most important technical developments to speed up MRI. Commonly, they are categorized into image- or k-space-based methods. The most widely used image-based approach is sensitivity encoding (SENSE), and k-space-based technique is generalized autocalibrating partially parallel acquisition (GRAPPA).

For GRAPPA $c_n(r)$ is used as an additional spatial encoding term. For SENSE on the other hand, it describes a standard Fourier encoding of spatially weighted images. Both techniques address the same problem of reconstructing an image without aliasing artefacts from undersampled k-space data, but they utilize the spatially varying coil sensitivities in different ways [17].

In matrix notation, the above signal equation can now be written as

$$s = Em \text{ with } E = LFC \qquad (8.15)$$

where the encoding operator was extended by the coil sensitivity operator C.

8.6.1 SENSE

As mentioned above, the receiver coils have different spatial sensitivity and therefore provide a spatial localization of the recorded MR signal. In an image reconstructed from an undersampled k-space, each pixel is the sum of the true image information with aliased image information from different spatial locations. For each coil the aliased image information is now weighted differently with the different coil sensitivities. If these sensitivities are known, the true and the aliased image information can be unwrapped or unfolded. How well the true image information can be recovered depends on the coil sensitivities, the undersampling scheme, and the imaged object.

If we assume a simple example of a 2D Cartesian k-space with a regular undersampling factor of two (i.e., skipping every second k-space line) along the k_y direction, then reconstructing an image from this k-space with $\Delta k_x = \dfrac{1}{F_x}$ and with $\Delta k_y = \dfrac{2}{F_y}$ leads to aliasing artefacts along y. Due to the regular undersampling, these artefacts are copies of the original image m_{orig} shifted by $\dfrac{F_y}{2}$:

$$m(x,y) = m_{\text{orig}}(x,y) + m_{\text{orig}}\left(x,y + \frac{F_y}{2}\right) \tag{8.16}$$

If the MR data has been acquired not just with one but with two receiver coils, two images can be reconstructed each weighted by different coil sensitivities c:

$$m_{n=1}(x,y) = m_{\text{orig}}(x,y)c_{n=1}(x,y) + m_{\text{orig}}\left(x,y + \frac{F_y}{2}\right)c_{n=1}\left(x,y + \frac{F_y}{2}\right) \tag{8.17}$$

$$m_{n=2}(x,y) = m_{\text{orig}}(x,y)c_{n=2}(x,y) + m_{\text{orig}}\left(x,y + \frac{F_y}{2}\right)c_{n=2}\left(x,y + \frac{F_y}{2}\right) \tag{8.18}$$

leading to two equations for the two pixel intensities $m_{\text{orig}}(x,y)$ and $m_{\text{orig}}\left(x,y + \dfrac{F_y}{2}\right)$. If these two equations are linearly independent, then the pixel intensities can be recovered accurately, and an undersampling artefact-free image can be reconstructed. A graphical representation of the above equations is given in Fig. 8.7.

The above equations can also be written more generally as

$$m = Cm_{\text{orig}} \tag{8.19}$$

The artefact-free image m_{orig} can therefore be reconstructed by inverting the matrix C. For a regularly undersampled Cartesian k-space, such as the one considered in the above example, this matrix can be directly inverted because only very few pixels contribute to the aliasing artefacts at a certain pixel position. For irregularly undersampled k-space data or non-Cartesian trajectories, C is commonly too large to be directly inverted, and iterative approaches using, for example, conjugate gradient (CG), techniques are used to remove undersampling artefacts [18].

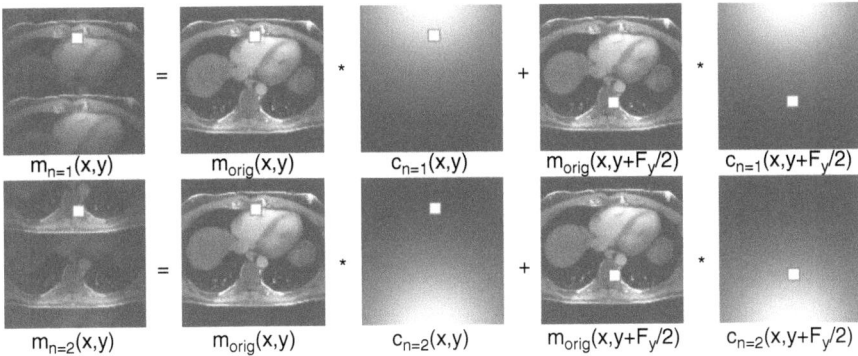

Fig. 8.7 Example of a 2D Cartesian acquisition with a regular undersampling factor of two using two receiver soils ($n = 1$ and $n = 2$). Each receiver coil leads to images with aliasing artefacts ($m_{n=1}$ and $m_{n=2}$), but the image content (m_{orig}) and thus the undersampling artefacts are weighted by the respective spatially varying coil sensitivity maps ($c_{n=1}$ and $c_{n=2}$). This allows for the removal of the aliasing artefacts (so-called unfolding)

Coil sensitivity information can be obtained from a separate so-called reference scan which is carried out prior to the actual image data acquisition. Self-reference techniques have also been developed, which obtain $c_n(\boldsymbol{r})$ from the image data itself [19].

8.6.2 GRAPPA

GRAPPA does not remove the undersampling artefacts in image space, but utilizes the spatially varying sensitivity of the receiver coils to fill in the missing k-space information and then reconstructs artefact-free images. The term $c_n(\boldsymbol{r})$ is therefore seen as spatial encoding in addition to the spatial encoding carried out by the field gradients. The main idea of GRAPPA is that missing k-space data $S(\boldsymbol{k}_m)$ can be estimated as a linear combination of acquired k-space data $S(\boldsymbol{k}_a)$ recorded with N_c receiver coils:

$$S_n\left(\boldsymbol{k}_m\right) = \sum_{j=1}^{N_c} w_j^n S_j\left(\boldsymbol{k}_a\right) \tag{8.20}$$

The weights w for each coil n are determined from autocalibration lines acquired in the center of k-space. If the acquired k-space points lie on a regular lattice (i.e., Cartesian sampling scheme), then the weights learned from the k-space center can be used throughout the entire k-space. Non-Cartesian trajectories can require multiple sets of weights to ensure missing k-space information is accurately calculated [20, 21].

Multiple k-space points in each coil can be used to calculate one missing k-space point which determines the size of the GRAPPA kernel. In theory, the entire acquired k-space can be used for each missing k-space point, but in practice less than ten data points are sufficient for an accurate reconstruction.

GRAPPA yields a fully sampled k-space for each coil, and thus an undersampling artefact-free image can be reconstructed for each coil. SENSE on the other hand does not recover the missing k-space information and provides only a single artefact-free image.

Any inaccuracies during the estimation of the coil sensitivities for SENSE or the calibration of the GRAPPA kernel directly affect the obtained image quality.

Even if the coil sensitivities are perfectly known, parallel imaging leads to a reduced SNR compared to a fully sampled data acquisition:

$$\text{SNR}_{\text{parallel}}(r) = \frac{\text{SNR}_{\text{full}}}{\sqrt{R}g(r)} \qquad (8.21)$$

R is the undersampling factor, and $g(r)$ depends on the conditioning of C and is determined by the spatial distribution of the coil sensitivities, the undersampling pattern, and the direction of undersampling. These parameters can vary as a function of space, and therefore, parallel imaging leads to a spatially dependent SNR with locally varying noise amplification compared to the fully sampled data acquisition.

8.7   Compressed Sensing

The concept of compressed sensing, which was originally proposed in the early 2000s by Donoho [22] and Candes et al. [23] and was soon translated to MRI by Lustig et al., [24] represents another powerful approach for increasing imaging speed in MRI by exploiting image redundancy in a different way. Compressed sensing takes advantage of the fact that an image **m** is usually sparse in some appropriate transform basis (Ψ) and enables reconstruction from a reduced number of k-space samples s if they are acquired in an incoherent fashion. Incoherence is a key component that aims to break the usual regularity in sampling patterns and enables the use of sparsity-based reconstructions.

A true representation of the image can then be reconstructed by using a nonlinear reconstruction approach maximizing the sparsity of signal subject to data consistency:

$$\min_{m} \left\| \Psi m \right\|_1 \text{ subject to } \left\| Em - s \right\|_2 \le \epsilon. \qquad (8.22)$$

Compressed sensing allows for high acceleration and has found wide application in MRI [25–29].

8.8   Temporal Acceleration

Fast imaging is especially important if temporal processes such as the beating of the heart or the uptake of a contrast agent in the vasculature need to be resolved. Although real-time MR imaging is possible, it is mainly limited to 2D imaging with

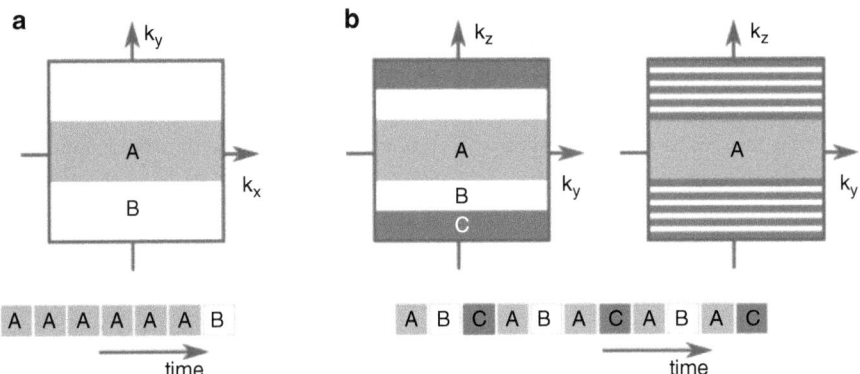

Fig. 8.8 Temporal acceleration. (**a**) Original keyhole techniques separate k-space in an inner (A) and outer (B) regions. A is acquired repeatedly over time but B is only acquired once. (**b**) Similar techniques separate k-space into more than two regions and vary the sampling of these regions to ensure high temporal resolution for A and sufficient update of B, C, ... to avoid artefacts

limited spatial resolution. In order to reduce scan times and improve both temporal and spatial resolution, temporal acceleration schemes have been developed.

The main assumption of many imaging approaches is that a contrast agent mainly affects the image contrast which is captured mainly by low k-space frequencies. The surrounding anatomy, which requires also high k-space frequency information to be accurately imaged, does not change over time. Therefore, a central k-space region is acquired with a higher temporal resolution than the outer k-space. For the image reconstruction at a certain time point t, the central k-space region acquired at t is combined with the outer k-space acquired closest to t. This of course leads to inconsistencies in k-space, but as long as the initial assumption holds true, images with an apparent high temporal resolution can be reconstructed.

There is a wide range of acquisition techniques which follow this scheme but differ in the way the k-space is separated into different blocks and how these blocks are acquired over time (Fig. 8.8). The original keyhole imaging technique separates the k-space into an inner and outer region, the inner region is repeatedly acquired over time, and the outer region is only acquired once [30]. Other approaches such as BRISK (block regional interpolation scheme for k-space) [31], TRICKS (time-resolved imaging of contrast kinetics) [32], or TWIST (time-resolved imaging with stochastic trajectories) [33] separate the k-space into an inner region and multiple outer regions. The outer regions are also repeatedly acquired but with a lower temporal frequency as the inner k-space region.

8.9   *K–t* Acceleration Techniques

Temporal acceleration techniques such as keyhole imaging or TRICKS discussed in the previous section adapt k-space sampling to achieve higher temporal resolution but still assume that a fully sampled image is reconstructed for each time step.

K–t acceleration techniques achieve a scan acceleration by reconstructing undersampled images for each time point. The sampling pattern changes over time and thus leads to temporally varying undersampling artefacts.

The separation between undersampling artefacts and true image content is done in *x–f* space, where *f* is the temporal frequency of the dynamic image sequence (Fig. 8.9). If the temporal changes in image content (i.e., the change of the anatomy

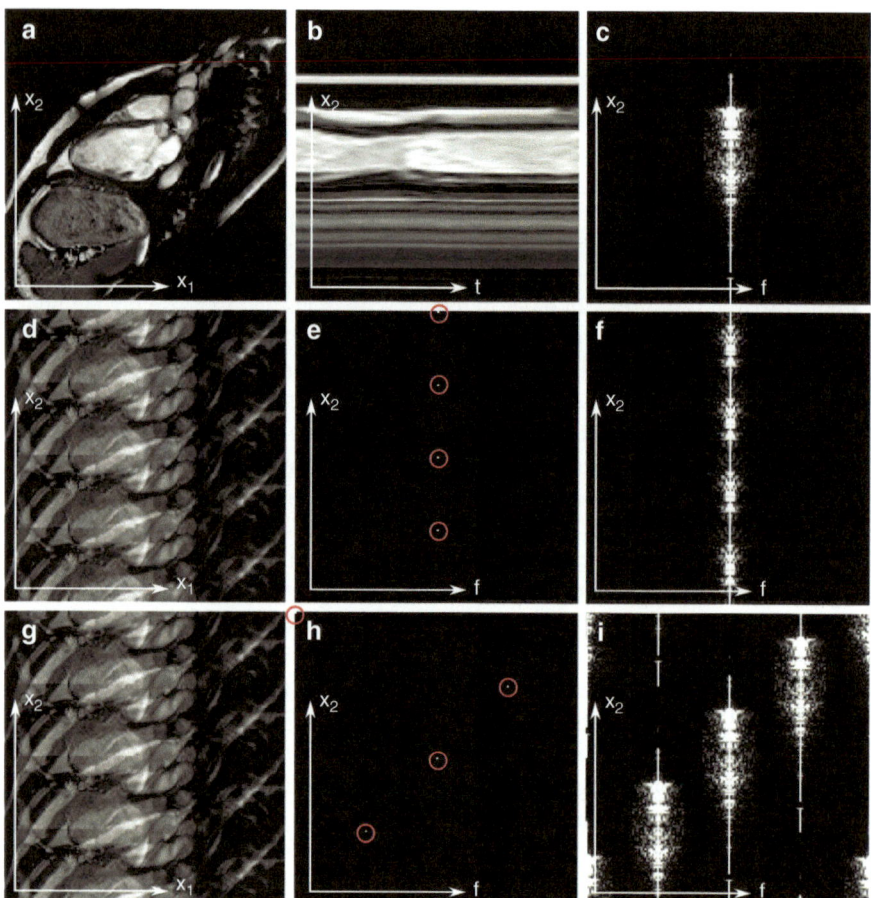

Fig. 8.9 *K–t* sampling pattern. (**a**) First frame of a dynamic fully sampled cardiac MR sequence. (**b**) Temporal profile showing the changes of the heart during the cardiac cycle. (**c**) *X–f* space of (**b**). (**d**) First frame of a regular undersampled data (i.e., constant undersampling pattern at each time point). (**e, f**) Point spread function in *x–f* space and temporal profile in *x–f* space for a regular undersampling pattern. (**g**) First frame of *k–t* undersampled data (i.e., varying undersampling pattern over time). (**h, i**) Point spread function in *x–f* space and temporal profile in *x–f* space for a *k–t* undersampling pattern. Due to the temporally varying undersampling patterns, undersampling artefacts have a higher frequency than the slowly varying anatomical information and thus can be separated in *x–f* space. For regular undersampled data, the undersampling artefacts do not vary over time and therefore overlap in the *x–f* space and cannot be separated

of the heart during the cardiac cycle) are slow compared to the temporal resolution of the dynamic image sequence, anatomic information is confined to low frequencies in $x–f$ space. If the undersampling pattern varies with each time frame, undersampling artefacts mainly occur in the high frequencies of the $x–f$ space.

A straightforward approach to suppress undersampling artefacts is to apply a simple low-pass filter in $x–f$ space (UNFOLD) [34] which can also be combined with parallel imaging for further acceleration (TSENSE) [35]. These techniques make assumptions about the distribution of image information and undersampling artefacts in $x–f$ space. $K–t$ BLAST and $k–t$ SENSE obtain a training data set with low spatial resolution but high temporal resolution in order to better distinguish between image information and undersampling artefacts in $x–f$ space [36]. The training data can be obtained from an additional scan or from the data itself using for example non-Cartesian sampling schemes such as radial k-space trajectories [37].

The information of the training data set can then be used as a regularization in $x–f$ space during an iterative image reconstruction approach. Instead of the Fourier transform a long time, other transformations such as principal component analysis or Karhunen-Loeve transform can be used to separate image content and undersampling artefacts [38, 39].

The $k–t$ reconstruction can also be formulated as a L_1-norm minimization problem and can then be solved with a compressed sensing approach to enforce sparsity in $x–f$ space [40, 41]. This does not require any training data and is thus very well suited for nonperiodic dynamic processes, such as myocardial perfusion imaging [42].

8.10 Outlook

Making MRI faster is still a very active field of research. Especially the combination of different reconstruction approaches [43–45] including information about physiological motion into the encoding operator [28, 46, 47] promises to allow for even higher undersampling factors without losing image quality. One of the main limitations of undersampling is SNR, because SNR is proportional to the square root of the number of acquired data points. Therefore, for higher undersampling factors additional regularization terms become more and more important for image reconstruction to ensure high image quality.

The temporal behavior of the magnetization during an MRI scan can be described by the Bloch equations. This can be used, for example, to obtain quantitative MR images (i.e., T_1- or T_2-maps) from qualitative MR images acquired with different MR sequence parameters [48, 49]. Other techniques utilize this additional information to compensate for blurring effects due to T_2-relaxation [50]. The main idea of all of these techniques is to keep MR sequence parameters constant and ensure a consistent magnetization for each T_R.

Recently, a new type of MR approach has been introduced which uses different MR sequence parameters for each T_R. These so-called MR fingerprinting techniques create a magnetization signal which varies over time, and the time curves for each voxel depend on known sequence parameters (e.g., T_E, T_R, flip angle), unknown scanner parameters (e.g., inhomogeneities of the main magnetic field B_0), and unknown diagnostic parameters (e.g., T_1, T_2, spin density) [51]. By comparing the measured signal time curves to a known signal library calculated from the Bloch equations, T_1-, T_2-, and B_0-, and spin density maps can be determined from very few acquired k-space samples.

MR fingerprinting utilizes knowledge about the underlying spin physics to separate MR image information and undersampling artefacts and allows for multiparametric quantitative MRI from a single scan in a very short time holding great promises for future clinical applications.

References

1. Hennig J. Echoes – how to generate, recognize, use or avoid them in MR-imaging sequences. Part I: fundamental and not so fundamental properties of spin echoes. Concepts Magn Reson. 1991;3:125–43. https://doi.org/10.1002/cmr.1820030402.
2. Hennig J. Echoes – how to generate, recognize, use or avoid them in MR-imaging sequences. Part II: echoes in imaging sequences. Concepts Magn Reson. 1991;3:179–92. https://doi.org/10.1002/cmr.1820030402.
3. Hargreaves B. Rapid gradient-echo imaging. J Magn Reson Imaging. 2012;36:1300–13. https://doi.org/10.1002/jmri.23742.
4. Markl M, Leupold J. Gradient echo imaging. J Magn Reson Imaging. 2012;35:1274–89. https://doi.org/10.1002/jmri.23638.
5. Tsai C-M, Nishimura DG. Reduced aliasing artifacts using variable-density k-space sampling trajectories. Magn Reson Imaging. 2000;43:452–8.
6. Stadler A, Schima W, Ba-Ssalamah A, Kettenbach J, Eisenhuber E. Artifacts in body MR imaging: their appearance and how to eliminate them. Eur Radiol. 2007;17:1242–55.
7. Lauzon ML, Rutt BK. Effects of polar sampling in k-space. Magn Reson Med. 1996;36:940–9.
8. Lauzon ML, Rutt BK. Polar sampling in k-space: reconstruction effects. Magn Reson Med. 1998;40:769–82. https://doi.org/10.1002/mrm.1910400519.
9. Haacke EM, Lindskogj ED, Lin W. A fast, iterative, partial-fourier technique capable of local phase recovery. J Magn Reson. 1991;92:126–45. https://doi.org/10.1016/0022-2364(91)90253-P.
10. Sodickson DK, Manning WJ. Simultaneous acquisition of spatial harmonics (SMASH): fast imaging with radiofrequency coil arrays. Magn Reson Imaging. 1997;38:591–603.
11. Pruessmann KP, Weiger M, Scheidegger MB, Boesiger P. SENSE: sensitivity encoding for fast MRI. Magn Reson Imaging. 1999;42:952–62.
12. Griswold MA, Jakob PM, Heidemann RM, Nittka M, Jellus V, Wang J, Kiefer B, Haase A. Generalized autocalibrating partially parallel acquisitions (GRAPPA). Magn Reson Med. 2002;47:1202–10. https://doi.org/10.1002/mrm.10171.
13. Tsao J, Kozerke S. MRI temporal acceleration techniques. J Magn Reson Imaging. 2012;36:543–60. https://doi.org/10.1002/jmri.23640.
14. Davies M, Puy G, Vandergheynst P, Wiaux Y. A compressed sensing framework for magnetic resonance fingerprinting. SIAM J Imaging Sci. 2014;7:2623–56. https://doi.org/10.1137/130947246.
15. Roemer PB, Edelstein WA, Hayes CE, Souza SP, Mueller OM. The NMR phased array. Magn Reson Imaging. 1990;16:192–225.
16. Blaimer M, Breuer F, Mueller M, Heidemann RM, Griswold MA, Jakob PM. SMASH, SENSE, PILS, GRAPPA. Top Magn Reson Imaging. 2004;15:223–36. https://doi.org/10.1097/01.rmr.0000136558.09801.dd.

17. Pruessmann KP. Encoding and reconstruction in parallel MRI. NMR Biomed. 2006;19:288–99. https://doi.org/10.1002/nbm.1042.
18. Pruessmann KP, Weiger M, Boernert P, Boesiger P. Advances in sensitivity encoding with arbitrary k-space trajectories. Magn Reson Med. 2001;46:638–51.
19. Uecker M, Lai P, Murphy MJ, Virtue P, Elad M, Pauly JM, Vasanawala SS, Lustig M. ESPIRiT-an eigenvalue approach to autocalibrating parallel MRI: where SENSE meets GRAPPA. Magn Reson Imaging. 2013. https://doi.org/10.1002/mrm.24751.
20. Seiberlich N, Breuer FA, Blaimer M, Barkauskas K, Jakob PM, Griswold MA. Non-Cartesian data reconstruction using GRAPPA operator gridding (GROG). Magn Reson Imaging. 2007;58:1257–65. https://doi.org/10.1002/mrm.21435.
21. Seiberlich N, Breuer F, Blaimer M, Jakob P, Griswold M. Self-calibrating GRAPPA operator gridding for radial and spiral trajectories. Magn Reson Med. 2008;59:930–5. https://doi.org/10.1002/mrm.21565.
22. Donoho DL. Compressed sensing. IEEE Trans Inf Theory. 2006;52:1289–306. https://doi.org/10.1109/TIT.2006.871582.
23. Candes EJ, Romberg J, Tao T. Robust uncertainty principles: exact signal reconstruction from highly incomplete frequency information. IEEE Trans Inf Theory. 2006;52:489–509. https://doi.org/10.1109/TIT.2005.862083.
24. Lustig M, Donoho D, Pauly JM. Sparse MRI: the application of compressed sensing for rapid MR imaging. Magn Reson Imaging. 2007;58:1182–95. https://doi.org/10.1002/mrm.21391.
25. Gamper U, Boesiger P, Kozerke S. Compressed sensing in dynamic MRI. Magn Reson Imaging. 2008;59:365–73. https://doi.org/10.1002/mrm.21477.
26. Akçakaya M, Rayatzadeh H, Basha TA, Hong SN, Chan RH, Kissinger KV, Hauser TH, Josephson ME, Manning WJ, Nezafat R. Accelerated late gadolinium enhancement cardiac MR imaging with isotropic spatial resolution using compressed sensing: initial experience. Radiology. 2012;264:691–9. https://doi.org/10.1148/radiol.12112489.
27. Li W, Griswold M, Yu X. Fast cardiac T1 mapping in mice using a model-based compressed sensing method. Magn Reson Med. 2012;68:1127–34. https://doi.org/10.1002/mrm.23323.
28. Usman M, Atkinson D, Odille F, Kolbitsch C, Vaillant G, Schaeffter T, Batchelor PG, Prieto C. Motion corrected compressed sensing for free-breathing dynamic cardiac MRI. Magn Reson Med. 2013;70:504–16. https://doi.org/10.1002/mrm.24463.
29. Lustig M, Donoho DL, Santos JM, Pauly JM. Compressed sensing MRI: a look at how CS can improve on current imaging techniques. IEEE Signal Process Mag. 2008;25:72–82. https://doi.org/10.1109/MSP.2007.914728.
30. Van Vaals JJ, Brummer ME, Thomas Dixon W, Tuithof HH, Engels H, Nelson RC, Gerety BM, Chezmar JL, Den Boer JA. "Keyhole" method for accelerating imaging of contrast agent uptake. J Magn Reson Imaging. 1993;3:671–5. https://doi.org/10.1002/jmri.1880030419.
31. Doyle M, Walsh EG, Blackwell GG, Pohost GM. Block regional interpolation scheme for k-space (BRISK): a rapid cardiac imaging technique. Magn Reson Med. 1995;33:163–70. https://doi.org/10.1002/mrm.1910330204.
32. Korosec FR, Frayne R, Grist TM, Mistretta CA. Time-resolved contrast-enhanced 3D MR angiography. Magn Reson Imaging. 1996;36:345–51.
33. Lim RP, Shapiro M, Wang EY, et al. 3D time-resolved MR angiography (MRA) of the carotid arteries with time-resolved imaging with stochastic trajectories: comparison with 3D contrast-enhanced Bolus-Chase MRA and 3D time-of-flight MRA. AJNR Am J Neuroradiol. 2008;29:1847–54. https://doi.org/10.3174/ajnr.A1252.
34. Madore B, Glover GH, Pelc NJ. Unaliasing by fourier-encoding the overlaps using the temporal dimension (UNFOLD), applied to cardiac imaging and fMRI. Magn Reson Imaging. 1999;42:813–28.
35. Kellman P, Epstein FH, Mcveigh ER. Adaptive sensitivity encoding incorporating temporal filtering (TSENSE). Magn Reson Med. 2001;45:846–52. https://doi.org/10.1002/mrm.1113.
36. Tsao J, Boesiger P, Pruessmann KP. k-t BLAST and k-t SENSE: dynamic MRI with high frame rate exploiting spatiotemporal correlations. Magn Reson Imaging. 2003;50:1031–42. https://doi.org/10.1002/mrm.10611.

37. Hansen MS, Baltes C, Tsao J, Kozerke S, Pruessmann KP, Eggers H. k-t BLAST reconstruction from non-Cartesian k-t space sampling. Magn Reson Imaging. 2006;55:85–91. https://doi.org/10.1002/mrm.20734.
38. Pedersen H, Kozerke S, Ringgaard S, Nehrke K, Won YK. K-t PCA: temporally constrained k-t BLAST reconstruction using principal component analysis. Magn Reson Med. 2009;62:706–16. https://doi.org/10.1002/mrm.22052.
39. Ding Y, Chung Y-C, Jekic M, Simonetti OP. A new approach to autocalibrated dynamic parallel imaging based on the Karhunen-Loeve transform: KL-TSENSE and KL-TGRAPPA. Magn Reson Med. 2011;65:1786–92. https://doi.org/10.1002/mrm.22766.
40. Jung H, Ye JC, Kim EY. Improved k – t BLAST and k – t SENSE using FOCUSS. Phys Med Biol. 2007;52:3201–26. https://doi.org/10.1088/0031-9155/52/11/018.
41. Jung H, Sung K, Nayak KS, Kim EY, Ye JC. K-t FOCUSS: a general compressed sensing framework for high resolution dynamic MRI. Magn Reson Med. 2009;61:103–16. https://doi.org/10.1002/mrm.21757.
42. Otazo R, Kim D, Axel L, Sodickson DK. Combination of compressed sensing and parallel imaging for highly accelerated first-pass cardiac perfusion MRI. Magn Reson Imaging. 2010;64:767–76. https://doi.org/10.1002/mrm.22463.
43. Liang D, Liu B, Wang J, Ying L. Accelerating SENSE using compressed sensing. Magn Reson Imaging. 2009;62:1574–84. https://doi.org/10.1002/mrm.22161.
44. Lustig M, Pauly JM. SPIRiT: iterative self-consistent parallel imaging reconstruction from arbitrary k-space. Magn Reson Imaging. 2010;64:457–71. https://doi.org/10.1002/mrm.22428.
45. Uecker M, Zhang S, Frahm J. Nonlinear inverse reconstruction for real-time MRI of the human heart using undersampled radial FLASH. Magn Reson Imaging. 2010;63:1456–62. https://doi.org/10.1002/mrm.22453.
46. Feng L, Axel L, Chandarana H, Block KT, Sodickson DK, Otazo R. XD-GRASP: golden-angle radial MRI with reconstruction of extra motion-state dimensions using compressed sensing. Magn Reson Med. 2015;0:1–14. https://doi.org/10.1002/mrm.25665.
47. Royuela-del-Val J, Cordero-Grande L, Simmross-Wattenberg F, Martín-Fernández M, Alberola-López C. Nonrigid groupwise registration for motion estimation and compensation in compressed sensing reconstruction of breath-hold cardiac cine MRI. Magn Reson Med. 2016;75:1525–36. https://doi.org/10.1002/mrm.25733.
48. Look DC, Locker DR. Time saving in measurement of NMR and EPR relaxation times. Rev Sci Instrum. 1970;41:250.
49. Roujol S, Weingärtner S, Foppa M, Chow K, Kawaji K, Ngo LH, Kellman P, Manning WJ, Thompson RB, Nezafat R. Accuracy, precision, and reproducibility of four T1 mapping sequences: a head-to-head comparison of MOLLI, ShMOLLI, SASHA, and SAPPHIRE. Radiology. 2014;272:683–9. https://doi.org/10.1148/radiol.14140296.
50. Block KT, Uecker M, Frahm J. Model-based iterative reconstruction for radial fast spin-echo MRI. IEEE Trans Med Imaging. 2009;28:1759–69.
51. Ma D, Gulani V, Seiberlich N, Liu K, Sunshine JL, Duerk JL, Griswold MA. Magnetic resonance fingerprinting. Nature. 2013;495:187–92. https://doi.org/10.1038/nature11971.

4D Flow MRI

9

Sebastian Schmitter and Susanne Schnell

Abstract
4D flow is a unique technique to measure the in vivo blood velocity vector spatially and temporally resolved. It allows deriving important hemodynamic parameters such as peak velocity, net flow, wall shear stress, and others, which are important parameters for a large range of diseases. This chapter will cover the basic techniques of MR velocity encoding, current acquisition strategies and practical aspects, post-processing of the acquired data, current limitations, and an overview of current and future research directions.

9.1 Introduction

It was discovered more than a decade before the acquisition of the first human MR images that magnetic resonance allows the assessment of blood flow in vivo [1]. Several years later, in 1982, Moran [2] gave a mathematical and technical description for an experiment allowing to spatially resolve the blood velocity. The work demonstrated that velocity encoding can be realized in addition to spatial encoding using velocity sensitive gradients, which still is the basis of today's velocity mapping techniques. In 1986, time-resolved two-dimensional phase-contrast (cine 2D PC) MR imaging was introduced [3], a technique that has been established as a

S. Schmitter (✉)
Physikalisch-Technische Bundesanstalt, Berlin, Germany
e-mail: sebastian.schmitter@ptb.de

S. Schnell
Department of Radiology, Northwestern University, Feinberg School of Medicine, Chicago, IL, USA
e-mail: susanne.schnell@northwestern.edu

© Springer International Publishing AG 2018
I. Sack, T. Schaeffter (eds.), *Quantification of Biophysical Parameters in Medical Imaging*, https://doi.org/10.1007/978-3-319-65924-4_9

routine tool for a variety of clinical applications. Usually, only the through-plane component of the velocity vector is measured requiring accurate orientation of the imaging slice. Later, PC-MRI advanced toward a time-resolved PC imaging technique with full three-dimensional spatial coverage and with three-directional velocity encoding. It is often termed "4D flow MRI" [4] with "4D" referring to the 3D spatial and the temporal dimension and "flow" referring to the three-directional velocity encoding. 4D flow MRI offers the opportunity for retrospective assessment of blood flow in different regions in a volume, in contrast to cine 2D PC velocity imaging where the position of the slice needs to be planned carefully to the area of interest. It also allows the derivation of advanced parameters such as wall shear stress, pressure differences, and others.

Today, velocity-encoded MR imaging is being used in a large range of cardiovascular applications. At present, however, cine 2D PC velocity imaging is predominantly applied in clinical routine. This is mostly explained by the longer scan time, which is a limitation of this technique as discussed later in more detail.

The following sections will provide an introduction to 4D flow MRI and its applications. First, a detailed description of encoding and velocity compensation using MR gradients will be given followed by practical realization of 4D flow MRI and common acquisition schemes. The post-processing required for subsequent quantification of various hemodynamic parameters will be covered prior to selected applications of 4D flow MRI. Finally, we will cover present limitations of this technique and provide an outlook of current research topics and future directions of 4D flow MRI.

9.2 Flow Compensation and Flow Encoding

Moving spins in a spatially heterogeneous magnetic field, such as produced by a gradient coil, accumulate a phase along their path relative to a stationary spin. Such phases, which are not fully rewound by switching a second gradient with opposite polarity, are blessing and curse at the same time: they can create unwanted artifacts in standard imaging techniques, but they also enable the quantification of the blood velocity.

9.2.1 Moving Spins in a Magnetic Gradient Field

In an MRI acquisition, the spins' local precession frequency $\omega(\mathbf{r})$ in the rotating frame, i.e., a coordinate system rotating with precession frequency γB_0, is given by:

$$\omega(\mathbf{r}) = \gamma \left(\mathbf{Gr} + \Delta B_0 (\mathbf{r}) \right) \tag{9.1}$$

Here, B_0 denotes the main magnetic field, $\mathbf{G} = (G_x, G_y, G_z)^T$ the switched linear magnetic field gradient realized by the three gradient coils, $\mathbf{r} = (x, y, z)^T$ the spatial coordinate vector measured from isocenter, and $\Delta B_0(\mathbf{r})$ denotes local magnetic field deviations from B_0, for example, caused by local susceptibility differences (see Chap. 15).

For convenience, $G=G_x$ describes a magnetic field gradient along the x-axis, and ΔB_0 is assumed to be zero. Due to the application of the gradient, the spin's local precession frequency depends on the position along the x-axis. Thus, the spins accumulate a spatially dependent phase that is given by the integration of the precession frequency:

$$\phi(t) = \gamma \int_{t_0}^{t} G_x(t') x(t') dt' \tag{9.2}$$

Figure 9.1 illustrates a bipolar gradient waveform consisting of two rectangular gradient pulses of duration Δt and $2\Delta t$ applied along the x-axis. If static spins at position x_0 are considered, the accumulated phase at the end of the waveform is given by $\phi(t) = \gamma x_0 m_0$ with $m_0 = \int_{0}^{t} G_x(t') dt'$ being the zeroth gradient moment, which is the area under the gradient waveform in Fig. 9.1a. If moving spins are considered, such as flowing blood within a vessel, the spatial coordinate changes during applications of the gradient, and the phase will not only depend on the coordinate, but also on the velocity, acceleration, and higher-order terms as illustrated in the following.

The spatial position of the spin can be expressed by a Taylor series around an expansion time point, which is here set to zero:

$$x(t) = x_0 + tv_0 + \frac{1}{2}t^2 a_0 + \cdots + \frac{1}{n!}t^n \left.\frac{dx(t)}{dt}\right|_{t=0} \tag{9.3}$$

x_0, v_0, and a_0 denote the initial spatial coordinate, the velocity, and the acceleration at $t = 0$. Thus, the accumulated phase for a gradient switched in x direction is given by:

$$\phi(t) = \gamma x_0 \int_{0}^{t} G_x(t') dt' + \gamma v_0 \int_{0}^{t} G_x(t')t' dt' + \frac{\gamma}{2} a_0 \int_{0}^{t} G_x(t')t'^2 dt'$$

$$+ \cdots + \frac{\gamma}{n!}\left.\frac{dx(t)}{dt}\right|_{t=0} \int_{0}^{t} G_x(t')t'^n dt' = \gamma x_0 m_0 + \gamma v_0 m_1 + \frac{\gamma}{2}a_0 m_2 + \cdots + \frac{\gamma}{n!}\left.\frac{dx(t)}{dt}\right|_{t=0} m_n \tag{9.4}$$

Here, $m_n = \int_{0}^{t} G_x(t')t'^n dt'$ denotes the nth gradient moment. In the following discussion, only the first two terms of the Taylor series are considered, while acceleration, jerk, and higher terms are neglected. In this case, the phase is linearly dependent on the spatial coordinate weighted by the zeroth gradient moment as well as on the velocity weighted by the first gradient moment (see Eq. (2.66) in Chap. 2):

$$\phi(t) \cong \gamma x_0 m_0 + \gamma v_0 m_1 \tag{9.5}$$

Fig. 9.1 Bipolar gradient
waveform (**a**) with
corresponding zeroth (**b**)
and first (**c**) gradient
moment. The resulting
m_1-induced phase shown in
(**d**) is linearly dependent
on the velocity

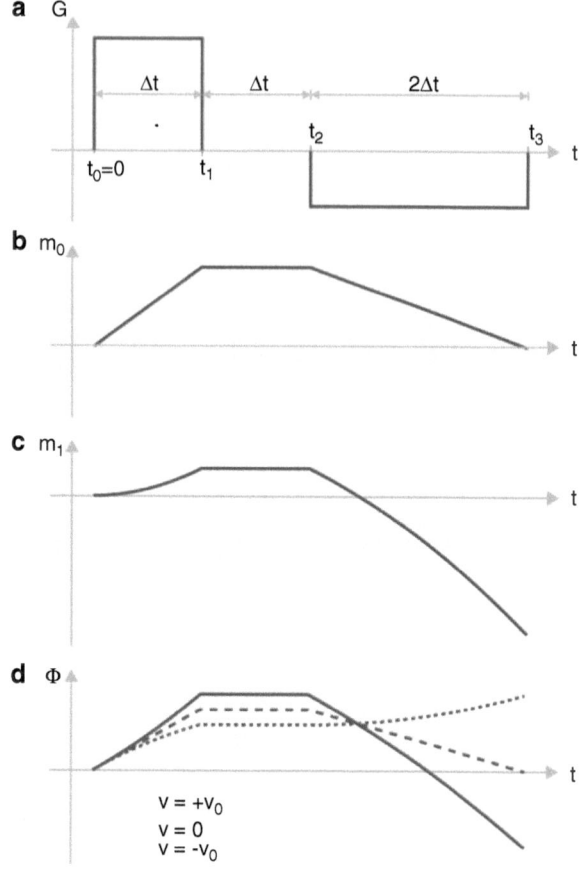

Fig. 9.1 Bipolar gradient waveform (**a**) with corresponding zeroth (**b**) and first (**c**) gradient moment. The resulting m_1-induced phase shown in (**d**) is linearly dependent on the velocity

Figure 9.1c illustrates $m_1(t)$ for the gradient waveform given in Fig. 9.1a, and the corresponding phases for spins with constant velocities of v_0, 0, and $-v_0$ starting at the same initial coordinate x_0 are shown in Fig. 9.1d. At the end of the bipolar gradient pulse (t_3), a residual phase $\phi \neq 0$ remains that scales linearly with the velocity of the moving spins.

When calculating the gradient moments, the time origin and the starting time of the gradient pulse are of importance. In slice-selective imaging, the time origin ($t=0$) can be set to the isodelay time point of the excitation pulse. For symmetric SINC-shaped RF pulses of duration T_{RF} that are often used in MRI, the isodelay time point is considered to be $T_{RF}/2$. Gradient waveforms switched on an axis other than slice selection, such as the phase-encoding gradient, typically do not start at $t = 0$ but with a delay δt. In general, if a gradient pulse starting at $t=0$ with moments m_0 and m_1 is shifted by $\delta t > 0$, the first moment m_1' of the shifted waveform at time $t + \delta t$ is given by $m_1' = m_1 + m_0 \cdot \delta t$. In other words, the first moment is invariant with respect to temporal delays if m_0 vanishes. This is the case for the bipolar gradient pulse in Fig. 9.1a at time point $t = t_3$; shifting of the entire gradient waveform in time by δt will lead to unchanged m_1 values at the end of the gradient waveform. It can be shown more generally that similar relationships exist for higher gradient moments: the nth resulting gradient moment m_n' is given by $m_n' = m_n + m_{m-1} \cdot \delta t$ if the waveform is shifted by δt.

9.2.2   Flow Artifacts and Flow Compensation

Non-compensated first gradient moments ($m_1 \neq 0$) typically cause artifacts in MR images. For example, a constant velocity in phase-encoding direction causes mislocalization in that direction if conventional phase-encoding gradients with $m_1 \neq 0$ are being used as shown in Fig. 9.2. In this case, not only m_0 changes incrementally (Fig. 9.2b) with phase-encoding line number, as intended for phase-encoding purposes, but also m_1 as displayed in Fig. 9.2c. Therefore, assuming constant blood velocity within the voxel, such m_1-induced linear phase in k-space generates a shift in image space. In vessels with pulsatile blood flow, the flow-induced phase varies over k-space resulting in more complex artifacts, often seen as multiple replication artifacts as shown in Fig. 9.2d. To avoid such artifacts, *velocity-compensated* gradient waveforms using bipolar gradient waveforms with $m_1 = 0$ and constant m_0 can be applied as illustrated in Fig. 9.3. In case of the slice selection gradient, both m_0 and m_1 vanish at the end of the gradient pulses, and therefore the resulting phase is independent of the blood velocity. Velocity compensation can also be realized for

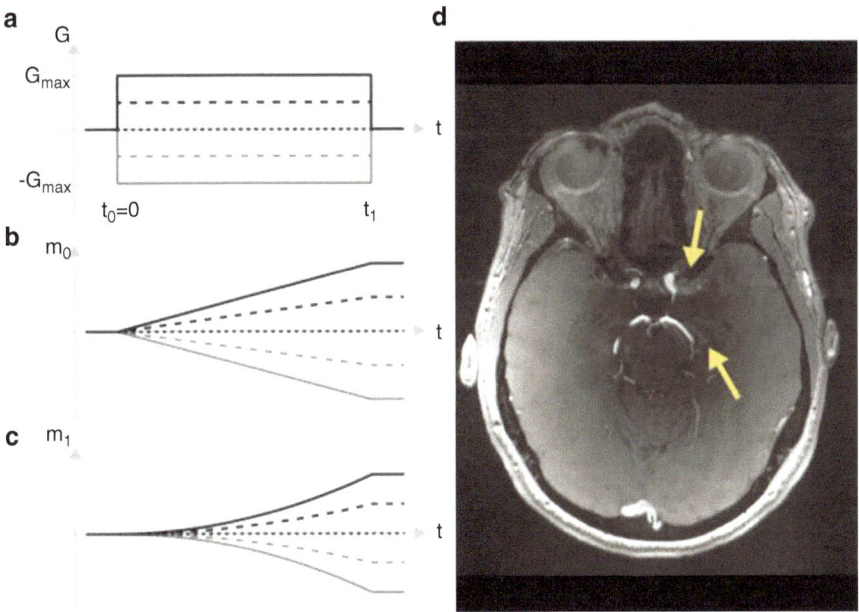

Fig. 9.2 (**a–c**) Phase-encoding gradient and respective moments $\boldsymbol{m_0}$ and $\boldsymbol{m_1}$ as a function of time. (**d**) Typical artifacts (yellow arrows) observed in non-flow-compensated time-of-flight angiography images caused by pulsatile blood flow

Fig. 9.3 Velocity-compensated slice and phase-encoding gradient waveform. For the slice selection gradient, m_0 and m_1 vanish at the end of the three gradient pulses. For the phase-encoding gradient, m_0 changes for different lines (indicated by different colors), but m_1 vanishes at the end of the bipolar phase-encoding gradient for all lines

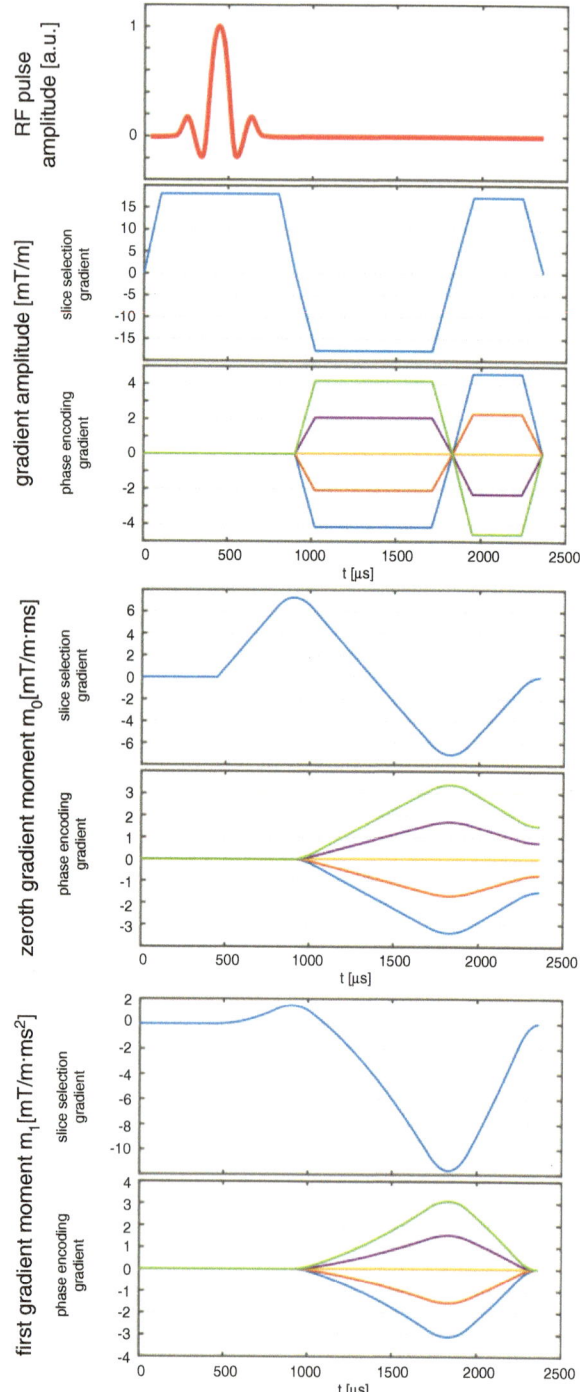

phase-encoding gradients with $m_1 = 0$ but $m_0 \neq 0$ as shown in Fig. 9.3. Because both gradient amplitude and gradient duration impact m_0 and m_1, the total duration is typically minimized while keeping the amplitude within gradient limits [5].

9.2.3 Encoding the Blood Velocity

Besides *compensating* velocity induced phase shifts, gradient pulses with $m_1 \neq 0$ offer the ability to *encode* the velocity through the MR phase. It is the basis of velocity-encoded MR imaging, often also termed flow encoding, that allows for quantification of the blood velocity vector in vivo. For velocity encoding along an arbitrary spatial direction ("encoding direction"), two acquisitions with velocity-encoding gradients generating two different m_1 values along that direction are being applied, yielding two images with different velocity sensitivities of the MR phase. Its phase difference is proportional to the velocity along the encoding direction, and the sensitivity of such phase difference with respect to the velocity encoding can be controlled by the difference of the first moments of the velocity-encoding gradients Δm_1. In practice, it is the maximum measurable velocity, also termed the *encoding velocity* (v_{enc}), that needs to be set in the sequence protocol and that is given by:

$$v_{enc} = \frac{\pi}{\gamma |\Delta m_1|} \tag{9.6}$$

The velocity range from $-v_{enc}$ to $+v_{enc}$ is mapped onto the $\pm\pi$ range in the phase-difference images. Typical v_{enc} values are about 100 cm/s for brain imaging and 150 cm/s or more for the thoracic aorta. Actual velocities larger than v_{enc} will be aliased as outlined later in this chapter; therefore, v_{enc} is typically set above the maximum expected velocity. However, the v_{enc} value should not be set unreasonably high as it increases the velocity noise σ_v:

$$\sigma_v = \frac{2}{\pi} \frac{v_{enc}}{SNR} \tag{9.7}$$

Here SNR denotes the signal-to-noise ratio in the magnitude data.

So far, only velocity encoding along a single encoding direction was considered. This is typically the case for slice-selective imaging, where only the through-plane velocity is encoded. Three-directional encoding of the velocity vector can be performed by applying m_1 values successively along different directions (x, y, z). The respective v_{enc} values do not need to match for all directions; a lower v_{enc} can be set along directions in which lower velocities are expected.

If three-directional encoding is needed, at least four acquisitions ("four-point encoding") with different velocity-encoding moments and directions are required to fully resolve the velocity vector. According to Eq. (9.6), v_{enc} depends on Δm_1 and not on the individual absolute m_1 values chosen for each encoding step. Thus, for a given v_{enc} value, several different solutions exist for how to choose m_1 for each encoding step. A straightforward approach consists in choosing one flow-compensated acquisition with $m_1 = 0$ for all gradient axes and subsequently three acquisitions with flow encoding enabled with gradient moments

$$m_1 = \frac{\pi}{\gamma v_{enc}}$$ for one of the three orthogonal directions i = u, v, w, respectively. The

directions u, v, and w correspond to readout-, phase-, and slice-encoding direction, but they do not necessarily coincide with the scanner's coordinate axes, x, y, and z. Shorter echo times while applying the same v_{enc} values can be achieved if the first acquisition applies an encoding moment of $\Delta m_1^i / 2$ in all three directions, and the same encoding is repeated for acquisition 2 to 4, but with inverted m_1 for one direction at a time. Beside these methods other encoding techniques such as balanced four-point encoding as well as five-point encoding have been investigated, which are explained in more detail in [6, 7].

9.3 4D Flow Acquisition

The three-directional velocity-encoding technique can be extended to acquire the data in a time-resolved manner in a 3D volume. This method is often termed "4D flow" MRI, with "4D" denoting the three-dimensional spatial imaging volume and the temporal dimension while the term "flow" is used synonymously for the three-directional encoding of the velocity vector. Note that in order to properly distinguish between the dimensions of velocity encoding and imaging volume, the term "dimension" refers to the spatial dimension of the imaging volume such as two dimensional for slice-selective imaging and three dimensional for volume-selective imaging, while the term "direction" refers to the velocity-encoding directions, such as one directional or three directional.

4D flow MRI requires data acquisition in synchrony with the cardiac cycle, which is typically realized using an ECG, a pulse oximeter (be aware that the detected R-wave using pulse oximetry ocurs later in time than the R-wave in your imaging volume, therefore ECG is the recommended triggering method), or other physiological units that detect the patient's heart beat. Most 4D flow sequences are based on a gradient echo acquisition with Cartesian or non-Cartesian readout.

9.3.1 Cartesian 4D Flow Acquisition Sequence

Figure 9.4 illustrates the acquisition scheme of a Cartesian 4D flow sequence using a basic four-point velocity encoding [8] in combination with ECG trigger and respiratory navigator. The cardiac cycle is divided into N_p cardiac phases, typically between 10 and 30 phases, and each phase covers a time frame of t_p = 20–50 ms. Each phase is subdivided into four velocity-encoding blocks: assuming a simple four-point velocity encoding, the first block acquires a reference signal with $m_1 = 0$ for all three axes at echo time. The same phase-encoding and partition line is acquired again in the following three blocks, but in each block a different first

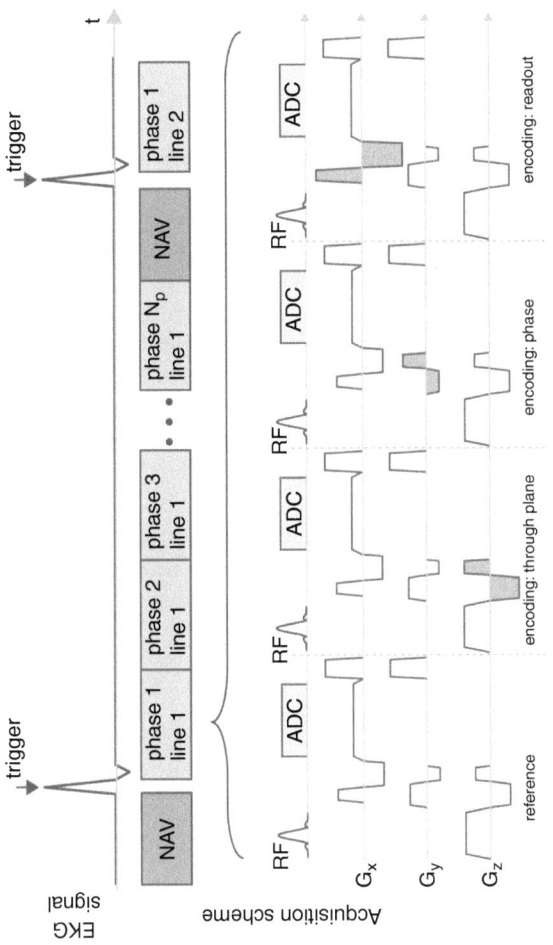

Fig. 9.4 4D flow acquisition scheme with corresponding physiological signal

moment m_1 is set for one of the axis G_x, G_y, and G_z according to the preset v_{enc} value for that axis (see gray-shaded gradient pulses in Fig. 9.4). After the acquisition of the fourth block are acquired again for the next cardiac phase, and this scheme is repeated until phase N_p is reached. If needed, a respiratory navigator can be acquired after phase N_p as discussed later in the following paragraph. The acquisition is then stopped until the beginning of the next cardiac cycle is detected by the MR scanner. The entire acquisition is repeated but with different phase-encoding gradients for spatial encoding of the volume in k-space (line 1, line 2, etc.).

9.3.2 Non-Cartesian Acquisition

4D flow MRI is not limited to Cartesian imaging. Non-Cartesian PC-based velocity-encoded imaging has been successfully applied, for example, using the phase-contrast "vastly undersampled isotropic projection reconstruction" (PC-VIPR) technique, which is based on a 3D radial acquisition scheme [6]. Here, the k-space is sampled along isotropically distributed radial spokes that pass through the k-space origin. In radial imaging the angle between neighboring spokes is typically given by the spatial resolution of the 3D image according to the Nyquist theorem. An important feature of radial imaging, however, is that the artifacts are more acceptable than in a Cartesian acquisition if fewer lines are acquired. Although the undersampling will introduce streaking artifacts, the object can still be resolved, and it has been shown that moderate streaking artifacts in PC-MRI are tolerable and that substantial acceleration is feasible. Therefore, the method can provide full 3D spatial coverage with high spatial isotropic resolution. Furthermore, radial acquisitions are less susceptible to motion, and navigator information can be derived from the radial data itself. On the other side, non-Cartesian acquisitions are typically more prone to eddy current effects and off-resonances. However, the opportunity to achieve high acceleration factors is highly attractive for future applications, particularly for high-resolution scans.

9.3.3 Respiration Control

When 4D flow MRI is applied for thoracic targets, respiratory motion will cause artifacts and blurring especially in Cartesian implementations. Most prominent methods to address respiratory motion are external respiration bellows or navigator techniques [9]. In both cases a respiratory acceptance window is defined typically around the end-expiratory or end-inspiratory phase, such that only acquisitions are being accepted if the current respiratory position falls into the defined acceptance window (Fig. 9.5). In case of the respiratory bellow, an elastic strap is wrapped around the upper abdomen that fixates the bellow to the body. Changes in the respiratory position will cause pressure changes within the bellow, which is detected by the MR scanner. In case of the respiratory navigator, the position of the diaphragm dome along the head-feet direction is recorded by a

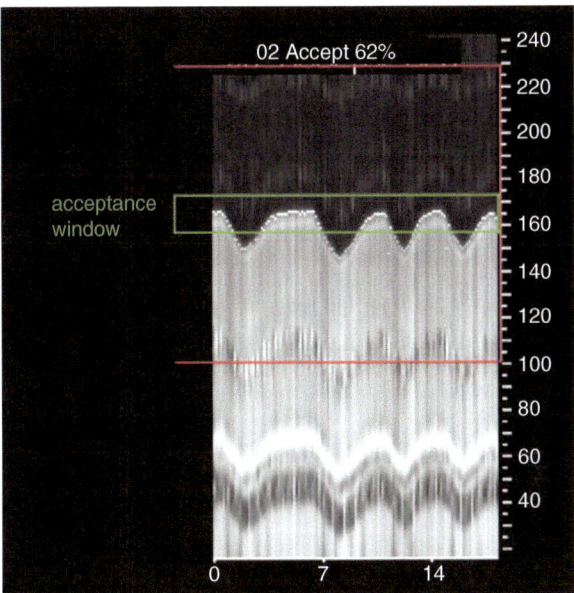

Fig. 9.5 Navigator signal displaying the motion of the diaphragm. The peaks of the curve reflect the exhale position, while the valleys indicated the inhale position of the diaphragm. Accepted positions for the image reconstruction are marked by the green window. The red window indicates the search region for the detection algorithm

one-dimensional acquisition in combination with a pencil beam excitation. This pencil beam is created either by 2D selective RF pulse or by a pair of 90° and 180° slice-selective pulses where only the intersection of the two slices generates signal. The navigator is typically obtained after the acquisition of the last cardiac phase N_p. Only if the diaphragm position falls within the defined acceptance window as illustrated in Fig. 9.5, the acquired data is accepted; otherwise it is rejected and automatically reacquired. Respiration control impacts the scan efficiency, which is defined as the number of accepted scans divided by the total number of accepted and rejected scans. Typical values are between 60 and 80% depending on several factors such as the acceptance width.

9.4   Resolution, Timing, and Practical Trade-Offs

Figure 9.4 shows the case where only a single segment ($N_{seg} = 1$), thus a single phase-encoding line, is acquired during each block in each cardiac cycle. This allows for high temporal resolution with a temporal spacing of $T_{res} = 4 \cdot TR$, with TR being the repetition time from RF pulse to RF pulse. Minimum TR values of 4–7 ms are feasible in 4D flow MRI, thus allowing for temporal resolutions of about $4 \times TR = 16$–28 ms. Note that even higher temporal resolutions could be achieved if the four encodings are acquired in different cardiac cycles. The disadvantage of high temporal resolutions is the resulting long scan time: assuming that no acceleration techniques are being applied, a dataset with 128×32 phase-encoding lines (along y and z) will require 4096 cardiac cycles, thus ~68 min assuming a heart rate of 60 beats/min and 100% scan efficiency. One way to reduce the total acquisition time (TA) consists in increasing the number of acquired k-space lines per block, which are often termed segments (N_{seg}). For example, $N_{seg} = 3$ will reduce TA approximately by a factor of 3, but to the cost of a reduced minimum temporal resolution with temporal spacing of $T_{res} = N_{seg} \times 4 \times TR = 48{-}84$ ms. Other factors that impact TA are the spatial resolution and spatial coverage particularly in phase-encoding and partition-encoding direction as well as the adjusted v_{enc}. Therefore, a trade-off between temporal resolution, spatial resolution, and spatial coverage needs to be found in practice in order to limit TA.

9.5   Acceleration Techniques

The total duration of 4D flow acquisitions may still be beyond 20 min despite increasing the number of segments to N_{seg}, as illustrated in the previous paragraph. Such TA is unacceptably long for clinical use and further acceleration is needed. MRI acceleration strategies are explained in detail in Chap. 8. Common acceleration methods such as GRAPPA or SENSE [10, 11] with acceleration factors between 2 and 3 are often included in the scans. Such acceleration techniques acquire less k-space lines as required by the Nyquist theorem and reconstruct the missing information via multiple receive coil elements. Higher acceleration factors can be achieved by combining GRAPPA or SENSE with acceleration in the temporal dimension, using techniques such as kt-GRAPPA [12, 13] or kt-SENSE [14, 15]. Acceleration factors of R = 5–8 have been used without significantly altering velocity curves in the clinical setting [16, 17]. Recently, also compressed sensing approaches have been investigated for higher acceleration exploiting the divergence-free properties of the flow vector field [18].

9.6   Post-processing, Visualization, and Quantification

Many studies over the last three decades have systematically validated the quantitative and qualitative analysis of blood flow using in vitro and in vivo experiments. It was shown that there are a number of sources of inaccuracies in PC-MRI, which can results in errors of measured velocities. The major sources of errors include eddy currents [19], Maxwell terms [20], gradient field distortions [21], and velocity aliasing [8]. Appropriate correction strategies implemented in the post-processing pipeline should thus be applied to ensure accurate flow quantification using PC-MRI data. Additional sources of error as a result of complex flow and inadequate timing of the flow encoding include acceleration effect and spatial displacement [22, 23]. Once data is corrected for errors, the complex data is reviewed by a radiologist, which is done qualitatively by inspecting the blood flow pattern and quantitatively based on derived hemodynamic parameters.

9.6.1   Maxwell Terms

4D flow MRI relies on switching magnetic field gradients for spatial encoding and for encoding the velocity. Switching a field gradient G_i always generates magnetic field components oriented in x- or y-direction, which cause the resulting B-field vector to not co-align with the z-axis except in isocenter. Due to these additional transverse components, the length of the residual B-field vector that determines the precession frequency of the spins exceeds the ideal B-field B_{ideal} $=B_0+\mathbf{Gr}=B_0+G_x x+G_y y+G_z z$ by an additional amount B_c, termed "concomitant field" or "Maxwell term" [20]. These concomitant fields create phase errors, which increase with distance from isocenter and thus affect the velocity measurements especially when protocols with large field of views are acquired. A prominent example is 4D flow MRI of the thoracic aorta. The concomitant magnetic field B_c can be derived using a Taylor expansion, and to the lowest order, it is given by:

$$Bc\left(x,y,z,t\right)=\frac{1}{2B_0}\left\{G_x^2 z^2+G_y^2 z^2+G_z^2 z^2\frac{x^2+y^2}{4}-G_x G_z xz-G_y G_z yz\right\} \quad (9.8)$$

The concomitant magnetic field of Eq. (9.8) comprises five concomitant gradient terms or Maxwell terms. The first three terms are the so-called "self-squared" type, since they involve the product of a gradient with itself. The last two terms are the concomitant gradient cross-terms since they require the simultaneous activation of the longitudinal and a transverse gradient. Note that B_c scales with $1/B_0$ and with G_i^2 and thus decreases with increasing magnetic field strength, but strongly increases with gradient field strength. The resulting accumulated phase due to the concomitant gradient terms in phase-contrast imaging can be calculated by:

$$\phi_c\left(x,y,z\right)=\gamma\int B_c\left(x,y,z,t\right)dt \quad (9.9)$$

The concomitant field errors can be described by a polynomial with the following equation in magnet-based spatial coordinates:

$$\Delta\phi\left(x,y,z\right)=Az^2+B\left(x^2+y^2\right)+Cxz+Dyz \quad (9.10)$$

The coefficients A–D can be calculated directly in the sequence program based on the knowledge of the applied gradient waveforms and passed to the reconstruction program, which applies the phase correction according to Eq. (9.10) in the image domain [20].

9.6.2 Gradient Field Nonlinearities

In addition to the aforementioned concomitant fields, there are secondary deviations of the magnetic field gradients with nonlinear behavior also increasing with distance from isocenter. These gradient field nonlinearities can cause image warping and require correction. In PC-MRI, this introduces errors in velocity measurements by affecting the first gradient moments used to encode velocity [21, 24], which can lead to considerable deviations between the designed and the actual velocity encoding. This does not only affect the magnitude of encoded velocities but also the velocity-encoding direction. Errors due to gradient field nonlinearities depend on spatial location and increase with distance to isocenter. In velocity magnitude these errors can be as high as 60%, while errors in the velocity-encoding direction can be up to 45°. Correction for gradient field nonlinearities can be performed during image reconstruction, based on the knowledge of the gradient field nonlinearities.

9.6.3 Eddy Currents

Switching of imaging and velocity-encoding gradients in PC-MRI results in changes in magnetic flux and thereby induces eddy currents in the conducting parts of the scanner system. These eddy currents can cause alterations of the desired gradient strengths and duration and thus result in spatially varying phase errors in the MR images [25]. For PC-MRI, the different gradient waveforms used for the subsequent velocity encodings lead to different eddy current-induced phase changes in the phase images of each velocity-encoded acquisition. As a result, subtraction of phase images from the reference phase does not take eddy current-related errors into account, and additional post-processing correction schemes are needed.

Several correction strategies have been proposed and are typically based on the subtraction of an estimation of the spatially varying eddy current-induced phase effects. The spatial variation of the phase difference in regions containing static tissue is used to calculate the eddy current-induced offset for the entire image. By fitting a spatially linear function or higher-order polynomial to the detected static tissue of the phase-difference data, the phase offset characteristics can be estimated and subsequently used to correct the entire image by subtraction (Fig. 9.6) [19]. Automatic correction algorithms have been reported to correct for eddy currents; however, for a reliable estimation of background offsets, user interaction may often be required to correctly identify static background signal. An alternative approach is the use of a dedicated magnetic field monitoring device to measure the phase evolution during the sequence and to correct for phase effects in the reconstruction [26]. However, this comes at the cost of additional hardware and advanced reconstruction techniques.

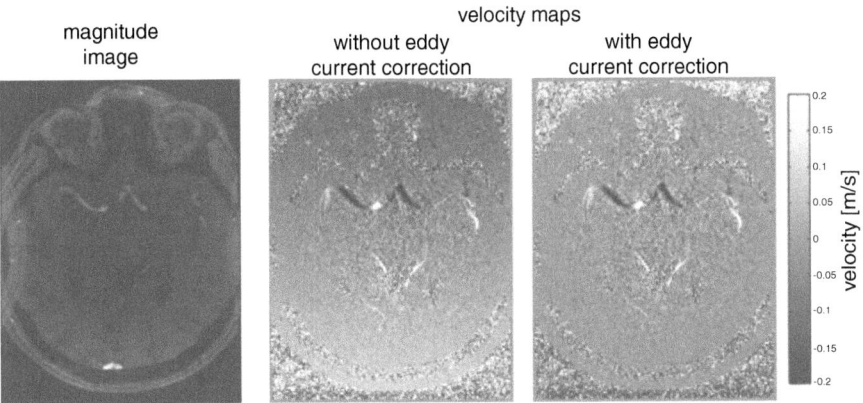

magnitude
image

velocity maps
without eddy
current correction

with eddy
current correction

Fig. 9.6 Impact of eddy currents on velocity maps. The left image shows the signal magnitude in axial orientation through parts of the middle cerebral arteries. The middle image shows uncorrected velocity maps with velocity encoding in anterior-posterior direction. Successful eddy current correction by fitting a spatially linear phase to the data that is subsequently subtracted from the phase data results in the right image

Fig. 9.7 Velocity aliasing for a parabolic flow profile. For an adjusted velocity sensitivity v_{enc} that is smaller than the highest occurring flow, velocities get wrapped to be inverted

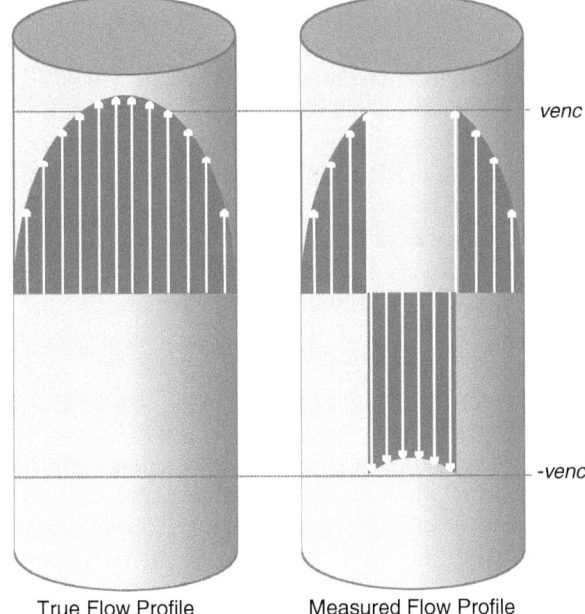

True Flow Profile   Measured Flow Profile

9.6.4   Velocity Aliasing and Noise

For each PC-MRI acquisition, the v_{enc} is set prior to acquisition to some maximum value at, for example, 50 or 100 cm/s, translating to an encoded phase difference of π. If, however, velocities in the acquired slice or volume exist with values

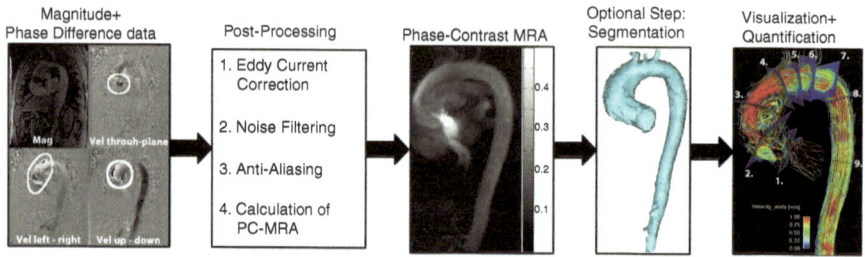

Fig. 9.8 Post-processing workflow example of a patient with a unicuspid aortic valve. Starting from the phase-difference and magnitude data, the images are first corrected for eddy currents, noise, and aliasing. The example shown here shows aliasing artifacts in the unprocessed phase-difference data (white circles in left panel). Then a PC-MRA is calculated, which is then either further segmented or directly used to mask the velocity vectors within vessel boundaries only. Resulting images are used to visualize the blood flow (here stationary streamlines at peak systole with color coding the speed from blue = slow to red = fast) and quantify the data. Quantification of net flow, peak velocity, and backward flow is typically done using analysis planes positioned perpendicular to the vessel (right panel). In this patient nine planes at (1) the aortic root, (2) the sino-tubular junction, (3) mid-ascending aorta, (4) distal ascending aorta, (5) proximal aortic arch at the brachiocephalic trunk, (6) mid aortic arch between brachiocephalic trunk and left common carotid artery, (7) distal aortic arch between left common carotid artery and left subclavian artery, (8) proximal descending aorta, and (9) mid-thoracic descending aorta

greater than this maximum, then the phase will be aliased, meaning it falls outside the $[-\pi, \pi]$ interval and is wrapped back onto itself (Fig. 9.7) [8]. Phase wrapping causes problems in flow quantification and in determining the flow direction (in vivo example in left panel). If the maximum likely velocity is known a priori, then the v_{enc} can be adjusted to prevent aliasing. However, phase-contrast velocity images suffer from noise that can lead to errors in the acquired velocities. Noise in the velocity-encoded images (σ_v) is inversely related to the signal-to-noise ratio (SNR) in the corresponding magnitude images ($\sigma_v \sim 1/SNR$, compare Eq. (9.7)). For a given SNR, the velocity noise is thus determined by the user selected v_{enc}, resulting in a trade-off between the minimum v_{enc} needed to detect velocities without aliasing and with low noise. For optimal noise performance, v_{enc} should therefore always be selected as small as possible. Such an approach sometimes does result in velocity aliasing, which needs to be corrected in another post-processing step. Several algorithms and strategies have been introduced over the last couple of years, for example, a technique based on the Poisson solution to unwrap the phase [27]. The method is based on the assumption that the magnitude of the phase change between adjacent voxels is less than π per voxel. From this an estimate of the phase gradient can be obtained by wrapping the gradient of the original phase image. The remaining problem is then to obtain the absolute phase given the estimate of the phase gradient. The least-squares solution to this problem is a solution of the Poisson equation and can be solved with a Poisson solver. The absolute phase can then be obtained by mapping the least-squares phase to the nearest multiple of 2π from the measured phase. Recently, the usage of dual- or multi-v_{enc} approaches to correct for aliasing while increasing SNR gained popularity as discussed in the outlook [28–33].

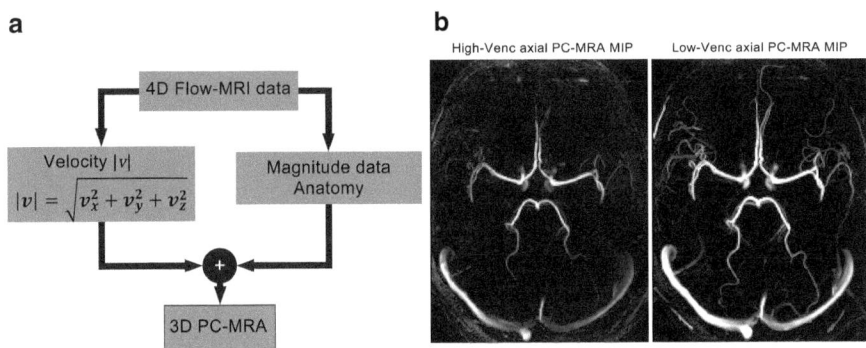

Fig. 9.9 (**a**) Calculation of a 3D PC-MRA from 4D Flow MRI data. (**b**). Axial views of maximum intensity projections of a PC-MRA in the head with two different v_{enc} (left, high v_{enc} = 140 cm/s; right, low v_{enc} = 45 cm/s) showing the effect of increased sensitivity to slow velocities and decreased noise a low v_{enc}

9.6.5   Post-Processing Workflow and Phase-Contrast Angiogram

Once the phase-difference and magnitude images are reconstructed, a typical post-processing workflow starts with eddy current correction using either the abovementioned additional measurement of the non-flow situation or the estimation of the offset using a linear or polynomial fit on the detected static tissue [19] (Fig. 9.8). This is followed by anti-aliasing of wrapped velocities with either an automatic method based on the assumption that adjacent voxel velocities in temporal or slice direction should not differ by more than v_{enc} or by manually selecting aliased voxels and adding v_{enc}, until the velocity is visually similar to the adjacent voxels. In a following step, noise masking is typically performed by thresholding the signal intensity in the magnitude images to exclude regions with low signal intensity such as air in the lungs or outside the vessels [19]. This step could also be omitted, since too aggressive noise filtering could result in falsely deleting voxels within the vessels.

Beyond measuring the three-dimensional velocity vector field using 4D flow MRI, the technique can also be used to calculate a 3D phase-contrast MR angiogram (PC-MRA). A PC-MRA provides detailed information on the vascular geometry [34, 35]. Several strategies exist for calculating a PC-MRA from the flow MRI data [34–36]. The principle of PC-MRA is to combine the magnitude image to suppress image background with the velocity magnitude to highlight the vessels (Fig. 9.9a). It is important to note that the depiction of the vessels is determined by the choice of the velocity sensitivity v_{enc}. The maximum intensity projections (MIP) of the cranial vessels in Fig. 9.9b demonstrate the effect of the velocity sensitivity on PC-MRA data. Small vessels with slow flow are more clearly visible using a smaller v_{enc} factor (right panel Fig. 9.9b). Larger vessels with higher blood flow velocities demonstrate a decreased signal due to aliasing of the signal phase. For increased v_{enc} factors, the visualization of larger arterial vessels is improved, and venous or slow flow signal is suppressed. The quality of the depicted vessels is best with an as low as possible selected velocity sensitivity, which still allows recovering aliased voxels using unwrapping strategies.

9.6.6   Visualization and Quantification of 4D Flow MRI Data

A first qualitative analysis of the data requires the visualization of the blood flow. Visualization and quantification of blood flow have been widely done in a number of applications and have proven to be useful tools for the assessment of blood flow within the cardiac- and neurovascular system [37–45].

Traditionally, MRI imaging of flow is accomplished using individual 2D slices with one-directional (through-plane) velocity-encoding direction. Such methods are typically used for blood flow quantification in the heart and great vessels. Applications include the assessment of left ventricular performance (e.g., cardiac output), regurgitation volumes in case of valve insufficiency, or evaluation of flow acceleration in stenotic regions (e.g., aortic valve stenosis). Data analysis is done using semiautomatic segmentation of the vascular lumen of interest and calculation of time-resolved blood flow from mean flow velocities and vascular cross-sectional area.

Alternatively, 3D spatial encoding offers the possibility of high spatial, isotropic resolution and the ability to measure and visualize the temporal evolution of complex flow and motion patterns in a 3D volume [38, 46–49]. This is typically done with third party commercial software. To visualize the blood flow, a first step is to restrict the velocities within the vessel boundaries for which the PC-MRA is employed. For some applications a thorough segmentation of the vessel boundaries using the PC-MRA as source images is advantageous (optional segmentation step, Fig. 9.8). This, however, means an additional step of image processing, and the usage of additional software with advanced segmentation capability is often necessary. However, many clinical questions can already be answered by just using the boundaries from the calculated PC-MRA as is. Visualization of blood flow can be achieved by displaying the three-dimensional vector field, time-resolved pathlines within the 3D volume or stationary streamlines for each individual time frame. Basic parameters such as net flow (ml/cycle), peak or mean velocity (m/s), and regurgitant fraction can be determined at specific locations by positioning analysis planes (visualization + quantification) [50–52]. In specific clinical applications, only the evaluation of peak velocity may be of interest, for example, for the evaluation of aortopathy for patients with bicuspid aortic valves. In this case a quick MIP analysis with automatic detection of peak velocity in addition to blood flow visualization could be sufficient for clinical purposes [53]. Since 4D flow MRI data reflect the true underlying time-resolved blood flow velocity vector field, it is possible to calculate advanced parameters from the data such as pressure difference maps [54, 55] between two points within a vessel computed from Navier-Stokes equations (see Eq. (2.50) in Chap. 2) or wall shear stress [56, 57] describing the tangential drag force on the vessel wall induced from flowing blood (more reading is provided in Chap. 3). Furthermore, parameters have been proposed to characterize turbulent flow and kinetic energy [28, 58, 59]. All these advanced parameters require abovementioned thorough segmentation of the vessel wall as well as sufficient spatial (at least 5–6 voxels across the vessel, ideally in the submillimeter range) [60, 61] and temporal resolution of the acquired data set (below 50 ms allowing the error to be below 10%) [61].

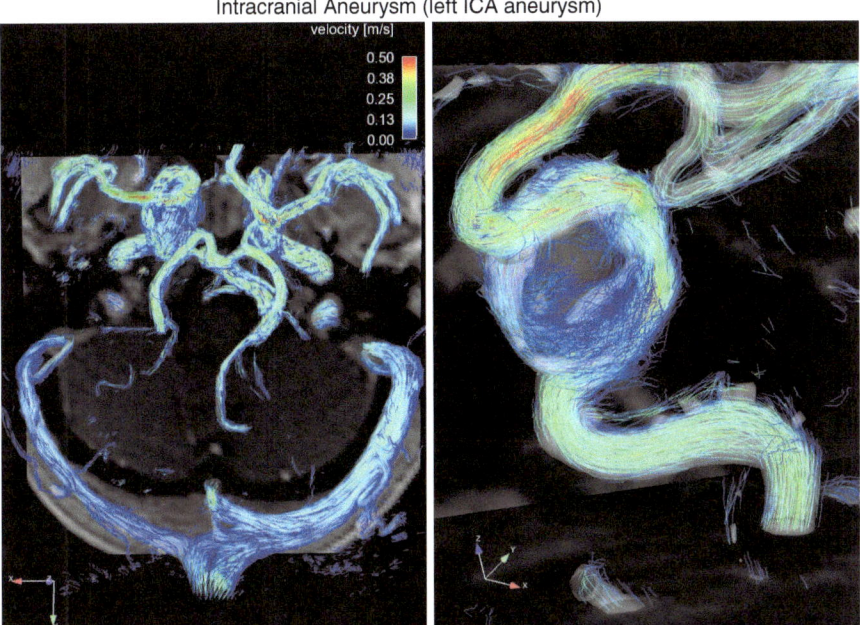

Fig. 9.10 Flow-sensitive 4D MRI in the brain. 3D flow visualization in the large intracranial arteries in the circle of Willis using streamlines at peak systole (color scale from blue to red, with red being fast velocities). The patient presented with a giant intracranial aneurysm at the ICA (internal carotid artery). The magnitude is shown in the background for anatomical reference. The tortuous routes of the different vascular segments and changes in regional velocities can clearly be appreciated

9.7   Clinical Application for Intracranial Aneurysms

Many vascular regions have been investigated with 4D flow MRI. In this section however, we concentrate on one specific application, which uses many of described concepts. Figure 9.10 illustrates the potential of 4D flow MRI to assess, visualize, and quantify flow characteristics intracranially.

Intracranial aneurysms (IA) are focal dilations of arteries in the cerebral vasculature (Fig. 9.10. The worldwide prevalence of IA without adjustment for IA risk factors is about 3%, [62] and the annual incidence of IA rupture resulting in subarachnoid hemorrhage is about 1% [63]. The mortality rate of subarachnoid hemorrhage resulting from IA rupture without intervention ranges from about 35% to 65%, and many survivors experience moderate to severe lifelong disabilities. Treatment for IA involving surgical clipping or endovascular coiling of the aneurysm has an associated 1-year mortality of about 3% and morbidity.

Research using 4D flow MRI to study the hemodynamics of IA may yield more accurate patient risk stratification schemes by studying the fundamental characteristics of IA that make them more prone to rupture. Many studies investigated, for example, flow pattern within the aneurysm [64, 65], flow residence time and its influence on thrombus formation within the aneurysm [66], as well as mean and peak velocity of different aneurysm morphologies (saccular versus fusiform shaped) [36]. Many studies also investigated wall shear stress (WSS, frictional force exerted by blood flow along the artery wall) as potential additional risk or growth factor to predict aneurysms rupture [36, 67]. Steiger et al. [68] showed that aneurysm growth could be understood as a passive yield to blood pressure and reactive healing and thickening of the wall with increasing aneurysm diameter. There are contradicting theories about the effects of WSS. The high-flow theory indicates WSS elevation results in aneurysm growth [69–72] and the low-flow theory with low WSS at the aneurysm wall assumes progressive thinning until rupture [73, 74]. At present, 4D flow MRI applied to intracranial aneurysms is still a research tool until hemodynamic factors are completely understood. It constitutes an important tool to develop new imaging biomarkers and has promising potential to serve as predictive tool in the future.

9.8   Limitations

Most imaging techniques have limitations, which is also the case for 4D flow MRI. Major limitations are limited spatial resolution, long scan times, and limited SNR. The combination of these issues limits the wide use of 4D flow MRI in clinical routine. Improving one limitation may result in worsening another, since all imaging parameters are dependent. For example, increasing spatial resolution will extend scan time and decrease SNR. However, even if low spatial resolutions are chosen to shorten the acquisition to scan times below 10 min (considered as being clinically acceptable), such scan times are typically not achievable without using advanced acceleration of image acquisition. Therefore, a key challenge is to push current techniques to achieve higher acceleration factors or to develop novel, fast flow quantification methods. New acquisition methods based on the combination of phase-contrast MRI and fast sampling strategies such as spiral or radial imaging with advanced acceleration methods have been proposed and shown promising results [6, 16, 75–81].

Another limitation is given by the fact that 4D flow MRI acquires data over a period of several minutes. The acquisition will therefore reflect hemodynamic properties averaged over the entire acquisition time and may miss temporally changing features of one heart beat versus another or introduce imaging artifacts. In many cases, ECG triggering is used instead of gating hindering the acquisition during the R-wave. Although real-time phase-contrast MRI has been demonstrated for slice-selective velocity imaging [82–84], 4D flow MRI in real time is not feasible at present.

Furthermore, the wide use of 4D flow MRI in clinical routine is also limited by the amount of (semi-) automatic post-processing steps required for deriving the parameters of interest from the phase-contrast data. Many of such steps are still semiautomatic and require user interaction. Vendors are continuously working on streamlining this process to improve the post-processing pipeline for clinical routine applications.

Limited SNR and therefore high velocity noise are particularly challenging in applications requiring high dynamic velocity ranges. This is the case in some pathologies, such as aneurysms and stenosis or if the venous and the arterial system is of interest at the same time. Such pathologies are characterized by high velocity jets adjacent to regions with slow blood velocity. High SNR is needed to reduce the velocity noise far enough to simultaneously capture slow and fast velocities simultaneously.

Addressing the abovementioned limitations is subject of numerous research projects worldwide conducted by an active research community in collaboration with the MR vendors.

9.9 Conclusions and Outlook

4D flow MR imaging offers the unique ability to directly quantify the blood velocity vector field in vivo. The technique has great potential for a large range of vascular diseases, which is being investigated in many studies.

Recently, initial studies have demonstrated that 4D flow MRI benefits from 7 T ultrahigh field MR imaging [67, 85]. The gain compared to clinical field strengths of 1.5 T and 3 T is multifold. A 2.2–2.6-fold increase in SNR compared to 3 T has been reported, which translates into a 5.5-fold increase of SNR compared to 1.5 T [86]. Note that such increases achieved by averaging multiple acquisitions would require a 4.8-fold (3 T) and more than 30-fold (1.5 T) increase in measurement time. Another benefit is given by increased T_1 relaxation constant values at 7 T [87], which results in a stronger suppression of the background tissue and thus in lower background steady-state signal, while the blood flowing into the imaging slab provides strong signal. The resulting increased vessel-to-background contrast is expected to benefit the segmentation process. In addition to previous advantages of ultrahigh field MRI, the increased field strength also improves parallel imaging performance and therefore allows for higher acceleration factors [88, 89].

According to Eq. (9.7), lower velocity noise can also be achieved by reducing v_{enc} values to the cost of potential phase wraps in the velocity data. In order to prevent phase wraps in the velocity data, dual-v_{enc}, or multi-v_{enc}, approaches have been proposed and are presently under investigation to facilitate both high SNR and a high dynamic range [28, 32, 33]. The high-v_{enc}, dataset can then be utilized to unwrap the low-v_{enc}, data, which allows low velocity noise for the entire dataset [31, 32]. The trade-off of such approaches is longer scan time.

The present chapter has outlined the fundamental techniques of 4D flow MR imaging, the post-processing pipeline, clinical applications, as well as the limitations. It should be noted, however, that a large range of advanced techniques beyond those presented here exist, such as the ability of decoding acceleration and higher orders, the decoding of velocity using Fourier-based methods, or the calculation of further parameters such as pressure differences. For such advanced topics, the reader is referred to current literature.

Acknowledgments We would like to thank Dr. Kelvin Chow (Siemens Inc.) for providing the figure of the Scout Navigator, Carmen Blanken for the analysis and segmentation of the patient with the unicuspid aortic valve for the figure explaining the post-processing workflow, and Simon Schmidt, German Cancer Research Center, Heidelberg, Germany for providing the figure explaining the flow compensation. We also acknowledge the following funding sources: AHA Scientist Development Grant 16SDG30420005; NIH NHLBI R01HL115828.

References

1. Singer JR. Blood flow rates by nuclear magnetic resonance measurements. Science. 1959;130(3389):1652–3.
2. Moran PR. A flow velocity zeugmatographic interlace for NMR imaging in humans. Magn Reson Imaging. 1982;1(4):197–203.
3. Nayler GL, Firmin DN, Longmore DB. Blood flow imaging by cine magnetic resonance. J Comput Assist Tomogr. 1986;10(5):715–22.
4. Markl M, Frydrychowicz A, Kozerke S, Hope M, Wieben O. 4D flow MRI. J Magn Reson Imaging JMRI. 2012;36(5):1015–36.
5. Bernstein MA, Shimakawa A, Pelc NJ. Minimizing TE in moment-nulled or flow-encoded two- and three-dimensional gradient-echo imaging. J Magn Reson Imaging. 1992;2(5):583–8.
6. Johnson KM, Lum DP, Turski PA, Block WF, Mistretta CA, Wieben O. Improved 3D phase contrast MRI with off-resonance corrected dual echo VIPR. Magn Reson Med. 2008;60(6):1329–36.
7. Markl M, Wallis W, Brendecke S, Simon J, Frydrychowicz A, Harloff A. Estimation of global aortic pulse wave velocity by flow-sensitive 4D MRI. Magn Reson Med. 2010;63(6):1575–82.
8. Pelc NJ, Bernstein MA, Shimakawa A, Glover GH. Encoding strategies for three-direction phase-contrast MR imaging of flow. J Magn Reson Imaging. 1991;1(4):405–13.
9. McConnell MV, Khasgiwala VC, Savord BJ, Chen MH, Chuang ML, Edelman RR, et al. Comparison of respiratory suppression methods and navigator locations for MR coronary angiography. Am J Roentgenol. 1997;168(5):1369–75.
10. Griswold MA, Jakob PM, Heidemann RM, Nittka M, Jellus V, Wang J, et al. Generalized auto-calibrating partially parallel acquisitions (GRAPPA). Magn Reson Med. 2002;47(6):1202–10.
11. Pruessmann KP, Weiger M, Scheidegger MB, Boesiger P. SENSE: sensitivity encoding for fast MRI. Magn Reson Med. 1999;42(5):952–62.
12. Huang F, Akao J, Vijayakumar S, Duensing GR, Limkeman M. k-t GRAPPA: a k-space implementation for dynamic MRI with high reduction factor. Magn Reson Med. 2005;54(5):1172–84.
13. Jung B, Ullmann P, Honal M, Bauer S, Hennig J, Markl M. Parallel MRI with extended and averaged GRAPPA kernels (PEAK-GRAPPA): optimized spatiotemporal dynamic imaging. J Magn Reson Imaging. 2008;28(5):1226–32.
14. Tsao J, Boesiger P, Pruessmann KP. k-t BLAST and k-t SENSE: dynamic MRI with high frame rate exploiting spatiotemporal correlations. Magn Reson Med. 2003;50(5):1031–42.
15. Giese D, Schaeffter T, Kozerke S. Highly undersampled phase-contrast flow measurements using compartment-based k-t principal component analysis. Magn Reson Med. 2013;69(2):434–43.
16. Schnell S, Markl M, Entezari P, Mahadewia RJ, Semaan E, Stankovic Z, et al. k-t GRAPPA accelerated four-dimensional flow MRI in the aorta: effect on scan time, image quality, and quantification of flow and wall shear stress. Magn Reson Med. 2014;72(2):522–33.
17. Giese D, Wong J, Greil GF, Buehrer M, Schaeffter T, Kozerke S. Towards highly accelerated Cartesian time-resolved 3D flow cardiovascular magnetic resonance in the clinical setting. J Cardiovasc Magn Reson. 2014;16:42.
18. Santelli C, Loecher M, Busch J, Wieben O, Schaeffter T, Kozerke S. Accelerating 4D flow MRI by exploiting vector field divergence regularization. Magn Reson Med. 2016;75(1):115–25.
19. Walker PG, Cranney GB, Scheidegger MB, Waseleski G, Pohost GM, Yoganathan AP. Semiautomated method for noise reduction and background phase error correction in MR phase velocity data. J Magn Reson Imaging. 1993;3(3):521–30.

20. Bernstein MA, Zhou XJ, Polzin JA, King KF, Ganin A, Pelc NJ, et al. Concomitant gradient terms in phase contrast MR: analysis and correction. Magn Reson Med. 1998;39(2):300–8.

21. Markl M, Bammer R, Alley MT, Elkins CJ, Draney MT, Barnett A, et al. Generalized reconstruction of phase contrast MRI: analysis and correction of the effect of gradient field distortions. Magn Reson Med. 2003;50(4):791–801.

22. Thunberg P, Wigstrom L, Wranne B, Engvall J, Karlsson M. Correction for acceleration-induced displacement artifacts in phase contrast imaging. Magn Reson Med. 2000;43(5):734–8.

23. Thunberg P, Wigstrom L, Ebbers T, Karlsson M. Correction for displacement artifacts in 3D phase contrast imaging. J Magn Reson Imaging. 2002;16(5):591–7.

24. Peeters JM, Bos C, Bakker CJ. Analysis and correction of gradient nonlinearity and B0 inhomogeneity related scaling errors in two-dimensional phase contrast flow measurements. Magn Reson Med. 2005;53(1):126–33.

25. McCauley TR, Pena CS, Holland CK, Price TB, Gore JC. Validation of volume flow measurements with cine phase-contrast MR imaging for peripheral arterial waveforms. J Magn Reson Imaging. 1995;5(6):663–8.

26. Giese D, Haeberlin M, Barmet C, Pruessmann KP, Schaeffter T, Kozerke S. Analysis and correction of background velocity offsets in phase-contrast flow measurements using magnetic field monitoring. Magn Reson Med. 2012;67(5):1294–302.

27. Song SMH, Napel S, Pelc NJ, Glover GH. Phase unwrapping of MR Phase images using Poisson equation. IEEE Trans Image Process. 1995;4(5):667–76.

28. Binter C, Knobloch V, Manka R, Sigfridsson A, Kozerke S. Bayesian multipoint velocity encoding for concurrent flow and turbulence mapping. Magn Reson Med. 2013;69(5):1337–45.

29. Callaghan FM, Kozor R, Sherrah AG, Vallely M, Celermajer D, Figtree GA, et al. Use of multi-velocity encoding 4D flow MRI to improve quantification of flow patterns in the aorta. J Magn Reson Imaging. 2015;43(2):352–63.

30. Ha H, Kim GB, Kweon J, Kim YH, Kim N, Yang DH, et al. Multi-VENC acquisition of four-dimensional phase-contrast MRI to improve precision of velocity field measurement. Magn Reson Med. 2015;75(5):1909–19.

31. Nett EJ, Johnson KM, Frydrychowicz A, Del Rio AM, Schrauben E, Francois CJ, et al. Four-dimensional phase contrast MRI with accelerated dual velocity encoding. J Magn Reson Imaging. 2012;35(6):1462–71.

32. Schnell S, Ansari SA, Wu C, Garcia J, Murphy IG, Rahman OA, Rahsepar AA, Aristova M, Collins JD, Carr JC, Markl M. Accelerated dual-venc 4D flow MRI for neurovascular applications. J Magn Reson Imaging. 2017;46(1):102–14.

33. Lee AT, Pike GB, Pelc NJ. Three-point phase-contrast velocity measurements with increased velocity-to-noise ratio. Magn Reson Med. 1995;33(1):122–6.

34. Bernstein MA, Grgic M, Brosnan TJ, Pelc NJ. Reconstructions of phase contrast, phased array multicoil data. Magn Reson Med. 1994;32(3):330–4.

35. Dumoulin CL. Phase contrast MR angiography techniques. Magn Reson Imaging Clin N Am. 1995;3(3):399–411.

36. Schnell S, Ansari SA, Vakil P, Wasielewski M, Carr ML, Hurley MC, et al. Three-dimensional hemodynamics in intracranial aneurysms: influence of size and morphology. J Magn Reson Imaging. 2014;39(1):120–31.

37. Hangiandreou NJ, Rossman PJ, Riederer SJ. Analysis of MR phase-contrast measurements of pulsatile velocity waveforms. J Magn Reson Imaging. 1993;3(2):387–94.

38. Kilner PJ, Yang GZ, Mohiaddin RH, Firmin DN, Longmore DB. Helical and retrograde secondary flow patterns in the aortic arch studied by three-directional magnetic resonance velocity mapping. Circulation. 1993;88(5 Pt 1):2235–47.

39. Mohiaddin RH, Kilner PJ, Rees S, Longmore DB. Magnetic resonance volume flow and jet velocity mapping in aortic coarctation. J Am Coll Cardiol. 1993;22(5):1515–21.

40. Bogren HG, Mohiaddin RH, Kilner PJ, Jimenez-Borreguero LJ, Yang GZ, Firmin DN. Blood flow patterns in the thoracic aorta studied with three-directional MR velocity mapping: the effects of age and coronary artery disease. J Magn Reson Imaging. 1997;7(5):784–93.

41. Kilner PJ, Yang GZ, Wilkes AJ, Mohiaddin RH, Firmin DN, Yacoub MH. Asymmetric redirection of flow through the heart. Nature. 2000;404(6779):759–61.
42. Korperich H, Gieseke J, Barth P, Hoogeveen R, Esdorn H, Peterschroder A, et al. Flow volume and shunt quantification in pediatric congenital heart disease by real-time magnetic resonance velocity mapping: a validation study. Circulation. 2004;109(16):1987–93.
43. Alaraj A, Amin-Hanjani S, Shakur SF, Aletich VA, Ivanov A, Carlson AP, et al. Quantitative assessment of changes in cerebral arteriovenous malformation hemodynamics after embolization. Stroke. 2015;46(4):942–7.
44. Wu C, Honarmand AR, Schnell S, Kuhn R, Schoeneman SE, Ansari SA, et al. Age-related changes of normal cerebral and cardiac blood flow in children and adults aged 7 months to 61 years. J Am Heart Assoc. 2016;5(1):e002657.
45. Rivera-Rivera LA, Turski P, Johnson KM, Hoffman C, Berman SE, Kilgas P, et al. 4D flow MRI for intracranial hemodynamics assessment in Alzheimer's disease. J Cereb Blood Flow Metab. 2016;36:1718–30.
46. Wigstrom L, Sjoqvist L, Wranne B. Temporally resolved 3D phase-contrast imaging. Magn Reson Med. 1996;36(5):800–3.
47. Kozerke S, Hasenkam JM, Pedersen EM, Boesiger P. Visualization of flow patterns distal to aortic valve prostheses in humans using a fast approach for cine 3D velocity mapping. J Magn Reson Imaging. 2001;13(5):690–8.
48. Bogren HG, Buonocore MH. 4D magnetic resonance velocity mapping of blood flow patterns in the aorta in young vs. elderly normal subjects. J Magn Reson Imaging. 1999;10(5):861–9.
49. Markl M, Draney MT, Hope MD, Levin JM, Chan FP, Alley MT, et al. Time-resolved 3-dimensional velocity mapping in the thoracic aorta: visualization of 3-directional blood flow patterns in healthy volunteers and patients. J Comput Assist Tomogr. 2004;28(4):459–68.
50. Schnell S, Entezari P, Mahadewia RJ, Malaisrie SC, McCarthy PM, Collins JD, et al. Improved semiautomated 4D flow MRI analysis in the aorta in patients with congenital aortic valve anomalies versus tricuspid aortic valves. J Comput Assist Tomogr. 2016;40(1):102–8.
51. Frydrychowicz A, Harloff A, Jung B, Zaitsev M, Weigang E, Bley TA, et al. Time-resolved, 3-dimensional magnetic resonance flow analysis at 3 T: visualization of normal and pathological aortic vascular hemodynamics. J Comput Assist Tomogr. 2007;31(1):9–15.
52. Clough RE, Waltham M, Giese D, Taylor PR, Schaeffter T. A new imaging method for assessment of aortic dissection using four-dimensional phase contrast magnetic resonance imaging. J Vasc Surg. 2012;55(4):914–23.
53. Rose MJ, Jarvis K, Chowdhary V, Barker AJ, Allen BD, Robinson JD, et al. Efficient method for volumetric assessment of peak blood flow velocity using 4D flow MRI. J Magn Reson Imaging. 2016;44(6):1673–82.
54. Tyszka JM, Laidlaw DH, Asa JW, Silverman JM. Three-dimensional, time-resolved (4D) relative pressure mapping using magnetic resonance imaging. J Magn Reson Imaging. 2000;12(2):321–9.
55. Ebbers T, Wigstrom L, Bolger AF, Engvall J, Karlsson M. Estimation of relative cardiovascular pressures using time-resolved three-dimensional phase contrast MRI. Magn Reson Med. 2001;45(5):872–9.
56. Potters WV, van Ooij P, Marquering H, vanBavel E, Nederveen AJ. Volumetric arterial wall shear stress calculation based on cine phase contrast MRI. J Magn Reson Imaging. 2015;41(2):505–16.
57. Stalder AF, Russe MF, Frydrychowicz A, Bock J, Hennig J, Markl M. Quantitative 2D and 3D phase contrast MRI: optimized analysis of blood flow and vessel wall parameters. Magn Reson Med. 2008;60(5):1218–31.
58. Dyverfeldt P, Sigfridsson A, Kvitting JP, Ebbers T. Quantification of intravoxel velocity standard deviation and turbulence intensity by generalizing phase-contrast MRI. Magn Reson Med. 2006;56(4):850–8.
59. Wong J, Chabiniok R, deVecchi A, Dedieu N, Sammut E, Schaeffter T, et al. Age-related changes in intraventricular kinetic energy: a physiological or pathological adaptation? Am J Physiol Heart Circ Physiol. 2016;310(6):H747–55.

60. Hofman MB, Visser FC, van Rossum AC, Vink QM, Sprenger M, Westerhof N. In vivo valida-
    tion of magnetic resonance blood volume flow measurements with limited spatial resolution in
    small vessels. Magn Reson Med. 1995;33(6):778–84.
61. Cibis M, Potters WV, Gijsen FJ, Marquering H, van Ooij P, Vanbavel E, et al. The effect of
    spatial and temporal resolution of cine phase contrast MRI on wall shear stress and oscillatory
    shear index assessment. PLoS One. 2016;11(9):e0163316.
62. Wiebers DO, Whisnant JP, Huston J 3rd, Meissner I, Brown RD Jr, Piepgras DG, et al.
    Unruptured intracranial aneurysms: natural history, clinical outcome, and risks of surgical and
    endovascular treatment. Lancet. 2003;362(9378):103–10.
63. Rinkel GJ, Djibuti M, Algra A, van Gijn J. Prevalence and risk of rupture of intracranial aneu-
    rysms: a systematic review. Stroke. 1998;29(1):251–6.
64. Hollnagel DI, Summers PE, Poulikakos D, Kollias SS. Comparative velocity investigations in
    cerebral arteries and aneurysms: 3D phase-contrast MR angiography, laser Doppler velocim-
    etry and computational fluid dynamics. NMR Biomed. 2009;22(8):795–808.
65. Marquering H, Van Ooij P, Streekstra G, Schneiders J, Majoie C, Vanbavel E, et al. Multi-
    scale flow patterns within an intracranial aneurysm phantom. IEEE Trans Biomed Eng.
    2011;58(12):3447–50.
66. Rayz VL, Boussel L, Ge L, Leach JR, Martin AJ, Lawton MT, et al. Flow residence time
    and regions of intraluminal thrombus deposition in intracranial aneurysms. Ann Biomed Eng.
    2010;38(10):3058–69.
67. van Ooij P, Zwanenburg JJ, Visser F, Majoie CB, vanBavel E, Hendrikse J, et al. Quantification
    and visualization of flow in the Circle of Willis: time-resolved three-dimensional phase con-
    trast MRI at 7 T compared with 3 T. Magn Reson Med. 2013;69(3):868–76.
68. Steiger HJ, Aaslid R, Keller S, Reulen HJ. Strength, elasticity and viscoelastic properties of
    cerebral aneurysms. Heart Vessel. 1989;5(1):41–6.
69. Nakatani H, Hashimoto N, Kang Y, Yamazoe N, Kikuchi H, Yamaguchi S, et al. Cerebral
    blood flow patterns at major vessel bifurcations and aneurysms in rats. J Neurosurg.
    1991;74(2):258–62.
70. Fukuda S, Hashimoto N, Naritomi H, Nagata I, Nozaki K, Kondo S, et al. Prevention
    of rat cerebral aneurysm formation by inhibition of nitric oxide synthase. Circulation.
    2000;101(21):2532–8.
71. Hara A, Yoshimi N, Mori H. Evidence for apoptosis in human intracranial aneurysms. Neurol
    Res. 1998;20(2):127–30.
72. Sho E, Sho M, Singh TM, Xu C, Zarins CK, Masuda H. Blood flow decrease induces apop-
    tosis of endothelial cells in previously dilated arteries resulting from chronic high blood flow.
    Arterioscler Thromb Vasc Biol. 2001;21(7):1139–45.
73. Griffith TM. Modulation of blood flow and tissue perfusion by endothelium-derived relaxing
    factor. Exp Physiol. 1994;79(6):873–913.
74. Liepsch DW. Flow in tubes and arteries – a comparison. Biorheology. 1986;23(4):395–433.
75. Pike GB, Meyer CH, Brosnan TJ, Pelc NJ. Magnetic resonance velocity imaging using a fast
    spiral phase contrast sequence. Magn Reson Med. 1994;32(4):476–83.
76. Thompson RB, McVeigh ER. Flow-gated phase-contrast MRI using radial acquisitions. Magn
    Reson Med. 2004;52(3):598–604.
77. Hutter J, Schmitt P, Aandal G, Greiser A, Forman C, Grimm R, et al. Low-rank and sparse
    matrix decomposition for compressed sensing reconstruction of magnetic resonance 4D phase
    contrast flow imaging (loSDeCoS 4D-PCI). Med Image Comput Comput Assist Interv.
    2013;16(Pt 1):558–65.
78. Hutter J, Schmitt P, Saake M, Stubinger A, Grimm R, Forman C, et al. Multi-dimensional
    flow-preserving compressed sensing (MuFloCoS) for time-resolved velocity-encoded phase
    contrast MRI. IEEE Trans Med Imaging. 2015;34(2):400–14.
79. Dyvorne H, Knight-Greenfield A, Jajamovich G, Besa C, Cui Y, Stalder A, et al. Abdominal
    4D flow MR imaging in a breath hold: combination of spiral sampling and dynamic com-
    pressed sensing for highly accelerated acquisition. Radiology. 2015;275(1):245–54.

80. Carlsson M, Toger J, Kanski M, Bloch KM, Stahlberg F, Heiberg E, et al. Quantification and visualization of cardiovascular 4D velocity mapping accelerated with parallel imaging or k-t BLAST: head to head comparison and validation at 1.5 T and 3 T. J Cardiovasc Magn Reson. 2011;13:55.
81. Tao YH, Rilling G, Davies M, Marshall I. Carotid blood flow measurement accelerated by compressed sensing: validation in healthy volunteers. Magn Reson Imaging. 2013;31(9):1485–91.
82. Nezafat R, Kellman P, Derbyshire J, McVeigh E. Real time high spatial-temporal resolution flow imaging with spiral MRI using auto-calibrated SENSE. Conf Proc IEEE Eng Med Biol Soc. 2004;3:1914–7.
83. Joseph AA, Merboldt KD, Voit D, Zhang S, Uecker M, Lotz J, et al. Real-time phase-contrast MRI of cardiovascular blood flow using undersampled radial fast low-angle shot and nonlinear inverse reconstruction. NMR Biomed. 2012;25(7):917–24.
84. Joseph A, Kowallick JT, Merboldt KD, Voit D, Schaetz S, Zhang S, et al. Real-time flow MRI of the aorta at a resolution of 40 msec. J Magn Reson Imaging. 2014;40(1):206–13.
85. Schmitter S, Schnell S, Ugurbil K, Markl M, Van de Moortele PF. Towards high-resolution 4D flow MRI in the human aorta using kt-GRAPPA and B1+ shimming at 7T. J Magn Reson Imaging. 2016;44(2):486–99.
86. Lotz J, Doker R, Noeske R, Schuttert M, Felix R, Galanski M, et al. In vitro validation of phase-contrast flow measurements at 3 T in comparison to 1.5 T: precision, accuracy, and signal-to-noise ratios. J Magn Reson Imaging JMRI. 2005;21(5):604–10.
87. Rooney WD, Johnson G, Li X, Cohen ER, Kim SG, Ugurbil K, et al. Magnetic field and tissue dependencies of human brain longitudinal 1H2O relaxation in vivo. Magn Reson Med. 2007;57(2):308–18.
88. Wiesinger F, Van de Moortele PF, Adriany G, De Zanche N, Ugurbil K, Pruessmann KP. Parallel imaging performance as a function of field strength – an experimental investigation using electrodynamic scaling. Magn Reson Med. 2004;52(5):953–64.
89. Ohliger MA, Grant AK, Sodickson DK. Ultimate intrinsic signal-to-noise ratio for parallel MRI: electromagnetic field considerations. Magn Reson Med. 2003;50(5):1018–30.

CEST MRI

10

Martin Kunth and Leif Schröder

Abstract

Magnetic resonance imaging (MRI) with indirect detection through chemical exchange saturation transfer (CEST) offers unique features for quantitative imaging as many biophysical and biochemical tissue parameters are linked to exchange-coupled magnetisation pools. The CEST approach also greatly improves the sensitivity of MRI and represents an emerging technique in diagnostic imaging. It uses an induced signal loss in an abundant spin pool to capture information from a dilute spin pool through an actively driven saturation transfer process and achieves much better signal than direct readout of the magnetisation from the low concentration analyte. This chapter provides an introduction into the concepts of CEST and its recent developments considering an extended library of agents, quantitative evaluation for contributing physico-chemical parameters, and advanced encoding techniques. CEST applications nowadays rely on endogenous and synthetic agents that are both discussed here. The inclusion of spin hyperpolarisation as another signal amplification concept illustrates special features that come with the increased dynamic range and extended timescale that magnetisation from hyperpolarised nuclei provide.

10.1 Quantitative MRI Using Exchange-Coupled Magnetisation Pools

Image contrast based on chemical exchange saturation transfer (CEST) is a relative new aspect of magnetic resonance imaging (MRI). It is an indirect detection technique that uses the abundant water signal as a proxy for obtaining information

M. Kunth • L. Schröder (✉)
Molecular Imaging, Leibniz-Forschungsinstitut für Molekulare Pharmakologie (FMP),
Berlin, Germany
e-mail: lschroeder@fmp-berlin.de

© Springer International Publishing AG 2018   213
I. Sack, T. Schaeffter (eds.), *Quantification of Biophysical Parameters in Medical Imaging*, https://doi.org/10.1007/978-3-319-65924-4_10

about a more dilute molecular species. Both endogenous and exogenous substances can be used to generate CEST contrast. An important difference compared to conventional T_1 and T_2 relaxivity agents is that the signal contrast is not just "passively" building up during the respective evolution intervals for the magnetisation by choosing a certain repetition (T_R) or echo time (T_E). Instead, the CEST MRI acquisition sequence with its initial saturation pulse actively drives the evolution of the CEST effect, thus generating an adjustable contrast. This makes it a unique technique with regard to quantitative imaging as the signal response of the tissue or the sample is linked to various molecular parameters of diagnostic interest. The CEST approach allows for either a simplified quantification to compare different tissues based on the observed tuneable CEST effect as collective measure for certain tissue conditions, or it can be used to further elucidate certain physico-chemical parameters that govern the exchange process and that are linked to disease onset and progression.

The structure of this chapter was motivated by the idea of making the reader familiar with saturation transfer experiments of both proton and xenon NMR/MRI. It is an emerging field as illustrated by its dedicated conferences [1–4]; hence, this work is far from claiming to be complete. The CEST mechanism comes with numerous aspects that contribute to quantitative imaging, and in many cases, details of the technical implementation go beyond the scope of this book. This chapter should rather be considered as an introduction that brings attention to the core concepts while referring to the original publications for further details. There are several extensive reviews available for proton CEST like those by van Zijl and Yadav [5], Woods et al. [6], De Leon-Rodriguez et al. [7], Wu et al. [8], Terreno et al. [9], Vinogradov et al. [10], Zaiss and Bachert [11], McMahon and Gilad [12], and Wu et al. [13] as well as a dedicated book by McMahon et al. [14] and a web site [15]. These are complemented with regard to CEST using hyperpolarised xenon (HyperCEST) by the works of Schröder [16, 17], Schnurr et al. [18], Wemmer [19], and Wang and Dmochowski [20].

Since protons represent by far the more widely used species, we utilise this system to present the general concepts of CEST and its underlying theoretical framework based on the Bloch-McConnell equations. These are the basis to describe the evolution of magnetisation during a CEST experiment. Next, we briefly discuss early applications and the related limitations that were subsequently addressed to perform meaningful quantitative CEST experiments. In particular, we look at methods to compensate for saturation spillover and magnetic field inhomogeneities, and how exchange parameters can be quantified. This is important knowledge when analysing any kind of saturation transfer data. Section 10.3 introduces the main classes of proton CEST agents which can be grouped into exchangeable nuclei, exchangeable water molecules, and compartmental exchange. We discuss how these different classes of agents carry diagnostic information and look at their different benefits. With regard to quantitative imaging, the design of agents that are responsive to their physico-chemical environment is one of the great potentials of the CEST technique. This concept has been implemented for different types of CEST agents that will be briefly reviewed.

The fourth section is dedicated to HyperCEST as a technique that combines the inherent amplification of CEST with hyperpolarised xenon with its artificially increased spin polarisation. In contrast to proton CEST, the dilute pool is generated by the addition of a suitable xenon host. We review the different hosts that have been used, with a particular focus on the most common family of xenon hosts, cryptophanes. The HyperCEST technique offers additional aspects of quantitative imaging, in particular for systems in which xenon is in exchange between lipidic and aqueous environments, as well as in and out of host molecules in both of these environments. We discuss how these systems, in combination with unique quantitative analysis techniques, have been used to investigate model membrane systems and discriminate their membrane fluidity.

While proton CEST and HyperCEST are very different systems, due to their conceptual similarity many aspects are equally applicable to both of them. The choice of a certain saturation scheme is of particular importance when performing experiments on clinical MRI scanners that cannot apply continuous wave (cw) saturation pulses. In addition, different pulse schemes can be used to reduce unwanted effects or may be less prone to particular artefacts. The last section deals with concepts that are of importance for data acquisition during the CEST magnetisation preparation itself or with respect to the subsequent spatial encoding. We review some recent advances to accelerated acquisition techniques which are of particular importance when working with hyperpolarised xenon. However these are also valuable for proton CEST since including the spectral dimension into imaging series always imposes time issues on the acquisition protocol.

10.1.1 Clinical MRI Sensitivity Limitations and Origins of CEST

MRI typically detects very weak magnetic moments, and consequently low sensitivity is an important issue. At physiological temperature and clinically available magnetic field strengths, these magnetic moments show only a tiny net alignment with the applied external magnetic field. It is both an advantage and a disadvantage that the energy transition of a spin-1/2 system as it manifests in water protons and as it is used in NMR experiments usually falls into the regime of radiofrequency (RF) wavelengths (nuclei with larger spin numbers are not discussed here). RF pulses are harmless (compared to, e.g. X-rays for atomic resolution) and come with excellent penetration depth. At the same time, this weak energy splitting is related to almost equal population densities according to the Boltzmann distribution. This results in a nearly vanishing net magnetisation and weak signals.

The laws governing the detectable macroscopic magnetisation leave only some limited approaches to increase the detectable signal based on Faraday induction. Regarding spin-1/2 systems, the Langevin equation for paramagnetism ($M \sim \tanh x$, with $x \sim \gamma B_0/k_B T$) contains only the sample temperature T and external magnetic field inductivity B_0 as measures that can be controlled by the experimentalist. Due to the nature of the law and the small coupling energy of the magnetic moments to

B_0 compared to thermal energy $k_B T$, even for protons with their high gyromagnetic ratio, temperature only has a noticeable impact for very low values below 1 K. Cooling the patient or animal is therefore not an option. Increasing the external field is an alternative approach but is usually impaired by the high costs and only yields limited improvements: The strongest permanent magnetic field inductivities achieved today reach ca. 45 T which provides, despite the tremendous technical challenge and efforts undertaken in various national high field laboratories [21], not much more than a 15-fold enhancement compared to the field inductivities of current clinical MRI scanners of 3 T.

Nevertheless, MRI is based on our excellent ability to manipulate the macroscopic magnetisation, and this leaves also some room for indirect sensitivity enhancements. In its most simple form, i.e. an isolated spin pool, the MRI signal somewhat reflects a snapshot of the spin system during the acquisition period and contains only information about the (immediate) evolution history governed by the direct interaction of the RF (and gradient) pulses with the detected spin species (usually the water signal). This concept can be extended for the readout of another spin pool with its own chemical environment characterised by a distinct chemical shift. Introducing chemical exchange between two pools extends the possibilities tremendously since one magnetisation can be used as a storage pool for the evolution history of the other one. An example is the exchange between amide protons and water protons as illustrated in Fig. 10.1a.

Such detection goes far beyond the snapshot information that can be extracted from a single pool and comes with the inherent capacity for signal amplification. This concept dates back to the early achievements of NMR introduced by Forsén and Hoffman [22] where chemical exchange of labile atoms between distinct molecules could be detected by manipulating the spins in one molecular environment and observing changes in the RF signal from another molecular signature: The manipulation is usually done by driving the indirectly detected pool into

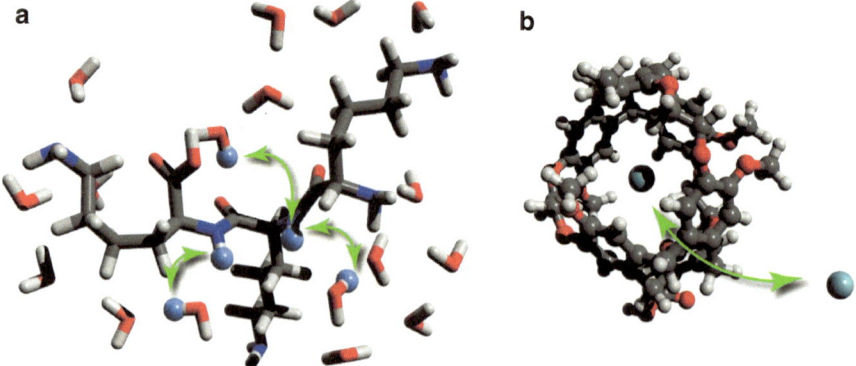

Fig. 10.1 Example of exchange-connected magnetisation pools. (**a**) NH protons of lysine (Lys) residues in a protein are in chemical exchange with protons from nearby water molecules. Shown is a short amino acid sequence (Lys)$_3$ as a DiaCEST agent, and the protons that participate in exchange are highlighted in blue. (**b**) Cryptophane-A-monoacid molecule with two exchanging Xe atoms as it is used in HyperCEST NMR

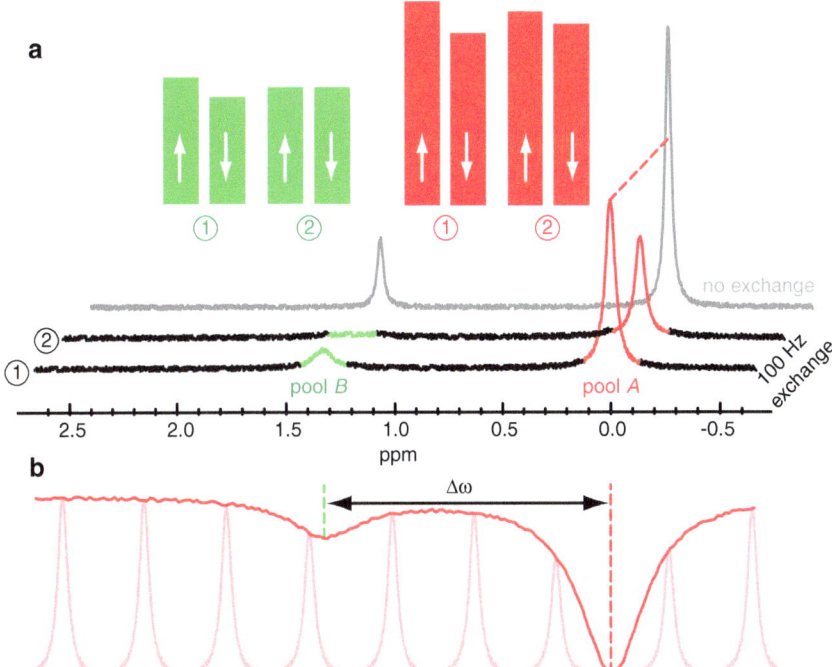

Fig. 10.2 Principle of saturation transfer experiments. (**a**) Nuclei from two spin pools A and B yield two sharp resonances (line width 10 Hz) when they are isolated from each other (grey spectrum in background). Chemical exchange causes line broadening (front spectrum) but facilitates saturation transfer. The solid bars indicate the relative population of the different spin states under different conditions. Without saturation, there is a difference between the population densities for both pools A and B. The application of a cw saturation pulse to resonance B drives pool B into equal population densities. Due to chemical exchange, the vanishing net magnetisation of pool B eventually also reduces the magnetisation of pool A, as its population difference is diminished (middle spectrum; bars representing the population densities not to scale; spectra simulated with WinDNMR [214] with pool A and B separated by $\Delta\omega = 400$ Hz at 300 MHz proton NMR frequency, 80% of the signal in pool A). (**b**) Collecting a whole series of spectra from pool A with application of the RF pulse at different carrier frequencies around resonance B yields an array of intensities of A as a function of the saturation frequency offset. This type of data is called a z-spectrum

saturation, i.e. cancelling its net magnetisation (see Fig. 10.2a). This lack of detectable magnetisation eventually manifests as a reduced signal of the other NMR resonance that was not directly affected by the saturation pulse. In principle, saturation is easily achieved in a spin-½ system with only two energy levels. Such double-resonance experiments are known as CEST experiments and initially focused on the chemical exchange between two spin pools of similar concentration. This approach became a valuable tool to study reaction rates since the magnetic properties can be used for spin labelling without affecting the chemical reaction to be studied; it also found many applications for investigating biomedical systems and focused initially on determination of conversion rates rather than on a signal amplification strategy [23].

10.1.2 Capabilities of CEST Detection

The CEST technique has become an important approach for addressing the MRI sensitivity problem and represents a special case of indirect or "remote" detection [24] because information from a dilute pool is transferred to and stored in a much more abundant and easily detected spin pool. This transfer requires a moderately fast chemical exchange of NMR-detectable nuclei. Besides revealing the sheer presence of such a dilute pool, with careful design, such saturation transfer experiments can reveal quantitative information about the exchange dynamics and size of participating spin pools. For certain cases, this exchange is sensitive to physiological parameters like pH or temperature. Moreover, the information about the chemical shift of the exchanging species can be recovered by applying the saturation pulse at different frequencies: The magnetisation of the abundant pool is recorded for each different saturation frequency to build up a spectrum (see Fig. 10.2b). Such a data set is referred to as a z-spectrum or nowadays as a CEST spectrum (when there are no additional components, see Sect. 10.2.1). This approach is useful when multiple dilute pools are present that are successively encoded in an abundant spin ensemble such as water.

In its most simple case of two exchanging pools, quantification of the intensity of the CEST effect is based on differential equations describing the dynamics of a two-site exchange system as introduced by McConnell in terms of modifications to the Bloch equations [25] which will be discussed in the next section. Despite being a very simple two-pulse NMR experiment (evolution under saturation conditions, followed by detection), it can reveal a surprising amount of information while preserving the chemical shift dimension of NMR. Although the pulse sequences used for such saturation transfer experiments are usually quite simple and can be implemented as easy extensions of already existing methods (especially for MRI), they can be quite demanding for the hardware because of the length of the saturation pulses. For human MRI studies, limitations for applied RF power also have to be considered [26].

10.1.3 Exchanging Nuclei and the Bloch-McConnell Equations

For identifying two separate spin pools, they must have distinct chemical shifts. Though CEST detection is not strictly limited to the case of slow exchange (this would be the case where the difference in chemical shift between the two pools is larger than the rate of exchange in Hz) such exchange conditions do represent the most common case in (pre-)clinical MRI applications. Whereas the exchange rate is intrinsic to the physico-chemical properties of the system (with some control by the experimentalist through temperature and potentially pH of the sample or chemical design of the agent), the difference in resonance frequency depends linearly on the external field B_0. Systems with faster exchange rates are in general candidates for more efficient saturation transfer onto the detected pool. Stronger external fields B_0 therefore provide favourable experimental conditions to resolve the exchanging resonances at sufficient concentration for identifying the right saturation frequency of not yet characterised systems. Higher fields also often come with the added benefit of decelerated T_1 relaxation. Such relaxation is in general an effect that diminishes the theoretically achievable saturation transfer (both for thermally polarised and hyperpolarised systems) as T_1 drives the z-magnetisation of the detected pool towards the thermal polarisation, independent of the applied saturation pulses. Hence, the build-up of any detectable saturation effect must be dominant over the T_1 dynamics [27].

The Bloch-McConnell (BM) equations (a coupled system of first-order linear differential equations) that describe two exchanging spin pools under saturation are given by

$$\frac{dM_x^A(t)}{dt} = -R_2^A M_x^A(t) \quad - \Delta\omega_A M_y^A(t) \quad - k_{AB}M_x^A(t) + k_{BA}M_x^B(t)$$

$$\frac{dM_y^A(t)}{dt} = -R_2^A M_y^A(t) \quad + \Delta\omega_A M_x^A(t) + \omega_1 M_z^A(t) - k_{AB}M_y^A(t) + k_{BA}M_y^B(t)$$

$$\frac{dM_z^A(t)}{dt} = -R_1^A(M_z^A(t) - M_0^A) - \omega_1 M_y^A(t) \quad - k_{AB}M_z^A(t) + k_{BA}M_z^B(t)$$

$$\frac{dM_x^B(t)}{dt} = -R_2^B M_x^B(t) \quad - \Delta\omega_B M_y^B(t) \quad - k_{BA}M_x^B(t) + k_{AB}M_x^A(t)$$

$$\frac{dM_y^B(t)}{dt} = -R_2^B M_y^B(t) \quad + \Delta\omega_B M_x^B(t) + \omega_1 M_z^B(t) - k_{BA}M_y^B(t) + k_{AB}M_y^A(t)$$

$$\frac{dM_z^B(t)}{dt} = -R_1^B(M_z^B(t) - M_0^B) - \omega_1 M_y^B(t) \quad - k_{BA}M_z^B(t) + k_{AB}M_z^A(t)$$

These can be summarised as the following matrix notation originally proposed by Woessner et al. [28]:

$$\frac{d}{dt}\underbrace{\begin{pmatrix} M_x^A \\ M_y^A \\ M_z^A \\ M_x^B \\ M_y^B \\ M_z^B \end{pmatrix}}_{=:\,\mathbf{M}} = \underbrace{\begin{pmatrix} -(R_2^A+k_{AB}) & -\Delta\omega_A & 0 & +k_{BA} & 0 & 0 \\ +\Delta\omega_A & -(R_2^A+k_{AB}) & +\omega_1 & 0 & +k_{BA} & 0 \\ 0 & -\omega_1 & -(R_1^A+k_{AB}) & 0 & 0 & +k_{BA} \\ +k_{AB} & 0 & 0 & -(R_2^B+k_{BA}) & -\Delta\omega_B & 0 \\ 0 & +k_{AB} & 0 & +\Delta\omega_B & -(R_2^B+k_{BA}) & +\omega_1 \\ 0 & 0 & +k_{AB} & 0 & +\omega_1 & -(R_1^B+k_{BA}) \end{pmatrix}}_{=:\,\hat{A}} \cdot \begin{pmatrix} M_x^A \\ M_y^A \\ M_z^A \\ M_x^B \\ M_y^B \\ M_z^B \end{pmatrix} + \underbrace{\begin{pmatrix} 0 \\ 0 \\ R_1^A M_0^A \\ 0 \\ 0 \\ R_1^B M_0^B \end{pmatrix}}_{=:\mathbf{B}}$$

with the formal solution

$$\mathbf{M}(t) = \left(\mathbf{M}_0 + \frac{\mathbf{B}}{\hat{A}} \right) \exp(\hat{A}t) - \frac{\mathbf{B}}{\hat{A}}. \tag{10.1}$$

This allows for powerful and fast solution of the BM equations. An even faster numerical solution has been proposed by Murase and Tanki [29] by reordering the matrix with the aim to remove the inhomogeneity of the equation:

$$\frac{d}{dt}\underbrace{\begin{pmatrix} M_x^A \\ M_x^B \\ M_y^A \\ M_y^B \\ M_z^A \\ M_z^B \\ 1 \end{pmatrix}}_{=:\,\mathbf{M}} = \underbrace{\begin{pmatrix} -(R_2^A+k_{AB}) & +k_{BA} & -\Delta\omega_A & 0 & 0 & 0 & 0 \\ +k_{AB} & -(R_2^B+k_{BA}) & 0 & -\Delta\omega_B & 0 & 0 & 0 \\ +\Delta\omega_A & 0 & -(R_2^A+k_{AB}) & +k_{BA} & +\omega_1 & 0 & 0 \\ 0 & +\Delta\omega_B & +k_{AB} & -(R_2^B+k_{BA}) & 0 & +\omega_1 & 0 \\ 0 & 0 & -\omega_1 & 0 & -(R_1^A+k_{AB}) & +k_{BA} & R_1^A M_0^A \\ 0 & 0 & 0 & -\omega_1 & +k_{AB} & -(R_1^B+k_{BA}) & R_1^B M_0^B \\ 0 & 0 & 0 & 0 & 0 & 0 & 0 \end{pmatrix}}_{=:\,\hat{A}} \cdot \underbrace{\begin{pmatrix} M_x^A \\ M_x^B \\ M_y^A \\ M_y^B \\ M_z^A \\ M_z^B \\ 1 \end{pmatrix}}_{=:\mathbf{M}}$$

with solution $\mathbf{M}(t) = \mathbf{M}_0 \exp(\hat{A}t)$. Herein, the parameters in the rotating frame are $\Delta\omega_{A/B}$ for the saturation frequency offset in Hz with respect to the Larmor frequency of each pool A/B, and ω_1 is the power of saturation pulse of $\vec{B}_1 = (\omega_1/\gamma, 0, 0)$, γ being the gyromagnetic ratio (for ^{1}H: $\gamma^H = 2\pi \cdot 42.577$ MHz/T and for ^{129}Xe: $\gamma^{Xe} = -11.777$ MHz/T). The intrinsic longitudinal and transverse relaxation rates in Hz of both pools are $R_{1/2,A/B}$, and the chemical exchange rates k_{AB} from $A{\to}B$ and vice versa in Hz obey the rate equation in steady state $k_{AB} = \dfrac{M_B^0}{M_A^0} k_{BA} = f_B \cdot k_{BA}$ where f_B is the relative size of pool B. The green terms are the magnetisations in thermal equilibrium, the red terms describe the impact of the RF saturation field, and the blue terms describe the chemical exchange dynamics. Without going into the details of deriving these equations, a few useful statements can already be concluded from the different matrix elements. In order to achieve a strong saturation transfer, the terms including ω_1 and k_{AB} should be the dominant ones on the timescale of longitudinal relaxation of the detection pool A. The saturation pulse will convert z-magnetisation, i.e. spin polarisation, temporarily into transverse components. For small sizes of pool B and if the saturation pulse duration is about or shorter than T_{2A}, ripples around the solution resonance appear in the z-spectrum [29]. A commonly accepted convention in CEST experiments is that the dominant pool, here referred to as pool A, is also denoted by the index w (for bulk water protons). This is the detection pool. The dilute pool, here referred to as pool B, is accordingly denoted by the index s (for solute protons of the CEST pool, i.e. the exchanging labile protons). The observed longitudinal magnetisation, M_{zA}, decreases upon irradiation on the resonance frequency of M_{zB}. The parameters to quantify the chemical exchange k_{AB} will be further investigated in Sect. 10.2.

An analytical solution of the BM equations $M(t) = M(t{=}0) \cdot e^{A \cdot t}$ including the above-mentioned matrix A has later been used to gain insight into the build-up of CEST contrast and the design of improved contrast agents [28].

10.1.4 Basic CEST Quantification and Limitations of Original CEST Agents

The term CEST for chemical exchange saturation transfer was first coined by Ward et al. [30]. They classified saturation transfer effects for OH, NH$_2$, and NH groups of a set of sugars, amino acids, nucleosides, imino acids, as well as barbituric acid and some of its derivatives. The reported CEST effects ranged from 7% (250 mM sorbitol, saturation of OH group) to 67% (125 mM L-alanine, saturation of NH_2 group). Although the applied saturation pulses did not exceed 3.5 µT, it was obvious right from the beginning that the small chemical shift range of protons imposes limitations for achieving strong CEST effects. Based on these early observations, exchange-related parameters had been investigated in initial studies with deriving the analytic expressions for water exchange filter spectroscopy, CEST effect, and amide proton transfer [31].

Increasing the amplitude B_1 of the saturation pulses affects spins over a wider spectral range (for long pulses the width of the pulse in the spectral domain is no longer given by the inverse of the pulse length in the time domain; the Fourier transform of a pulse only predicts the excitation profile in a reasonable way for small perturbations). Hence, efficient saturation schemes are usually not very selective and often result in unwanted direct saturation of pool A while applying the saturation pulse on-resonant to pool B at the offset $\Delta\omega$, i.e. the so-called spillover effect. This manifests itself in a z-spectrum through the observation that the baseline between the two resonance responses does not completely recover to the full z-magnetisation of pool A. Such effects were already reported by Ward et al. when introducing the CEST technique [30], especially when accelerating the exchange by increasing pH. In order to eliminate this effect, reference data has to be taken with a pulse applied symmetrically on the opposite side of pool A (see Fig. 10.3). This second saturation at offset $-\Delta\omega$ is then called the off-resonant (with respect to pool B) measurement. Many CEST effects just show up in the z-spectrum as an additional dip on the shoulder of the saturation response from the detected water pool. An alternative evaluation of the experiment is therefore defined by the magnetisation transfer ratio MTR = $(1 - S_{on}/S_{off})$ that yields the parameter MTR_{asym} by quantifying the spectral asymmetry as follows: instead of just determining one pair of signal values (at $\Delta\omega$ and $-\Delta\omega$ symmetrically around pool A), a whole range of symmetric pairs S_{on}/S_{off} around the detected signal (intensity S_0 at thermal equilibrium) is evaluated (see Fig. 10.3). The saturation response under investigation is then identified as the asymmetric shape of the broad dip in the z-spectrum and by plotting $(S_{on} - S_{off})/S_0$.

$$\mathrm{MTR}_{asym} = \mathrm{MTR}\left(\Delta\omega\right) - \mathrm{MTR}\left(-\Delta\omega\right) = S_{on}\left(-\Delta\omega\right)/S_{off} - S_{on}\left(\Delta\omega\right)/S_{off} \quad (10.2)$$

However, this parameter is not always exclusively determined by chemical exchange. Additional effects following the saturation of one species may also impact the signal of the detected pool.

Z-spectra are actually often obtained from an MR image series and subsequent pixel-wise evaluation or the definition of regions of interest (ROIs) for further quantification. Since the CEST information is stored in terms of z-magnetisation, a series of MRI scans with saturation pulses applied at different frequencies prior to the spatial encoding of the image data maps this information and enables extraction of the spectral information when plotting the intensity of the z-magnetisation versus the variable carrier frequency of the saturation pulse. This evaluation of CEST data is illustrated in Fig. 10.4 with an example of dissolved xenon

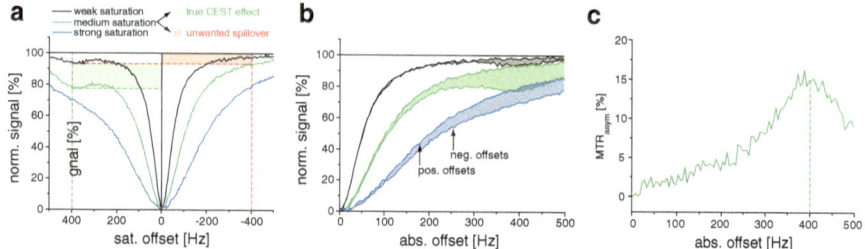

Fig. 10.3 Evaluation of true CEST effect and spillover contribution for the spin system from Fig. 10.2. (**a**) A weak saturation pulse (black line) may cause only a small CEST response. Increasing the saturation power (green line) improves the signal decrease at 400 Hz offset from the observed pool but also broadens all resonances and causes an increasing contribution of direct saturation of this pool. Hence, a reference measurement is performed with 400 Hz offset on the opposite side (−400 Hz, red dashed line). The signal decrease at that position yields the unwanted spillover effect. Excessive saturation (blue line) can cause substantial line broadening in the CEST spectrum that makes the CEST pool almost invisible as the spillover is even more pronounced. (**b**) Comparison of the signals from positive offset with negative offset. The shaded separation of the curves illustrates the true CEST effect and how clear the CEST response manifests within the spectrum. (**c**) In order to isolate the true CEST effect, the parameter MTR_{asym} can be defined (see text). The medium saturation (green line in (**a**, **b**)) yields a clear maximum of ca. 15% CEST effect in this example

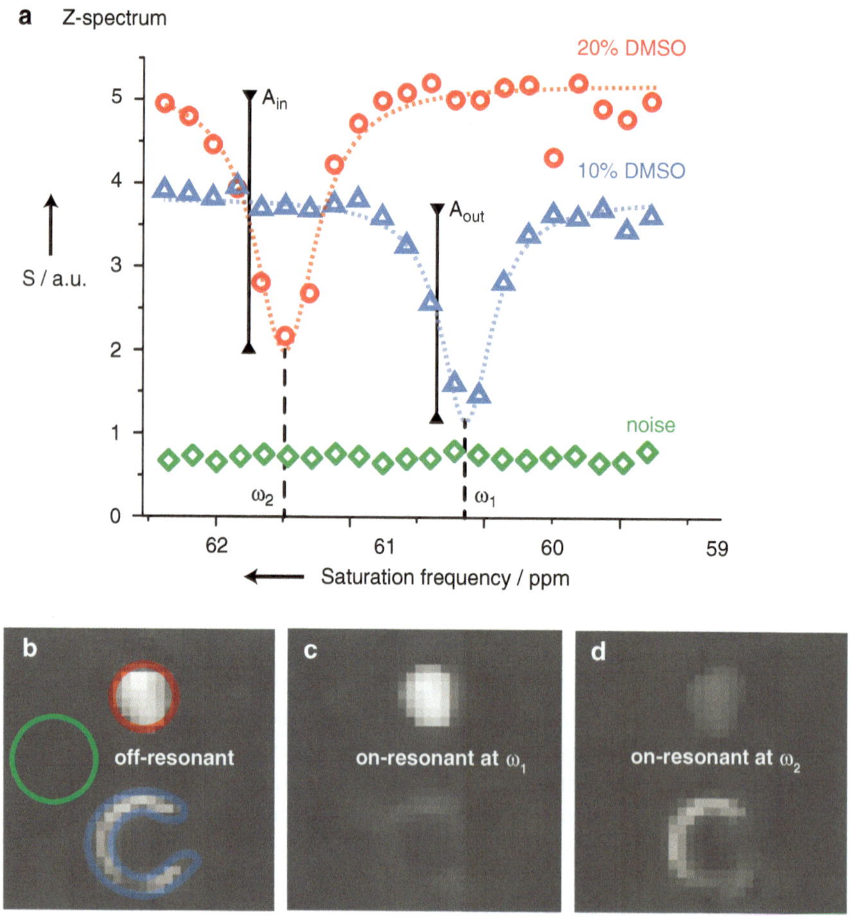

a   Z-spectrum

that reversibly binds to a molecular cage and hence forms a CEST agent. Such systems will be discussed in more detail in Sect. 10.4.

10.2 Detailed CEST Analysis and Quantification

In general, quantitative MRI can already be performed using the above-mentioned MTR_{asym} parameter as a relative collective measure for signal contrast between different tissues. The experimentalist should always be aware of the fact that the achieved saturation transfer is usually not a universal number but rather a collective effect that depends on each experimental setting including the hardware, the selected pulse sequence, and the sample/tissue conditions. This imposes certain challenges when actual physical parameters shall be derived from the induced signal loss and the community developed a multitude of quantitative evaluation approaches. Some of them will be discussed here, while there is also a recent review that focuses on exchange quantification [32]. It should also be mentioned that there is a related approach based on spin-lock experiments. For a comparison of CEST and related spin-lock techniques, the reader is referred to other publications [33–37].

10.2.1 CEST Information and Interfering Effects

A considerable effort has been undertaken to ensure proper analysis and quantification of the CEST effect and its related parameters. This holds especially true for studies where there is a potential strong impact from other effects, spillover as mentioned above being one of them. The separation from magnetisation transfer effects [38] (MT; based on saturating a broad spectral component of semi-solid macromolecular protons) is another important aspect [5, 11]. One technique is transformation of z-spectra into the time domain, followed by filtering of fast-decaying components attributed to MT effects [39]. Another one addresses MT effects that are asymmetric around the water resonance: Applying two different saturation frequencies flattens out MT asymmetry when both frequency components lie within the spectrum of an MT pool and allows separation of the CEST effect [40]. Line-shape analysis also helps to isolate such MT contributions and spillover from the desired saturation

---

Fig. 10.4 z-spectrum derived from a CEST imaging series illustrating the link between imaging and z-spectroscopy (modified from the work by Kunth et al. [159] with permission). (**a**) z-spectra from three different regions of interest in the ^{129}Xe MRI scans shown in (**b**). The Xe host (CrA) is dissolved in water with different admixtures of DMSO (20% in the top compartment, 10% in the bottom compartment). (**b–d**) Axial Xe MRI scans of a setup with two different compartments and definition of the three ROIs. Off-resonant saturation yields signal from both compartments, saturation at the resonance frequency of Xe inside the host in either of the solvent mixtures yields selective signal decrease in the bottom compartment (**c**) or the top compartment (**d**)

transfer [41]. Another approach to combined elimination of diluting effects from spillover and MT for better quantification of the underlying biochemical exchange processes is based on the evaluation of the inverse z-spectrum [42]. This method reveals the exchange-dependent relaxation rate in the rotating frame and the inherent exchange rate. A combined analytical solution describing z-spectra of a water pool in the presence of a semi-solid MT pool and multiple CEST pools applicable to both transient- and steady-state saturation transfer experiments is a more recent progress [43].

For some responses in a z-spectrum, it was also realised that because of their assignment in the aliphatic region, they actually originate from nonexchangeable protons. The observed signal contribution rather comes from the nuclear Overhauser effect (NOE) [44]. This is another contribution that complicates asymmetry analysis when it is used to derive pure exchange-related parameters. However, it can be correlated with information about the folding of proteins that are the source of the signalling protons [45]. The NOE contribution has also been investigated in the context of tumour imaging [46] and for acquisition schemes that avoid extensive data sampling for full asymmetry analysis [47].

Keeping such interfering contributions in mind, a quantitative CEST experiment often focuses on revealing the pure net exchange. This is also described as the proton transfer ratio (PTR) that is a normalised quantity with respect to the number of protons involved. Under the assumption of negligible spillover at the end of the saturation time t_{sat}, this quantity is eventually based on the Bloch-McConnell equations and is given by

$$\text{PTR} = x_s \alpha k_{sw} \, T_{1w} \left(1 - e^{-t_{sat}/T_{1w}} \right) \tag{10.3}$$

with $x_s = \dfrac{[\text{solute protons}]}{[\text{water protons}]}$ being the normalised concentration of the solute (CEST agent) protons, T_{1w} as the water longitudinal relaxation time, k_{sw} the exchange rate from the solute pool into the water pool, and $\alpha = \dfrac{(\gamma B_1)^2}{(\gamma B_1)^2 + (k_{sw})^2}$ the saturation transfer rate for a certain saturation amplitude B_1. For negligible MT and NOE effects, the PTR is obtained from the observed normalised signal loss:

$$\text{PTR} = 1 - S_{on}(\Delta\omega) / S_{off}(-\Delta\omega) \tag{10.4}$$

It illustrates how the induced signal loss is related to the exchange parameters k_{sw} and x_s but is of limited value as long as the interfering effects are not isolated.

CEST applications for MRI are also prone to distortions of the spectral data caused by field inhomogeneities. As demonstrated above, the MTR_{asym} analysis is based on referencing the saturation effect symmetrically around the water signal. Hence, changes in the water resonance frequency between different regions require individual referencing for each pixel. An illustration for the necessity of such corrections is given in Fig. 10.5 where direct saturation of the water signal is evaluated.

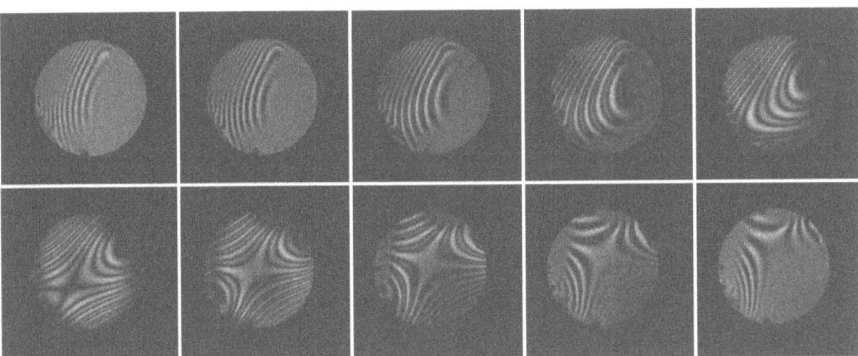

Fig. 10.5 Influence of magnetic field inhomogeneities on CEST imaging series of 10 mm NMR tube filled with water (axial scan). Using a 100 ms pulse with $B_1 = 15$ μT, small frequency offsets cause already large variation in direct saturation of the water signal in neighbouring areas. The patterns move through the sample as the saturation pulse frequency changes

The effect of spatially varying saturation is emphasised in this case where only a short saturation pulse was used (100 ms pulse duration, 15 μT pulse amplitude). Pulse limitations of clinical scanners might require the use of relative short but strong saturation pulses. Here, wiggles in the z-spectrum [29] make the direct saturation effect very susceptible to small field deviations. A small frequency offset causes large variation in achieved saturation. Hence, neighbouring regions with slightly different resonance frequencies show large contrast, and the patterns seem to move through the sample as the saturation pulse frequency is varied. A fast mapping and correction of MTR_{asym} is achieved using the WASSR method [48] that relies on weak direct water saturation for identifying the true reference frequency in each pixel of the localised MTR_{asym} data set. Field inhomogeneities are not only an issue for B_0 but also for the applied saturation field B_1. Various correction approaches have been published, one of them relying on multiple acquisitions with different B_1-values [49].

Methods for general improvement of z-spectra are also useful for noisy data that can be an issue from imaging series with high resolution. Spline-based methods that smooth the data for fitting and correcting z-spectra can compensate for this [50]. A related approach called enhanced and integral saturation transfer aims to better exploit all the information contained in the z-spectrum [51] and achieves significant improvements for certain in vitro and in vivo detections.

10.2.2 Quantification of Exchange Parameters

As mentioned above, PTR is a measure for the overall exchange-mediated signal transfer and is linked to the underlying exchange kinetics determined by k_{sw} and x_s. These quantities can be obtained by analysing the response of the system to well-defined changes in saturation time or power. In many experiments, it is assumed that a steady state of the water signal is reached instantaneously after saturating the

CEST agent pool. The observed PTR can then be described by the following equation in which the former T_{1w} relaxation time is replaced by the water exchange rate $R_{1w} = 1/T_{1w}$ plus a term $x_s k_{sw}$ that accounts for back-exchange of saturated water back into the CEST pool:

$$\text{PTR}\left(t_{\text{sat}}, \alpha\right) = 1 - S_{\text{on}}\left(\Delta\omega\right) / S_{\text{off}}\left(-\Delta\omega\right) = \frac{x_s \alpha k_{sw}}{R_{1w} + x_s k_{sw}}\left(1 - e^{-t_{\text{sat}}\left(R_{1w} + x_s k_{sw}\right)}\right) \quad (10.5)$$

The sum $R_{1w} + x_s k_{sw}$ in fact counteracts the desired signal loss as relaxation is restoring already saturated magnetisation and back-exchange is a null-contribution, i.e. "wasted" exchange capacity of the system. This equation illustrates that the observed signal loss increases with saturation time t_{sat} and saturation power B_1 that drives the saturation transfer rate α. Evaluating a whole set of saturation conditions therefore in principle yields access to k_{sw} and x_s. The corresponding QUEST and QUESP methods (quantifying exchange rates in chemical exchange saturation transfer agents using the saturation time/power dependencies) were first applied to pH calibration for poly-L-lysine and a starburst dendrimer [52]. A simplified quantification of the exchange rate is also possible with the QUESTRA method [53] (quantification of labile proton concentration-weighted chemical exchange rate with RF saturation time-dependent ratiometric analysis).

Various other evaluation techniques have been compared by Randtke et al. [54] and are also discussed in other reviews [10, 55]. The problem is that the CEST effect depends on both the exchange rate k_{sw} and the size of the CEST pool relative to the water pool, x_s. Unfortunately, the two parameters appear as a product term and can compensate for each other while still producing the same CEST response. These saturation parameter-derived approaches therefore require knowledge of the agent concentration for proper evaluation of k_{sw} through the Bloch-McConnell equations. Different studies focused on obtaining the two parameters separately. A concentration-independent determination of the exchange rate can be achieved by so-called omega plots [56] that yield the water exchange rate from the x-intercept of a plot of steady-state CEST intensity divided by reduction in signal caused by CEST irradiation versus $(\omega_1)^{-2}$. Further studies used such omega plots to obtain both proton concentration and the exchange rate [57, 58]. Sun developed another method based on the fact that there is an optimum saturation power that depends much more on k_{sw} than on x_s [59]. This allows the simultaneous determination of the two parameters [60].

10.3 Proton CEST Agents

The field of CEST MRI has brought up quite a variety of molecular probes in the last decade. Most of them are still based on ^1H NMR detection since water provides an obvious dominant pool to encode information from other molecules. In the following, we use the classification of CEST agents as introduced by van Zijl and Yadav [5], i.e. we will discuss exchangeable nuclei, exchangeable water molecules, and compartmental exchange.

10.3.1 Exchangeable Nuclei: DiaCEST and APT

Despite the above-mentioned limitations of agents relying on exchangeable protons within a limited chemical shift range, numerous applications have been realised with such systems. The term DiaCEST was coined to have a clear distinction from agents containing a paramagnetic shift centre which later became the ParaCEST class. Based on the molecules initially studied by Ward et al. [30] and the successful observation of a CEST effect in vivo, the molecular explanation for the origin of such contrast was often assigned to tissue rich in amide-containing compounds, i.e. endogenous mobile proteins and peptides. In such cases, the term amide proton transfer, APT, was established. MRI scans with and without signal saturation of such compounds have found applications in diagnostic imaging of, e.g. brain tumours [61], pH mapping [62], and acute ischemia [63]. The effect is usually not very pronounced; a maximum water signal reduction of 2.9 ± 0.3% between ischemic and normal brain regions was reported for 0.75 µT in the latter case.

More specific mapping of endogenous molecules due to their identified CEST effect has been performed in terms of glycoCEST [64] (detecting glycogen) and gagCEST [44] (revealing glycosaminoglycan, GAG). The first one has been used for monitoring glycogen metabolism in isolated, perfused mouse livers before and after administration of glucagon. The results correlated with ^{13}C NMR spectroscopy which is usually much more challenging to perform and cannot be that easily translated into clinical diagnostics. GagCEST is of interest since GAG loss in cartilage is linked to osteoarthritis and intervertebral disc degeneration. Again, the CEST approach allows a more direct access to the molecular tracer itself than indirect monitoring through delayed Gd contrast agent uptake or ^{23}Na MRI. Quantitative GAG investigation also found applications in detecting disease-related pH changes intervertebral discs [65, 66]. Related to the search for biocompatible DiaCEST agents are also protocols that administer a bolus of a nonmetabolisable derivative of glucose such as 3-O-methyl-D-glucose (3OMG) [67]. The substance is taken up rapidly and preferentially by tumours, followed by complete excretion by the kidneys. The concept is similar to fluorodeoxyglucose (FDG) as the most common tracer in positron emission tomography (PET; see Chap. 11). Another approach for visualising the increased energy consumption in tumours is based on glucosamine (GlcN). Enhanced accumulation of GlcN is postulated for all tumours that overexpress the glucose transporters and could be demonstrated for lung metastasis in mice as illustrated in Fig. 10.6 with additional evaluation yielding quantitative CEST kinetic measurements [68].

Another endogenous compound is creatine (Cr) where the related CEST detection has been coined CrEST. This metabolite has been studied extensively in conjunction with phosphocreatine (PCr) in in vivo ^{1}H and ^{31}P NMR spectroscopy. The proton signals of PCr and Cr are difficult to separate, and ^{31}P NMR only detects PCr but with even lower sensitivity. The CEST imaging approach for these metabolites that are linked through the important creatine kinase (CK) reaction is of great potential: Quantification of the exchange rates of the amine protons suggests a ca. seven-fold faster dissociation of Cr amine protons when compared to PCr and ATP [69].

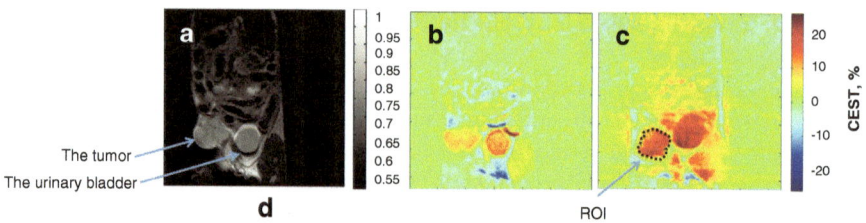

The tumor

The urinary bladder

Fig. 10.6 GlcN CEST MRI results for mice bearing 4T1 breast tumours at B_0 = 7 T. (**a**) A T_2-weighted image for a mouse bearing a 4T1 breast tumour (before administration of the agent). (**b**) A CEST image before administration of the agent; at frequency offset of 1.2 ppm, B_1 = 2.4 µT, 10.4% CEST was obtained in the tumour. (**c**) A CEST image 48 min after PO treatment with GlcN, 1.1 g/kg; at frequency offset of 1.2 ppm, B_1 = 2.4 µT, 20.2% CEST was obtained in the tumour. The CEST calculation was made for the marked ROI (Reproduced from Rivlin and Navon [68] under the Creative Commons license)

Appropriate selection of the saturation parameters yields CEST contrast that mainly originates from Cr with negligible contribution from PCr and ATP and is the basis for high-resolution proton imaging of Cr. Initial in vitro experiments then initiated studies for mapping conversion of PCr to Cr by spatiotemporal detection of Cr in human calf muscle under exercise with potential translation to studying myocardial disorders [70]. However, a tissue Cr concentration of ca. 10 mM causes only ca. 8% CEST effect. The CEST effect of the creatine NH and NH_2 protons was also used to reveal the energy landscape for proton exchange with water with two distinct regions corresponding to the zwitterionic creatine structures and deprotonated creatine [71]. Another metabolite known from ^1H NMR spectroscopy is myo-inositol (mIns) which has been shown to increase at early stages of Alzheimer's disease (AD). Saturation transfer using its hydroxyl groups is done in the MICEST technique [72] that revealed ca. 50% higher CEST effect in the AD mouse model compared to wild type controls.

The second generation of DiaCEST agents comprises compounds with high numbers of exchangeable protons such as cationic polymers that are reported to cause water signal changes of 40–50% at concentrations of ~100 µM [73]. Poly-L-Lysine (PLL) with more than 7000 exchangeable sites yielded an enhancement factor of more than 486,000. Polyuridilic acid, a polymer of 2000 uridine units, was observed at 10 µM with an enhancement factor of ~10.8 × 10^6 via its imino protons [74]. Related CEST agents have subsequently been used in the context of implementing MRI reporter genes [75]. The common principle is to monitor proteins providing an overall large number of exchangeable protons. A first application was based on lysine-rich protein [76] (LRP, 200 lysine residues) and saturation of its OH resonance with CEST effects of up to 8% in LRP-expressing tumour xenografts. Arginine-rich proteins such as hPRM1 (47% arginine residues) also cause a detectable CEST effect [77]. For its CEST characterisation, microwave-assisted peptide synthesis was used to obtain the 51 amino acid-long hPRM1. Both bacterial and human cells could be engineered such that they express an hPRM1 gene to achieve higher CEST contrast compared to controls. In a non-polymer-based approach for gene expression MRI, imaging of the herpes simplex virus type 1 thymidine kinase (HSV1-*tk*) reporter gene expression in rodents could be achieved by CEST detection of 5-methyl-5,6-dihydrothymidine (5-MDHT) [78].

Despite the limited chemical shift range, the spectral dimension was recognised early as valuable parameter and led to the idea of multiplexing and multicolour imaging with DiaCEST agents [79]. The different imaging "channels" correspond to the CEST signatures of NH, NH_2, and OH groups. Due to spillover effects, the suitable subtraction of some data sets has to be done to isolate the different components. An increased chemical shift range can eliminate such unwanted direct saturation of the observed water signal. This concept ultimately triggered the development of ParaCEST agents which will be discussed in the next section. But also without a paramagnetic shift, it is possible to design molecules with well-separated responses in the z-spectrum. Salicylic acid analogues containing a 2-hydroxybenzoic acid scaffold provide highly shifted protons for CEST MRI contrast [80]. An in vivo application was demonstrated for quantitative monitoring the opening of the blood–brain barrier [81].

10.3.2 Exchangeable Water: ParaCEST

It soon became clear that labile protons of the early CEST applications do not have sufficient spectral separation from the water signal to completely avoid unwanted spillover effects during saturation. The use of paramagnetic substances was therefore the obvious approach to increase the chemical shift separation between the saturated and observed pool. The chelates used in conventional contrast agents provide only very short life times for the coordinated water and do not allow for observing a second resonance. In fact, a fast exchange is usually wanted for conventional Gd contrast agents to ensure efficient relaxation of the bulk water pool that yields the relaxivity contrast of the image [82]. Sherry and co-workers realised that slow exchange could be achieved [83, 84] for enabling the observation of a second pool that introduces the chemical shift dimension for encoding of information. Slower exchange dynamics of coordinated water allowed the first ParaCEST agent introduced by Zhang et al. [85] for which coordinated water exhibited a chemical shift separation of 49.7 ppm relative to free water due to coordination to Eu^{3+} in a DOTAtetra(amide) derivative.

Studying the system at 4.7 T (hence 9.8 kHz chemical shift separation) allowed application of relative strong saturation pulses to achieve a CEST effect of 61% which is a significant increase compared to the molecules that were initially investigated [30]. Since serious spillover occurs only for extreme conditions, some studies report saturation pulse amplitudes of up to 250 µT (~10.6 kHz) at 7 T [86]. A summary about the early development of ParaCEST can be found in earlier reviews [6, 87] that discuss lanthanide(III) complexes of DOTA with decelerated exchange of water at the inner-sphere coordination sites. Nevertheless, the coordination lifetime is still relative short and represents a problem that causes broad saturation responses in the z-spectrum. This is illustrated in

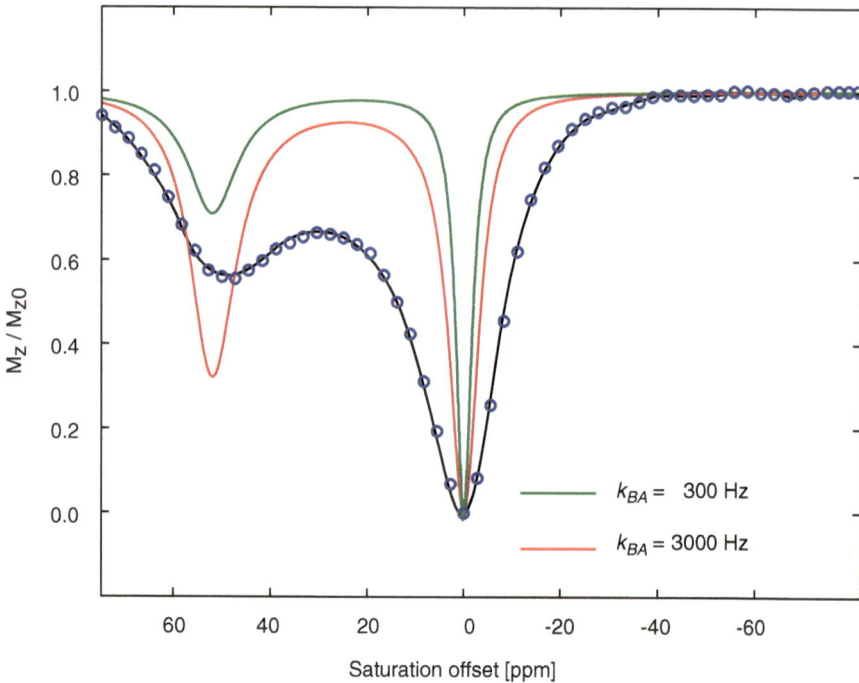

Fig. 10.7 Typical ParaCEST z-spectrum and simulation for decelerated chemical exchange. Data obtained with a EuDOTA complex (blue circles) yields an exchange rate of ca. 30,000 Hz at room temperature when analysing the spectrum with the Bloch-McConnell equations (black line). This is compared with simulated data where the exchange rate k_{BA} is reduced to 3000 Hz (red line) or 300 Hz (green line), respectively. The largest CEST effect is found for medium exchange

Fig. 10.7 where data from a EuDOTA ParaCEST agent with its true exchange parameters is compared with simulated spectra for slower exchange. Previous studies have demonstrated that for field strengths below 10 T only complexes containing the lanthanide ions Eu^{3+}, Tb^{3+}, Dy^{3+}, and Ho^{3+} have the right combination of water exchange time and induced chemical shift to justify further investigations [87]. Detection limits of ca. 10 µM have been reported for chelates that show up to 500 ppm shift for bound water.

Increasing the available chemical shift range also enabled the further development of the above-mentioned multiplexing approach in which several agents can be detected simultaneously at different frequencies. This approach can either be realised by using complexes of different lanthanides [86] (dotamGly complex of Eu^{3+} with +50 ppm induced shift for bound water and a Tb^{3+} version with a −600 ppm shift) or by using the same central ion (Eu^{3+}) with different ligands [88]. The former approach allowed selective in vitro cell imaging of two marked cell populations incubated with either of the ParaCEST probes at 40 mM concentration. The multiplexing concept also applies in the context of HyperCEST but with usually much lower agent concentrations (see Sect. 10.4).

A way to increase the sensitivity is to link several ParaCEST agents in a polymeric approach to have multiple CEST sites per molecule [89]. In doing so, the sensitivity was enhanced ca. tenfold. The general concept behind this idea is a nanoparticle approach that has been used in various contexts [90]. Liposomes (see Sect. 10.3.3) are a special case with an impressive sensitivity when referring to the particle concentration instead of the CEST site concentration.

The concept of ParaCEST agents comes with a side effect due to the large chemical shift separation between the two pools involved: Protons hopping between different frequencies can experience significant T_2 shortening due to chemical exchange rather than due to conventional transverse relaxation effects [91]. Using a set of different Eu^{3+} complexes with water exchange rates ranging from 0 to 5×10^6 s^{-1} revealed that intermediate water exchange rates can cause exchange relaxivity of ca. 0.5 s^{-1} mM^{-1}, whereas fast or very slow exchange does not contribute significantly to transverse relaxation. Impact of the water exchange kinetics in different Eu complexes is also still used for further optimisation of ParaCEST agents. Introduction of carboxyl groups or carboxyl ethyl ester groups on the amide substituents of DOTA-tetraamide ligands modifies water exchange in such complexes depending on the axial versus equatorial orientation of the carboxyl groups relative to the inner-sphere Eu(III)-bound water molecule [92]. Such insights are helpful for designing even more slowly water exchanging systems which will be necessary for in vivo applications.

10.3.3 Compartmental Exchange: LipoCEST

The basic idea behind this class of agents is to generate two different compartments of water with exchange between them. The resonance frequency for spins in one compartment is shifted relative to the other using a paramagnetic shift reagent (SR). The separation is realised by liposomes that are loaded with the SR. With liposome diameters of approximately 270 nm, about 300 million water molecules are encapsulated in the core of each liposome that contribute to the CEST effect. Such agents can be imaged at concentrations as low as 90 pM [93].

The chemical shift between the two compartments is of particular importance as large shifts allow to benefit from fast exchange rates and allow better separation between the exogenous CEST agent and endogenous proteins with exchangeable protons. Early designs were limited to chemical shifts of approximately ±5 ppm, which was mainly determined by the maximum concentration of the SR that could be encapsulated in the liposomes. Further increases in the chemical shift exploited the effects from bulk magnetic susceptibility on the chemical shift. This effect depends on the shape and orientation of the liposomes as well as the concentration and magnetic moment of the SR but does not require a direct chemical interaction between the paramagnetic SR and the water molecules. For spherical liposomes, this contribution is zero, but for nonspherical liposomes, it can be the dominant contribution. Such nonspherical liposomes can be prepared by osmotically shrinking them [94]; this increases the effective chemical shift range to approximately

±12 ppm [90]. Further improvement came from incorporating amphiphilic SRs into the liposomes in addition to encapsulating them [94]. With chemical shifts between the two compartments as large as +30/−45 ppm, two different LipoCEST agents could be co-localised [95].

The other major parameter that can be modified is the exchange rate between the water in the two compartments. Phospholipid composition of the liposomes and incorporation of amphiphiles such as cholesterol alter the water permeability of the membrane [96]. Alternatively the exchange rate can be controlled by altering the size of the nanoparticles [97, 98]. By modelling the effect of liposome size, it was found that smaller liposomes led to greater CEST contrast [97]. This needs to be balanced against the influence of size on bio distribution and uptake of liposomes. An optimal diameter of 200 nm was found, and additionally the liposome should be PEGylated to improve the pharmacokinetic properties of the liposomes.

Liposomes themselves are also of particular interest as they are well-studied drug carrier systems and CEST opens the door to investigating the delivery of the payload to its desired location in vivo. Such theranostic applications look for characteristic changes in the CEST signal upon breakup of the liposomes. Delli Castelli et al. [99] were able to investigate differences in payload release between stealth and pH-sensitive liposomes in vivo using LipoCEST. Dual ^{19}F/CEST liposomes, triggered by changes in temperature, allowed the monitoring of carrier localisation using CEST and quantitative drug release using ^{19}F MRI [100]. Recent work has used the conversion of LipoCEST contrast to ParaCEST contrast upon breakup of the liposomes to determine the selective release of liposomal content triggered by either pH or the application of ultrasound [101].

Rather than using liposomes that release their contents upon changes in their environment, these SR-loaded carriers that are sensitive to their environment are an alternative method to make responsive LipoCEST agents. This has been used to create pH- and temperature-sensitive LipoCEST agents [102]. Targeted LipoCEST agents have also been developed. In particular, $\alpha v \beta 3$ receptors are strongly expressed during angiogenesis in tumours, and, using an RGD peptide, liposomes were targeted to these receptors and imaged [103].

It should also be mentioned that encapsulation of CEST agents using liposomes is not limited to paramagnetic SRs. The idea of compartmentalisation also works with DiaCEST agents and has been reviewed by Chan et al. [104]. By loading the liposomes with different diamagnetic agents, multicolour CEST images could be acquired in live mice [105]. This has also been used in a novel sensor for cell death of implanted cells. Cells and pH-sensitive LipoCEST agents (L-arginine in liposomes) were incorporated into hydrogel microcapsules. By monitoring the CEST effect, the apoptosis of the subcutaneously transplanted xenogeneic hepatocyte cells could be observed [106]. In addition other devices can be used to partition water into two exchanging pools. As alternative to liposomes, red blood cells [107] and block copolymer vesicles [108] have both been successfully used to encapsulate paramagnetic SRs and observed to function as contrast agents.

10.3.4 Chemo-responsive CEST Imaging

Since exchange rates and availability of coordination sites depend for many compounds on parameters such as temperature, pH, the presence of competing ligands, etc., the design of agents that are responsive to their physico-chemical environment is one of the great potentials of the CEST approach. ParaCEST and the related LipoCEST agents with their large range in chemical shifts and exchange rates are of particular interest in this context. An early application was thermometry [109], but such agents have also been used to detect the presence of glucose [110, 111], lactate [112], singlet oxygen [113], or zinc ions [114]. The activity of the enzyme transglutaminase has been detected by its catalytic activity to form a covalent bond between Tm-DO3A-cadaverine and the side chain of a glutamine amino acid residue [115] (therefore termed catalyCEST). This approach shows both changes in CEST peak positions (at -9.2 ppm for the chelate unit and $+4.6$ ppm for the albumin unit) and modified exchange rates when analysing the data with BM equations. The catalyCEST approach was also applied to quantitatively detect β-galactosidase and β-glucuronidase activities [116], the enzyme sulfatase [117], and activity of γ-glutamyl transferase (GGT) within mouse models of human ovarian cancer [118].

Apart from the exchangeable water molecules, some ParaCEST agents also contain labile protons from NH groups. The combination of exchangeable water and amide protons is very valuable to establish a ratiometric pH measurement. It was described early on that the CEST effect of NH is pH sensitive [119]. However, changes in CEST amplitude may also be concentration related and not unambiguously assigned to a certain pH difference. An internal concentration reference (from the metal-coordinated water) is therefore needed to delineate changes in local concentration from changes in pH-mediated exchange rates [120, 121].

A somewhat different idea was realised with a Eu(III) complex that had a single phenolic proton as pH-sensitive unit [122] to induce a ~5 ppm shift in the water exchange CEST peak for $6 < \text{pH} < 7.6$. Acquiring CEST data at two slightly different saturation frequencies yields the solution pH without the need of a concentration marker. ParaCEST pH measurements achieve a precision of 0.21 pH units and an accuracy of 0.09 pH units [123]. An additional pH imaging application for monitoring cell death has been demonstrated with hydrogel-embedded LipoCEST sensors [106]. Because this conventional ratiometric approach works only with certain CEST agents, a generalised ratiometric pH MRI method has been developed based on the analysis of CEST effects under different RF irradiation power levels [124]. The sensitivity of this RF power-based ratiometric CEST ratio (rCESTR) is ca. 0.15–0.20 pH units [125]. With respect to biocompatibility, a series of intramolecular hydrogen bonded imidazoles has demonstrated promising pH mapping in the mouse kidney [126]. It should be mentioned that a recent approach was able to demonstrate pH monitoring via the exchange rates of the ^{15}N-bound protons in a heteronuclear CEST experiment that is initiated by inversion of neighbouring protons that are scalar coupled to ^{15}N [127].

Redox-activated ParaCEST agents have been implemented in different approaches: either by changing the oxidation state of the side groups of a lanthanide complex [128] or by acting directly on the paramagnetic ion [129]. In the first case, two N-methylquinolinium moieties on a Eu^{3+} complex are reduced by β-NADH and become CEST-active. The other example makes use of switching between the paramagnetic Co(II) and an NMR-invisible Co(III) complex. Another class of Co(II)-responsive CEST agents also relies on the paramagnetic properties of the central ion but are not considered as ParaCEST agents in the above-mentioned definition since the saturation transfer relies on labile protons of NH groups of such Co(II) complexes [130]. This approach includes an example with multiple CEST peaks that is suitable for ratiometric pH measurements as initially demonstrated by Wu et al. [122]. A switchable ParaCEST agent can also be generated by influencing the properties of the detected water, i.e. by using the agent to alter the T_1 properties of the bulk pool. By introducing nitroxide free radical groups to a model Eu(DOTA-tetraamide) complex, the T_1 of the bulk water protons is substantially shortened. This quenches the CEST effect which is turned on upon reduction of the paramagnetic nitroxide moieties to diamagnetic −NOH units by L-ascorbate [131]. In addition to the sensing abilities for biomedical imaging applications, the CEST effect has also been used to screen for parameters that influence water exchange in ParaCEST complexes [132]. Such an MRI assay in turn allows rapid screening of libraries for identifying the most sensitive imaging agents.

Despite their smaller chemical shift range, DiaCEST agents can also show a responsive behaviour. Amine-based block copolymers show no CEST effect when they are intact around physiological pH but become detectable through saturation transfer upon dissociation at acidic pH [133]. Enzyme detection of cytosine deaminase (CDase, catalysing the deamination of cytosine to uracil) has been demonstrated for two different substrates, namely, cytosine and 5-fluorocytosine (5FC). Deamination by recombinant CDase causes the CEST contrast to disappear when applying saturation pulses to protons at +2 ppm and +2.4 ppm downfield from water [134]. This allowed visualisation of different expression levels of CDase in different cell lines with CEST MRI.

A somewhat unique case of CEST detection is a recent approach by Bar-Shir et al. based on ^{19}F NMR [135]. Although the detected nucleus does not itself exchange nor is it part of an exchanging molecule, a saturation transfer between different conformations of the 5,5′-difluoro derivative of 1,2-bis(oaminophenoxy) ethane-N,N,N',N'-tetraacetic acid (5F-BAPTA) occurs in the presence of different divalent metal central ions that are bound by 5F-BAPTA. This agent allowed detection of 500 nM of Ca^{2+} ions in ^{19}F MRI.

10.4 HyperCEST

This technique is a special combination of two kinds of amplification processes, namely CEST and hyperpolarised (hp) nuclei. We will first briefly introduce the concept of hyperpolarisation itself before making the link to CEST applications that are possible under particular conditions.

10.4.1 Hyperpolarised Nuclei

Improving the intrinsically poor sensitivity of NMR that is governed by the Boltzmann distribution has been addressed in numerous approaches. One of them is to artificially increase the difference of the spin state populations. There are different ways to achieve this but all of them utilise either polarisation or high spin order of a precursor that is eventually converted into a hyperpolarisation (hp) of the detected spin system [136]. The achieved enhancement factors can be quite high ($>10^4$). However, the systems return to the thermal equilibrium through longitudinal relaxation which occurs on the timescales of seconds to many hours. This can limit the available time frame, but it is in many cases enough to apply many of the established NMR encoding techniques and detect spins at much lower concentrations than usually required.

Xenon is of particular interest for CEST experiments because of its large chemical shift range [137]. In fact, the absolute chemical shift separations in xenon systems can be comparable to ParaCEST agents despite the 3.6-fold lower gyromagnetic ratio compared to protons. The isotope ^{129}Xe is a spin-½ nucleus, and its 2-level spin system can easily be saturated for implementing the combination of CEST and hyperpolarised nuclei as introduced by Schröder et al. [138]. A related version is xenon transfer contrast (XTC) for lung imaging [139–141] by Mugler and co-workers.

Hyperpolarised Xe also comes with the advantage that T_1 relaxation is reasonable slow in many applications. This has two consequences: (a) it provides a large dynamic range to encode CEST information and (b) it allows studying fairly low concentrations of a CEST agent where the build-up of the signal loss of the detected pool takes several 10 s of seconds. Moreover, weak saturation with $B_1 \approx 1$ μT at a host concentration of 50 μM can cause ongoing CEST build-up for ca. 1 min but is a very selective saturation [27]. This situation is conceptually different from proton CEST where the desired saturation competes with counteracting T_1 relaxation on the same timescale and the maximum achievable CEST effect is usually limited to a fraction of the initial thermal magnetisation and also reaches a steady state after few seconds—everything that could not be detected within that timeframe remains invisible. Consequently, the dynamic range for HyperCEST is much better than for proton CEST applications. The saturation pulse parameters have to be adjusted accordingly when compartments of high and low concentrations of HyperCEST agents appear in the same object side by side. Otherwise, either "overexposure" of the first compartment or "underexposure" of the second one occurs [142].

A challenge of hp Xe is the fact that the valuable spin polarisation does not recover by itself but rather requires a redelivery of fresh hp Xe into the sample before the next measurement with identical starting conditions can be acquired. The repetition cycles for such redelivery into sample solutions can be much longer than those usually applied in ^1H CEST for which they are governed by the corresponding T_1. Optimised shared use of the Xe magnetisation for multiple encodings to partially circumvent this problem will be discussed later. In the conventional approach, accurate HyperCEST measurements depend on very stable repetitive Xe delivery. Shot-to-shot noise of more than 10% can be an issue [143] but can be minimised to less than 1% with careful design of sample preparation with mass flow controllers that are triggered by the spectrometer [144] and a constant flow through the pumping cell [145].

10.4.2 Suitable Host-Guest Systems

Although Xe participates only in some very few covalent bonds, it forms inclusion complexes through hydrophobic cores of host structures. Often the interaction time is short, and the only detectable effect is a shift of the Xe solution peak or a change in the line shape that is verified by repeating measurements at different concentrations of the host [146]. Dedicated host-guest systems, however, can provide excellent CEST conditions.

Cryptophanes [147] have been investigated in many studies since they have been introduced as host systems for Xe [148] because they show favourable binding properties for the noble gas with binding constants up to 42,000 M^{-1} depending on the solvent [149]. They are by far the most extensively characterised group of hosts for NMR of reversibly bound Xe. Cucurbit[n]urils (CB[n]) are alternative host molecules and have been investigated in terms of CB[5] [150], CB[6] [151, 152], and CB[7] [153]. Nanoemulsions, such as perfluorooctyl bromide (PFOB), that are known to have favourable gas-binding properties provide another approach to convey a separate chemical shift to exchangeable xenon [154, 155].

The high interest in cryptophanes as very potent host structures for CEST experiments with hp xenon is based on their ability to induce a change in chemical shift of xenon of more than 100 ppm. The host-guest interaction between Xe and cryptophanes can be very efficient and is dominated by van der Waals (vdW) forces, an aspect that makes the size of the cavity relative to the size of the noble gas (42 $Å^3$ vdW volume for Xe) an important parameter. Optimising this cavity size was initially performed with the aim of increasing the binding constant [156]. However, important in the context of CEST detection is also a relative fast exchange to achieve efficient saturation transfer. In general, cryptophanes come in a variety of sizes and possess different Xe-binding properties, leaving quite some parameter space for optimisation concerning reversible binding of xenon [157, 158]. Many studies have been performed with cryptophane-A (CrA, also known as Cr-222), especially for demonstrations of HyperCEST imaging [159, 160]. This host has a cavity size of 95 $Å^3$, and its derivative with one carboxylic acid group for further synthesis steps to attach functional groups is shown in Fig. 10.1b.

It has been demonstrated that despite a lower Xe solubility in water at increased temperature, the CEST sensitivity for exchangeable hp Xe increases with temperature [161, 162]. In general, fairly low total host concentrations have been detected at elevated temperature such as 250 nM for imaging [159] and 1.4 pM for unlocalised detection [163]. This illustrates that the signal transfer in HyperCEST is extremely efficient, especially when keeping in mind that the fractional occupancy of the hosts leaves room for improvement (for cryptophane-A conjugates in water, ca. 30% of the hosts are assumed to be occupied [152]). For the imaging application by Kunth et al. [159], a signal contrast of ~40% was achieved with only ~30 nM concentration of NMR-active sensor and ~95 μM of dissolved, detectable xenon. This is in contrast to CEST experiments with thermally polarised protons that usually require agent concentrations of at least ~10 μM to achieve signal changes of a few percent and are based on significantly higher concentrations of detectable nuclei (factor ~42 or ~72 for protons at 7 or 11 T, respectively). Though the above-mentioned PLL agent [73] caused ca. 50% CEST effect, one has to keep in mind that the concentration of active CEST sites was 700 mM in that case. HyperCEST works at much lower concentrations and can reduce acquisition times by a factor of ~10^9 compared to conventional biosensor detection implemented by Hilty et al. [164]. These impressive numbers are related to the fact that the adjusted acquisition time is inverse proportional to the square of the gain in sensitivity. The nanomolar detection by Kunth et al. represented already a ca. 1500-fold sensitivity enhancement compared to the original HyperCEST paper (20-fold reduced sensor concentration, 13-fold reduced acquisition time, better spatial resolution) which itself represented already a ~57-fold increased sensitivity compared to direct detection. Further sensitivity enhancement can be achieved by grafting multiple cages onto a carrier structure. This follows a similar principle as used for ParaCEST agents [89] or the PLL example and has been demonstrated by using viral capsids [165, 166], bacteriophages [167, 168], or liposomal carriers [155].

The so-called multiplexing option that comes with the large chemical shift range of Xe in different hosts was already discussed in early applications and supported the idea of designing a whole class of "biosensors" [148]. Multichannel imaging has also been addressed for Dia- and ParaCEST agents, but for protons it often suffers from spectral overlap of the different agents (see Sect. 10.3.1). For Xe NMR applications, such detection is easier to achieve, and it has been shown with different approaches: (a) different cages in different liquid phases [169], (b) identical cages exchanging between different environments [170] (see also next section), (c) identical cages in two physically separated solutions using different solvents [159], (d) different cages in the same solvent [145], and (e) an approach with two different types of hosts [155]. While (a) employed selective direct detection with extensive signal averaging, HyperCEST was used in (b)–(e) with excellent spectral resolution and clearly demonstrates the feasibility to selectively address different reporters like in fluorescence imaging. Kunth et al. used an adjustable system for chemical shifts based on variable DMSO concentrations (see also Fig. 10.4) with two Xe@host peaks separated by only 130 Hz [159]. Such high selectivity is important since some sensors show only small chemical shift changes upon enzymatic activity [171] or small chemical modifications [145].

Similar to the responsive agents discussed in Sect. 10.2.4, caged xenon has also been used to detect the presence of other particular molecules or the change of parameters that influence the exchange rate or binding properties. Impacts of temperature changes on HyperCEST were discussed early [161, 162, 172] and also led to the "transpletor" concept [162] in which the cages are used as controllable depolarisation gates with signal amplification similar to a transistor. Binding to different targets has been demonstrated with the biotin-avidin system [148] and much weaker DNA hybridisation [173] of certain nucleotides but also for the detection of ions like zinc [174], Pb, and Cd [175]. Enzyme detection was investigated for matrix metalloproteinase [171] (MMP-7) and human carbonic anhydrase [176] (HCA). A sensitivity to pH has also been reported for two different concepts [177, 178]. Cellular internalisation has been addressed both in targeted approaches (through the receptors for transferrin [179] and folate [180] or cell-penetrating peptides [181, 182]) or just based on the hydrophobic properties of CrA that enables Xe MRI cell tracking [160].

The combination of HyperCEST and fluorescence readout, partially based on corresponding building blocks [183], has turned out useful to quantify cell labelling efficiency through optical methods and then in an opaque sample through the high sensitivity of the HyperCEST technique. It could be used to demonstrate novel applications with labelling concentrations in the nanomolar range that are otherwise inaccessible with conventional relaxivity-based contrast agents. Sensors have been designed to detect inflammation marker CD14 with Xe MRI [184] as well as cell surface glycans [185] or to highlight cells of the blood–brain barrier [186].

10.4.3 Nested Exchanging Hyperpolarised Systems

The large chemical shift range of Xe enables investigations of more complex chemical exchange systems similar to the LipoCEST approach. The spectral dimension can contain information about the molecular environment of the Xe host which might be itself in chemical exchange between different environments. This was first described by Meldrum et al. [170] using a solution of mixed micelles that provide good solubility conditions for hydrophobic CrA. The partitioning of CrA from aqueous solution into lipids was then studied in more detail with liposomes of pure phospholipids and modified CrA molecules using HyperCEST and fluorescence-based methods [187]. Characteristically, the relatively broad NMR resonance attributed to CrA in lipid environment shows up 8–12 ppm shifted downfield from the resonance of CrA in aqueous solution. This typical chemical shift difference between the two cage pools is reduced at higher temperature. The large separation of the resonances is of particular interest since unbound Xe swapping between aqueous and lipid environment only shows a small chemical shift difference which is due to relative fast exchange and only becomes measurable at temperatures below 277 K [170]. Thus the distinct chemical shift difference can be helpful to identify that CrA is cell associated [160, 179]. It confirms the uptake even without dedicated

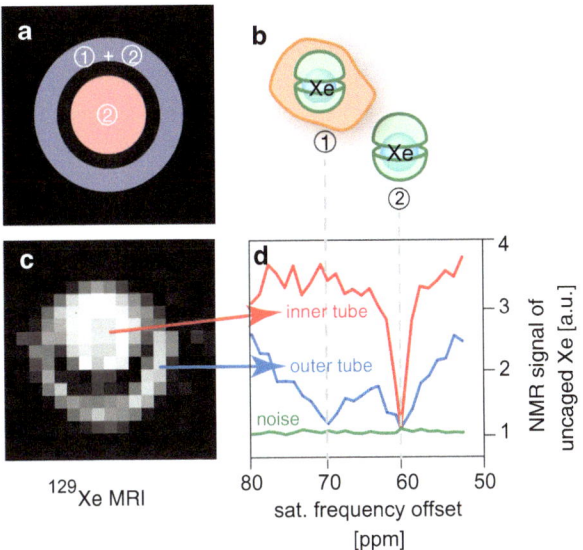

Fig. 10.8 HyperCEST spectroscopy of caged xenon in phospholipid environments. (**a**) Schematics for axial view of two NMR tubes nested inside each other. The inner one contains only free CrA, the outer one also cells labelled with CrA. (**b**) Schematics for free CrA as Xe host and cell-associated CrA. (**c**) Xe MRI scan with off-resonant saturation showing the Xe distribution inside the sample. (**d**) CEST spectra obtained from an MRI series showing two distinct CEST responses for cell-associated and free CrA (Figs. (**b**–**d**) modified from supplementary material of work by Klippel et al. [160])

cell-penetrating peptides as they were used in earlier studies [181]. An example of HyperCEST identification of CrA-labelled cells is shown in Fig. 10.8. The broadening of the lipid-based CrA resonance is presumably due to fast exchanging Xe atoms inside the lipid environment. Hence, in the lipid environment, the HyperCEST effect is more efficient, which also benefits from a higher local concentration of both CrA and Xe atoms which show high solubility in lipid environments [188].

10.4.4 DeLTA Spectroscopy

The large dynamic range of HyperCEST enables the investigation of the build-up of the CEST effect and thus introduces an additional dimension. It can resolve different HyperCEST pools even if they have the same chemical shift and offers a quantitative tool to discriminate different micro-environments. One example is different lipid environments that consist of different types of phospholipids and thus possess different membrane fluidity. Membrane fluidity is determined by the packing and ordering of the membrane's components. Low membrane fluidity hampers the permeation of Xe atoms in and out of the membrane as well as the diffusion of CrA and

Xe atoms inside the membrane. That hampering entails a slow build-up time of the HyperCEST effect. Schnurr et al. investigated the CEST-driven depolarisation of Xe atoms that interact with CrA nested in lipid environment using a Laplace transform analysis [189, 190] known as depolarisation Laplace transform analysis (DeLTA). The concept of using this operation to investigate saturation transfer was previously applied for amide protons [191]. The Laplace transform yields the depolarisation times which—in that case—can be assigned to highly fluid and less fluid model membranes, namely, 1-palmitoyl-2-oleoyl-sn-glycero-3-phosphocholine (POPC) and 1,2-dipalmitoyl-sn-glycero-3-phosphocholine (DPPC). In POPC the depolarisation time was sevenfold faster than in DPPC at 303 K. Combining HyperCEST encoding schemes developed for Xe MRI [159] with variable saturation times and performing a pixel-wise DeLTA yields an MRI contrast that represents the fluidity of the model membranes.

Beside model membranes consisting of different phospholipids, the cholesterol level of model membranes can be determined using HyperCEST with DeLTA [190]. The saturation transfer is less pronounced when membrane fluidity is decreased due to the incorporation of cholesterol. Thus, also the build-up of the depolarisation is decelerated and can be further investigated with DeLTA. This approach complements existing methods which detect membrane fluidity using fluorescence-based methods. Moreover, as DeLTA is based on NMR and HyperCEST, it is not limited by penetration depth and is able to detect low concentrations CrA.

10.4.5 Quantitative HyperCEST

The ability to derive quantitative results from HyperCEST data is of equal interest as for proton CEST. However, the special starting conditions with hyperpolarised nuclei are favourable for eliminating some of the limitations mentioned in Sect 10.2. It is common practice that some of the parameters from proton CEST are renamed in a more general way when water is not the abundant pool. The fractional size of the CEST pool B (formerly x_s) becomes f_B, and the exchange rate out of the CEST pool (formerly k_{sw}) is generalised as k_{BA}. Zaiss et al. derived a simplified solution for the driven depolarisation of hp nuclei [192] in a HyperCEST experiment. It benefits from the fact that the steady state magnetisation is typically much smaller than the initial magnetisation of the hyperpolarised system. The observed normalised magnetisation of the solution pool is then given by

$$S_{on}\left(\Delta\omega,t_{sat},B_1\right) / S_{off}\left(-\Delta\omega,t_{sat},B_1\right) = e^{-t_{sat}\lambda_{depol}\left(\Delta\omega,B_1\right)} \qquad (10.6)$$

where λ_{depol} is the depolarisation rate that depends on the saturation offset $\Delta\omega$, the saturation power B_1, and the exchange parameters f_B and k_{BA} as follows:

$$\lambda_{depol}\left(\Delta\omega,B_1\right) = f_B k_{BA} \frac{\left(\gamma B_1\right)^2}{\left(\gamma B_1\right)^2 + \left(k_{BA}\right)^2 + \left(\Delta\omega - \omega_B\right)^2} \qquad (10.7)$$

Herein, ω_B is the on-resonance frequency of the pool of caged xenon. Kunth et al. used this theoretical framework to quantify the Xe exchange with cryptophane in different solvents and demonstrated that f_B and k_{BA} can be determined independently from different saturation conditions [193]. Hence, a HyperCEST system is much more simple to characterise than proton CEST systems. Knowing the concentration of free Xe in solution through the Ostwald solubility coefficient and the applied partial pressure, it is possible to derive further parameters from f_B such as the binding constant K_a of the system and the fraction of occupied hosts, β. These can be mapped with MRI to characterise inhomogeneous samples or to determine unknown concentrations next to a reference sample [194]. Quantitative HyperCEST also allows straightforward prediction of optimum saturation conditions for maximum signal contrast once the exchange kinetics of a system have been characterised [27].

The comprehensive characterisation of labile Xe complexes also allows for classification of Xe hosts in terms of their HyperCEST efficiency. It is possible to define the product βk_{BA} as the gas turnover rate, i.e. the host efficiency in terms of maximal ^{129}Xe depolarisation rate per μM host concentration at a given Xe concentration. This quantitative approach revealed that CB[6] is a ca. 100-fold more potent HyperCEST agent than CrA [152].

10.5   Advanced Encoding and Evaluation

Proton CEST and HyperCEST face different challenges due to their respective starting conditions and exchange dynamics. Although recent developments for both techniques addressed different problems, they can be grouped into two categories according to the sequence of the experiment: saturation preparation and subsequent spatial/spectral encoding.

10.5.1 Improved Saturation Schemes

As mentioned earlier, some CEST experiments can be quite demanding for the hardware of a spectrometer and especially for clinical MRI scanners. One limitation on such systems is often that no continuous wave (cw) irradiation can be applied. In order to circumvent this limitation, the cw saturation can be decomposed into a pulse train [195], which can be efficiently implemented with multiple transmitters known from parallel imaging [196]. The decomposition into a pulse train also introduces the inter-pulse delay as an additional parameter. By acquiring two or more datasets with different delays between the pulses, effects of direct water saturation can be removed [47]. This method eliminates the need to acquire whole z-spectra and perform asymmetry analysis as usually done with DiaCEST agents.

Other new saturation schemes focus on separating the pure CEST effect from other contributions. Two-frequency RF irradiation can be used to uniformly saturate the protons in macromolecules [40], allowing for an isolation of the CEST contrast from

MT asymmetry, whereas a technique termed LOVARS [197] (length and offset varied saturation) modulates the exchange contrast from different pools (e.g. MT, CEST) with different frequencies and phases by systematically varying the saturation parameters. LOVARS phase maps furthermore show better contrast-to-noise ratios than conventional CEST maps and are less prone to B_0 inhomogeneities. However, they are also less sensitive to concentration and exchange rate of the CEST agent.

A different approach is frequency-labelled exchange transfer (FLEX) [198], where the modulation of the bulk water signal is observed as a function of the chemical shift evolution time of the dilute pool. In the resulting FLEX FID, rapidly decaying components such as MT effects can be removed by time domain analysis, which has been successfully demonstrated for ParaCEST agents in vitro [199] and in vivo [200]. FLEX has also been employed with on-resonant frequency labelling [201], detecting rapidly exchanging protons that resonate close to the bulk water frequency. However, FLEX is a rather slow technique since the FID data is acquired point-wise before reconstructing its spectral components.

CEST contrast is traditionally based on a signal difference after signal depletion, hence a negative contrast mechanism. A hybrid approach between CEST and a spin-lock preparation implemented a positive signal (pCEST) [202]. Although the absolute pCEST effect is smaller than the absolute CEST effect, the positive CEST contrast shows improved background suppression and hence a better dynamic range.

Regarding ParaCEST agents, one drawback is the severe T_2 shortening that comes with the paramagnetic effect. When conventional imaging sequences are employed, this can hamper the detection of the CEST effect. The problem has been addressed in a method called SwiftCEST [203], where almost zero echo time imaging ($TE < 10\ \mu s$) is achieved by swept RF excitation with nearly simultaneous acquisition.

10.5.2 Accelerated Acquisition

Once the saturation transfer into the bulk pool is generated, there is usually the demand to do further encoding for either the spatial distribution of the magnetisation or for the spectral dispersion of the CEST effect. Particularly the latter aspect can make CEST imaging very time-consuming when it goes beyond CEST MRI in its most simple form with just two preparations of magnetisation. Including the spectral dimension (with or without further spatial encoding) is more demanding and conventionally requires a repetitive starting magnetisation which undergoes saturation at different frequencies. One of the limitations is obviously that the z-spectrum has to be acquired point by point with long TR for either (a) full recovery of ^1H magnetisation or (b) redelivery of hp Xe.

In order to accelerate the characterisation of multiple CEST agents, an initial approach was proposed using spectral CEST imaging of a whole bundle of test tubes [204]. The main achievement was to eliminate field inhomogeneities for proper realignment of the CEST data from different parts of the array. More efficient k-space sampling for spectral CEST images can be achieved by using the keyhole technique [205]. Here, only the central k-space lines are acquired per frequency encoding step and then reconstructed using a fully sampled reference scan, allowing for an approximately fourfold reduction in scan time. Principal component analysis (PCA) and constrained reconstruction of undersampled CEST image data rely on correlations between data points in the spectral domain [206]. For hp Xe, this technique is not only useful to reduce the acquisition time (this effect is marginal for HyperCEST) but mainly for having more magnetisation per encoding step which results in an up to five times increased SNR.

Other approaches also aim to avoid multiple acquisitions that are performed using a selective saturation pulse with variable carrier frequency. While in the original experiment, this scheme is combined with a homogeneous magnetic field throughout the sample the problem can be inverted by using a single saturation frequency and a linear field gradient along one dimension of the sample. This requires that the available magnetisation is sufficiently large to allow for 1D projection measurements. The result encodes the spectral information along the gradient's direction, thus rendering the use of multiple saturation frequencies redundant. Such fast z-spectroscopy can be applied to samples that are homogeneous along the dimension that is used for the field gradient [207]. Multiple samples can be investigated at once when incorporating slice selection [208], and the gradient-based approach has meanwhile also been extended to localised CEST spectroscopy [209] and the quantitative investigation of tissue samples with solid state NMR techniques under magic angle spinning [210].

Hp nuclei should be used as efficient as possible for avoiding time-consuming redelivery of fresh magnetisation into the sample. Any longitudinal magnetisation that has been converted into detectable transverse components does not recover as is the case for thermally polarised magnetisation. For HyperCEST, it is therefore helpful that T_2 in aqueous solutions of cryptophane systems has been proven to be long enough to acquire entire echo trains, either based on spin echoes (RARE imaging [152, 160, 206] or z-spectroscopy applications [211]) or on gradient echoes (e.g. for EPI-based imaging [159]). A complementary approach for hp nuclei is to share the available magnetisation for multiple applications by using variable flip angles [212]. Since CEST requires a minimum of two acquisitions (reference and saturation encoding), a $45°/90°$ excitation scheme has been implemented in the smashCEST approach [159] (using shared magnetisation after single hyperpolarisation). With careful calibration of the pulses, this method can acquire MRI scans with a single delivery of hp nuclei and encode the entire information in less than 1 s.

Based on this gradient-encoded z-spectroscopy for proton CEST, the idea has been further developed and implemented for applications with hp Xe by Döpfert et al. [211] and Boutin et al. [213]. To make maximum use of available magnetisation, both studies included spin echo refocussing to increase SNR. The smashCEST approach [159] was also utilised in both papers to further accelerate data acquisition, requiring only a single hp Xe delivery for the entire experiment. However, smashCEST results were only shown by Döpfert et al.

Conclusion

The CEST signal amplification is a powerful tool with a variety of different aspects for quantitative MRI. It tackles the NMR sensitivity problem when exchange-coupled spin systems are available and the technique has experienced promising refinement over the last decade with an increasing number of (responsive) agents and improved encoding and evaluation protocols. Quantitative CEST MRI yields access to exchange-related parameters such as pH and pool sizes of certain dilute substances. Synthetic CEST agents extend the toolbox of endogenous diamagnetic substances to benefit from paramagnetic and hyperpolarised compounds. They also can be designed as chemo-responsive agents for a variety of stimuli. Their in vivo applications are still limited, but they have shown a great potential for further studies and may also prove valuable for NMR in vitro diagnostics. Applications cover various markers for medical imaging and illustrate an unprecedented sensitivity that will stimulate the design of further MRI sensors for molecular diagnostics.

Acknowledgements Part of this work has been supported by the Human Frontiers Science Program (MK funded through programme grant no. RGP0050/2016) and a Koselleck Grant of the German Research Foundation (SCHR 995-1/1 to LS).

References

1. Terreno E, et al. Conference abstracts. Contrast Media Mol Imaging. 2010;5:333–45.
2. Abstracts from the 2nd international workshop on CEST and PARACEST techniques, Dallas, USA, June 1–2, 2011. Contrast Media Mol Imaging. 2012;7:101–20.
3. Mani T, et al. Proceedings for October CEST, the third international workshop on CEST imaging, 15–17 October 2012. Contrast Media Mol Imaging. 2013;8:293–331.
4. PENN-CEST Symposium 2015. http://mediasite.med.upenn.edu/Mediasite/Catalog/Full/2d77 3ae853fb4a2f953af8c0bdffada621. Accessed 27 Nov 2016.
5. van Zijl PCM, Yadav NN. Chemical exchange saturation transfer (CEST): what is in a name and what isn't? Magn Reson Med. 2011;65:927–48.
6. Woods M, Woessner DE, Sherry AD. Paramagnetic lanthanide complexes as PARACEST agents for medical imaging. Chem Soc Rev. 2006;35:500–11.
7. De Leon-Rodriguez LM, et al. Responsive MRI agents for sensing metabolism in vivo. Acc Chem Res. 2009;42:948–57.
8. Wu Y, Evbuomwan M, Melendez M, Opina A, Sherry AD. Advantages of macromolecular to nanosized chemical-exchange saturation transfer agents for MRI applications. Future Med Chem. 2010;2:351–66.
9. Terreno E, Castelli DD, Aime S. Encoding the frequency dependence in MRI contrast media: the emerging class of CEST agents. Contrast Media Mol Imaging. 2010;5:78–98.
10. Vinogradov E, Sherry AD, Lenkinski RE. CEST: from basic principles to applications, challenges and opportunities. J Magn Reson. 2013;229:155–72.
11. Zaiss M, Bachert P. Chemical exchange saturation transfer (CEST) and MR Z-spectroscopy in vivo: a review of theoretical approaches and methods. Phys Med Biol. 2013;58:R221.
12. McMahon MT, Gilad AA. Cellular and molecular imaging using chemical exchange saturation transfer. Top Magn Reson Imaging. 2016;25:197–204.
13. Wu B, et al. An overview of CEST MRI for non-MR physicists. EJNMMI Phys. 2016;3:19.
14. McMahon MT, Gilad AA, Bulte JWM, van Zijl PCM, editors. Chemical exchange saturation transfer imaging - advances and applications. Singapore: Pan Stanford Publishing; 2016.
15. cest-sources.org. Available at: http://www.cest-sources.org/doku.php. Accessed 27 Nov 2016.

16. Schröder L. Xenon for NMR biosensing – inert but alert. Phys Med. 2013;29:3–16.
17. Schröder L. In: Hane FT, editor. Hyperpolarized and inert gas MRI. Cambridge: Academic; 2017. p. 263–77.
18. Schnurr M, Witte C, Schröder L. Hyperpolarized xenon-129 magnetic resonance: concepts, production, techniques and applications. Cambridge: RSC Publishing; 2015.
19. Wemmer DE. Hyperpolarized xenon-129 magnetic resonance: concepts, production, techniques and applications. Cambridge: RSC Publishing; 2015.
20. Wang Y, Dmochowski IJ. An expanded palette of xenon-129 NMR biosensors. Acc Chem Res. 2016;49:2179–87.
21. Meet the 45 Tesla hybrid magnet - MagLab. National High Magnetic Field Laboratory - Meet the 45 Tesla hybrid magnet. Available at: https://nationalmaglab.org/about/around-the-lab/meet-the-magnets/meet-the-45-tesla-hybrid-magnet. Accessed 1 Dec 2016.
22. Forsén S, Hoffman RA. Study of moderately rapid chemical exchange reactions by means of nuclear magnetic double resonance. J Chem Phys. 1963;39:2892.
23. Kuchel PW. Spin-exchange NMR spectroscopy in studies of the kinetics of enzymes and membrane transport. NMR Biomed. 1990;3:102–19.
24. Harel E, Schröder L, Xu S. Novel detection schemes of nuclear magnetic resonance and magnetic resonance imaging: applications from analytical chemistry to molecular sensors. Annu Rev Anal Chem. 2008;1:133–63.
25. McConnell H, Reaction M. Rates by nuclear magnetic resonance. J Chem Phys. 2004;28:430–1.
26. Sun PZ, Benner T, Kumar A, Sorensen AG. Investigation of optimizing and translating pH-sensitive pulsed-chemical exchange saturation transfer (CEST) imaging to a 3T clinical scanner. Magn Reson Med. 2008;60:834–41.
27. Kunth M, Witte C, Schröder L. Continuous-wave saturation considerations for efficient xenon depolarization. NMR Biomed. 2015;28:601–6.
28. Woessner DE, Zhang S, Merritt ME, Sherry AD. Numerical solution of the Bloch equations provides insights into the optimum design of PARACEST agents for MRI. Magn Reson Med. 2005;53:790–9.
29. Murase K, Tanki N. Numerical solutions to the time-dependent Bloch equations revisited. Magn Reson Imaging. 2011;29:126–31.
30. Ward KM, Aletras AH, Balaban RS. A new class of contrast agents for MRI based on proton chemical exchange dependent saturation transfer (CEST). J Magn Reson. 2000;143:79–87.
31. Zhou J, Wilson DA, Sun PZ, Klaus JA, Van Zijl PCM. Quantitative description of proton exchange processes between water and endogenous and exogenous agents for WEX, CEST, and APT experiments. Magn Reson Med. 2004;51:945–52.
32. Kim J, Wu Y, Guo Y, Zheng H, Sun PZ. A review of optimization and quantification techniques for chemical exchange saturation transfer MRI toward sensitive in vivo imaging. Contrast Media Mol Imaging. 2015;10:163–78.
33. Yuan J, Zhou J, Ahuja AT, Wang Y-XJ. MR chemical exchange imaging with spin-lock technique (CESL): a theoretical analysis of Z-spectrum using a two-pool $R1\rho$ relaxation model beyond the fast-exchange limit. Phys Med Biol. 2012;57:8185–200.
34. Zaiss M, Bachert P. Exchange-dependent relaxation in the rotating frame for slow and intermediate exchange -- modeling off-resonant spin-lock and chemical exchange saturation transfer. NMR Biomed. 2013;26:507–18.
35. Zaiss M, Bachert P. Equivalence of spin-lock and magnetization transfer NMR experiments. arXiv:1203.2067; 2012.
36. Cobb JG, Li K, Xie J, Gochberg DF, Gore JC. Exchange-mediated contrast in CEST and spin-lock imaging. Magn Reson Imaging. 2014;32:28–40.
37. Roeloffs V, Meyer C, Bachert P, Zaiss M. Towards quantification of pulsed spinlock and CEST at clinical MR scanners: an analytical interleaved saturation–relaxation (ISAR) approach. NMR Biomed. 2015;28:40–53.
38. Balaban RS, Ceckler TL. Magnetization transfer contrast in magnetic resonance imaging. Magn Reson Q. 1992;8:116–37.

39. Yadav NN, Chan KWY, Jones CK, Mcmahon MT, van Zijl PCM. Time domain removal of irrelevant magnetization in chemical exchange saturation transfer Z-spectra. Magn Reson Med. 2013;70:547–55.
40. Lee J-S, Regatte RR, Jerschow A. Isolating chemical exchange saturation transfer contrast from magnetization transfer asymmetry under two-frequency rf irradiation. J Magn Reson. 2012;215:56–63.
41. Zaiss M, Schmitt B, Bachert P. Quantitative separation of CEST effect from magnetization transfer and spillover effects by Lorentzian-line-fit analysis of z-spectra. J Magn Reson. 2011;211:149–55.
42. Zaiss M, et al. Inverse Z-spectrum analysis for spillover-, MT-, and T1-corrected steady-state pulsed CEST-MRI – application to pH-weighted MRI of acute stroke. NMR Biomed. 2014;27:240–52.
43. Zaiss M, et al. A combined analytical solution for chemical exchange saturation transfer and semi-solid magnetization transfer. NMR Biomed. 2015;28:217–30.
44. Ling W, Regatte RR, Navon G, Jerschow A. Assessment of glycosaminoglycan concentration in vivo by chemical exchange-dependent saturation transfer (gagCEST). Proc Natl Acad Sci U S A. 2008;105:2266–70.
45. Zaiss M, Kunz P, Goerke S, Radbruch A, Bachert P. MR imaging of protein folding in vitro employing nuclear-Overhauser-mediated saturation transfer. NMR Biomed. 2013;26:1815–22.
46. Xu J, et al. On the origins of chemical exchange saturation transfer (CEST) contrast in tumors at 9.4 T. NMR Biomed. 2014;27:406–16.
47. Xu J, et al. Variable delay multi-pulse train for fast chemical exchange saturation transfer and relayed-nuclear overhauser enhancement MRI: variable delay multi-pulse CEST. Magn Reson Med. 2014;71:1798–812.
48. Kim M, Gillen J, Landman BA, Zhou J, van Zijl PCM. Water saturation shift referencing (WASSR) for chemical exchange saturation transfer (CEST) experiments. Magn Reson Med. 2009;61:1441–50.
49. Windschuh J, et al. Correction of B1-inhomogeneities for relaxation-compensated CEST imaging at 7 T. NMR Biomed. 2015;28:529–37.
50. Stancanello J, et al. Development and validation of a smoothing-splines-based correction method for improving the analysis of CEST-MR images. Contrast Media Mol Imaging. 2008;3:136–49.
51. Terreno E, et al. Methods for an improved detection of the MRI-CEST effect. Contrast Media Mol Imaging. 2009;4:237–47.
52. McMahon MT, et al. Quantifying exchange rates in chemical exchange saturation transfer agents using the saturation time and saturation power dependencies of the magnetization transfer effect on the magnetic resonance imaging signal (QUEST and QUESP): Ph calibration for poly-L-lysine and a starburst dendrimer. Magn Reson Med. 2006;55:836–47.
53. Sun PZ. Simplified quantification of labile proton concentration-weighted chemical exchange rate (k(ws)) with RF saturation time dependent ratiometric analysis (QUESTRA): normalization of relaxation and RF irradiation spillover effects for improved quantitative chemical exchange saturation transfer (CEST) MRI. Magn Reson Med. 2012;67:936–42.
54. Randtke EA, Chen LQ, Corrales LR, Pagel MD. The Hanes-Woolf linear QUESP method improves the measurements of fast chemical exchange rates with CEST MRI. Magn Reson Med. 2014;71:1603–12.
55. Liu G, Song X, Chan KWY, McMahon MT. Nuts and bolts of chemical exchange saturation transfer MRI. NMR Biomed. 2013;26:810–28.
56. Dixon WT, et al. A concentration-independent method to measure exchange rates in PARACEST agents. Magn Reson Med. 2010;63:625–32.
57. Sun PZ, Wang Y, Dai Z, Xiao G, Wu R. Quantitative chemical exchange saturation transfer (qCEST) MRI - RF spillover effect-corrected omega plot for simultaneous determination of labile proton fraction ratio and exchange rate: RF spillover effect-corrected omega plot analysis. Contrast Media Mol Imaging. 2014;9:268–75.

58. Meissner J-E, et al. Quantitative pulsed CEST-MRI using Ω-plots. NMR Biomed. 2015;28:1196–208.

59. Sun PZ. Simultaneous determination of labile proton concentration and exchange rate utilizing optimal RF power: radio frequency power (RFP) dependence of chemical exchange saturation transfer (CEST) MRI. J Magn Reson. 2010;202:155–61.

60. Sun PZ, Wang Y, Xiao G, Wu R. Simultaneous experimental determination of labile proton fraction ratio and exchange rate with irradiation radio frequency power-dependent quantitative CEST MRI analysis. Contrast Media Mol Imaging. 2013;8:246–51.

61. Jones CK, et al. Amide proton transfer imaging of human brain tumors at 3T. Magn Reson Med. 2006;56:585–92.

62. Zhou J, Payen J-F, Wilson DA, Traystman RJ, van Zijl PCM. Using the amide proton signals of intracellular proteins and peptides to detect pH effects in MRI. Nat Med. 2003;9: 1085–90.

63. Sun PZ, Zhou J, Huang J, van Zijl P. Simplified quantitative description of amide proton transfer (APT) imaging during acute ischemia. Magn Reson Med. 2007;57:405–10.

64. van Zijl PCM, Jones CK, Ren J, Malloy CR, Sherry AD. MRI detection of glycogen in vivo by using chemical exchange saturation transfer imaging (glycoCEST). Proc Natl Acad Sci U S A. 2007;104:4359–64.

65. Melkus G, Grabau M, Karampinos DC, Majumdar S. Ex vivo porcine model to measure pH dependence of chemical exchange saturation transfer effect of glycosaminoglycan in the intervertebral disc. Magn Reson Med. 2014;71:1743–9.

66. Zhou Z, et al. Quantitative chemical exchange saturation transfer MRI of intervertebral disc in a porcine model. Magn Reson Med. 2016;76:1677–83.

67. Rivlin M, Tsarfaty I, Navon G. Functional molecular imaging of tumors by chemical exchange saturation transfer MRI of 3-O-methyl-D-glucose. Magn Reson Med. 2014;72: 1375–80.

68. Rivlin M, Navon G. Glucosamine and N-acetyl glucosamine as new CEST MRI agents for molecular imaging of tumors. Sci Rep. 2016;6:32648.

69. Haris M, et al. Exchange rates of creatine kinase metabolites: feasibility of imaging creatine by chemical exchange saturation transfer MRI. NMR Biomed. 2012;25:1305–9.

70. Haris M, et al. A technique for in vivo mapping of myocardial creatine kinase metabolism. Nat Med. 2014;20:209–14.

71. Ivchenko O, et al. Proton transfer pathways, energy landscape, and kinetics in creatine–water systems. J Phys Chem B. 2014;118:1969–75.

72. Haris M, et al. MICEST: a potential tool for non-invasive detection of molecular changes in Alzheimer's disease. J Neurosci Methods. 2013;212:87–93.

73. Goffeney N, Bulte JW, Duyn J, Bryant LH Jr, van Zijl PC. Sensitive NMR detection of cationic-polymer-based gene delivery systems using saturation transfer via proton exchange. J Am Chem Soc. 2001;123:8628–9.

74. Snoussi K, Bulte JWM, Guéron M, van Zijl PCM. Sensitive CEST agents based on nucleic acid imino proton exchange: detection of poly(rU) and of a dendrimer-poly(rU) model for nucleic acid delivery and pharmacology. Magn Reson Med. 2003;49:998–1005.

75. Liu G, Gilad AA. MRI of CEST-based reporter gene. Methods Mol Biol. 2011;771:733–46.

76. Gilad AA, et al. Artificial reporter gene providing MRI contrast based on proton exchange. Nat Biotechnol. 2007;25:217–9.

77. Bar-Shir A, et al. Human protamine-1 as an MRI reporter gene based on chemical exchange. ACS Chem Biol. 2014;9:134–8.

78. Bar-Shir A, Liu G, Greenberg MM, Bulte JWM, Gilad AA. Synthesis of a probe for monitoring HSV1-tk reporter gene expression using chemical exchange saturation transfer MRI. Nat Protoc. 2013;8:2380–91.

79. McMahon MT, et al. New 'multicolor' polypeptide diamagnetic chemical exchange saturation transfer (DIACEST) contrast agents for MRI. Magn Reson Med. 2008;60:803–12.

80. Yang X, et al. Salicylic acid and analogues as diaCEST MRI contrast agents with highly shifted exchangeable proton frequencies. Angew Chem Int Ed. 2013;52:8116–9.

81. Song X, et al. Salicylic acid analogues as chemical exchange saturation transfer MRI contrast agents for the assessment of brain perfusion territory and blood-brain barrier opening after intra-arterial infusion. J Cereb Blood Flow Metab. 2016. https://doi.org/10.1177/02716 78X16637882.

82. Caravan P. Strategies for increasing the sensitivity of gadolinium based MRI contrast agents. Chem Soc Rev. 2006;35:512.

83. Zhang S, Wu K, Sherry AD. Gd3+ complexes with slowly exchanging bound-water molecules may offer advantages in the design of responsive MR agents. Investig Radiol. 2001;36:82–6.

84. Sherry AD, Wu Y. The importance of water exchange rates in the design of responsive agents for MRI. Curr Opin Chem Biol. 2013;17:167–74.

85. Zhang S, Winter P, Wu K, Sherry AD. A novel europium(III)-based MRI contrast agent. J Am Chem Soc. 2001;123:1517–8.

86. Aime S, Carrera C, Delli Castelli D, Geninatti Crich S, Terreno E. Tunable imaging of cells labeled with MRI-PARACEST agents. Angew Chem Int Ed Engl. 2005;44:1813–5.

87. Zhang S, Merritt M, Woessner DE, Lenkinski RE, Sherry AD. PARACEST agents: modulating MRI contrast via water proton exchange. Acc Chem Res. 2003;36:783–90.

88. Viswanathan S, et al. Multi-frequency PARACEST agents based on europium(III)-DOTA-tetraamide ligands. Angew Chem Int Ed Engl. 2009;48:9330–3.

89. Wu Y, et al. Polymeric PARACEST agents for enhancing MRI contrast sensitivity. J Am Chem Soc. 2008;130:13854–5.

90. Castelli DD, Terreno E, Longo D, Aime S. Nanoparticle-based chemical exchange saturation transfer (CEST) agents. NMR Biomed. 2013;26:839–49.

91. Soesbe TC, Merritt ME, Green KN, Rojas-Quijano FA, Sherry AD. T2 exchange agents: a new class of paramagnetic MRI contrast agent that shortens water T2 by chemical exchange rather than relaxation. Magn Reson Med. 2011;66:1697–703.

92. Mani T, et al. The stereochemistry of amide side chains containing carboxyl groups influences water exchange rates in EuDOTA-tetraamide complexes. J Biol Inorg Chem. 2014;19:161–71.

93. Aime S, Delli Castelli D, Terreno E. Highly sensitive MRI chemical exchange saturation transfer agents using liposomes. Angew Chem Int Ed Engl. 2005;44:5513–5.

94. Terreno E, et al. From spherical to osmotically shrunken paramagnetic liposomes: an improved generation of LIPOCEST MRI agents with highly shifted water protons. Angew Chem. 2007;119:984–6.

95. Terreno E, et al. First ex-vivo MRI co-localization of two LIPOCEST agents. Contrast Media Mol Imaging. 2008;3:38–43.

96. Terreno E, et al. Determination of water permeability of paramagnetic liposomes of interest in MRI field. J Inorg Biochem. 2008;102:1112–9.

97. Zhao JM, et al. Size-induced enhancement of chemical exchange saturation transfer (CEST) contrast in liposomes. J Am Chem Soc. 2008;130:5178–84.

98. Terreno E, et al. Osmotically shrunken LIPOCEST agents: an innovative class of magnetic resonance imaging contrast media based on chemical exchange saturation transfer. Chemistry. 2009;15:1440–8.

99. Delli Castelli D, et al. In vivo MRI multicontrast kinetic analysis of the uptake and intracellular trafficking of paramagnetically labeled liposomes. J Control Release. 2010;144:271–9.

100. Langereis S, et al. A temperature-sensitive liposomal 1H CEST and 19F contrast agent for MR image-guided drug delivery. J Am Chem Soc. 2009;131:1380–1.

101. Castelli DD, Boffa C, Giustetto P, Terreno E, Aime S. Design and testing of paramagnetic liposome-based CEST agents for MRI visualization of payload release on pH-induced and ultrasound stimulation. J Biol Inorg Chem. 2014;19:207–14.

102. Opina ACL, et al. TmDOTA-tetraglycinate encapsulated liposomes as pH-sensitive LipoCEST agents. PLoS One. 2011;6:e27370.

103. Flament J, et al. In vivo CEST MR imaging of U87 mice brain tumor angiogenesis using targeted LipoCEST contrast agent at 7 T. Magn Reson Med. 2013;69:179–87.

104. Chan KWY, Bulte JWM, McMahon MT. Diamagnetic chemical exchange saturation transfer (diaCEST) liposomes: physicochemical properties and imaging applications. Wiley Interdiscip Rev Nanomed Nanobiotechnol. 2014;6:111–24.

105. Liu G, et al. In vivo multicolor molecular MR imaging using diamagnetic chemical exchange saturation transfer liposomes. Magn Reson Med. 2012;67:1106–13.
106. Chan KWY, et al. MRI-detectable pH nanosensors incorporated into hydrogels for in vivo sensing of transplanted-cell viability. Nat Mater. 2013;12:268–75.
107. Ferrauto G, et al. Lanthanide-loaded erythrocytes as highly sensitive chemical exchange saturation transfer MRI contrast agents. J Am Chem Soc. 2014;136:638–41.
108. Grüll H, et al. Block copolymer vesicles containing paramagnetic lanthanide complexes: a novel class of T1- and CEST MRI contrast agents. Soft Matter. 2010;6:4847–50.
109. Zhang S, Malloy CR, Sherry AD. MRI thermometry based on PARACEST agents. J Am Chem Soc. 2005;127:17572–3.
110. Zhang S, Trokowski R, Sherry AD. A paramagnetic CEST agent for imaging glucose by MRI. J Am Chem Soc. 2003;125:15288–9.
111. Ren J, Trokowski R, Zhang S, Malloy CR, Sherry AD. Imaging the tissue distribution of glucose in livers using a PARACEST sensor. Magn Reson Med. 2008;60:1047–55.
112. Aime S, Delli Castelli D, Fedeli F, Terreno E. A paramagnetic MRI-CEST agent responsive to lactate concentration. J Am Chem Soc. 2002;124:9364–5.
113. Song B, et al. A europium(III)-based PARACEST agent for sensing singlet oxygen by MRI. Dalton Trans Camb Engl 2003. 2013;42:8066–9.
114. Trokowski R, Ren J, Kálmán FK, Sherry AD. Selective sensing of zinc ions with a PARACEST contrast agent. Angew Chem Int Ed Engl. 2005;44:6920–3.
115. Hingorani DV, Randtke EA, Pagel MD. A catalyCEST MRI contrast agent that detects the enzyme-catalyzed creation of a covalent bond. J Am Chem Soc. 2013;135:6396–8.
116. Fernández-Cuervo G, Tucker KA, Malm SW, Jones KM, Pagel MD. Diamagnetic imaging agents with a modular chemical design for quantitative detection of β-galactosidase and β-glucuronidase activities with catalyCEST MRI. Bioconjug Chem. 2016;27:2549–57.
117. Sinharay S, Fernández-Cuervo G, Acfalle JP, Pagel MD. Detection of sulfatase enzyme activity with a catalyCEST MRI contrast agent. Chemistry. 2016;22:6491–5.
118. Sinharay S, et al. Noninvasive detection of enzyme activity in tumor models of human ovarian cancer using catalyCEST MRI. Magn Reson Med. 2016;77(5):2005–14. https://doi.org/10.1002/mrm.26278.
119. Ward KM, Balaban RS. Determination of pH using water protons and chemical exchange dependent saturation transfer (CEST). Magn Reson Med. 2000;44:799–802.
120. Aime S, Delli Castelli D, Terreno E. Novel pH-reporter MRI contrast agents. Angew Chem Int Ed. 2002;41:4334–6.
121. Aime S, et al. Paramagnetic lanthanide(III) complexes as pH-sensitive chemical exchange saturation transfer (CEST) contrast agents for MRI applications. Magn Reson Med. 2002;47:639–48.
122. Wu Y, Soesbe TC, Kiefer GE, Zhao P, Sherry AD. A responsive europium(III) chelate that provides a direct readout of pH by MRI. J Am Chem Soc. 2010;132:14002–3.
123. Sheth VR, Liu G, Li Y, Pagel MD. Improved pH measurements with a single PARACEST MRI contrast agent. Contrast Media Mol Imaging. 2012;7:26–34.
124. Longo DL, et al. A general MRI-CEST ratiometric approach for pH imaging: demonstration of in vivo pH mapping with iobitridol. J Am Chem Soc. 2014;136:14333–6.
125. Wu R, Longo DL, Aime S, Sun PZ. Quantitative description of radiofrequency (RF) power-based ratiometric chemical exchange saturation transfer (CEST) pH imaging. NMR Biomed. 2015;28:555–65.
126. Yang X, et al. Developing imidazoles as CEST MRI pH sensors. Contrast Media Mol Imaging. 2016;11:304–12.
127. Zeng H, et al. 15N heteronuclear chemical exchange saturation transfer MRI. J Am Chem Soc. 2016;138:11136–9.
128. Ratnakar SJ, et al. Europium(III) DOTA-tetraamide complexes as redox-active MRI sensors. J Am Chem Soc. 2012;134:5798–800.
129. Tsitovich PB, Spernyak JA, Morrow JR. A redox-activated MRI contrast agent that switches between paramagnetic and diamagnetic states. Angew Chem. 2013;125:14247–50.

130. Dorazio SJ, Olatunde AO, Spernyak JA, Morrow JR. CoCEST: cobalt(II) amide-appended paraCEST MRI contrast agents. Chem Commun (Camb). 2013;49:10025–7.
131. Ratnakar SJ, et al. Modulation of CEST images in vivo by T1 relaxation: a new approach in the design of responsive PARACEST agents. J Am Chem Soc. 2013;135:14904–7.
132. Napolitano R, Soesbe TC, De León-Rodríguez LM, Sherry AD, Udugamasooriya DG. On-bead combinatorial synthesis and imaging of chemical exchange saturation transfer magnetic resonance imaging agents to identify factors that influence water exchange. J Am Chem Soc. 2011;133:13023–30.
133. Zhang S, et al. A novel class of polymeric pH-responsive MRI CEST agents. Chem Commun (Camb). 2013;49:6418–20.
134. Liu G, et al. Monitoring enzyme activity using a diamagnetic chemical exchange saturation transfer MRI contrast agent. J Am Chem Soc. 2011;133:16326–9.
135. Bar-Shir A, et al. Metal ion sensing using ion chemical exchange saturation transfer 19F magnetic resonance imaging. J Am Chem Soc. 2013;135:12164–7.
136. Witte C, et al. Magnetic resonance imaging of cell surface glycosylation using Hyper-CEST xenon biosensors. In: Proceedings of the WMIC; 2014.
137. Goodson BM. Nuclear magnetic resonance of laser-polarized noble gases in molecules, materials, and organisms. J Magn Reson. 2002;155:157–216.
138. Schroder L, Lowery TJ, Hilty C, Wemmer DE, Pines A. Molecular imaging using a targeted magnetic resonance hyperpolarized biosensor. Science. 2006;314:446–9.
139. Ruppert K, Brookeman JR, Hagspiel KD, Mugler JP. Probing lung physiology with xenon polarization transfer contrast (XTC). Magn Reson Med. 2000;44:349–57.
140. Ruppert K, et al. XTC MRI: sensitivity improvement through parameter optimization. Magn Reson Med. 2007;57:1099–109.
141. Dregely I, et al. Multiple-exchange-time xenon polarization transfer contrast (MXTC) MRI: initial results in animals and healthy volunteers. Magn Reson Med. 2012;67:943–53.
142. Kunth, M., Döpfert, J., Witte, C. & Schröder, L. Simultaneous MRI monitoring of diffusion of multiple hyper-CEST contrast agents. In: 54th experimental nuclear magnetic resonance conference (ENC), Asilomar, CA, 2013.
143. Garcia S, et al. Sensitivity enhancement by exchange mediated magnetization transfer of the xenon biosensor signal. J Magn Reson. 2007;184:72–7.
144. Witte C, Kunth M, Döpfert J, Rossella F, Schröder L. Hyperpolarized xenon for NMR and MRI applications. J Vis Exp. 2012;67:e4268. https://doi.org/10.3791/4268.
145. Witte C, Kunth M, Rossella F, Schröder L. Observing and preventing rubidium runaway in a direct-infusion xenon-spin hyperpolarizer optimized for high-resolution hyper-CEST (chemical exchange saturation transfer using hyperpolarized nuclei) NMR. J Chem Phys. 2014;140:084203.
146. Lowery TJ, et al. Distinguishing multiple chemotaxis Y protein conformations with laser-polarized 129Xe NMR. Protein Sci. 2005;14:848–55.
147. Brotin T, Dutasta J-P. Cryptophanes and their complexes—present and future. Chem Rev. 2009;109:88–130.
148. Spence MM, et al. Functionalized xenon as a biosensor. Proc Natl Acad Sci U S A. 2001;98:10654–7.
149. Jacobson DR, et al. Measurement of radon and xenon binding to a cryptophane molecular host. Proc Natl Acad Sci U S A. 2011;108:10969–73.
150. Huber G, et al. Interaction of xenon with cucurbit[5]uril in water. ChemPhysChem. 2011;12:1053–5.
151. Kim BS, et al. Water soluble cucurbit[6]uril derivative as a potential Xe carrier for 129Xe NMR-based biosensors. Chem Commun. 2008;2008:2756–8. https://doi.org/10.1039/B805724A.
152. Kunth M, Witte C, Hennig A, Schröder L. Identification, classification, and signal amplification capabilities of high-turnover gas binding hosts in ultra-sensitive NMR. Chem Sci. 2015;6:6069–75.

153. Schnurr M, Sloniec-Myszk J, Döpfert J, Schröder L, Hennig A. Supramolecular assays for mapping enzyme activity by displacement-triggered change in hyperpolarized 129Xe magnetization transfer NMR spectroscopy. Angew Chem Int Ed. 2015;54:13444–7.

154. Stevens TK, Ramirez RM, Pines A. Nanoemulsion contrast agents with sub-picomolar sensitivity for xenon NMR. J Am Chem Soc. 2013;135:9576–9.

155. Klippel S, Freund C, Schröder L. Multichannel MRI labeling of mammalian cells by switchable nanocarriers for hyperpolarized xenon. Nano Lett. 2014;14:5721–6.

156. Fogarty HA, et al. A cryptophane core optimized for xenon encapsulation. J Am Chem Soc. 2007;129:10332–3.

157. Huber G, et al. Cryptophane-xenon complexes in organic solvents observed through NMR spectroscopy. J Phys Chem A. 2008;112:11363–72.

158. Kotera N, et al. Design and synthesis of new cryptophanes with intermediate cavity sizes. Org Lett. 2011;13:2153–5.

159. Kunth M, Döpfert J, Witte C, Rossella F, Schröder L. Optimized use of reversible binding for fast and selective NMR localization of caged xenon. Angew Chem Int Ed. 2012;51:8217–20.

160. Klippel S, et al. Cell tracking with caged xenon: using cryptophanes as MRI reporters upon cellular internalization. Angew Chem Int Ed. 2014;53:493–6.

161. Schröder L, et al. Temperature response of Xe129 depolarization transfer and its application for ultrasensitive NMR detection. Phys Rev Lett. 2008;100:257603.

162. Schröder L, et al. Temperature-controlled molecular depolarization gates in nuclear magnetic resonance. Angew Chem Int Ed. 2008;47:4316–20.

163. Bai Y, Hill PA, Dmochowski IJ. Utilizing a water-soluble cryptophane with fast xenon exchange rates for picomolar sensitivity NMR measurements. Anal Chem. 2012;84:9935–41.

164. Hilty C, Lowery TJ, Wemmer DE, Pines A. Spectrally resolved magnetic resonance imaging of a xenon biosensor. Angew Chem Int Ed. 2006;45:70–3.

165. Meldrum T, et al. A xenon-based molecular sensor assembled on an MS2 viral capsid scaffold. J Am Chem Soc. 2010;132:5936–7.

166. Jeong K, et al. Targeted molecular imaging of cancer cells using MS2-based 129Xe NMR. Bioconjug Chem. 2016;27:1796–801.

167. Palaniappan KK, et al. Molecular imaging of cancer cells using a bacteriophage-based 129Xe NMR biosensor. Angew Chem Int Ed Engl. 2013;52:4849–53.

168. Stevens TK, et al. HyperCEST detection of a 129Xe-based contrast agent composed of cryptophane-a molecular cages on a bacteriophage scaffold. Magn Reson Med. 2013;69:1245–52.

169. Berthault P, Bogaert-Buchmann A, Desvaux H, Huber G, Boulard Y. Sensitivity and multiplexing capabilities of MRI based on polarized 129Xe biosensors. J Am Chem Soc. 2008;130:16456–7.

170. Meldrum T, Schröder L, Denger P, Wemmer DE, Pines A. Xenon-based molecular sensors in lipid suspensions. J Magn Reson. 2010;205:242–6.

171. Wei Q, et al. Designing ^{129}Xe NMR biosensors for matrix metalloproteinase detection. J Am Chem Soc. 2006;128:13274–83.

172. Schilling F, et al. MRI thermometry based on encapsulated hyperpolarized xenon. ChemPhysChem. 2010;11:3529–33.

173. Roy V, et al. A cryptophane biosensor for the detection of specific nucleotide targets through xenon NMR spectroscopy. Chemphyschem. 2007;8:2082–5.

174. Kotera N, et al. A sensitive zinc-activated 129Xe MRI probe. Angew Chem Int Ed Engl. 2012;51:4100–3.

175. Tassali N, et al. Smart detection of toxic metal ions - Pb2+ and Cd2+ - using a 129Xe NMR-based sensor. Anal Chem. 2014;86:1783–8. https://doi.org/10.1021/ac403669p.

176. Chambers JM, et al. Cryptophane xenon-129 nuclear magnetic resonance biosensors targeting human carbonic anhydrase. J Am Chem Soc. 2009;131:563–9.

177. Berthault P, et al. Effect of pH and counterions on the encapsulation properties of xenon in water-soluble cryptophanes. Chemistry. 2010;16:12941–6.

178. Riggle BA, Wang Y, Dmochowski IJ. A 'smart' [129]Xe NMR biosensor for pH-dependent cell labeling. J Am Chem Soc. 2015;137:5542–8.
179. Boutin C, et al. Cell uptake of a biosensor detected by hyperpolarized 129Xe NMR: the transferrin case. Bioorg Med Chem. 2011;19:4135–43.
180. Khan NS, Riggle BA, Seward GK, Bai Y, Dmochowski IJ. Cryptophane-folate biosensor for 129Xe NMR. Bioconjug Chem. 2015;26:101–9.
181. Seward GK, Wei Q, Dmochowski IJ. Peptide-mediated cellular uptake of cryptophane. Bioconjug Chem. 2008;19:2129–35.
182. Seward GK, Bai Y, Khan NS, Dmochowski IJ. Cell-compatible, integrin-targeted cryptophane-129Xe NMR biosensors. Chem Sci. 2011;2:1103–10.
183. Rossella F, Rose HM, Witte C, Jayapaul J, Schröder L. Design and characterization of two bifunctional cryptophane A-based host molecules for xenon magnetic resonance imaging applications. ChemPlusChem. 2014;79:1463–71.
184. Rose HM, et al. Development of an antibody-based, modular biosensor for 129Xe NMR molecular imaging of cells at nanomolar concentrations. Proc Natl Acad Sci U S A. 2014;111:11697–702.
185. Witte C, et al. Live-cell MRI with xenon hyper-CEST biosensors targeted to metabolically labeled cell-surface glycans. Angew Chem Int Ed. 2015;54:2806–10.
186. Schnurr M, Sydow K, Rose HM, Dathe M, Schröder L. Brain endothelial cell targeting via a peptide-functionalized liposomal carrier for xenon hyper-CEST MRI. Adv Healthc Mater. 2015;4:40–5.
187. Sloniec J, et al. Biomembrane interactions of functionalized cryptophane-a: combined fluorescence and 129Xe NMR studies of a bimodal contrast agent. Chemistry. 2013;19:3110–8.
188. Booker RD, Sum AK. Biophysical changes induced by xenon on phospholipid bilayers. Biochim Biophys Acta. 2013;1828:1347–56.
189. Schnurr M, Witte C, Schröder L. Functionalized 129Xe as a potential biosensor for membrane fluidity. Phys Chem Chem Phys. 2013;15:14178.
190. Schnurr M, Witte C, Schröder L. Depolarization laplace transform analysis of exchangeable hyperpolarized 129Xe for detecting ordering phases and cholesterol content of biomembrane models. Biophys J. 2014;106:1301–8.
191. Koskela H, Heikkinen O, Kilpeläinen I, Heikkinen S. Rapid and accurate processing method for amide proton exchange rate measurement in proteins. J Biomol NMR. 2007;37:313–20.
192. Zaiss M, Schnurr M, Bachert P. Analytical solution for the depolarization of hyperpolarized nuclei by chemical exchange saturation transfer between free and encapsulated xenon (HyperCEST). J Chem Phys. 2012;136:144106–10.
193. Kunth M, Witte C, Schröder L. Quantitative chemical exchange saturation transfer with hyperpolarized nuclei (qHyper-CEST): sensing xenon-host exchange dynamics and binding affinities by NMR. J Chem Phys. 2014;141:194202.
194. Kunth M, Witte C, Schröder L. Determination of Absolute xenon-host concentration by quantitative 129Xe MRI using hyper-CEST. In: Proceedings of the WMIC; 2014.
195. Sun PZ, et al. Simulation and optimization of pulsed radio frequency irradiation scheme for chemical exchange saturation transfer (CEST) MRI-demonstration of pH-weighted pulsed-amide proton CEST MRI in an animal model of acute cerebral ischemia. Magn Reson Med. 2011;66:1042–8.
196. Keupp J, Baltes C, Harvey PR, van den Brink J. Parallel RF transmission based MRI technique for highly sensitive detection of amide proton transfer in the human brain at 3T. Proc Intl Soc Magn Reson Med. 2011;19:710.
197. Song X, et al. CEST phase mapping using a length and offset varied saturation (LOVARS) scheme. Magn Reson Med. 2012;68:1074–86.
198. Friedman JI, McMahon MT, Stivers JT, Van Zijl PCM. Indirect detection of labile solute proton spectra via the water signal using frequency-labeled exchange (FLEX) transfer. J Am Chem Soc. 2010;132:1813–5.
199. Lin C-Y, et al. Using frequency-labeled exchange transfer to separate out conventional magnetization transfer effects from exchange transfer effects when detecting ParaCEST agents. Magn Reson Med. 2012;67:906–11.

200. Lin C-Y, Yadav NN, Ratnakar J, Sherry AD, van Zijl PCM. In vivo imaging of paraCEST agents using frequency labeled exchange transfer MRI. Magn Reson Med. 2014;71:286–93.

201. Yadav NN, et al. Detection of rapidly exchanging compounds using on-resonance frequency-labeled exchange (FLEX) transfer. Magn Reson Med. 2012;68:1048–55.

202. Vinogradov E, Soesbe TC, Balschi JA, Sherry AD, Lenkinski RE. pCEST: positive contrast using chemical exchange saturation transfer. J Magn Reson. 2012;215:64–73.

203. Soesbe TC, Togao O, Takahashi M, Sherry AD. SWIFT-CEST: a new MRI method to overcome T_2 shortening caused by PARACEST contrast agents. Magn Reson Med. 2012;68:816–21.

204. Liu G, Gilad AA, Bulte JWM, van Zijl PCM, McMahon MT. High-throughput screening of chemical exchange saturation transfer MR contrast agents. Contrast Media Mol Imaging. 2010;5:162–70.

205. Varma G, Lenkinski RE, Vinogradov E. Keyhole chemical exchange saturation transfer. Magn Reson Med. 2012;68:1228–33.

206. Döpfert J, Witte C, Kunth M, Schröder L. Sensitivity enhancement of (hyper-)CEST image series by exploiting redundancies in the spectral domain. Contrast Media Mol Imaging. 2014;9:100–7.

207. Xu X, Lee J-S, Jerschow A. Ultrafast scanning of exchangeable sites by NMR spectroscopy. Angew Chem Int Ed Engl. 2013;52:8281–4.

208. Döpfert J, Witte C, Schröder L. Slice-selective gradient-encoded CEST spectroscopy for monitoring dynamic parameters and high-throughput sample characterization. J Magn Reson. 2013;237:34–9.

209. Wilson NE, D'Aquilla K, Debrosse C, Hariharan H, Reddy R. Localized, gradient-reversed ultrafast z-spectroscopy in vivo at 7T. Magn Reson Med. 2016;76:1039–46.

210. Zhou IY, et al. Tissue characterization with quantitative high-resolution magic angle spinning chemical exchange saturation transfer Z-spectroscopy. Anal Chem. 2016;88:10379–83.

211. Döpfert J, Witte C, Schröder L. Fast gradient-encoded CEST spectroscopy of hyperpolarized xenon. ChemPhysChem. 2014;15(2):261–4. https://doi.org/10.1002/cphc.201300888.

212. Zhao L, et al. Gradient-echo imaging considerations for hyperpolarized 129Xe MR. J Magn Reson B. 1996;113:179–83.

213. Boutin C, Léonce E, Brotin T, Jerschow A, Berthault P. Ultrafast Z-spectroscopy for 129Xe NMR-based sensors. J Phys Chem Lett. 2013;4:4172–6.

214. Reich HJ. WinDNMR: dynamic NMR spectra for windows. J Chem Educ. 1995;72:1086.

Innovative PET and SPECT Tracers

11

Ulrich Abram

Abstract
Although suitable radiotracers are available for many clinical applications in emission tomography such as SPECT and PET, improved tracers are required for advanced nuclear medical imaging. But their development remains a challenge. The demand for new tracers arises from recent progress in the development of imaging techniques and innovations in isotope production. Particularly the latter has triggered an interest in hitherto less favored radionuclides for routine applications of nuclear medical imaging. Since most new tracers are radioactive metals, the coordination chemistry of some of the elements of interest needs to be extended to meet the specific requirements that apply to the preparation of radioactive pharmaceuticals.

New tracers for SPECT and PET are typically developed for one or more of the following reasons: (a) to optimize (frequently multidentate) ligand structures for specific radiometal ions, (b) to develop multiuse ligands for the complexation of multiple metal ions, (c) to search for convenient coupling strategies for bioconjugates, and (d) to provide agents for theranostic solutions. This chapter gives an overview of strategies in radiochemistry and tracer development for nuclear medical imaging applications.

11.1 Introduction

Despite significant advances in the development of medical imaging modalities that do not use ionizing radiation, computed tomography (CT) and nuclear medical techniques (SPECT, PET) still take a prominent place in the portfolio of methods

U. Abram
Freie Universität Berlin, Institute of Chemistry and Biochemistry, Berlin, Germany
e-mail: Ulrich.Abram@fu-berlin.de

© Springer International Publishing AG 2018
I. Sack, T. Schaeffter (eds.), *Quantification of Biophysical Parameters in Medical Imaging*, https://doi.org/10.1007/978-3-319-65924-4_11

currently in clinical use. The provision of healthcare worldwide relies on the availability of diagnostic tests wherever needed and at reasonable costs for individual patients and national healthcare systems.

The probes used in SPECT and PET studies are key to meeting these demands. They are the chemical compounds that contain the radioactive tracer isotope and provide the desired organ distribution for the administered radiopharmaceutical. Classically, small molecules are preferred, since they frequently possess favorable pharmacokinetics and distribution patterns due to their small size [1]. More specific studies, however, also require specific distribution mechanisms, which also stimulate the related chemistry. Imaging a biological target with a reasonable signal-to-noise ratio and with high specificity requires the development of molecules optimized for the specific imaging modality. The molecular probes must also follow the specific biological transport and uptake mechanisms of the target organ. This can be accomplished by direct radioactive labelling of target-seeking biomolecules or by the synthesis of bioconjugates in which radioactive tracers are attached to target-seeking molecular fragments via a covalent spacer. The first approach is limited to radioactive isotopes of the native elements of organic compounds (carbon, hydrogen, nitrogen, oxygen, sulfur, phosphorus, selenium) or halides (fluorine, iodine), which may replace H, OH, or CH_3 groups in organic molecules without drastically changing their biological properties. For the majority of radioactive metal ions, however, direct labelling is less favorable or impossible, as is explained in the following sections. Thus, the development of suitable synthetic routes for coordination chemistry-based bioconjugates remains a major task for the upcoming years. The recent progress in isotope production, which has made additional radiometals available for diagnostic mass applications, and the goal of coupling imaging tools with metal-based radiotherapy in what is known as theranostics represent new challenges for the chemistry behind nuclear medical imaging.

11.2   SPECT and PET: The Methods and Classic Tracers

Single photon emission computed tomography (SPECT) and positron emission tomography (PET) are well-established imaging techniques in nuclear medicine. Both rely on the ability of medium-energy gamma radiation to penetrate living organisms and to deliver spatially resolved images by means of a suitable detector system. However, these two imaging techniques use different gamma radiation sources. While most SPECT applications exploit gamma quants emitted as accompanying radiation of ß$^-$ decay or (preferably) arising from metastable nuclear isomers, the γ-quants used for PET are formed by the annihilation of an electron-positron pair. The latter process produces two γ-quants, which are coincidently emitted in an angle of exactly 180° to each other (Fig. 11.1) and can be used for a coincident detection of the events in a tube-shaped detector system, PET normally produces images of higher quality and resolution. On the other hand, the instrumentation required for the generation of high-quality PET images and particularly the high production costs of classic PET tracers such as ^{18}F or ^{11}C still limit the mass application of positron emitters. Unfortunately, there exist no radioactive isotopes of

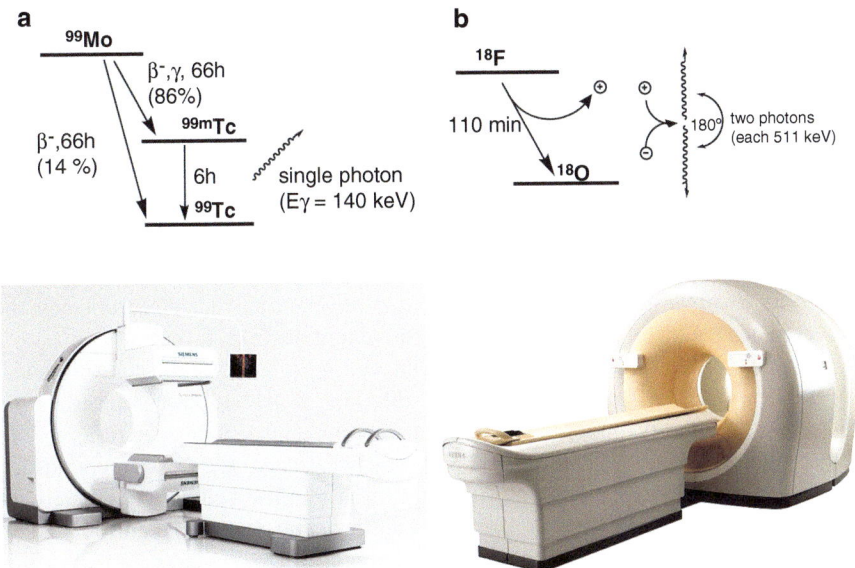

Fig. 11.1 The formation of (**a**) photons from the metastable ^{99m}Tc and a typical SPECT/CT system (courtesy of Siemens Healthcare GmbH) and (**b**) the formation of annihilation photons from the positron decay of ^{18}F and a PET TF/CT system (courtesy of Philips Healthcare) with the typical detector ring

"organic" chemical elements such as carbon, nitrogen, oxygen, or sulfur with optimal nuclear properties in terms of decay scheme, half-life, availability, and photon energy.

Thus, a rather unusual nuclide, the metastable nuclear isomer ^{99m}Tc, has become the "workhorse" in diagnostic nuclear medicine. It is estimated that between 80 and 90% of all studies in diagnostic nuclear medicine are currently performed with this nuclide, amounting to a total of about 35 million clinical studies per year [2–5]. The dominating role of ^{99m}Tc in diagnostic nuclear medicine is mainly attributable to its favorable nuclear properties. Its radiation is almost free of accompanying particle radiation, the γ-energy of 140 keV is in the optimal range for collimation, and the half-life of 6 h is long enough to acquire diagnostic scans, but short enough to preclude a considerable radiation burden to the patient. The fact that no organ accumulates technetium ensures rapid excretion of the non-targeted technetium compound from the human body. Another important argument for the use of ^{99m}Tc is its permanent availability at clinical sites from ^{99}Mo/^{99m}Tc nuclide generators [6].

Chemically, technetium, which is the first man-made element, is not an optimal candidate for the synthesis of novel tracers for medical imaging procedures. We find it in the 7th group of the Periodic Table of Elements, together with manganese and rhenium. This means that the chemistry behind almost all nuclear medical applications of ^{99m}Tc is the coordination chemistry of the artificial transition metal technetium. This causes a number of problems in the synthesis of biologically active tracers, since it is chemically impossible to simply attach a ^{99m}Tc label to a small organic molecule without significantly changing its biological properties. Recent

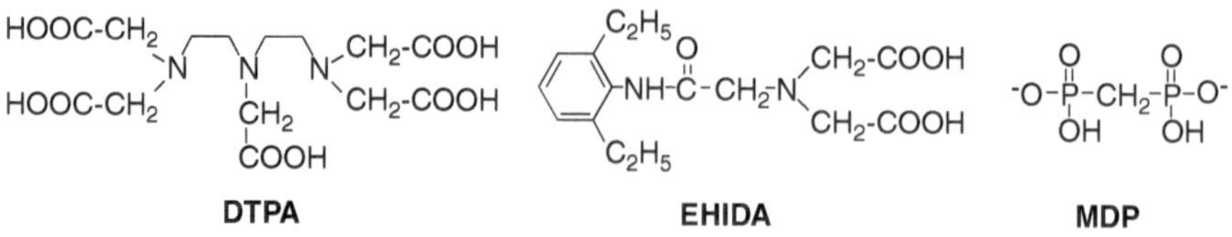

HOOC-CH₂   CH₂-COOH   C₂H₅ O   CH₂-COOH   O   O
        N N N   NH-C-CH₂N   ⁻O-P-CH₂-P-O⁻
HOOC-CH₂   CH₂ CH₂-COOH   CH₂-COOH   OH   OH
            COOH   C₂H₅

       DTPA   **EHIDA**   **MDP**

Fig. 11.2 Ligands of the first generation of ^{99m}Tc radiopharmaceuticals used for kidney (DTPA), liver (EHIDA), and bone scintigraphy (MDP)

attempts to solve such problems at least for larger biomolecules resulted in the development of bioconjugates and will be treated in Sect. 11.4.

Most of the classic ^{99m}Tc imaging agents belong to a group of relatively small molecules ($M < 2000$) whose biological distribution patterns are determined by their charge, polarity, lipophilicity, and other parameters. A first generation of ^{99m}Tc radiopharmaceuticals was developed almost 40 years ago, and the compounds are still in regular clinical use. Many ligands currently in use belong to the family of complexones. The periphery of the related aminoacetic acids can readily be modified, and via the resulting changes in the lipophilicity of the anionic technetium complexes, they can be directed to specific biological excretion pathways. For instance, the technetium complex with the hydrophilic DTPA (diethylenetriamine pentaacetic acid) ligand is directed to renal excretion, while hepatobiliary excretion is preferred by the more lipophilic EHIDA (N-(2,6-diethylacetanilido)iminodiacetic acid) derivative, and consequently the compounds are used for kidney and liver scintigraphy, respectively (Fig. 11.2). It is interesting to note that the exact chemical structures of these classic technetium radiopharmaceuticals are still not known [6]. The same holds true for the ^{99m}Tc complex with MDP (methylene diphosphonate) (Fig. 11.2). The compound (like related complexes with 1-hydroxyethylene diphosphonate or 1-hydroxy-4-aminobutylidene-1,1-diphosphonate) has been in routine use for bone imaging for a long time. It is highly probable that the diphosphonates act as bidentate ligands for technetium and calcium and, thus, can incorporate ^{99m}Tc into the mechanism of the formation of the inorganic components during bone growth.

Much more structural information is available for a second generation of ^{99m}Tc radiopharmaceuticals, since most of these heart- and brain-seeking ^{99m}Tc molecules, shown in Fig. 11.3, were specifically designed taking into account fundamental pharmacological parameters and rational structure-activity relationships. This means that the sum parameters of the entire ^{99m}Tc complex molecules have been optimized. Thus, the term "labelling," which means the replacement of a nonradioactive atom in a target-finding molecule by a radioactive one of the same element of by another group chemically resembling the replaced atom as closely as possible, should not be used for such metal-based radiopharmaceuticals. The biological distribution pathway of metal-based, small-molecule SPECT (and also PET) tracers is determined by the nature of the ligand (donor atoms, polarity, lipophilicity), charge, size and shape of the resulting complex molecule, and the inherent polarity of a potential metal core.

Following the fundamental concepts of pharmacology, cationic compounds became interesting for the replacement of ^{201}Tl in myocardial imaging. Two of the widely used ^{99m}Tc complexes are shown in Fig. 11.3: [99m]Tc sestamibi and ^{99m}Tc-tetrofosmin. Both compounds are cationic and contain ligands with peripheral ether substituents. These functionalities ensure retention of the radioactive tracers in myocardial cells by (partial) hydrolysis and the associated change in the charges of the complex molecules [6]. Neutral molecules with a balanced lipophilicity and a relatively low molecular weight are able to cross the blood-brain barrier and are thus potential candidates for brain imaging. ^{99m}Tc-HMPAO, an optimized member of a whole series of technetium complexes with tetradentate amineoxime ligands, was the first technetium-containing standard drug for this application. Retention of this compound in the brain is explained by an enzymatic conversion into a more hydrophilic complex, which can no longer pass the blood-brain barrier. This trapping mechanism is more effective for the complex with the *d,l*-isomer of the chiral ligand, while its *meso*-form shows much faster redistribution and is, thus, not suitable for brain imaging. This example nicely illustrates the problems that may occur when chiral molecules are involved in the development of pharmaceuticals. Such problems were avoided with the second brain-seeking agent shown in Fig. 11.3, ^{99m}Tc-ECD. This agent contains the ethylene cysteine ester dimer as a nonchiral, tetradentate ligand. It belongs to the extensively studied diaminodithiolato family, and only one of the two amino groups deprotonates during coordination with the $\{TcO\}^{3+}$ core. Brain retention of the resulting neutral technetium(V) complex is effected by (partial) hydrolysis of the peripheral ester functionalities, a mechanism already discussed for tissue retention of the heart-seeking technetium radiopharmaceuticals [6].

While ^{99m}Tc diagnostic agents are widely used, routine application of other SPECT tracers is comparatively rare. They are used in special cases or when the availability of [99m]TcO$_4^-$ is (locally) restricted. Typical nuclides and applications are [123,131]I-iodohippurate for renal imaging, [47]Ca for studies of calcium metabolism, ^{201}Tl for myocardial perfusion, ^{75}Se-selenomethionin for pancreas studies, ^{133}Xe for pulmonary perfusion, and ^{111}In-oxine for the detection of infections [2].

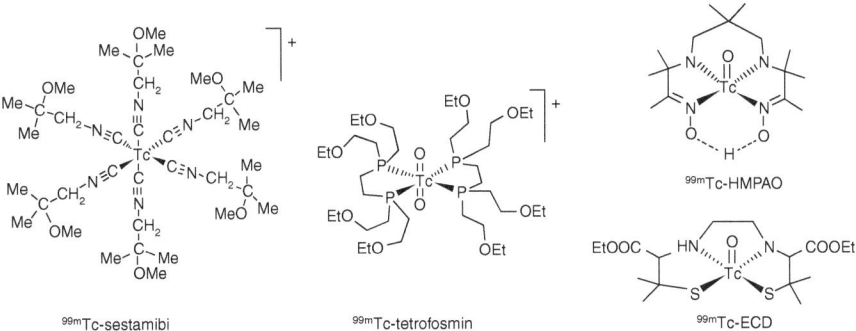

Fig. 11.3 ^{99m}Tc complexes for routine myocardial studies (^{99m}Tc-sestamibi and ^{99m}Tc-tetrofosmin) and brain perfusion (^{99m}Tc-HMPAO and ^{99m}Tc-ECD)

Similar to the situation with SPECT tracers, there is one positron emitter that dominates routine applications: ^{18}F. It can be prepared in cyclotron reactions starting from ^{18}O or ^{20}Ne targets. For the majority of ^{18}F applications, a (p,n) reaction with a $H_2^{18}O$ target is used, which yields $^{18}F^-$ ions in aqueous solution. This method delivers the isotope in a high specific activity (typically in the range of 100 GBq/μmol). The fluoride ions are suitable for a large number of nucleophilic reactions. Electrophilic reactions require production of F_2. This can also be made from 18-oxygen using an $^{18}O_2$ gas target with the same nuclear reaction or by a $^{20}Ne(d,\alpha)^{18}F$ reaction [7, 8]. The availability of ^{18}F for both nucleophilic and electrophilic reactions gives access to a large variety of ^{18}F-containing molecules that can be used in clinical imaging and as valuable tools in the development and/or evaluation of (conventional) drugs [9].

^{18}F emits positrons with β_{max} = 0.630 MeV, which corresponds to a maximum range of the positrons of 2.4 mm in water and yields images with good spatial resolution. The workhorse of clinical ^{18}F applications is ^{18}F-desoxyglucose (FGD). As a modified sugar, this molecule can be used for imaging of a large variety of malignant tumors and a number of metabolic diseases (see Chap. 18). Some examples of ^{18}F-labelled organic molecules, including amino acids, nucleosides, and neurotransmitters, are presented in Fig. 11.4. A special case is 5-fluorouracil, which is frequently used for the (conventional) treatment of several forms of cancer, and the degree of uptake of the ^{18}F-labelled compound in tumor tissue helps to evaluate the response to 5-fluorouracil chemotherapy.

The half-lives of the "organic" positron emitters ^{11}C, ^{13}N, and ^{15}O are very short (20 min, 10 min, and 2 min, respectively), representing a considerable obstacle to the use of these PET nuclides in routine applications. Naturally, all processes related to the synthesis of these pharmaceuticals, their administration, and their biodistribution and/or metabolism must be fast [2]. As a result of these restrictions, the two-minute nuclide ^{15}O can only be used as $C^{15}O_2$, $H_2^{15}O$, or $^{15}O_2$ for fast blood flow or oxygen extraction studies (see Chap. 21). Some more specific organic compounds can be prepared for ^{13}N and particularly for ^{11}C. $^{11}CO_2$ and $^{11}CH_4$ are common precursors for the synthesis of ^{11}C-labelled compounds. Finally, some amino acids,

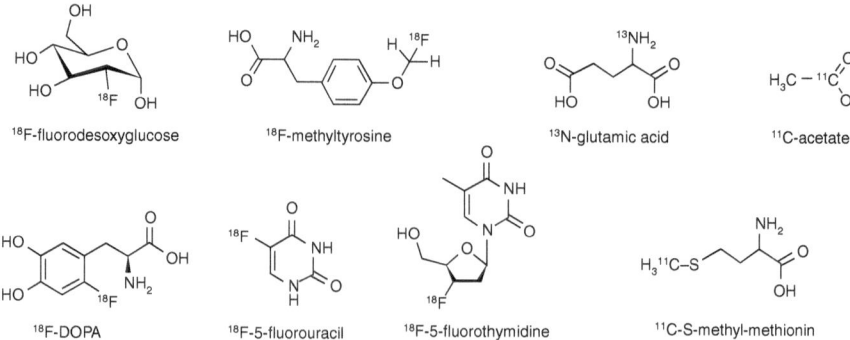

Fig. 11.4 PET tracers for various diagnostic applications and treatment monitoring

nucleosides, or drugs can be prepared and employed for imaging of blood flow, tumor viability, and neuroreceptor characterization. Some frequently used compounds are shown in Fig. 11.4.

11.3 Prospective SPECT and PET Radionuclides and Tracers

Recent developments in isotope production, manufacturing, and supply, but also progress made in imaging modalities and related data processing, have drawn attention to nuclides hitherto difficult to access or neglected for other reasons. Tables 11.1 and 11.2 list suitable PET and SPECT tracers together with some basic information on their production and radiation properties.

As outlined in subchapter 11.2, radiopharmaceuticals based on ^{99m}Tc and ^{18}F currently dominate routine clinical applications. This domination is mainly attributable to the radiation properties and/or the availability of these nuclides and not their chemical properties or related favorable biological distribution patterns of readily

Table 11.1 PET nuclides

Isotope	Physical half-life	β^+ energy (MeV)[a]	Isotope production
^{18}F	6.0 h	0.634	Cyclotron
^{11}C	20.4 min	1.983	Cyclotron
^{13}N	9.9 min	2.220	Cyclotron
^{15}O	2.0 min	2.757	Cyclotron
^{68}Ga	1.1 h	1.899	^{68}Ge/^{68}Ga generator
^{62}Cu	9.6 min	2.910	^{62}Zn/^{62}Cu generator
^{64}Cu	78.2 h	0.653	Cyclotron
^{182}Rb	1.3 min	3.381	^{82}Sr/^{82}Rb generator
^{44}Sc	80.2 h	1.474	^{44}Ti/^{44}Sc generator
^{124}I	100.2 h	2.138, 1.532	Cyclotron
^{86}Y	14.7 h	1.221	Cyclotron

[a]Data taken from http://www.nucleide.org/DDEP_WG/DDEPdata.htm

Table 11.2 SPECT nuclides

Isotope	Physical half-life	Mode of decay (%)	γ Energy (MeV)[a]	Isotope production
^{99m}Tc	6.0 h	IT (99.99)	0.141	^{99}Mo/^{99m}Tc generator
^{131}I	8 days	β^- (100), γ	0.284, 0.365	Reactor
^{123}I	13.2 h	EC (100)	0.159	Cyclotron
^{111}In	67.4 h	EC (100)	0.171, 0.245	Cyclotron
^{201}Tl	73 h	EC (100)	0.135, 0.167	Cyclotron
^{67}Ga	78.2 h	EC (100)	0.093, 0.185, 0.300, 0.393	Cyclotron
^{133}Xe	5.2 days	β^-, γ	0.081	Reactor

[a]Data taken from http://www.nucleide.org/DDEP_WG/DDEPdata.htm
IT internal transition, *EC* electron capture

available compounds of these nuclides. Therefore, research aims not only at developing ^{99m}Tc and ^{18}F compounds with improved properties, but also at replacing them by agents with other isotopes, including those with relatively short half-lives such as ^{11}C. This has become possible by the development of novel and fast synthetic routes for important key compounds of the 11-carbon chemistry such as $^{11}C–CH_3I$, which can nowadays be produced from cyclotron-produced $^{11}CO_2$ with reaction times of less than 30 min [10]. This is commonly done using a kind of online procedure with computer-controlled synthesis modules. Such modules integrate a series of vials, solvent reservoirs, valves, pumps and heaters, and chromatographic columns. They allow remote, computer-controlled synthesis of radiotracers with high specific activities and reduce the radiation burden of the operator to a minimum. A typical synthesis module is shown in Fig. 11.5.

The modular design and the operation in a shielded cell grant access to synthetic procedures generally not recommended for use with highly radioactive compounds. Thus, $^{11}CO_2$ is reduced in a catalyzed gas phase reaction to give $^{11}CH_4$, which is finally converted to $^{11}CH_3I$ by free-radical iodination. The efficient and fast synthesis of methyl iodide gives access to a variety of innovative ^{11}C-methylated tracers. A collection of such compounds is shown in Fig. 11.6. It is clearly apparent that relatively "complicated" compounds, such as ^{11}C-raclopride, a synthetic antagonist on D_2 dopamine receptors; ^{11}C-doxepin, which binds selectively to histamine H1 receptors; and many other products, can be prepared in a rational, fast, and reproducible way. This approach also paves the way for other reactive secondary precursors such as ^{11}C-methanol, ^{11}C-formaldehyde, ^{11}CO, $H^{11}CN$, ^{11}C-carboxylates, $^{11}CS_2$, or other methylating agents such as ^{11}C-triflate or ^{11}C-nitromethane [10].

The development of innovative fluorination methods for (nonradioactive) organic compounds, particularly those with "late-stage" strategies [11, 12], has also stimulated the synthesis of ^{18}F tracers. Unfortunately, not all of the sophisticated

Fig. 11.5 Typical computer-controlled module for the automated synthesis of PET tracers (courtesy of GE Healthcare)

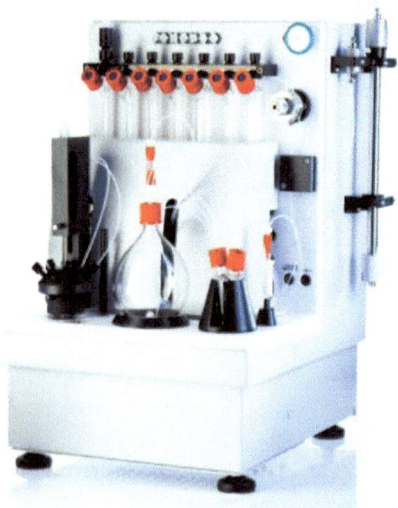

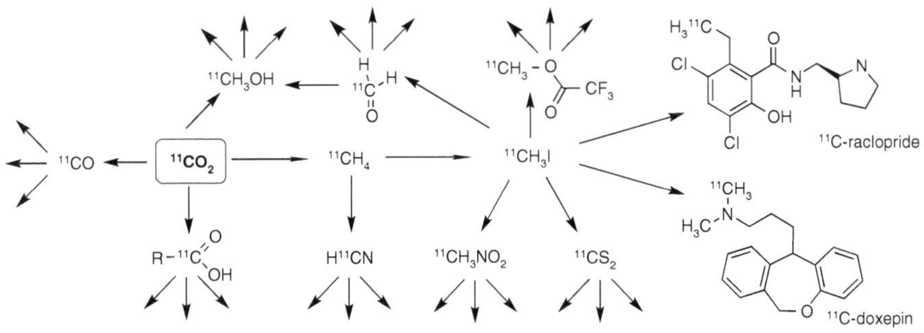

Fig. 11.6 ^{11}C precursors for the synthesis of innovative and flexible ^{11}C tracers

procedures can readily be transferred to a synthesis under "clinical conditions." Nevertheless, the established methods of cyclotron reactions provide access to ^{18}F in the form of "nucleophilic" and "electrophilic" precursors; however, it must be noted that batches of the nucleophilic $^{18}F^-$ are available with high specific activity (typically 100 GBq/μmol and higher), while the production and isolation of the electrophilic $^{18}F_2$ in corresponding gas targets lower the accessible specific activity significantly (typical values are lower than 1 GBq/μmol). This is mainly due to the need to add nonradioactive fluorine during the isolation of $^{18}F_2$ from the target, which in turn lowers the specific activity of ^{18}F (or increases the mass of the produced radiotracer) and thus strongly limits the use of ^{18}F compounds synthesized from electrophilic procedures in nuclear medical receptor binding studies due to receptor saturation or pharmacological effects resulting from the relatively high (chemical) concentration of the fluorinated compounds [13].

For both nucleophilic and electrophilic reactions, the ^{18}F from the target has to be converted into a chemical form that is suitable for the ongoing organic labelling procedures. Therefore, the nucleophilicity of $^{18}F^-$, which is drastically decreased by a number of strong hydrogen bonds in aqueous solutions, must be increased. This is commonly done by the addition of phase transfer agents such as tetramethylammonium or tetrabutylammonium ions or cryptants such as Kryptofix-222, which forms stable complexes with the accompanied potassium ions and provides transfer of F^- into aprotic solvents (acetonitrile, dimethylformamide, or dimethylsulfoxide). After removal of the water, which is commonly done by azeotropic distillation, the $^{18}F^-$ is highly reactive and can be used for numerous reactions following the S_N2 or S_NAr mechanisms. They allow aliphatic as well as aromatic fluorination. Again, the use of automated synthesis modules decreases the time required for the production and purification of novel ^{18}F radiotracers.

Some examples of the use of such an approach for aliphatic fluorination are presented in Fig. 11.7. They include the synthesis of ^{18}F-fluoro-17β-estradiol, which was performed in a cassette-type ^{18}F-FDG synthesizer [14], but also that of ^{18}F-fluoroethylflumazenil, a tracer for the imaging of human benzodiazepine receptors [15], and a series of phosphonium salts (MitoPhos), which are explored as candidates for the imaging of myocardial perfusion by targeting mitochondrial membrane potential depletion [16].

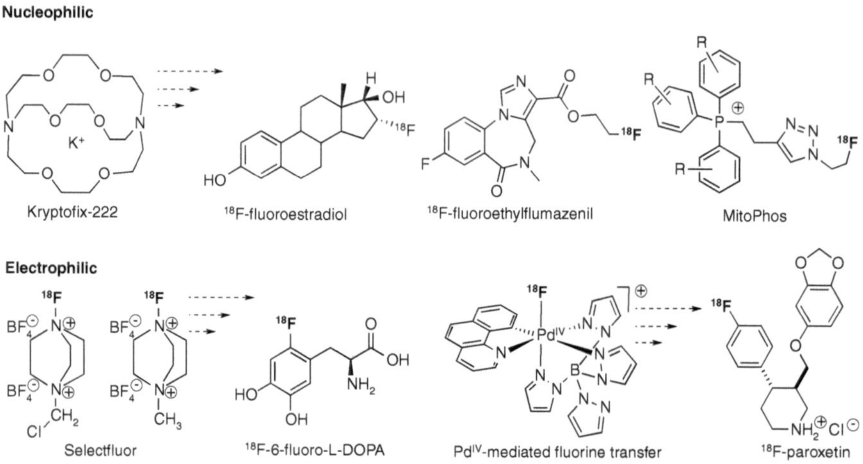

Nucleophilic

Kryptofix-222 ^{18}F-fluoroestradiol ^{18}F-fluoroethylflumazenil   MitoPhos

Electrophilic

Selectfluor ^{18}F-6-fluoro-L-DOPA   Pd^{IV}-mediated fluorine transfer ^{18}F-paroxetin

Fig. 11.7 ^{18}F transfer reagents for nucleophilic and electrophilic reactions together with typical products

Several attempts have been undertaken to make electrophilic fluorination more effective, particularly with regard to the achievable specific activity of the nuclide and the selectivity of the labelling procedures. An interesting approach is the conversion of "nucleophilic" ^{18}F$^-$ ions into ^{18}F$_2$ by means of an electric discharge chamber. The procedure, which involves the intermediate formation of ^{18}F-fluoromethane, results in a reported specific activity of the final ^{18}F-labelled radiopharmaceutical of 15 GBq/μmol[17, 18]. The selectivity of electrophilic fluorination can be increased by the intermediate conversion of ^{18}F$_2$ into less reactive (and thus more selective) intermediates such as XeF$_2$, N-fluorobenzenesulfonimides, or N-fluoropyridinium salts. An interesting, mild electrophilic fluorination reagent was introduced by V. Gouveneur and her group: Selectfluor (Fig. 11.7) [17]. This diamine can be prepared using the electric discharge method mentioned above and has been used in the synthesis of 6-[^{18}F]fluoro-L-DOPA [18]. Another interesting approach for electrophilic fluorination has been proposed by the group of Ritter [19] (Fig. 11.7). They use an organopalladium compound as fluoride transfer reagent. The palladium ion in the octahedral complex has the high formal oxidation state "+IV" and acts as oxidant, which transfers the substrate to the nucleophilic fluoride, while being reduced to a lower oxidation state. It has successfully been used, e.g. for the synthesis of ^{18}F-paroxetine. The transfer of this conceptually highly interesting two-step approach to an automated ^{18}F reaction for clinical application, however, revealed a number of problems [20]. Nevertheless, the use of transition metal complexes as transfer reagents in electrophilic fluorination is a promising innovative route, and we are likely to read about successful applications of this method in the near future. Similar reactions with nickel and copper complexes have already been applied for the ^{18}F labelling of aromatic systems [21–26].

The introduction of copper-catalyzed 1,3-dipolar cycloaddition—a so-called click reaction—by Huisgens and Sharpless [27, 28] has also influenced the development of organic and inorganic radiopharmaceuticals. A "click reaction" is a convenient option for orthogonal coupling of a target-seeking molecule to a radio-active label. The chemistry of corresponding metal complexes will be described in the following subchapters. "Click reactions" have also been used for the designing of ^{11}C and ^{18}F compounds [29]. An example using Mitophos derivatives is shown in Fig. 11.7 [16].

Although the dominance of ^{99m}Tc in classic SPECT imaging is based on the use of small molecules (some of them are shown in Fig. 11.3), current research in ^{99m}Tc tracer chemistry mainly focuses on bioconjugates, which will be discussed in sub-chapter 11.4. Nevertheless, some more small-molecule ^{99m}Tc tracers shall shortly be mentioned. Most of them were developed in parallel to or after the major products presented in Fig. 11.3. It is obvious that the cardiac imaging agents shown in Fig. 11.8 differ completely from the cationic reagents like ^{99m}Tc-sestamibi or ^{99m}Tc-tetrofosmin in terms of their structure. ^{99m}TcN-NOET and ^{99m}Tc-teboroxim are neu-tral compounds, and their myocardial uptake and retention are most probably due to membrane binding of the highly lipophilic compounds. On the other hand, the structure of ^{99m}Tc-MAG$_3$ is very similar to those of the brain perfusion reagents ^{99m}Tc-HMPAO and ^{99m}Tc-ECD shown in Fig. 11.3. All three compounds possess five-coordinate Tc(V) ions with a central {TcO}$^{3+}$ core and tetradentate ligands. Unlike the neutral, lipophilic brain-seeking compounds, however, ^{99m}Tc-MAG$_3$ is a hydrophilic anion with a peripheral carboxylic group. These examples underline the previous discussion that the biological distribution pattern is governed by the over-all charge, lipophilicity, and polarity of the entire molecule.

Although ^{99m}Tc radiopharmaceuticals are inexpensive, well established, and available for almost all routine SPECT studies, they have a considerable drawback: the production of the mother isotope, ^{99}Mo, which is commonly obtained from ^{235}U targets by fission with a high yield of approximately 6% in nuclear reactors. With regard to the huge number of nuclear reactors running worldwide, this sounds like an inexhaustible source. In fact, however, most ^{99}Mo is produced in only five research reactors, which are all about 40 years old, and planned or unplanned out-ages of these reactors cause considerable problems in ^{99}Mo production. A

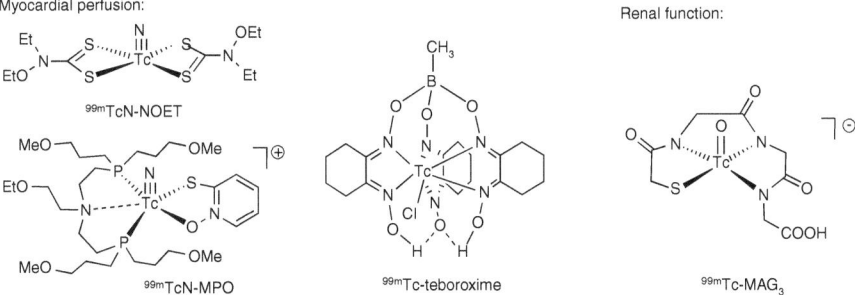

Fig. 11.8 Small-molecule ^{99m}Tc tracers for myocardial perfusion imaging and renal function imaging

considerable shortage of ^{99}Mo occurred in 2008–2009, and another shortage is expected during the scheduled outage in 2017 and 2018 of a Canadian reactor, which currently supplies about 20 percent of the world's ^{99}Mo demand [30]. Thus, there are good reasons to search for alternative agents for use in routine imaging modalities.

Additionally, new agents are desirable to replace or supplement traditional SPECT imaging procedures by PET solutions. ^{68}Ga is a nuclide with high potential for such a task. It can be eluted from ^{68}Ge/^{68}Ga generators. The germanium mother isotope is efficiently produced in a $(p,2n)$ reaction in high-energy accelerators [4, 5, 31]. Due to the long half-life of ^{68}Ge (271 d), such generators can produce ^{68}Ga at clinical sites over a period of months. The immobilization of ^{68}Ge in such commercial generators is accomplished by several solid carrier materials: SiO_2, TiO_2, SnO_2, or polymer-supported CeO_2 [32]. The resulting ^{68}Ga^{3+} ions are washed from such columns using diluted HCl. Another interesting nuclide in this context is ^{64}Cu. It decays in three different modes (EC, ß$^-$, and ß$^+$) and can, thus, potentially be used for imaging and therapy. The formation and disintegration characteristics of the two isotopes are shown in Fig. 11.9.

About the "radiopharmaceutical chemistry" of ^{64}Cu virtually nothing is known [33]. Studies have been conducted using ^{64}Cu complexes with diacetyl-2,3-bis(N^4-methyl-3-thiosemicarbazone) (^{64}Cu-ATSM, see Fig. 11.10) and some related thiosemicarbazones for PET imaging of hypoxia [34–36]. Most of other ^{64}Cu studies (e.g., as PET probe for cancer imaging) have been done with non-chelated Cu^{2+} ions [33].

Far more research has been spent on developing the nuclear medical coordination chemistry of gallium. Gallium is a nonphysiological metal and occurs in aqueous solutions exclusively in the oxidation state "+3." Ga^{3+} ions are relatively "hard" Lewis acids and form stable complexes, preferably with ligands having "hard" donor atoms such as oxygen or nitrogen. But stable Ga^{3+} complexes with relatively "soft" sulfur donor atoms also exist. The coordination chemistry of most Ga^{3+} ions is similar to that of the physiologically important Fe^{3+} ions. Both ions have the same charge, electronic configurations without ligand field stabilization energy, and similar radii (Ga^{3+}: 0.62 Å, Fe^{3+}: 0.65 Å) [4]. They form preferably octahedral complexes.

Typical ligand systems for the coordination of Ga^{3+} ions are summarized in Fig. 11.10. Most of them belong to the class of aminocarboxylic acids like the

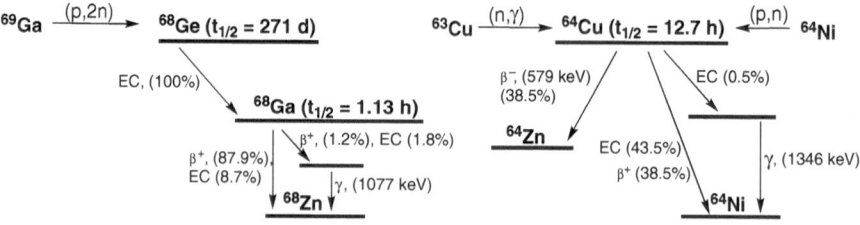

Fig. 11.9 Production and decay schemes of ^{68}Ga and ^{64}Cu

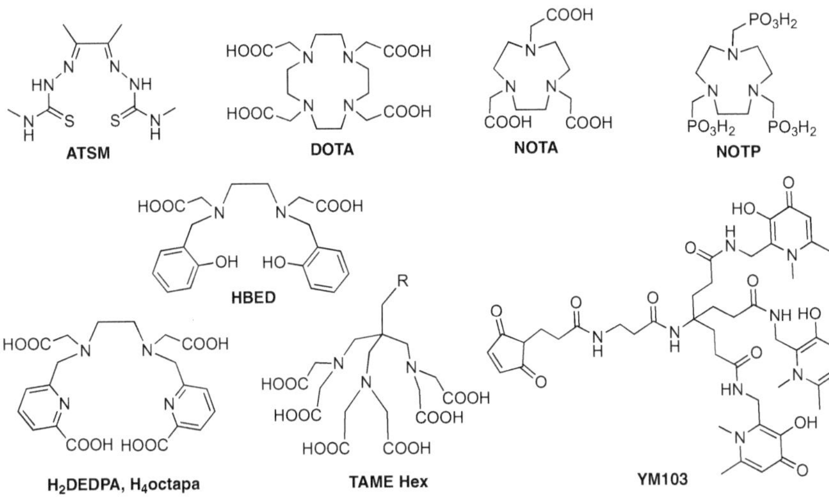

Fig. 11.10 Chelators for ^{64}Cu, ^{68}Ga, and/or ^{111}In complexation

DTPA ligand presented in Fig. 11.2. DTPA is applied for renal imaging as its ^{99m}Tc complex and can also be used for the complexation of Ga^{3+} ions. Most of the ligands shown in Fig. 11.10 are representatives of whole families, and substitutions in the periphery of the organic compounds allow modifications in their coordination modes, the number of donor atoms used for coordination, and lipophilicity. An example of such modifications is shown with the NOTA and NOTP ligands. Both compounds have a 1,4,7-triazacyclononane skeleton, while their side chains have different donor groups. Other representatives including those with thiolates or phosphonic acid esters have also been studied [37, 38]. The hitherto most popular representatives of the class of cyclic ligands are DOTA and NOTA. Both ligands form Ga^{3+} complexes with a thermodynamic stability that is high enough for clinical applications. Of the two derivatives, NOTA is clearly preferred because the "hard" Ga^{3+} ions are perfectly coordinated by the "N_3 unit" of the triazacyclononane ring. The "N_4 unit" of the corresponding DOTA ligand produces more steric strain in the octahedral coordination sphere. On the other hand, coordinated DOTA provides an "additional," non-coordinated carboxylic group for the formation of bioconjugates. Applications of the ligands presented in Fig. 11.10 for bifunctional labelling are discussed in subchapter 11.4.

For the NOTA derivatives, Ga^{3+} chelates form relatively fast when conditions are appropriate. Reaction times are short and no elevated temperatures are required. The same holds true for most of the open-chain chelators such as HBED [39] or TAME. Particularly the facial triamine moiety of the latter ligand system allows generation of ligands with a large structural variety and functional groups and donor atoms that can readily be modified. The ^{68}Ga complex with the tris(hydroxypyridinone) ligand YM103 is formed rapidly under mild conditions and with high yield, thus representing a promising candidate for future clinical applications of ^{68}Ga-based PET [40].

A ligand with valuable properties in terms of tracer chemistry is the open-chained, pyridine-based H_2DEDPA (or $H_4octapa$), which forms stable complexes with gallium and indium [41, 42]. The products are formed in facile procedures at room temperature in a short time and with good labelling yields, which is a clear advantage over the more abrasive conditions of related syntheses with cyclic DOTA, and they form complexes of high thermodynamic stability and kinetic inertness. The indium compound is 8-coordinate with one unbound picolinic acid group. The remaining carboxylic unit recommends H_2DEDPA also as chelator for bioconjugates with ^{68}Ga, ^{111}In, or other M^{3+} ions, which will be discussed in the following subchapter. ^{111}In is produced in a cyclotron from ^{111}Cd via a (p,n) reaction. It emits two photons (172 and 245 keV) and Auger electrons, which can be used for therapy. A classic application of this isotope is blood cell labelling with ^{111}In oxine.

The examples just outlined show that particularly ^{68}Ga has a high potential for assuming an important role in future nuclear medical imaging applications. Before it can become a serious complement or even competitor for the dominating isotope ^{99m}Tc, however, several issues must be addressed: (a) the costs of PET scanners and of routine clinical PET examinations must be significantly lowered to ensure wider use worldwide. (b) the $^{68}Ge/^{68}Ga$ generator systems must be improved to ensure consistent quality of ^{68}Ga that is independent of the supplier. (c) ^{68}Ga reagents must be developed to meet the demands arising from routine nuclear medical use in procedures including myocardial perfusion scans, cerebral imaging, bone scans, lung scintigraphy, and renal and hepatobiliary excretion studies and (d) kit-like formulations for such standard ^{68}Ga reagents should be provided. As long as these requirements are not fulfilled, there are good reasons to develop and/or to improve ^{68}Ga *and* ^{99m}Tc compounds for existing and novel applications. Some recent promising developments in this field will be described in the following sections.

11.4 Targeting and Multi-targeting by Bioconjugation

While real labelling of target-seeking molecules, that means the substitution of a nonradioactive atom by a radioactive one, has been accomplished for most ^{11}C, ^{18}F, and ^{123}I tracers, such a procedure is more or less impossible for radiometals, and much effort has been undertaken to develop the "small-molecule" tracers discussed above. Although many of these tracers have sophisticated biological distribution and retention mechanisms, they result from the optimization of molecular sum parameters and not from specific receptor binding. Attempts to mimic the shape and charge distribution of estradiol by inorganic building blocks have failed so far [43, 44]. More promising is the so-called bifunctional approach with bioconjugated tracer molecules (Fig. 11.11). In such units, the radioactive label (commonly a chelator with a tight bond to the corresponding metal ion) is clearly separated from the receptor-binding part of the molecule (commonly a real or modified biomolecule) by a spacer. Implementation of the spacers can be done in different ways. Frequently, it is an integral part of the chelator and consists of a small alkyl chain with a

Biomolecule
Spacer
Chelator
Radiometal
Labelled Bioconjugate

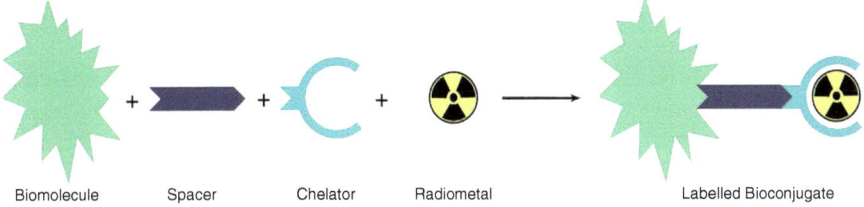

Fig. 11.11 Schematic representation of the bifunctional approach

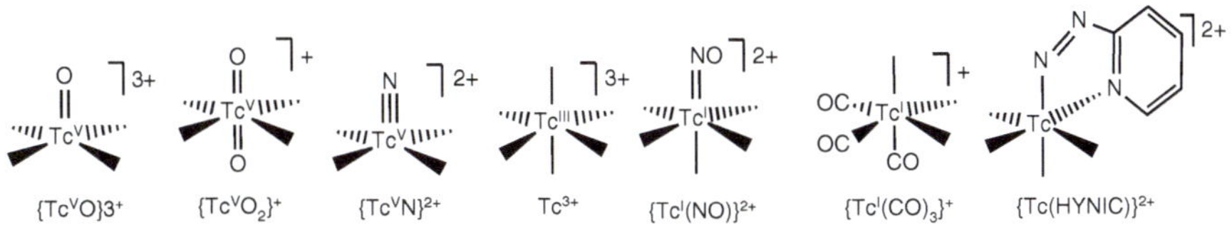

{TcVO}3+
{TcVO2}+
{TcVN}2+
Tc3+
{TcI(NO)}2+
{TcI(CO)3}+
{Tc(HYNIC)}2+

Fig. 11.12 Typical technetium cores

terminal reactive group (e.g., a carboxylic group), but it can also be established during the coupling procedure, e.g., a "click reaction." The latter is the favored method although it requires modification of both the chelator and the biological vector.

Targeting in such large bioconjugates is accomplished by the biological part of the molecule, while the formation of stable or inert chelates remains a domain of coordination chemistry. For technetium compounds, the well-established cores of the small-molecule chemistry are typically also used for the designing of bioconjugates. In conjunction with the chelator, they ensure stable binding of the radioactive metal and determine the net charge of the conjugate. Some frequently used cores are compiled in Fig. 11.12.

It becomes clear that technetium compounds with the metal in different oxidation states require chelators of different denticity and net charge for the formation of chelates with optimal stability. The most important oxidation states for such tracers are "+1," "+5," and (with less examples) "+3." Thus, the selection of ligands with appropriate donor atom constellations becomes important. Sulfur-containing ligands are well suitable for the relatively "soft" Tc^V cores, while oxygen and nitrogen donors are preferred to stabilize lower oxidation states. In most cases, ligands with different functional groups (amines, amides, carboxylates, aliphatic and aromatic hydroxylic groups, thiones, thiolates, or phosphines) are used. This allows optimization of the stability of the chelates and control over of the net charge of the complex. Fortunately, a large portfolio of technetium chelates is available from the development of small-molecules tracers, from which suitable candidates can be selected and equipped with an anchor group for binding the spacer and/or the biomolecule. Classic examples of reactive functionalities suitable for linking chelates to biological vectors are carboxylic groups, active esters, anhydrides, aromatic thiocyanates, and isothiocyanates. In addition, azides and alkynes for "click coupling" became popular for use in potential radiopharmaceuticals.

A small collection of Tc(V) chelates suitable for bioconjugation is shown in Fig. 11.13. The first three compounds possess the $\{TcO\}^{3+}$ core, which is frequently formed for Tc(V) compounds when pertechnetate is reduced using common reducing agents such as stannous chloride in the presence of a suitable ligand. Many oxidotechnetium(V) compounds have a coordination number of five, since the oxido ligand exerts a relatively strong *trans*-labilizing influence. This means that tetradentate ligands provide sufficient stability for the complex molecules, as has been shown for the DADT and tcb-4 ligands of Fig. 11.13 [45, 46]. More examples are presented in references [4–6, 47]. Pseudo-octahedral TcO complexes are frequently established when the position *trans* to the oxido ligand can be occupied by an oxygen donor. In such cases, some electron density is transferred from the Tc = O unit to the *trans* Tc-O bond, which results in higher stability of the complexes. A rare example of a highly stable TcO complex with a pentadentate, trinegative ligand is TcO-tcb-5-COOH [48]. A somewhat different bifunctional strategy is pursued with the HYNIC complexes [49, 50], which enable straightforward synthesis of bioconjugates with a wide range of targeting vectors [4]. The technetium-organohydrazino unit is a robust core, but its stability can be further improved by the use of 6-hydrazinonicotinic acid (HYNIC). A major drawback of the HYNIC compounds, however, is the uncertainty induced by the required co-ligands. They influence the course of the reactions, and some uncertainties in the choice of optimal coordination are inherent in this chemistry, particularly when the reactions are transferred from the macroscopic scale to the clinically relevant tracer chemistry with ^{99m}Tc. A variety of co-ligands have been tested including phosphine sulfonates, pyridine carboxylates, nicotinic acid, glucaric acid, glucamine, and tricine [49], but although bioconjugation, particularly with peptides, works well, several questions remain concerning the basic chemistry, which also pertain to (the important) structures of the HYNIC complexes at tracer level and even the oxidation state of the metal.

A facile precursor for ^{99m}Tc bioconjugates is the *fac*-$\{^{99m}Tc(CO)_3\}^+$ core. The development of a normal-pressure synthesis of the organometallic aqua complex $[^{99m}Tc(CO)_3(OH_2)_3]^+$ by Alberto [51] was a breakthrough not only for organometallic pharmaceutical chemistry in general but particularly for designing new bioconjugates. The basic arrangement of such complexes with three facially coordinated carbonyl ligands is very robust and can readily be completed with tripodal ligand systems of various donor atom constellations. The resulting complexes possess

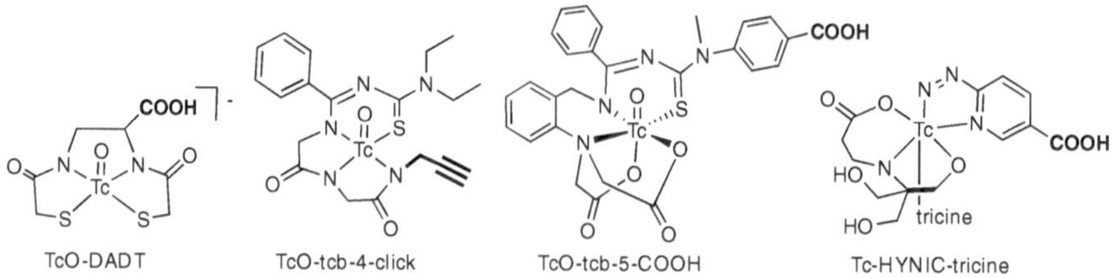

Tco-DADT   TcO-tcb-4-click   TcO-tcb-5-COOH   Tc-HYNIC-tricine

Fig. 11.13 Tc(V) chelates suitable for ^{99m}Tc bioconjugation (anchor groups in bold)

sufficient stability *in vivo*, while attempts with monodentate and bidentate ligands were less successful due to labilization of such ligands by carbonyls, which could not be compensated by the chelating effect in such ligands. Figure 11.14 shows only a small selection of neutral and cationic tricarbonyltechnetium(I) complexes with tripodal ligands designed for bioconjugation. Representatives with different *O,N,N* [52, 53], *S,N,N* [54], and *N,N,N* donor atoms [55] are shown, but a huge amount of other stable compounds with this core has been studied [4–6, 48, 52, 56–58]. {^{99m}Tc(CO)$_3$} compounds of particular interest are the cyclopentadienyl compounds [59, 60]. They provide by far the smallest labelling units, and the technetium atom is embedded deep inside a nonpolar unit. This means that such labelling units will most probably have no influence on the distribution patterns of the corresponding bioconjugates. The (substituted) cyclopentadienyl ligands are conveniently prepared by retro-Diels-Alder reactions from stable organic components during the labelling procedure or by using corresponding ferrocene units as cp$^-$ sources [59–63].

Technetium(III) compounds are relatively less explored, since this oxidation state has a kind of "border line behavior." This means that such compounds are stabilized by ligands with a balance of acceptor and donor properties, and Tc(III) compound occurs as paramagnetic octahedral complexes or as diamagnetic, trigonals of pentagonal bipyramids. Nevertheless, ^{99m}Tc(III) chemistry may become interesting with regard to ligands that are relevant for other M^{3+} ions such as In^{3+}, Ga^{3+}, and several lanthanide ions.

The development of ^{68}Ga agents follows the route outlined for ^{99m}Tc. Multidentate chelators, which form stable complexes with gallium ions, are provided with an anchor group for bioconjugation. The formation of amides, thioureas, or triazols follows common coupling strategies. Some prototype systems are shown in Fig. 11.15; a comprehensive collection of ligands tested for ^{68}Ga labelling has been compiled in a recent review [64].

A crucial issue in the labelling of biological vectors with ^{68}Ga is the search for mild reaction conditions. Depending on the pH of the solutions, "hard" and highly charged Ga^{3+} ions are surrounded by relatively large hydrate shells. Thus, some of

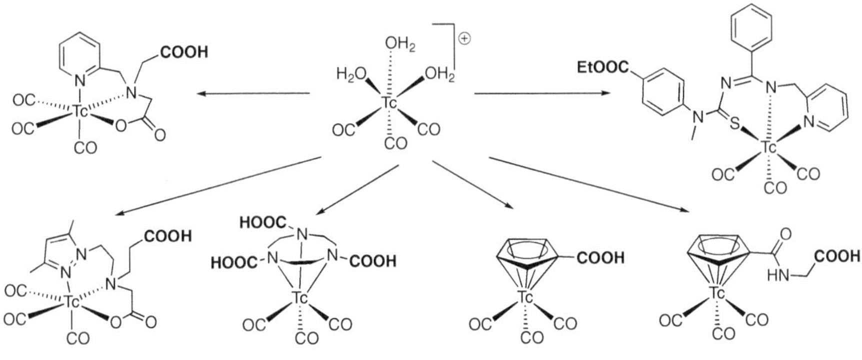

Fig. 11.14 {Tc^I(CO)$_3$} chelates suitable for ^{99m}Tc bioconjugation (anchor groups in bold)

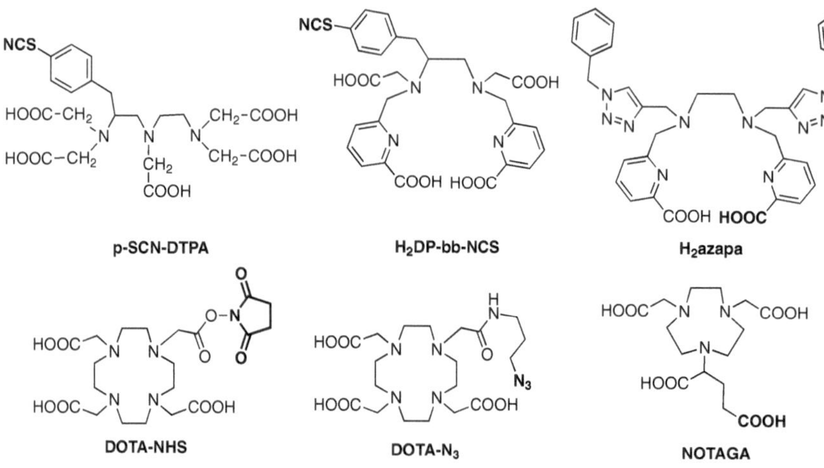

Fig. 11.15 Substituted chelators for ^{68}Ga bioconjugation (anchor groups in bold)

the complexation procedures require an elevated temperature and relatively long reaction times, which is unfavorable with regard to the peptide nature of many of the biological vectors, and work with a radioactive isotope with a half-life of 1.1 h. Ideally, ^{68}Ga labelling procedures should also be performed in a kit-like fashion, as is state of the art for ^{99m}Tc. Such kits should produce the ^{68}Ga radiopharmaceutical in a few minutes under mild conditions with regard to temperature and pH and in high yields. Ideally, no further purification steps are required, and the kit tolerates the presence of small amounts of competing metal ions (e.g., Zn^{2+}, which is the disintegration product of ^{68}Ga).

So far, some of the labelling procedures have been optimized and provide bio-conjugates with DOTA- and NOTA-type ligands within a reasonable time and quality [65–67]. Faster labelling under mild conditions has been achieved with open-chain chelators of the tris(hydroxypyridinone) type, which allow labelling procedures to be performed at ambient temperature within a short time [68, 69].

11.5 Multi-metal Solutions and Theranostic Approaches

The discussion in the previous subchapters has shown that for each nuclear medical routine application an individual tracer molecule was developed. This also holds true for the vast majority of the metal-based reagents and is clearly justified by the fact that the biological distribution patterns of ^{99m}Tc compounds are determined by the overall properties of the complex molecule and changes in both the periphery of the ligands and the metal core (see Fig. 11.12) have drastic consequences for their biological behavior.

With increasing interest in radiotherapy using β^-- or α-emitting metal isotopes, there grew also a need to visualize the in vivo distribution of these potentially hazardous particle emitters [70–73]. A compilation of radiometals with radiation characteristics suitable for *in vivo* therapy is presented in Table 11.3.

Table 11.3 Radiometals suitable for therapy[a]

Isotope	Physical half-life	Decay (%)	β^-_{max} (%) / α (MeV)	Isotope production
[186]Re	3.72 days	β^-	10.93 (22), 1.07 (71)	Reactor
[188]Re	17 h	β^-	1.97 (26), 2.12 (71)	[188]W/[188]Re generator
[90]Y	2.67 days	β^-	2.28 (100)	[90]Sr/^{90}Y generator
[177]Lu	6.65 days	β^-	0.18 (12), 0,39 (9), 0.50 (79)	Reactor
[153]Sm	1.9 days	β^-	0.64 (30), 0.70 (49), 0.81 (20)	Reactor
[225]Ac	10 days	α	5.90 (19), 5.94 (52)	[229]Th/[225]Ra/[225]Ac generator
[223]Ra	11.4 days	α	5.64 (11), 5.71 (26, 5.82 (50), 5.85 (10)	[227]Ac/[227]Th/[223]Ra generator
[213]Bi	46 min	α (2.1), β^- (97.9)	α: 5.65 (0.2), 5.98 (1.9); β^-: 0.98 (31), 1.42 (66)	[225]Ac/[213]Bi generator
[89]Sr	50.6 day	β^-	1.50 (100)	Reactor

[a]Data taken from http://www.nucleide.org/DDEP_WG/DDEPdata.htm

The chemical properties of some of the elements show a remarkable number of parallelisms to the behavior of some of the PET and SPECT tracers listed in Tables 11.1 and 11.2. For instance, technetium (^{99m}Tc) and rhenium ([186]Re, [188]Re) belong to the same group of elements in the periodic table and have similar chemical properties, and the coordination chemistry of the isotopes ^{111}In and ^{68}Ga closely matches that of the lanthanide ions [177]Lu^{3+}, [153]Sm^{3+}, and [90]Y^{3+}.

Although ^{99m}Tc and [186,188]Re have similar physicochemical properties, the hope of medical chemists to find in them a perfect matched pair of isotopes for imaging and therapy did not come true immediately. The two elements have different redox potentials and reaction kinetics, and the classic kit approaches starting from pertechnetate did not work with rhenium.

With the implementation of the tricarbonyltechnetium(I) chemistry [51, 52] and the development of a kit-like synthesis for $[M(CO)_3(OH_2)_3]^+$ complexes (M = [99m]Tc, [188]Re) [74–76], however, there exist two precursors which indeed behave like chemical twins. The extraordinary stability of the $\{M(CO)_3\}^+$ cores prevents oxidation, and the ligand exchange behavior with tridentate chelators shows no differences between the two elements. Thus, this core is particularly suitable for the synthesis of stable bioconjugates for targeting a variety of biological sites with both radiometals [77, 78]. This can be done by relatively small molecules such as the androgen and estrogen receptor tracers shown in Fig. 11.16 ([99m]TcCO-flutamide and [99m]TcCO-estradiol) or by larger conjugates with peptides or proteins [79–82].

The idea to use the chelators shown in Figs. 11.10 and 11.15 for metal ions other than ^{68}Ga or ^{111}In is straightforward, and the implementation of concerted solutions for M^{3+} radiometals suitable for imaging and those suitable for radiotherapy is a challenge for the future. An example already in use is shown in Fig. 11.16. M^{3+}-DOTATOC, which targets the somatostatin receptor of neuroendocrine tumors and is used for therapy with ^{90}Y or ^{177}Lu, while diagnostic imaging to monitor the

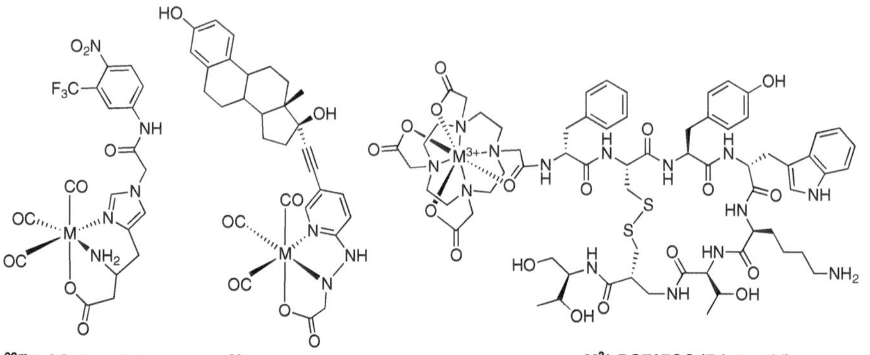

99mTcCO-flutamide   99mTcCO-estradiol   M3+-DOTATOC (Edotreotid)

Fig. 11.16 Bioconjugated 99mTc tricarbonyl compounds for receptor binding and the potential multi-metal bioconjugate M3+-DOTATOC

response to treatment has been done using the corresponding 68Ga or 111In derivatives [71]. This example impressively illustrates the high flexibility of the related macrocyclic ligand, which coordinates metal ions with ion radii ranging from 0.62 (Ga3+: coordination number 6) to 1.02 Å (Y3+: coordination number 8) [83]. On the other hand, such differences produce noticeable differences in the thermodynamic stability of the resulting complexes [71]. It also becomes clear that the DOTA ligand system is not optimized for the binding of multiple metal ions. To overcome this limitation, many efforts have been made in the designing of more cyclic and open-chain ligand systems. Their structures are manifold, and they comprise a variety of donor atom constellations. Details are summarized in some excellent reviews and original papers [5, 31, 64, 71–73, 84].

A very recent theranostic approach comes from optical labelling of metal-based drugs. The use of organic fluorophores or luminescent metal complexes seems to be an excellent method to keep drugs trackable in *in vitro* studies. Clinical applications, however, are strongly limited by the low tissue penetration of the radiation [85]. This drawback can be avoided when the fluorophores are replaced with radioactive labels-a strategy that can be used for radiometals suitable for both imaging and therapy. 18F, 64Cu, and 99mTc have been used for such experiments [86–88], and it is expected that this approach will receive more attention in the coming years.

Conclusions

The development of novel tracers for PET and SPECT is stimulated by innovations in radionuclide production, the use of new methods and principles of organic synthesis, and the advent of innovative technical tools such as modern synthesis modules. Such modules are computer-controlled and allow fast and reproducible syntheses with nuclides of short half-lives such as 11C or 18F. 68Ge/68Ga nuclide generators represent a tremendous progress in the development of novel metal-based radiotracers. 68Ga has the potential to become a very important isotope for routine radiopharmacy and, thus, an important alternative to 99mTc-currently the predominant isotope in medical imaging. The development of specific radiopharmaceuticals based on these two nuclides requires serious efforts in the development of novel chelators and/or fast and reliable techniques for the formation of the corresponding complexes. Bioconjugation will allow precise targeting of the radiotracers, and the combination with radionuclides for therapy is the basis for developing new theranostic approaches.

References

1. Rao J, Contag CH. More chemistry is needed for molecular imaging. Bioconjug Chem. 2016;27:265–6.
2. Kuntic V, Brboric J, Vujic Z, Uskokovic-Markovic S. Radioisotopes used as radiotracers for in vitro and in vivo diagnostics. Asian J Chem. 2016;28:235–41.
3. Alshaabi Y. Radioisotopes in medicine therapeutic techniques. http://www.world-nuclear.org/information-library/non-power-nuclear-applications/radioisotopes-research/radioisotopes-in-medicine.aspx, updated December 2016.
4. Bartholomä MD, Louie AS, Valliant JF, Zubieta J. Technetium and gallium derived radiopharmaceuticals: comparing and contrasting the chemistry of two important radiometals for the molecular imaging era. Chem Rev. 2010;110:2903–20.
5. Dilworth JR, Pascu SI. The radiopharmaceutical chemistry of technetium and rhenium. In: Long N, Wong W-T, editors. The chemistry of molecular imaging. New York: Wiley; 2015.
6. Abram U, Alberto R. Technetium and rhenium - coordination chemistry and nuclear medical applications. J Braz Chem Soc. 2006;17:1486–500.
7. Roeda D, Dolle F. Recent developments in the chemistry of [^{18}F]fluoride for PET. In: Long N, Wong W-T, editors. The chemistry of molecular imaging. New York: Wiley; 2015.
8. Jacobson O, Kiesewetter DO, Chen X. Fluorine-18 radiochemistry, labeling strategies and synthetic routes. Bioconjug Chem. 2015;26:1–18.
9. Preshlock S, Tredwell M, Gouverneur V. ^{18}F-Labeling of arenes and heteroarenes for applications in positron emission tomography. Chem Rev. 2016;115:719–66.
10. Miller PW, Kato K, Langstrom B. Carbon-11, nitrogen-13, and oxygen-15 chemistry: an introduction to chemistry with short-lived radioisotopes. In: Long N, Wong W-T, editors. The chemistry of molecular imaging. New York: Wiley; 2015.
11. Campbell MC, Ritter T. Modern carbon–fluorine bond forming reactions for aryl fluoride synthesis. Chem Rev. 2015;115:612–33.
12. Neumann CN, Ritter T. Late-stage fluorination: fancy novelty or useful tool? Angew Chem Int Ed. 2015;54:3216–21.
13. Cai L, Lu S, Pike VW. Chemistry with [^{18}F] fluoride. Eur J Org Chem. 2008;39:2853–73.
14. Mori T, Kasamatsu S, Mosdzianowski C, Welch MJ, Yonekura Y, Fujibayashi Y. Automatic synthesis of 16α-[^{18}F]-fluoro-17β-estradiol using a cassette-type [^{18}F]-fluorodeoxyglucose synthesizer. Nucl Med Biol. 2006;33:281–6.
15. Gründler G, Siessmeier T, Lange-Asschenfeld C, Vernaleken I, Buchholz HG, Stoeter P, Drzezga A, Lüddens H, Rösch F, Bartenstein P. [^{18}F]-Fluoroethylflumazenil: a novel tracer for PET imaging of human benzodiazepine receptors. Eur J Nucl Med. 2001;28:1463–70.
16. Haslop A, Wells L, Gee A, Plisson C, Long N. One-pot multi-tracer synthesis of novel ^{18}F-labeled PET imaging agents. Mol Pharm. 2014;11:3818–22.
17. Bergman J, Solin O. Fluorine-18-labeled fluorine gas for synthesis of tracer molecules. Nucl Med Biol. 1997;24:677–83.
18. Teare H, Robins EG, Kirjavainen A, Forsback S, Sandford G, Solin O, Luthra SK, Gouverneur V. Radiosynthesis and evaluation of [^{18}F]-Selectfluor bis(triflate). Angew Chem Int Ed Engl. 2010;49:6821–924.

19. Lee E, Kamlet AS, Powers DC, Neumann CN, Boursalian GB, Furuya T, Choi DC, Hooker JM, Ritter T. A fluoride-derived electrophilic late-stage fluorination reagent for PET imaging. Science. 2011;334:639–42.
20. Kamlet AS, Neumann CN, Lee E, Carlin SM, Moseley CK, Stephenson N, Hooker JM, Ritter T. Application of palladium-mediated (18)F-fluorination to PET radiotracer development: overcoming hurdles to translation. PLoS One. 2013;8(3):e59187.
21. Lee E, Hooker JM, Ritter T. Nickel-mediated oxidative fluorination for PET with aqueous [^{18}F] fluoride. J Am Chem Soc. 2012;134:17456–8.
22. Ren H, Wey H-Y, Strebl M, Neelamegam R, Ritter T, Hooker JM. Synthesis and imaging validation of [^{18}F]-MDL100907 enabled by Ni-mediated fluorination. ACS Chem Neurosci. 2014;5:611–5.
23. Zlatopolskiy BD, Zischler J, Urusova EA, Endepols H, Kordys E, Frauendorf H, Mattaghy FM, Neumaier BA. A practical one-pot synthesis of positron emission tomography (PET) tracers via nickel-mediated radiofluorination. Chemistry Open. 2015;4:457–62.
24. Tredwell M, Preshlock SM, Taylor NJ, Gruber S, Huiban M, Passchier J, Mercier J, Genicot C, Gouverneur VA. A general copper-mediated nucleophilic ^{18}F fluorination of arenes. Angew Chem Int Ed Engl. 2014;53:7751–5.
25. Niwa T, Ochiai H, Watanabe Y, Hosoya T. Ni/cu-catalyzed defluoroborylation of fluoroarenes for diverse C–F bond functionalizations. J Am Chem Soc. 2015;137:14313–8.
26. Mossine AV, Brooks AF, Makaravage KJ, Miller JM, Ichiishi N, Sanford MS, Scott PJH. Synthesis of [^{18}F]-arenes via the copper-mediated [^{18}F]-fluorination of boronic acids. Org Lett. 2015;17:5780–3.
27. Huisgens R. 1,3-Dipolare cycloadditions. Proc Chem Soc. 1961:357–96.
28. Kolb HC, Finn MG, Sharpless KB. Click chemistry: diverse chemical function from a few good reactions. Angew Chem Int Ed Engl. 2001;40:2004–21.
29. Meyer J-P, Adumeau P, Lewis JS, Zeglis BM. Click chemistry and radiochemistry: the first 10 years. Bioconjug Chem. 2016;27:2791–807.
30. Tollefson J. Reactor shutdown threatens world's medical-isotope supply. Nature. 2016. https://doi.org/10.1038/nature.2016.20577.
31. Price EW, Orvig C. Chemistry of inorganic nuclides (^{86}Y, ^{68}Ga, ^{64}Cu ^{89}Zr ^{124}I). In: Long N, Wong W-T, editors. The chemistry of molecular imaging. New York: Wiley; 2015.
32. Chakravarty R, Chakraborty S, Ram R, Vatsa R, Bhusari P, Shukla J, Mittal BR, Dash A. Detailed evaluation of different (68)Ge/(68)Ga generators: an attempt toward achieving efficient (68)Ga radiopharmacy. J Labelled Comp Radiopharm. 2016;59:87–94.
33. Chakravarty R, Chakraborty S, Dash A. ^{64}Cu^{2+} ions as PET probe: an emerging paradigm in molecular imaging of cancer. Mol Pharm. 2016;13:3601–12.
34. Bonnitcha PD, Vavere AL, Lewis JS, Dilworth JR. In vitro and in vivo evaluation of bifunctional bisthiosemicarbazone ^{64}Cu-complexes for the positron emission tomography imaging of hypoxia. J Med Chem. 2008;51:2985–91.
35. Vavere AL, Lewis JS. Cu-ATSM: a radiopharmaceutical for the PET imaging of hypoxia. Dalton Trans. 2007;4:4893–902.
36. Lopci E, Grassi I, Chiti A, Nanni C, Cicoria G, Toschi L, Fonti C, Lodi F, Mattioli S, Fanti. SPET. PET radiopharmaceuticals for imaging of tumor hypoxia: a review of the evidence. Am J Nucl Med Mol Imaging. 2014;4:365–84.
37. Yang DJ, Azhdarinia A, Kim EE. Tumor specific imaging using Tc-99m and Ga-68 labeled radiopharmaceuticals. Curr Med Imaging Rev. 2005;1:25–34.
38. Lazar I, Ramasamy R, Brucher E, Geraldes CFGC, Sherry AD. NMR and potentiometric studies of 1,4,7-triazacyclononane-N,N',N''-tris(methylenephosphonate monoethylester) and its complexes with metal ions. Inorg Chim Acta. 1992;195:89–93.
39. Sun Y, Anderson C, Pajeau T, Reichert D, Hancock R, Motekaitis R, Martell A, Welch M. Indium(III) and gallium(III) complexes of bis(aminoethanethiol) ligands with different denticities: stabilities, molecular modeling, and in vivo behavior. J Med Chem. 1996;39:458–70.
40. Berry DJ, Ma Y, Ballinger JR, Tavare R, Koers A, Sunassee K, Zhou T, Nawaz S, Mullen GED, Hider RC, Blower PJ. Efficient bifunctional gallium-68 chelators for positron emission tomography: tris(hydroxypyridinone) ligands. Chem Commun. 2011;47:7068–70.

41. Boros E, Ferreira CL, Cawthray JF, Price EW, Patrick PO, Wester DW, Adams MJ, Orvig C. Acyclic chelate with ideal properties for ^{68}Ga PET imaging agent elaboration. J Am Chem Soc. 2010;132:15726–33.

42. Price EW, Cawthray JF, Bailey GA, Ferreira CL, Boros E, Adam MJ, Orvig C. H$_4$octapa: an acyclic chelator for ^{111}In radiopharmaceuticals. J Am Chem Soc. 2012;134:8670–83.

43. Chi YD, Katzenellenbogen JA. Selective formation of heterodimeric bis-bidentate aminothiol-oxometal complexes or rhenium(V). J Am Chem Soc. 1993;115:7045–6.

44. Chi YD, O'Neil JP, Anderson CJ, Welch MJ, Katzenellenbogen JA. Homodimeric and heterodimeric bis(aminothiol) oxometal complexes with rhenium(V) and technetium(V). Control of heterodimeric complex formation and an approach to metal complexes that mimic steroid hormones. J Med Chem. 1994;37:928–37.

45. Rao TN, Adhikesavalu D, Camaraman A, Fritzberg AR. Technetium (V) and rhenium (V) complexes of 2,3-bis(mercaptoacetamido)propanoate. Chelate ring stereochemistry and influence on chemical and biological properties. J Am Chem Soc. 1990;112:5798–804.

46. Castillo Gomez JD, Hagenbach A, Abram U. Propargyl-substituted thiocarbamoylbenzamidines of technetium and rhenium: steps towards bioconjugation with use of click chemistry. Eur J Inorg Chem. 2016;2016:5427–34.

47. Nguyen HH, Pham CT, Abram U. Rhenium and technetium complexes with pentadentate thiocarbamoylbenzamidines: steps toward bioconjugation. Inorg Chem. 2014;54:5949–59.

48. Liu S, Chakraborty S. ^{99m}Tc-centered one-pot synthesis for preparation of ^{99m}Tc radiotracers. Dalton Trans. 2011;40:6077–86.

49. Meszaros LK, Dose A, Biagini SCG, Blower PJ. Hydrazinonicotinic acid (HYNIC) - coordination chemistry and applications in radiopharmaceutical chemistry. Inorg Chim Acta. 2010;363:1059–69.

50. Rennen H, van Eerd JE, Oyen WJG, Corstens FHM, Edwards DS, Boermann OC. Effects of coligand variation on the in vivo characteristics of Tc-99m-labeled interleukin-8 in detection of infection. Bioconjug Chem. 2002;13:370.

51. Alberto R, Schibli R, Egli A, Schubiger PA, Herrmann WA, Artus G, Abram U, Kaden TA. Metal carbonyl syntheses XXII. Low pressure carbonylation of [MOCl$_4$]$^-$ and [MO$_4$]$^-$: the technetium(I) and rhenium(I) complexes [NEt$_4$]$_2$[MCl$_3$(CO)$_3$]. J Organomet Chem. 1995;493:119–27.

52. Alberto R, Schibli R, Egli A, Schubiger PA. A novel organometallic aqua complex of technetium for the labeling of biomolecules: synthesis of [^{99m}Tc(OH$_2$)$_3$(CO)$_3$]$^+$ from [^{99m}TcO$_4$]$^-$ in aqueous solution and its reaction with a bifunctional ligand. J Am Chem Soc. 1998;120:7987–8.

53. Morais M, Paulo A, Gano L, Santos I, Correia JDG. Target-specific Tc(CO)$_3$-complexes for in vivo imaging. J Organomet Chem. 2013;744:125–39.

54. Oehlke E, Nguyen HH, Kahlke N, Deflon VM, Abram U. Tricarbonyltechnetium(I) and -rhenium(I) complexes with N'-thiocarbamoylpicolylbenzamidines. Polyhedron. 2012;40:153–8.

55. Braband H, Imstepf S, Benz M, Spingler B, Alberto R. Combining bifunctional chelator with (3 + 2)-cycloaddition approaches: synthesis of dual-function technetium complexes. Inorg Chem. 2012;51:4051–7.

56. Alberto R. The particular role of radiopharmacy within bioorganometallic chemistry. J Organomet Chem. 2007;692:1179–86.

57. Alberto R. Radiopharmaceuticals. In: Jaouen G, editor. Bioorganometallics: biomolecules, labeling, medicine. Weinheim: Wiley; 2005.

58. Soler M, Feliu L, Planas M, Ribas X, Costas M. Peptide-mediated vectorization of metal complexes: conjugation strategies and biomedical applications. Dalton Trans. 2016;45:12970–82.

59. Liu Y, Spingler B, Schmutz P, Alberto R. Metal-mediated retro Diels–Alder of dicyclopentadiene derivatives: a convenient synthesis of [(Cp-R)M(CO)$_3$] (M = ^{99m}Tc, Re) complexes. J Am Chem Soc. 2008;130(5):1554.

60. Wald J, Alberto R, Ortner K, Candreia L. One-pot synthesis of derivatized cyclopentadienyl–tricarbonyl complexes of ^{99m}Tc with an in situ CO source: application to a serotonergic receptor ligand. Angew Chem Int Ed Engl. 2001;40:3062–6.

61. N'Dongo HWP, Liu Y, Can D, Schmutz P, Spingler B, Alberto R. Aqueous syntheses of [(Cp-R)M(CO)₃] type complexes (Cp = cyclopentadienyl, M = Mn, ^{99m}Tc, Re) with bioactive functionalities. J Organomet Chem. 2009;694:981–7.

62. Sulieman S, Can D, Mertens J, N'Dongo HWP, Liu Y, Schmutz P, Bauwens M, Spingler B, Alberto R. Cyclopentadienyl-based amino acids (Cp-aa) as phenylalanine analogues for tumor targeting: syntheses and biological properties of [(Cp-aa)M(CO)₃](M = Mn, re, ^{99m}Tc). Organometallics. 2012;31:6880–6.

63. Ursilio S, Can D, N'Dongo HWP, Schmutz P, Spingler B, Alberto R. Cyclopentadienyl chemistry in water: synthesis and properties of bifunctionalized [(η⁵-C₅H₃{COOR}₂)M(CO)₃] (M = Re and ^{99m}Tc) complexes. Organometallics. 2014;33:6945–52.

64. Spang P, Herrmann C, Rösch F. Bifunctional gallium-68 chelators: past, present, and future. Sem Nucl Med. 2016;46:373–94.

65. Roosenburg S, Laverman P, Joosten L, Cooper MS, Kolenc-Peitl PK, Foster JM, Hudson C, Leyton J, Burnet J, Oyen WJG, Blower PJ, Mather SJ, Boerman OC, Sosabowski JK. PET and SPECT imaging of a radiolabeled minigastrin analogue conjugated with DOTA, NOTA, and NODAGA and labeled with ^{64}Cu, ^{68}Ga, and ^{111}In. Mol Pharm. 2014;11:3930–7.

66. Simecek J, Zemek O, Hermann P, Notni J, Wester H-J. Tailored gallium(III) chelator NOPO: synthesis, characterization, bioconjugation, and application in preclinical Ga-68-PET imaging. Mol Pharm. 2014;11:3893–903.

67. Notni J, Pohle K, Wester H. Be spoilt for choice with radiolabelled RGD peptides: preclinical evaluation of ^{68}Ga-TRAP(RGD)₃. Nucl Med Biol. 2013;40:33–41.

68. Ma MT, Cullinane C, Imberti C, Torres JB, Terry SYA, Roselt P, Hicks RJ, Blower PJ. New tris(hydroxypyridinone) bifunctional chelators containing isothiocyanate groups provide a versatile platform for rapid one-step labeling and pet imaging with ^{68}Ga^{3+}. Bioconjug Chem. 2016;27:309–18.

69. Ma MT, Cullinane C, Waldeck K, Roselt P, Hicks RJ, Blower PJ. Rapid kit-based (68) Ga-labelling and PET imaging with THP-Tyr(3)-octreotate: a preliminary comparison with DOTA-Tyr(3)-octreotate. EJNMMI Res. 2015;5:52.

70. Heeg MJ, Jurisson SS. The role of inorganic chemistry in the development of radiometal agents for cancer therapy. Acc Chem Res. 1999;32:1053–60.

71. Cutler CS, Hennkens HM, Sisay N, Huclier-Markai S, Jurisson SS. Radiometals for combined imaging and therapy. Chem Rev. 2013;113:858–83.

72. Bhattacharyya S, Dixon M. Metallic radionuclides in the development of diagnostic and therapeutic radiopharmaceuticals. Dalton Trans. 2011;40:6112–28.

73. Price EW, Orvig C. Matching chelators to radiometals for radiopharmaceuticals. Chem Soc Rev. 2014;43:260–90.

74. Causey PW, Besanger TR, Schaffer P, Valliant JF. Expedient multi-step synthesis of organometallic complexes of Tc and re in high effective specific activity. A new platform for the production of molecular imaging and therapy agents. Inorg Chem. 2008;47:8213–21.

75. Schibli R, Schwarzbach R, Alberto R, Ortner K, Schmalle H, Dumas C, Egli A, Schubiger PA. Steps toward high specific activity labeling of biomolecules for therapeutic application: preparation of precursor [^{188}Re(H₂O)₃(CO)₃]⁺ and synthesis of tailor-made bifunctional ligand systems. Bioconjug Chem. 2002;13:750–6.

76. Park SH, Seifert S, Pietzsch H-J. Novel and efficient preparation of precursor [^{188}Re(OH₂)₃(CO)₃]⁺ for the labeling of biomolecules. Bioconjug Chem. 2006;17:223–5.

77. Morais GR, Paulo A, Santos I. Organometallic complexes for SPECT imaging and/or radionuclide therapy. Organometallics. 2012;31:5693–714.

78. Jürgens S, Herrmann WA, Kühn FE. Rhenium and technetium based radiopharmaceuticals: Development and recent advances. J Organomet Chem. 2014;751:83–9.

79. Nayak TK, Hathaway HJ, Ramesh C, Arterburn JB, Dai D, Sklar LA, Norenberg JP, Prossnitz ER. Influence of charge on cell permeability and tumor imaging of GPR30-targeted ^{111}In-labeled non-steroidal imaging agents. J Nucl Med. 2008;49:978–86.

80. He HY, Morely JE, Silca-Lopez E, Bottenus B, Montajano M, Fugate GA, Twamley B, Benny PD. Synthesis and characterization of nonsteroidal-linked M(CO)₃⁺(M = 99ᵐTc, Re)

compounds based on the androgen receptor targeting molecule flutamide. Bioconjug Chem. 2009;20:78–86.

81. Binkley SL, Ziegler CJ, Herrick RS, Rowlett RS. Specific derivatization of lysozyme in aqueous solution with $Re(CO)_3(H_2O)_3^+$. Chem Commun. 2010;46:1203–5.

82. Binkley SL, Leeper TC, Rowlett RS, Herrick RS, Ziegler CJ. $Re(CO)_3$ $(H_2O)_3^+$ binding to lysozyme: structure and reactivity. Metallomics. 2011;3:909–16.

83. Shannon RD. Revised effective ionic radii and systematic studies of interatomic distances in halides and chalcogenides. Acta Cryst. 1976;A32:751–67.

84. Boros E, Lin Y-HS, Ferreira CL, Patrick BO, Häfli UO, Adam MJ, Orvig C. One to chelate them all: investigation of a versatile, bifunctional chelator for ^{64}Cu, ^{99m}Tc, Re and Co. Dalton Trans. 2011;40:6253–9.

85. Bertrand B, Doulain P-E, Goze C, Bodio E. Development of trackable metal-based drugs: new generation of therapeutic agents. Dalton Trans. 2016;45:13005–11.

86. Natarajan A, Türkcan S, Gambhir SS, Pratx G. Multiscale framework for imaging radiolabeled therapeutics. Mol Pharm. 2015;12:4556–60.

87. Lagisetty P, Vilekar P, Awashi V. Synthesis of radiolabeled cytarabine conjugates. Bioorg Med Chem Lett. 2009;19:4764–7.

88. Vineberg JG, Wang T, Zuniga ES, Ojima I. Design, synthesis, and biological evaluation of theranostic vitamin–linker–Taxoid conjugates. J Med Chem. 2015;58:2406–16.

Methods and Approaches in Ultrasound Elastography

12

Heiko Tzschätzsch

Abstract

Medical ultrasound is one of the most common medical imaging modalities. It allows real-time visualization of the morphology of soft tissues in the human body based on backscattered compression waves in the high-frequency (MHz) range. Special technologies are sensitive to organ functions such as blood flow and tissue strain. Over the past decade, ultrasound-based elastography (USE) has been developed and became widely applied in the clinic for the assessment of tissue stiffness in a variety of conditions including malignant tumors and liver fibrosis. The clinical benefit of USE has been overwhelmingly demonstrated in a large number of clinical studies and reviews. Unlike other sonographic methods, USE is not a single technique but rather a set of methodological ideas and approaches centered on the mechanical stimulation of soft tissues, deformation readout, and stiffness reconstruction. This chapter provides an overview of the key concepts of current USE methods and their potential clinical applications. Many of the methods discussed are still experimental, while others have already been replaced by more sophisticated quantitative and image-resolved methods. The aim of this chapter is to guide readers through the pros and cons of individual concepts, thereby helping them to gain insight into the basic principles of USE.

12.1 Introduction

Since the basic developments of medical ultrasound in the 1940s, sonography has become one of the most widely applied diagnostic imaging technique worldwide. The option of real-time data acquisition and intermediate image display, the high

H. Tzschätzsch
Department of Radiology, Charité – Universitätsmedizin Berlin, Berlin, Germany
e-mail: Heiko.Tzschaetzsch@charite.de

© Springer International Publishing AG 2018
I. Sack, T. Schaeffter (eds.), *Quantification of Biophysical Parameters in Medical Imaging*, https://doi.org/10.1007/978-3-319-65924-4_12

281

cost efficiency, the mobility of ultrasound devices, and the possibility of quantifying functional parameters such as flow and marked improvements of detail resolution and image quality over the past years have made sonography the diagnostic imaging modality with the widest dissemination. Since the early 1990s, elastography has been developed as a method to enhance the sensitivity of sonography to mechanical changes caused by pathologies such as benign nodules, malignant tumors, or fibrosis in the human body. The basic image contrast in sonography relates to the compression properties of the tissue, which are determined by the high water content of soft biological tissues and do not vary largely in the body. Elasticity, or more specifically shear elasticity, is measured by controlled imposition and readout of mechanical tissue deformations, followed by somewhat elaborate image processing to reconstruct stiffness maps or *elastograms*. Since the advent of ultrasound elastography (USE), many methods have been devised to address the three key points involved in USE—namely, mechanical stimulation, acquisition of deformation field data, and stiffness image reconstruction. This chapter is a brief introduction to medical ultrasound, the key concepts of USE, and the systems most commonly used to measure shear elasticity by ultrasound.

12.2   Medical Ultrasound

The basic principle of medical ultrasound includes the following steps: 1) generation of sound waves with a transducer, 2) propagation and interaction of these waves with the tissue, 3) acquisition of the reflected waves by the transducer, and 4) reconstruction of an anatomical image from the reflected waves. More insight is gained by considering the three fundamental ultrasound modes — A-mode, M-mode, and B-mode — which will be explained in the following. Additionally, a short introduction to functional ultrasound will be presented.

12.2.1 A-Mode

The *A-mode* (amplitude mode) is the basic and most simple ultrasound mode. Sound, or compression waves, is generated by the piezoelectric element in the head of the transducer (see Fig. 12.1). This element converts electric voltage into mechanical strain (indirect piezoelectric effect). The created compression wave pulse, which is commonly in the megahertz range, propagates from the transducer placed on the skin surface into deeper tissue (in positive x direction) at the speed of sound, $c_\| = 1540$ m/s (see Table 12.1). There are three main interactions between the propagating compression wave pulse and biological tissues in the body: reflection, scattering, and absorption. Reflection acts at the interface of two tissues with different impedance. The reflection coefficient, R, depends on the relative change of impedance. Due to energy conservation, the corresponding transmission coefficient, T, is given by $T = 1 + R$ (see Chap. 2, Eq. (2.35)). In contrast, scattering acts inside one tissue. The compression

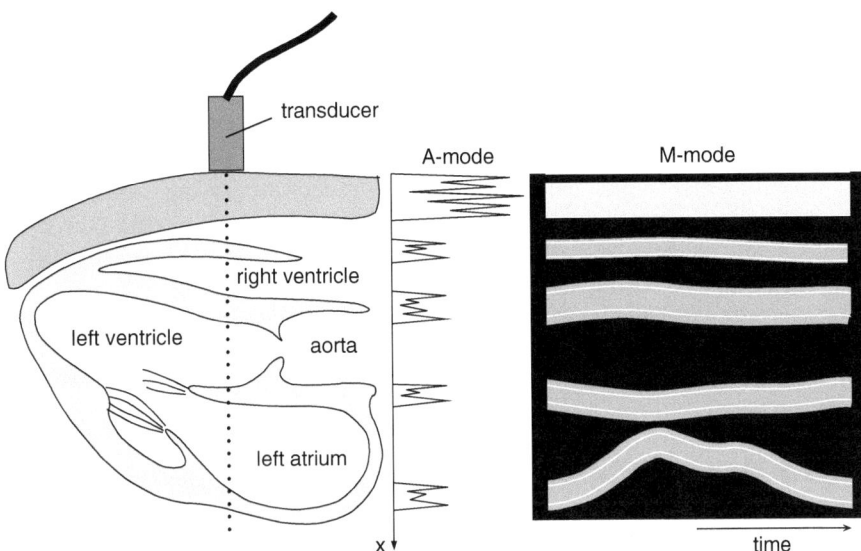

Fig. 12.1 (Left) Setup of single-element transducer and human heart; (middle) corresponding A-mode; (right) M-mode over one cardiac cycle [1]

Table 12.1 Comparison of compression wave speed $c_\parallel$ (Eq. (2.22) in Chap. 2) and shear wave $c_\perp$ speed (Eq. (2.23) in Chap. 2) of different media, organs, and tissues

Organ	$c_\parallel$ in m/s	$c_\perp$ in m/s
Air	331	–
Water	1492	–
Breast	1450–1570	1.1–3.5
Liver	1522–1623	0.9–3.0
Brain	1460–1580	1.0–4.6
Thrombus	1586–1597	0.3–1.3
Muscle	1500–1600	1.6–7.6
Bone	1630–4170	1420–3541

Values taken from Sarvazyan et al. [2]

wave scatters at microscopic tissue inhomogeneities. Both effects result in partial reflection and partial transmission of the wave. Absorption causes attenuation of the wave while it propagates through a viscous medium. Wave attenuation has to be compensated for by time-dependent amplification of the received high-frequency signals. Since it is merely an artifact rather than a contrast-generating interaction, absorption of compression waves will not be discussed further. The reflected wave pulse reaches the transducer and will be converted into an electric signal via the direct piezoelectric effect. The corresponding depth, x, of the reflection can be calculated as the time of flight, Δt, between sending and acquiring the echoes:

$$x = \frac{\Delta t \; c_{\parallel}}{2}. \tag{12.1}$$

To ensure that no interaction occurs between two wave packages consecutively emitted from the transducer, all echoes of the first pulse must return to the transducer before the next pulse is sent. Therefore, the *pulse repetition frequency* (PRF) is limited by the maximum penetration depth, x_{max}, through PRF = $c_{\parallel}/(2 \; x_{max})$. Typical values for PRF are in the range of a few thousands of hertz. The magnitude of the reflected wave pulses is related to the anatomical structure and is visualized over the spatial profile (see A-mode in Fig. 12.1).

12.2.2 M-Mode

The *M-mode* (motion mode) is an extension of the A-mode in time. The magnitude of the A-mode is displayed in gray scale over a certain time with the temporal resolution of 1/PRF (see M-mode in Fig. 12.1). High amplitudes of the reflected waves are displayed as hyperintense structures, while tissue regions with few scattering events appear as hypointense areas. Due to its high time resolution, the M-mode is mainly used in echocardiography.

12.2.3 B-Mode

The *B-mode* (brightness mode) is an extension of the A-mode in space. A two-dimensional image is created by the combination of certain A-mode lines with different sector orientation (see Fig. 12.2). For this the transducer consists of an array of piezoelectric elements. For each sector orientation, a so-called *line of sight* (LoS), a certain number of elements are activated simultaneously and act as one element (see Sect. 12.2.1). To acquire the next lateral LoS, another group of elements are activated. In common clinical ultrasound devices, the number n of LoS is in the order of 2^7–2^9. Therefore, the *frame rate* (FR) is reduced to FR = PRF/n and is typically in the order of several tens of hertz.

12.2.4 Functional Ultrasound

Unlike anatomical ultrasound, functional ultrasound measures the velocity of fluids like blood and moving tissues such as the cardiac wall. As such functional ultrasound relies on motion encoding based on the Doppler effect or by correlation techniques as explained in Chap. 2, Eq. (2.59).

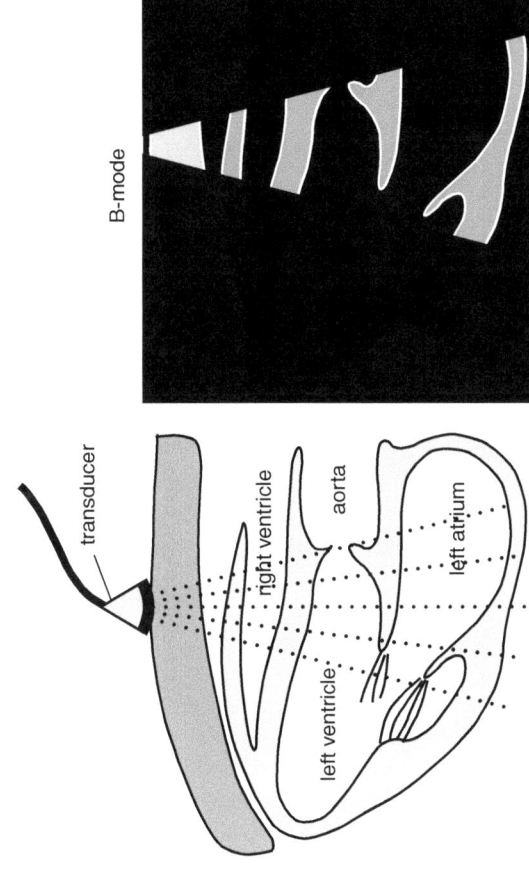

Fig. 12.2 (Left) Setup of multi-element transducer and human heart; (right) corresponding B-mode [1]

12.3   Principles of Ultrasound Elastography

The clinical motivation for developing elastography is lent from palpation. During palpation the physician uses his or her hands to apply a force to the surface of the body and senses the tissue response in terms of hard and soft. However, palpation is subjective and limited to the surface. To overcome these limitations, an elastography experiment uses three major techniques: 1) mechanical stimulation of deep tissue, 2) measurement of deep tissue deformation, and 3) quantification of viscoelastic parameters within the field of view. To understand the basic principles of USE, we briefly introduce, in the next section, four key concepts used by different elastography methods.

12.3.1 Qualitative Elastography

The physical background of palpation is the relationship between the amount of tissue deformation/strain and the applied force/stress. This relationship is in linear elasticity theory defined by Hooke's law (see Chap. 2, Eqs. (2.8) and (2.9)) by which the tissue's shear modulus is defined. Strain can be defined as relative tissue deformation or the gradient of tissue displacement (see Chap. 2, Eq. (2.6)). Strain measured by elastography can be generated by external force or intrinsic motion. The resulting tissue motion can be quantified by cross-correlation techniques similar to the approaches used in the color mode (see Sect. 12.2.4). Strain images are obtained by calculating the spatial derivative from the tissue motion image. The resulting strain map (*elastogram*) shows high strain for soft tissue portions and low strain for stiff tissue portions. Due to the unknown force/stress and boundary conditions, the strain map provides only qualitative information.

12.3.2 ARFI

Strain can be directly generated within the region of interest using a technique known as *acoustic radiation force impulse* (ARFI), which reduces artifacts due to ribs or ascites. ARFI generates a highly focused ultrasound pulse by a controlled delay of the electronic pulse for each single piezoelectric transducer element (see Fig. 12.3) [3]. The ultrasound pulse yields an axially orientated force, which deforms the tissue along the focus direction, thereby generating a laterally propagating shear wave. The resulting shear strain can be measured as described in Sect. 12.3.1. For quantitative elastography, the shear wave propagation speed has to be measured.

12.3.3 Quantitative Elastography

In contrast to strain, the shear wave velocity is independent of the amount of the applied stress and can thus be exploited as a quantitative parameter of the tissue's inherent stiffness (see Table 12.1). Shear waves can be generated in different ways

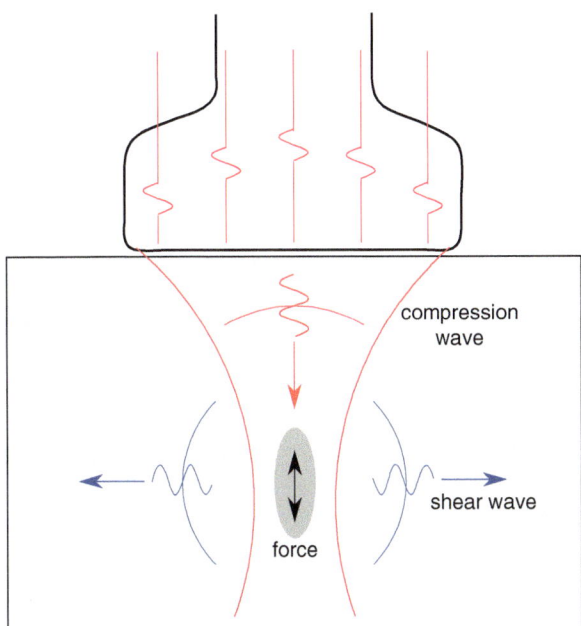

Fig. 12.3   The principle of ARFI. An electric delay between the transducer elements generates a highly focused compression wave, resulting in an axially orientated force within the focus region. The focused force is a new point source which generates a shear wave propagating in lateral direction

(see Fig. 12.4) such as by ARFI (see Fig. 12.3) or using external actuators. Most quantitative ultrasound elastography techniques track the shear wave at different time points and lateral positions. The velocity, which is typically in the order of several m/s, can be estimated using a time-of-flight algorithm. Depending on the specific method used, the shear wave speed is given either as mean value averaged over a small area or as a color-coded elastogram.

12.3.4 Coherent Plane-Wave Compounding

Common ultrasound frame rates are in the order of several tens of hertz. To observe rapid events such as fast shear waves, a higher frame rate is required. The technique of *plane-wave imaging* allows acquisition of B-mode images in a single shot (frame), yielding a frame rate equal to the PRF [4]. This is accomplished by the simultaneous activation of all piezoelectric elements, which generates an unfocused, plane compression wave. To achieve lateral resolution, the backscattered plane waves are acquired with a depth-dependent time delay between signal acquisitions in each piezoelectric element. This approach is called dynamic receive focusing and is similar to the focusing principle shown in Fig. 12.3 but in the receive mode. Due to the loss of focusing in the transmit mode, the improvement in frame rate is achieved at the cost of spatial resolution.

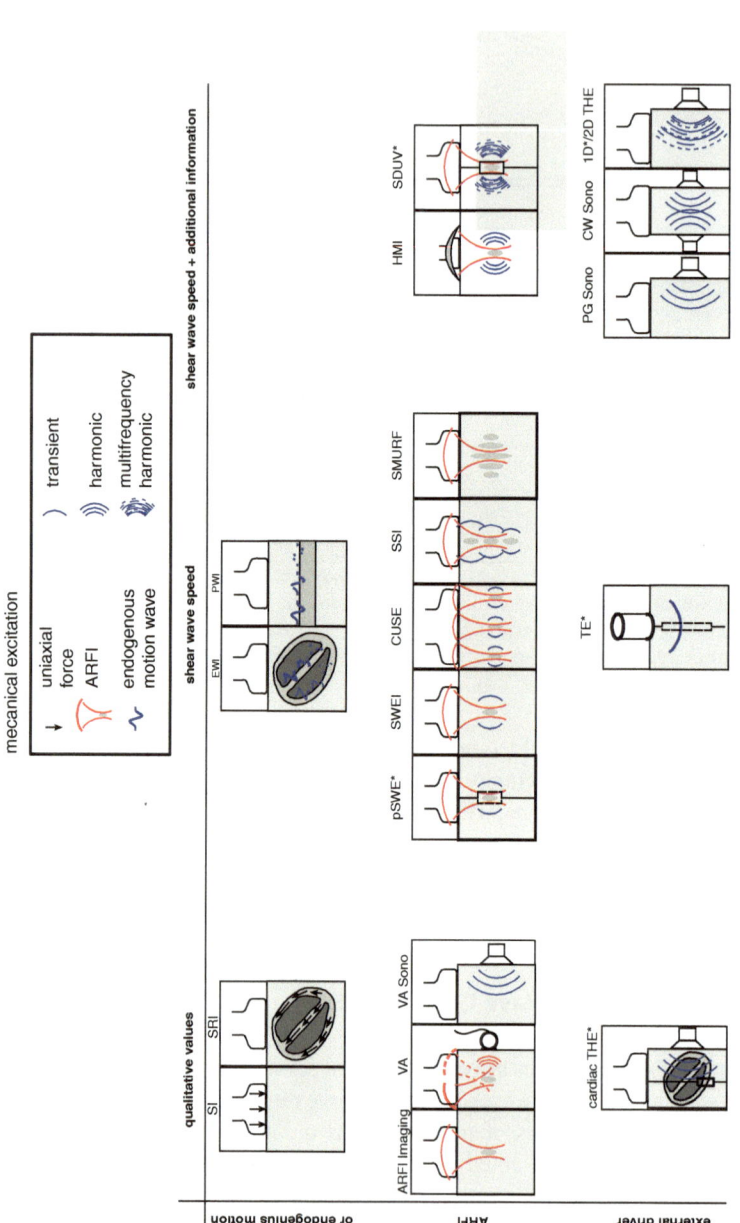

Fig. 12.4 Schematic overview of ultrasound elastography methods. Abbreviations of the names of USE methods correspond to section titles of Sect. 12.4, which explains the methods in greater detail. Compression waves are marked in red, whereas shear waves are marked in blue. Methods which provide a scalar parameter instead of an image are marked by "asterisk". Note that loudspeakers generate both shear waves and compression waves; however, for brevity we here assume that in USE only the shear wave speed is measured

The *coherent plane-wave compounding* method compensates for this drawback in image quality of plane-wave imaging by combining several plane-wave images of different tilting angles [5–7]. The final image quality is comparable to that of conventional B-mode ultrasound while the frame rate is one order of magnitude higher (in the order of a few hundreds of hertz). As a limitation, the coherent plane-wave compounding technique makes high demands on scanner hardware.

12.4   Ultrasound Elastography Methods

Many USE methods are reviewed in the literature [2, 8–14]. In this chapter we will only briefly summarize most of the USE methods and their (potential) application in diagnostic imaging. Some of the methods described below are commercially available, while others are still in an experimental state. Most USE methods can be classified by the values they measure, either strain, shear wave speed, or dispersion. Dispersion denotes the frequency dependence of the shear wave speed. Additional parameters such as the nonlinearity or anisotropy are also addressed, but, due to the highly experimental nature of these methods, we do not discuss them. A schematic overview of USE methods is given in Fig. 12.4.

12.4.1 Strain Imaging (SI)

Strain imaging (SI) was developed by Ophir et al. and first presented in 1991 [15]. An overview is given in Varghese [16].

SI is based on quasi-static tissue deformations with a dynamical change far below 10 Hz. Quasi-static deformations can be induced by manually applying compression forces to the body surface or by exploiting internal excitation such as respiratory movements, cardiac motion, or vascular pulsation. The resulting axial or lateral tissue displacement, u, can be detected using the color mode. This displacement is the tissue movement between two consecutive frames or over the temporal integration of several frames. Tissue strain ε (see Chap. 2, Eq. (2.6)) can be calculated using the spatial derivative of the displacement. In one dimension

$$\varepsilon = \frac{\partial u}{\partial x}. \tag{12.2}$$

The resulting strain images are visualized as shown in Fig. 12.5. In general, the amount of force/stress applied to the tissue is unknown, and therefore SI is a qualitative elastography method, and only the relative shear modulus can be reconstructed [17]. Due to the simple and palpation-related setup, SI requires no special hardware but longer training periods.

SI is used for evaluation of the breasts, prostate, thyroid, muscle, and lymph nodes, and in principle, all organs accessible to manual palpation can be examined [16].

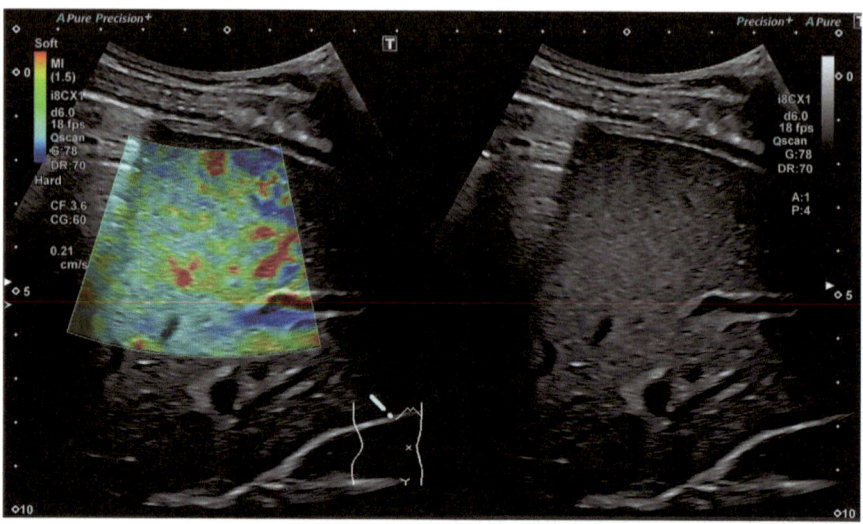

Fig. 12.5 Standard display of a liver examination by strain imaging using a Toshiba Aplio i900 ultrasound scanner (Toshiba, Otawara, Japan)

12.4.2 Strain Rate Imaging (SRI)

Strain rate imaging (SRI) is related to the cardiac Doppler measurement and therefore mainly used in echocardiography [18]. In contrast to SI, the basic measurement value is the tissue velocity $v = \dot{u}$, and therefore the strain rate $\dot{\varepsilon}$ is displayed:

$$\dot{\varepsilon} = \frac{\partial \dot{u}}{\partial x}.$$ (12.3)

Besides real-time tracking of the strain rate, the color mode-based values enable derivation of information on tissue dilatation ($\dot{\varepsilon} > 0$) or contraction ($\dot{\varepsilon} < 0$). SRI has been applied to intrinsic tissue deformation like in the heart, the muscle, or the gastrointestinal wall [8].

12.4.3 Acoustic Radiation Force Impulse (ARFI) Imaging

Acoustic radiation force impulse (ARFI) imaging is based on the work of Sarvazyan et al. [19] and was first published by Nightingale et al. [20] in 2002. An overview can be found in [21].

To avoid external tissue compression and to stimulate the tissue directly in the target region, ARFI uses highly focused ultrasound pulses to generate force/stress as explained above. The resulting deformation is in the order of tens of micrometers. Rapid acquisition of signals near the focus of the ARFI pulses provides information on strain, and the resulting images are comparable to SI. Due to the decreasing focus quality with increasing depth, ARFI is limited to approximately 8 cm depth.

ARFI imaging has been used to investigate abdominal organs, the breast, heart, vessels, nerve, and prostate and to monitor thermal ablation procedures [21].

12.4.4 Vibro-Acoustography (VA)

Vibro-acoustography (VA) was introduced by Fatemi and Greenleaf [22, 23] in 1998. An overview can be found in Urban et al. [24].

VA uses time-harmonic tissue excitation generated by two ARFI beams with slightly different ultrasonic frequencies. Within the focus region, a point source is generated with a difference frequency of typically 10–70 kHz. The amplitude of the emanating compression wave is detected by a hydrophone. The size of the compression wave amplitude is related to the tissue's stiffness. Spatial resolution in the field of view (5×5 cm^2) is achieved by sweeping the joint focus. Unlike ultrasound images, VA images are speckle free and highly sensitive to calcifications. It takes several minutes to scan a 5×5 cm^2 image; however, this time can be reduced to 1 min using a linear transducer.

VA has been applied to the breast, thyroid, liver, ex vivo prostate, and porcine arteries [24].

12.4.5 Vibration Amplitude Sonoelastography (VA Sono)

Vibration amplitude sonoelastography (VA Sono) was developed by Lerner et al. [25] in 1988. An overview can be found in Parker [26].

VA Sono was the first ultrasound elastography method based on time-harmonic vibration. For this an external driver generates mono- or multifrequency excitation in the range of 20–1000 Hz. The resulting tissue displacement is captured by Doppler techniques, and the corresponding vibration amplitude is calculated using Fourier-Bessel functions. Like in ARFI imaging and VA, this amplitude provides qualitative information on tissue stiffness.

VA Sono has been used in the liver, breast, kidney, prostate, and eye and to characterize lesions [26].

12.4.6 Cardiac Time-Harmonic Elastography (Cardiac THE)

Cardiac time-harmonic elastography (cardiac THE) was developed by Tzschätzsch et al. [27] and presented in 2012.

Cardiac THE is based on observations made by cardiac magnetic resonance elastography showing that externally induced shear wave amplitudes in the left ventricle vary in synchrony with alterations in myocardial stiffness [28, 29]. The technical principles are similar to vibration amplitude sonoelastography (see Sect. 12.4.5). An external loudspeaker generates a 30 Hz vibration, and the resulting myocardial wall vibration is captured in the A-mode. The vibration amplitude, which is displayed in real time, shows mechanical tissue contraction during systole and tissue relaxation during diastole by high and low values, respectively. Mechanical contraction and relaxation occur prior to the geometrical phases of the heart. The time delay between the wave amplitude alteration and heart geometry is a direct measure of isovolumetric times of the cardiac cycle [30]. Cardiac THE is currently under investigation for detecting cardiac diastolic dysfunction.

12.4.7 Electromechanical Wave Imaging (EWI)

Electromechanical wave imaging (EWI) was first published by Pernot and Konofagou [31] in 2005 and revised by Konofagou et al. [32].

EWI exploits the intrinsic electromagnetic activation of the heart muscle, which involves the generation of shear waves that propagate though the myocardium. These waves can be captured in the B-mode at a high frame rate of 500 frames/s. This is achieved by subdividing the color mode image into several sectors, which are consecutively acquired through the cardiac cycle. The isochrones are reconstructed in 3D based on the segmented strain image. Isochrones are hypothetical lines connecting sites where the activation signal at the tissue deflection due to electromechanical waves passes at the same time. In this way EWI visualizes the spreading of electromechanical activation across the myocardium. The feasibility of EWI has been demonstrated in canine tissue and humans [32]. An alternative approach based on the acoustic waves produced by heart sounds was developed by Kanai and Koiwa [33] in 2001. This method captures the shear wave generated by an aortic valve closure within the left ventricular septum. A high frame rate of 450 Hz is achieved by reducing the number of lines of sight to 16, and wave propagation is displayed in a time-distance plot whose slope indicates the shear wave speed [34].

12.4.8 Pulse Wave Imaging (PWI)

Pulse wave imaging (PWI) was published by Pernot et al. [35] in 2007, and an overview can be found in Konofagou et al. [32].

The pulse wave caused by cardiac contraction travels along the aorta as a bulk wave. The pulse wave velocity, c, is related to Young's modulus of the aortic wall, E, via the Moens-Korteweg equation [32]:

$$E = \frac{2 \, \rho \, R \, c^2}{b} \tag{12.4}$$

where ρ is the density of the aortic wall, R is the inner radius, and b is the thickness of the wall.

PWI is based on the findings of Kanai in 1994 [36] using Doppler ultrasound to measure the pulse wave displacement of the aortic wall at two lateral positions. From the displacements, the time of flight and finally Young's modulus are calculated.

To provide spatial resolution, Pernot et al. [35] used plane-wave imaging with up to 8000 frames/s to capture the pulse wave at multiple time points. Pulse wave velocity can be measured by a time-of-flight algorithm.

PWI has been applied to the aorta of mice and humans [32].

12.4.9 Point Shear Wave Elastography (pSWE)

Point shear wave elastography (pSWE) was developed by Palmeri et al. [37] in 2008. An overview is given in Bruno et al. [38].

The pSWE technique is also known as *ARFI quantification*. In contrast to ARFI quantification, tissue displacement is measured outside the excitation point (see Fig. 12.6). Furthermore, a time-of-flight algorithm estimates the mean speed from the spherically propagating shear wave, providing a quantitative value of intrinsic tissue properties.

pSWEI has been applied to the liver, spleen, kidney, pancreas, and thyroid [38].

12.4.10 Shear Wave Elasticity Imaging (SWEI)

Shear wave elasticity imaging (SWEI) was developed by Nightingale et al. [39] and presented in 2003. It is based on the principle of ARFI [19].

Tissue excitation in SWEI is similar to that in pSWEI except that a map of the shear wave speed (in the order of 3×4 cm^2) is obtained by repeated ARFI excitations at different lateral positions. Similar to pSWEI, the wave speed is calculated using wave peak detection and a time-of-flight algorithm (see Fig. 12.7).

SWEI has been used for investigation of the liver, spleen, breast, and lymph nodes [40–42].

12.4.11 Comb-Push Ultrasound Shear Elastography (CUSE)

Comb-push ultrasound shear elastography (CUSE) was developed by Song et al. [43] and presented in 2012.

CUSE uses simultaneous ARFI excitation to overcome the limited frame rate of SWEI, which is due to repeated scanning at different lateral positions. Using simultaneous excitations and plane-wave imaging, CUSE acquires multiple shear waves within a few milliseconds. The displacement is estimated by a 2D autocorrelation algorithm. Directional filtering combined with a time-of-flight algorithm is used to reconstruct the elastogram in smaller regions (approximately 4×4 cm^2).

The feasibility of CUSE was demonstrated in phantoms, in vivo thyroid, and breast tissue [44, 45].

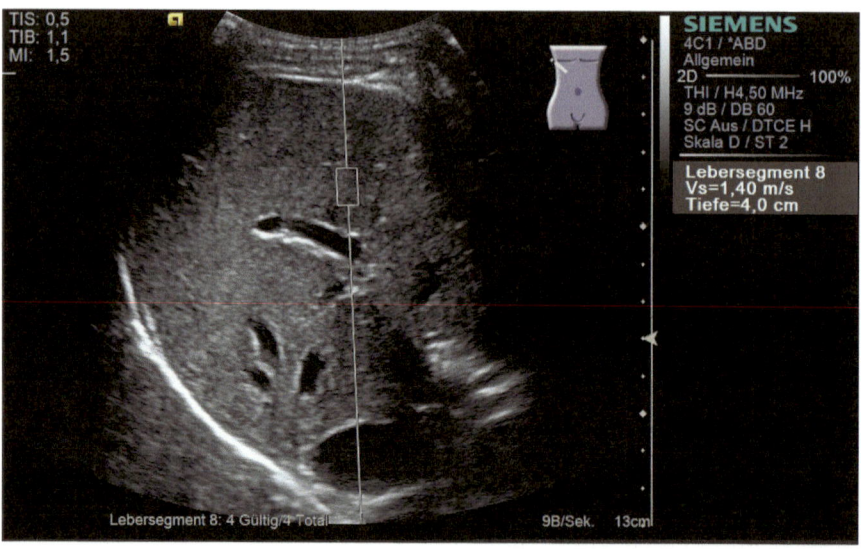

Fig. 12.6 Liver investigation by pSWEI using a Siemens Acuson S2000 ultrasound scanner (Siemens, Mountain View CA, USA). The ARFI focus and the corresponding shear wave speed of 1.40 m/s are obtained inside the rectangular region of interest

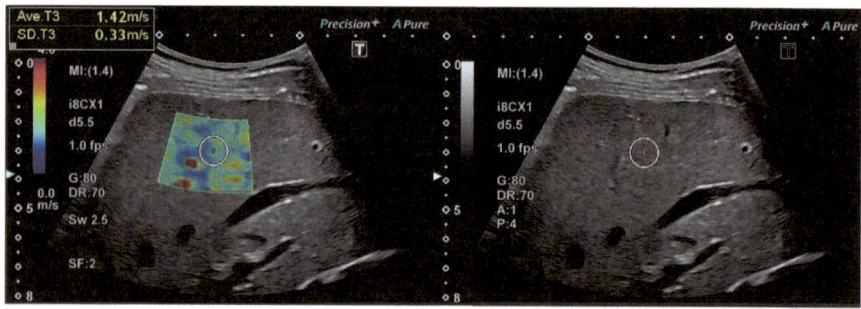

Fig. 12.7 Liver investigation by SWEI using a Toshiba Aplio i900 ultrasound scanner (Toshiba, Otawara, Japan). The mean shear wave speed within the selected ROI is 1.42 m/s

12.4.12 Supersonic Shear Imaging (SSI)

Supersonic shear imaging (SSI) was developed by Bercoff et al. [46] in 2004.

Spherically propagating shear waves from point sources as generated in pSWEI and SWEI can produce artifacts in the elastograms since most reconstruction methods assume plane-wave propagation. To generate plane waves, SSI applies repeated ARFI excitation at different focal depths. The foci are moved through the tissue with a higher speed than the shear wave speed, which gives rise to a Mach cone of nearly cylindrical geometry. The tissue displacement is captured with a frame rate of a few thousands of hertz by coherent plane-wave compounding. Shear wave speed is calculated using a time-of-flight algorithm, and the final elastogram is displayed with a frame rate of 3–4 Hz.

SSI has been applied to many organs and tissues including the breast, liver, spleen, thyroid, prostate, skeletal muscle, and transplanted kidneys [13].

12.4.13 Spatially Modulated Ultrasound Radiation Force (SMURF)

Spatially modulated ultrasound radiation force (SMURF) was developed by McAleavey et al. [47] and presented in 2007.

Most elastography methods control the temporal behavior of tissue motion and analyze the spatial behavior of the motion to reconstruct elastic properties. In contrast, SMURF attempts to control the spatial behavior of motion by a laterally modulated ARFI pattern. This pattern determines the wavelength, λ, of the shear wave, while its oscillating frequency, f, is measured by $c = \lambda \ f$. SMURF uses a high pulse repetition rate for each image line since the transiently induced shear waves are rapidly attenuated due to geometrical and viscous dispersion. SMURF was demonstrated in phantoms and an ex vivo porcine liver tissue [47, 48].

12.4.14 Transient Elastography (TE)

Transient elastography (TE) was developed by Sandrin et al. [49] for noninvasively staging liver fibrosis and presented in 2002.

In TE the tissue is excited by an external piston driver operated with a burst of a single 50 Hz vibration cycle. The generated shear wave propagates from the skin surface into deeper tissue and is captured in the M-mode. The shear wave fronts appear in the time-depth plot of the M-mode as parallel lines whose slope is the shear wave speed (see Fig. 12.8). TE provides a mean value for shear wave speed without spatial resolution. Similar to ARFI-based methods, TE is limited to a maximum depth of approx. 8 cm. TE was one of the first commercially available ultrasound elastography methods and is therefore best validated. Due to the lack of B-mode guidance, positioning the M-mode beam can be challenging.

The main application of TE is the staging of liver fibrosis. However, TE has also been tested in the breast, skeletal muscles, skin, and blood clots [13].

12.4.15 Harmonic Motion Imaging (HMI)

Harmonic motion imaging (HMI) was developed by Konofagou and Hynynen [50] and presented in 2003. An overview is given in Konofagou et al. [51].

The motivation for the development of HMI was the observation of changes in viscoelastic properties during thermal ablation with *high-intensity focused ultrasound* (HIFU). In contrast to other elastography techniques, HMI allows evaluation of tissue behavior during HIFU ablation in real time which is useful to control the

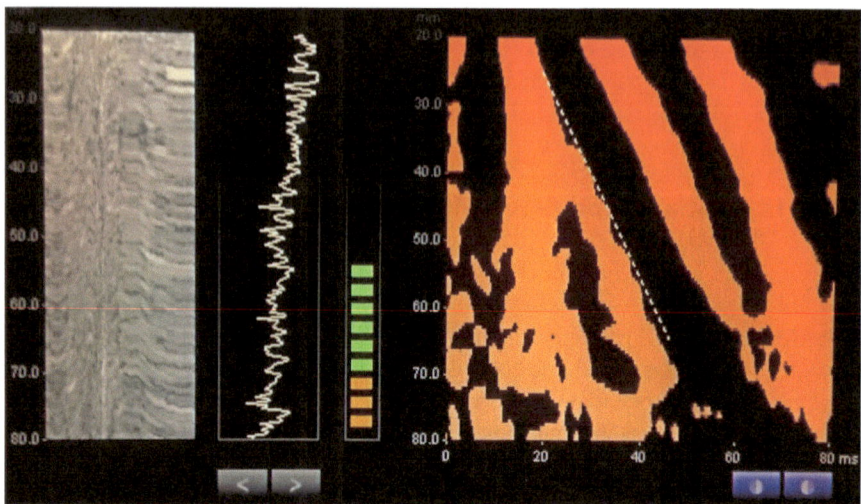

Fig. 12.8 Liver investigation by TE using a FibroScan® ultrasound scanner (Echosens, Paris, France). From left to right, M-mode, A-mode, feedback-color bar for contact pressure, and time-depth images of the shear wave with fitted slope (dashed white line)

duration of the ablation procedure. HMI is accomplished by amplitude modulation of the HIFU beam, which generates a harmonic vibration in the order of 50 Hz. Tissue deflection is captured using a second image transducer, embedded in the HIFU transducer. The necessary frame rate of a few hundreds of hertz is achieved by reducing the number of lines of sight. After temporal Fourier transformation, a phase gradient algorithm calculates the shear wave speed in the focus. As additional information, the phase shift between the vibration excitation and the measured tissue vibration is captured. This phase shift is identical to the phase angle of the complex-valued shear modulus. HMI is used for detection of breast cancer and other lesions and for real-time monitoring of thermal ablation [51].

12.4.16 Shear Wave Dispersion Ultrasound Vibrometry (SDUV)

Shear wave dispersion ultrasound vibrometry (SDUV) was developed by Chen et al. [52] and presented in 2004. An overview is given by Urban et al. [53].

Most ultrasound elastography methods measure only the shear wave phase velocity at one frequency or the group velocity of a transient burst. Additional information on viscosity can be obtained by evaluating the frequency dependence of shear wave speed (dispersion). Therefore SDUV excites harmonic vibration by ARFI at a base frequency and also at higher harmonics, which are typically in the range of 200–800 Hz. The superimposed tissue deflection is captured and frequency decomposed using Kalman filtering. For each frequency, the shear wave phase is evaluated at two lateral positions. The corresponding shear wave speed is calculated from this phase shift. Viscoelastic tissue properties can be reconstructed by fitting the Kelvin-Voigt model to the shear wave speed dispersion curve.

SDUV was used to investigate the skeletal muscle, liver, excised arteries, excised prostates, and excised porcine kidney [53].

A variant of SDUV, termed *Lamb wave dispersion ultrasound vibrometry* (LDUV), was used to investigate porcine myocardium during an open chest surgery [54].

12.4.17 Vibration Phase Gradient (PG) Sonoelastography

Vibration phase gradient (PG) sonoelastography is based on vibration amplitude sonoelastography and was developed by Yamakoshi et al. [55] and presented in 1990. An overview is given by Parker [26].

Tissue excitation and acquisition of the deflection are similar to vibration amplitude sonoelastography (see Sect. 12.4.5). In addition to the vibration amplitude, the vibration phase is captured, which allows quantification of shear wave speed using a phase gradient algorithm. When the tissue is stimulated by multifrequency excitation, PG sonoelastography can measure the dispersion of the shear wave speed.

PG sonoelastography was demonstrated in the skeletal muscle [26].

12.4.18 Crawling Waves (CW) Sonoelastography

Crawling waves (CW) *sonoelastography* was developed by Wu et al. [56] and presented in 2004. A review is given by Parker [26].

CW sonoelastography was developed to overcome limitations with regard to the frame rate of normal clinical scanners used for PG sonoelastography. Since the oscillation frequency f of time-harmonic shear waves in elastography is in the order of the frame rate of commercial ultrasound devices (approx. 80 Hz), the propagation of shear waves cannot be captured directly without aliasing artifacts. Therefore, CW sonoelastography uses two loudspeakers which are placed at opposite sides of the tissue and which have a slightly different vibration frequency, $f + \Delta f$ and $f - \Delta f$, with $\Delta f \ll f$. The resulting interference pattern is composed of a wave with two apparent wavelengths ($\lambda_1 = \Delta f/c$ and $\lambda_2 = f/c$) oscillating at two vibration frequencies (f and Δf). The apparent wave with wavelength λ_1 and oscillation frequency Δf meets the criteria to be captured by low frame rates while being suitable for inversion techniques since wavelengths are short ($\lambda_1 \ll \lambda_2$). Using multifrequency excitation CW sonoelastography is able to measure shear wave dispersion. CW sonoelastography with external loudspeakers was applied to ex vivo prostate and liver tissue [26] and demonstrated in vivo in the skeletal muscle [57–59].

12.4.19 Time-Harmonic Elastography (1D/2D THE)

Time-harmonic elastography (THE) is similar to vibration phase gradient sonoelastography except that it uses multiple harmonic drive frequencies and compound shear wave speed mapping [60].

THE uses a loudspeaker integrated into a patient bed and operated by a superposition of harmonic signals in the frequency range between 30 and 60 Hz. In 1D THE, the resulting tissue displacement is captured in the M-mode along multiple profiles corresponding to varying transducer positions. A fit-based algorithm automatically selects the most reliable wave speed values and evaluates the dispersion of shear wave speed.

2D THE also uses multifrequency wave stimulations. However, since motion is captured by a standard B-mode scanner with frame rates in the order of 80 Hz, vibrations slightly above the Nyquist limit are evaluated at aliasing frequency. Compound maps of wave speed are reconstructed by multifrequency inversion of directionally filtered wave images [61]. 2D THE is less limited in penetration depth than other methods and provides an elastogram which covers the entire B-mode image. A similar compounding reconstruction technique, *external vibration multidirectional ultrasound shear wave elastography* (EVMUSE), was proposed by Zhao et al. [62]. However, in EVMUSE, a single-frequency 50 Hz wave is observed a few milliseconds after vibration is stopped, giving rise to some transient wave effects.

1D THE was used in healthy volunteers [60, 63, 64], in liver fibrosis staging [65], and for evaluating liver decompression after shunt implantation [66]. Feasibility of 2D THE was demonstrated for liver and spleen examinations in volunteers [61].

Conclusion

Despite widespread clinical applications of USE, consistent thresholds and standardized stiffness-based diagnostic indices are still lacking. The discrepant results of USE studies are mainly attributable to the fact that so many different techniques are used. This is exactly the boon and bane of a rapidly developing and highly innovative field such as USE. On the one hand, new USE methods offer more insight into tissue mechanical parameters in health and disease. On the other hand, research groups and manufactures worldwide are inclined to validate their own methods for self-consistency instead of searching for modality-independent quantitative and tissue-inherent parameters. The importance of modality-independent and quantitative imaging biomarkers has been recognized and addressed by many alliances worldwide including the *Quantitative Imaging Biomarkers Alliance* (QIBA) or the *European Imaging Biomarkers Alliance* (EIBALL). The rapid progress made by USE promises that one day a gold standard of mechanical tissue properties will be developed, to which new methods can be compared. One promising approach is time-harmonic elastography (THE), which measures stiffness at well-defined excitation frequencies and which can thus be compared with reference methods such as MRI elastography. There is no doubt that well-documented standards in elastography will foster the dissemination of mechanical imaging biomarkers and thus contribute to higher precision of ultrasound-based diagnoses.

References

1. Tzschätzsch H. Entwicklung, anwendung und validierung der zeitharmonischen in vivo ultraschallelastografie an der menschlichen leber und am menschlichen herzen. Dissertation. Humboldt Universität Berlin. 2016.
2. Sarvazyan AP, Urban MW, Greenleaf JF. Acoustic waves in medical imaging and diagnostics. Ultrasound Med Biol. 2013;39(7):1133–46. https://doi.org/10.1016/j.ultrasmedbio.2013.02.006.
3. Sugimoto T, Ueha S, Itoh K. Tissue hardness measurement using the radiation force of focused ultrasound. In: IEEE symposium on ultrasonics. IEEE; 1990. p. 1377–1380. https://doi.org/10.1109/ULTSYM.1990.171591.
4. Lu JY. 2D and 3D high frame rate imaging with limited diffraction beams. IEEE Trans Ultrason Ferroelectr Freq Control. 1997;44(4):839–56. https://doi.org/10.1109/58.655200.
5. Cheng J, Lu J. Extended high-frame rate imaging method with limited-diffraction beams. IEEE Trans Ultrason Ferroelectr Freq Control. 2006;53(5):880–99. https://doi.org/10.1109/TUFFC.2006.1632680.
6. Montaldo G, Tanter M, Bercoff J, Benech N, Fink M. Coherent plane-wave compounding for very high frame rate ultrasonography and transient elastography. IEEE Trans Ultrason Ferroelectr Freq Control. 2009;56(3):489–506. https://doi.org/10.1109/TUFFC.2009.1067.
7. Song TK, Chang JH. Synthetic aperture focusing method for ultrasound imaging based on planar waves. 2004.
8. Bamber J, Cosgrove D, Dietrich C, Fromageau J, Bojunga J, Calliada F, et al. EFSUMB guidelines and recommendations on the clinical use of ultrasound elastography. Part 1: basic principles and technology. Ultraschall Med. 2013;34(2):169–84. https://doi.org/10.1055/s-0033-1335205.
9. Cosgrove D, Piscaglia F, Bamber J, Bojunga J, Correas J-M, Gilja O, et al. EFSUMB guidelines and recommendations on the clinical use of ultrasound elastography. Part 2: clinical applications. Ultraschall Med. 2013;34(3):238–53. https://doi.org/10.1055/s-0033-1335375.
10. Greenleaf JF, Fatemi M, Insana M. Selected methods for imaging elastic properties of biological tissues. Annu Rev Biomed Eng. 2003;5(1):57–78. https://doi.org/10.1146/annurev.bioeng.5.040202.121623.
11. Parker KJ, Doyley MM, Rubens DJ. Corrigendum: imaging the elastic properties of tissue: the 20 year perspective. Phys Med Biol. 2012;57(16):5359–60. https://doi.org/10.1088/0031-9155/57/16/5359.
12. Parker KJ, Taylor LS, Gracewski S, Rubens DJ. A unified view of imaging the elastic properties of tissue. J Acoust Soc Am. 2005;117(5):2705. https://doi.org/10.1121/1.1880772.
13. Sarvazyan A, Hall TJ, Urban MW, Fatemi M, Aglyamov SR, Garra BS. An overview of Elastography - an emerging branch of medical imaging. Curr Med Imaging Rev. 2011;7(4):255–82. https://doi.org/10.2174/157340511798038684.
14. Zaleska-Dorobisz U, Kaczorowski K, Pawluś A, Puchalska A, Inglot M. Ultrasound elastography - review of techniques and its clinical applications. Adv Clin Exp Med. 2014;23(4):645–55. Retrieved from http://www.ncbi.nlm.nih.gov/pubmed/25166452
15. Ophir J. Elastography: a quantitative method for imaging the elasticity of biological tissues. Ultrason Imaging. 1991;13(2):111–34. https://doi.org/10.1016/0161-7346(91)90079-W.
16. Varghese T. Quasi-static ultrasound elastography. Ultrasound Clin. 2009;4(3):323–38. https://doi.org/10.1016/j.cult.2009.10.009.
17. Oberai AA, Gokhale NH, Goenezen S, Barbone PE, Hall TJ, Sommer AM, Jiang J. Linear and nonlinear elasticity imaging of soft tissue in vivo: demonstration of feasibility. Phys Med Biol. 2009;54(5):1191.
18. Fleming AD, Xia X, McDicken WN, Sutherland GR, Fenn L. Myocardial velocity gradients detected by Doppler imaging. Br J Radiol. 1994;67(799):679–88. https://doi.org/10.1259/0007-1285-67-799-679.

19. Sarvazyan AP, Rudenko OV, Swanson SD, Fowlkes JB, Emelianov SY. Shear wave elasticity imaging: a new ultrasonic technology of medical diagnostics. Ultrasound Med Biol. 1998;24(9):1419–35. https://doi.org/10.1016/S0301-5629(98)00110-0.
20. Nightingale K, Soo MS, Nightingale R, Trahey G. Acoustic radiation force impulse imaging: in vivo demonstration of clinical feasibility. Ultrasound Med Biol. 2002;28(2):227–35. https://doi.org/10.1016/S0301-5629(01)00499-9.
21. Nightingale K. Acoustic radiation force impulse (ARFI) imaging: a review. Curr Med Imaging Rev. 2012;7(4):328–39. https://doi.org/10.2174/157340511798038657.
22. Fatemi M. Ultrasound-stimulated vibro-acoustic spectrography. Science. 1998;280(5360):82–5. https://doi.org/10.1126/science.280.5360.82.
23. Fatemi M, Greenleaf JF. Vibro-acoustography: an imaging modality based on ultrasound-stimulated acoustic emission. Proc Natl Acad Sci. 1999;96(12):6603–8. https://doi.org/10.1073/pnas.96.12.6603.
24. Urban MW, Alizad A, Aquino W, Greenleaf JF, Fatemi M. A review of vibro-acoustography and its applications in medicine. Curr Med Imaging Rev. 2011;7(4):350–9. https://doi.org/10.2174/157340511798038648.
25. Lerner RM, Parker KJ, Holen J, Gramiak R, Waag RC. Sono-elasticity: medical elasticity images derived from ultrasound signals in mechanically vibrated targets. Acoust Imaging. 1988;16:317–27. https://doi.org/10.1007/978-1-4613-0725-9_31.
26. Parker KJ. The evolution of vibration sonoelastography. Curr Med Imaging Rev. 2011;7(4):283–91. https://doi.org/10.2174/157340511798038675.
27. Tzschätzsch H, Elgeti T, Rettig K, Kargel C, Klaua R, Schultz M, et al. In vivo time harmonic elastography of the human heart. Ultrasound Med Biol. 2012;38(2):214–22. https://doi.org/10.1016/j.ultrasmedbio.2011.11.002.
28. Elgeti T, Beling M, Hamm B, Braun J, Sack I. Elasticity-based determination of isovolumetric phases in the human heart. J Cardiovasc Magn Reson. 2010;12(1):60. https://doi.org/10.1186/1532-429X-12-60.
29. Sack I, Rump J, Elgeti T, Samani A, Braun J. MR elastography of the human heart: noninvasive assessment of myocardial elasticity changes by shear wave amplitude variations. Magn Reson Med. 2009;61(3):668–77. https://doi.org/10.1002/mrm.21878.
30. Tzschätzsch H, Hättasch R, Knebel F, Klaua R, Schultz M, Jenderka K-V, et al. Isovolumetric elasticity alteration in the human heart detected by in vivo time-harmonic elastography. Ultrasound Med Biol. 2013;39(12):2272–8. https://doi.org/10.1016/j.ultrasmedbio.2013.07.003.
31. Pernot M, Konofagou EE. Electromechanical imaging of the myocardium at normal and pathological states. IEEE Ultrason Symp. 2005;2:1091–4. https://doi.org/10.1109/ULTSYM.2005.1603040.
32. Konofagou E, Lee W-N, Luo J, Provost J, Vappou J. Physiologic cardiovascular strain and intrinsic wave imaging. Annu Rev Biomed Eng. 2011;13(1):477–505. https://doi.org/10.1146/annurev-bioeng-071910-124721.
33. Kanai H, Koiwa Y. Myocardial rapid velocity distribution. Ultrasound Med Biol. 2001;27(4):481–98. https://doi.org/10.1016/S0301-5629(01)00341-6.
34. Kanai H. Propagation of vibration caused by electrical excitation in the normal human heart. Ultrasound Med Biol. 2009;35(6):936–48. https://doi.org/10.1016/j.ultrasmedbio.2008.12.013.
35. Pernot M, Fujikura K, Fung-Kee-Fung SD, Konofagou EE. ECG-gated, mechanical and electromechanical wave imaging of cardiovascular tissues in vivo. Ultrasound Med Biol. 2007;33(7):1075–85. https://doi.org/10.1016/j.ultrasmedbio.2007.02.003.
36. Kanai H, Kawabe K, Takano M, Murata R, Chubachi N, Koiwa Y. New method for evaluating local pulse wave velocity by measuring vibrations on arterial wall. Electron Lett. 1994;30(7):534–6. https://doi.org/10.1049/el:19940393.
37. Palmeri ML, Wang MH, Dahl JJ, Frinkley KD, Nightingale KR. Quantifying hepatic shear modulus in vivo using acoustic radiation force. Ultrasound Med Biol. 2008;34(4):546–58. https://doi.org/10.1016/j.ultrasmedbio.2007.10.009.
38. Bruno C, Minniti S, Mucell BA, Roberto P. ARFI: from basic principles to clinical applications in diffuse chronic disease—a review. Insights Imaging. 2016;7:735–46.

39. Nightingale K, McAleavey S, Trahey G. Shear-wave generation using acoustic radiation force: in vivo and ex vivo results. Ultrasound Med Biol. 2003;29(12):1715–23. https://doi.org/10.1016/j.ultrasmedbio.2003.08.008.

40. Cheng KL, Choi YJ, Shim WH, Lee JH, Baek JH. Virtual touch tissue imaging quantification shear wave elastography: prospective assessment of cervical lymph nodes. Ultrasound Med Biol. 2016;42(2):378–86. https://doi.org/10.1016/j.ultrasmedbio.2015.10.003.

41. Ianculescu V, Ciolovan LM, Dunant A, Vielh P, Mazouni C, Delaloge S, et al. Added value of virtual touch IQ shear wave elastography in the ultrasound assessment of breast lesions. Eur J Radiol. 2014;83(5):773–7.

42. Leschied JR, Dillman JR, Bilhartz J, Heider A, Smith EA, Lopez MJ. Shear wave elastography helps differentiate biliary atresia from other neonatal/infantile liver diseases. Pediatr Radiol. 2015;45(3):366–75. https://doi.org/10.1007/s00247-014-3149-z.

43. Song P, Zhao H, Manduca A, Urban MW, Greenleaf JF, Chen S. Comb-push ultrasound shear elastography (CUSE): a novel method for two-dimensional shear elasticity imaging of soft tissues. IEEE Trans Med Imaging. 2012;31(9):1821–32. https://doi.org/10.1109/TMI.2012.2205586.

44. Mehrmohammadi M, Song P, Meixner DD, Fazzio RT, Chen S, Greenleaf JF, et al. Comb-push ultrasound shear elastography (CUSE) for evaluation of thyroid nodules: preliminary in vivo results. IEEE Trans Med Imaging. 2015;34(1):97–106. https://doi.org/10.1109/TMI.2014.2346498.

45. Denis M, Mehrmohammadi M, Song P, Meixner DD, Fazzio RT, Pruthi S, et al. Comb-push ultrasound shear elastography of breast masses: initial results show promise. PLoS One. 2015;10(3):e0119398. https://doi.org/10.1371/journal.pone.0119398.

46. Bercoff J, Tanter M, Fink M. Supersonic shear imaging: a new technique for soft tissue elasticity mapping. IEEE Trans Ultrason Ferroelectr Freq Control. 2004;51(4):396–409. https://doi.org/10.1109/TUFFC.2004.1295425.

47. McAleavey SA, Menon M, Orszulak J. Shear-modulus estimation by application of spatially-modulated impulsive acoustic radiation force. Ultrason Imaging. 2007;104(2007):87–104.

48. McAleavey S, Menon M, Elegbe E. Shear modulus imaging with spatially-modulated ultrasound radiation force. Ultrason Imaging. 2009;31(4):217–34. https://doi.org/10.1177/016173460903100403.

49. Sandrin L, Tanter M, Gennisson J-L, Catheline S, Fink M. Shear elasticity probe for soft tissues with 1-D transient elastography. IEEE Trans Ultrason Ferroelectr Freq Control. 2002;49(4):436–46. https://doi.org/10.1109/58.996561.

50. Konofagou EE, Hynynen K. Localized harmonic motion imaging: theory, simulations and experiments. Ultrasound Med Biol. 2003;29(10):1405–13. https://doi.org/10.1016/S0301-5629(03)00953-0.

51. Konofagou EE, Maleke C, Vappou J. Harmonic motion imaging (HMI) for tumor imaging and treatment monitoring. Curr Med Imaging Rev. 2012;8:16. https://doi.org/10.2174/157340512799220616.

52. Chen S, Fatemi M, Greenleaf JF. Quantifying elasticity and viscosity from measurement of shear wave speed dispersion. J Acoust Soc Am. 2004;115(6):2781. https://doi.org/10.1121/1.1739480.

53. Urban MW, Chen S, Fatemi M. A review of shearwave dispersion ultrasound vibrometry (SDUV) and its applications. Curr Med Imaging Rev. 2012;8(1):27–36. https://doi.org/10.2174/157340512799220625.

54. Nenadic IZ, Urban MW, Mitchell SA, Greenleaf JF. Lamb wave dispersion ultrasound vibrometry (LDUV) method for quantifying mechanical properties of viscoelastic solids. Phys Med Biol. 2011;56(7):2245–64. https://doi.org/10.1088/0031-9155/56/7/021.

55. Yamakoshi Y, Sato J, Sato T. Ultrasonic imaging of internal vibration of soft tissue under forced vibration. IEEE Trans Ultrason Ferroelectr Freq Control. 1990;37(2):45–53. https://doi.org/10.1109/58.46969.

56. Wu Z, Taylor LS, Rubens DJ, Parker KJ. Sonoelastographic imaging of interference patterns for estimation of the shear velocity of homogeneous biomaterials. Phys Med Biol. 2004;49(6):911–22. https://doi.org/10.1088/0031-9155/49/6/003.

57. Hazard C, Hah Z, Rubens D, Parker K. Integration of crawling waves in an ultrasound imaging system. Part 1: system and design considerations. Ultrasound Med Biol. 2012;38(2):296–311. https://doi.org/10.1016/j.ultrasmedbio.2011.10.026.
58. Hah Z, Hazard C, Mills B, Barry C, Rubens D, Parker K. Integration of crawling waves in an ultrasound imaging system. Part 2: signal processing and applications. Ultrasound Med Biol. 2012;38(2):312–23. https://doi.org/10.1016/j.ultrasmedbio.2011.10.014.
59. Hoyt K, Castaneda B, Parker KJ. Two-dimensional sonoelastographic shear velocity imaging. Ultrasound Med Biol. 2008;34(2):276–88. https://doi.org/10.1016/j.ultrasmedbio.2007.07.011.
60. Tzschätzsch H, Ipek-Ugay S, Guo J, Streitberger K-J, Gentz E, Fischer T, et al. In vivo time-harmonic multifrequency elastography of the human liver. Phys Med Biol. 2014;59(7):1641–54. https://doi.org/10.1088/0031-9155/59/7/1641.
61. Tzschätzsch H, Nguyen Trong M, Scheuermann T, Fischer T, Schultz M, Braun J, Sack I. Two-dimensional time-harmonic elastography of the human liver and spleen. Ultrasound Med Biol. 2016;16:30163–6. https://doi.org/10.1016/j.ultrasmedbio.2016.07.004.
62. Zhao H, Song P, Meixner DD, Kinnick RR, Callstrom MR, Sanchez W, et al. External vibration multi-directional ultrasound shearwave elastography (EVMUSE): application in liver fibrosis staging. IEEE Trans Med Imaging. 2014;33(11):2140–8. https://doi.org/10.1109/TMI.2014.2332542.
63. Ipek-Ugay S, Tzschätzsch H, Braun J, Fischer T, Sack I. Physiological reduction of hepatic venous blood flow by Valsalva maneuver decreases liver stiffness. J Ultrasound Med. 2017;36:1305.
64. Ipek-Ugay S, Tzschätzsch H, Hudert C, Marticorena Garcia SR, Fischer T, Braun J, et al. Time harmonic elastography reveals sensitivity of liver stiffness to water ingestion. Ultrasound Med Biol. 2016;42(6):1289–94. https://doi.org/10.1016/j.ultrasmedbio.2015.12.026.
65. Tzschätzsch H, Ipek-Ugay S, Nguyen Trong M, Guo J, Eggers J, Gentz E, et al. Multifrequency time-harmonic elastography for the measurement of liver viscoelasticity in large tissue windows. Ultrasound Med Biol. 2015;41(3):724–33. https://doi.org/10.1016/j.ultrasmedbio.2014.11.009.
66. Tzschätzsch H, Sack I, Marticorena Garcia SR, Ipek-Ugay S, Braun J, Hamm B, Althoff CE. Time-harmonic elastography of the liver is sensitive to intrahepatic pressure gradient and liver decompression following transjugular intrahepatic portosystemic shunt (TIPS) implantation. Ultrasound Med Biol. 2017;43:595.

Photoacoustic Imaging: Principles and Applications

Jan Laufer

Abstract

Photoacoustic (PA) imaging is an emerging imaging technology with potential for preclinical biomedical research and clinical applications. PA imaging, which relies on the generation of broadband acoustic waves via the absorption of intensity-modulated light in tissue, offers the combination of strong optical contrast and high spatial resolution provided by ultrasound. For excitation wavelengths in the visible and near-infrared region, image contrast is predominately due to haemoglobin. Exogenous contrast agents, such as dyes or genetically expressed absorbers, can be used to obtain targeted molecular contrast. Over the past decade, PA imaging has rapidly evolved into different microscopy and tomography modalities, while novel methodologies have led to a variety of exciting applications. This chapter explains the basic principles of PA imaging, its implementation in the different modalities and provides examples of applications to morphological, functional and molecular imaging. Furthermore, the challenge of recovering quantitative information from PA image data sets is described.

13.1 Introduction

Photoacoustic (PA) imaging [1] relies on the generation of broadband ultrasound waves in biological tissue following the absorption of short, low energy optical pulses. By detecting the waves at multiple points on the tissue surface and by employing reconstruction algorithms, 3D images of the initial pressure are reconstructed from the recorded signals. Since the initial pressure is a function of the local abundance of tissue chromophores, PA images acquired using visible or

J. Laufer
Institut für Physik, Martin-Luther-Universität Halle-Wittenberg, Halle, Germany
e-mail: jan.laufer@physik.uni-halle.de

© Springer International Publishing AG 2018
I. Sack, T. Schaeffter (eds.), *Quantification of Biophysical Parameters in Medical Imaging*, https://doi.org/10.1007/978-3-319-65924-4_13

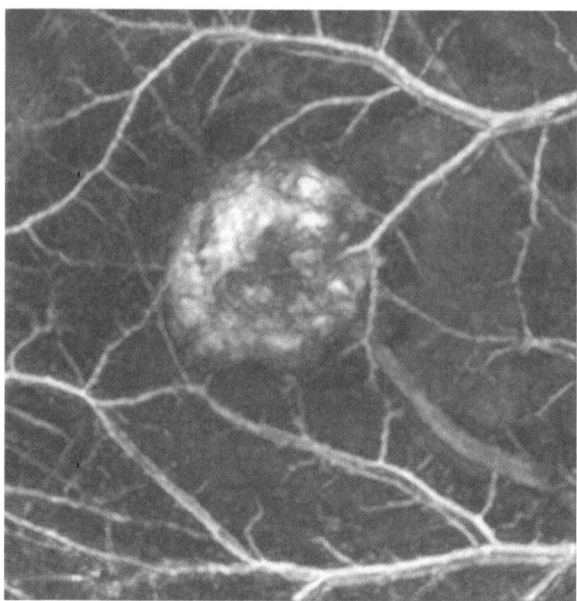

Fig. 13.1 *x-y* maximum intensity projection of a 3D photoacoustic image data set acquired in a mouse in vivo showing the vasculature of the skin and a subcutaneous tumour (reproduced with permission from [2])

near-infrared excitation wavelengths typically show the vasculature (Fig. 13.1). PA imaging combines a number of powerful attributes. It has multiscale imaging capabilities ranging from single cell resolution, achieved using PA microscopy (PAM), to micron resolution using PA tomography (PAT) (tens of micrometres at millimetre depths to hundreds of micrometres at centimetre depths). It provides strong contrast in vascularised soft tissues due to the absorption by haemoglobin where other modalities, such as MRI, X-ray CT and ultrasound lack sensitivity. However, its greatest strength arguably lies in the potential to recover the spatial distribution of the local abundance of tissue chromophores, such as oxyhaemoglobin (HbO_2) and deoxyhaemoglobin (Hhb), and exogenous contrast agents, by exploiting the differences in their wavelength-dependent optical absorption. From the spatial maps of HbO_2 and Hhb, derived parameters, such as blood oxygen saturation (sO_2) may then be obtained. PA imaging therefore provides the capability of functional and molecular imaging at depths and spatial resolutions that are inaccessible to most other purely optical imaging modalities, such as microscopy, optical coherence tomography and diffuse optical tomography. Importantly, PA imaging has the potential to allow spatially resolved *quantitative* measurements of chromophore concentrations and derived parameters. However, this potential has yet to be harnessed. Deep-tissue 3D quantitative PA tomography (qPAT) in particular remains challenging due to the large scale of the inverse problem, and remains an active area of research.

This chapter introduces the physical principles of the generation of PA signals, provides an overview of current ultrasound detectors used in PA scanners, discusses the main PA imaging modalities and image reconstruction algorithms, and gives examples of in vivo molecular and functional imaging applications. Finally, a brief introduction to qPAT is given.

13.2 PA Signal Generation

The generation of PA waves in tissue relies on the absorption of intensity-modulated light, i.e. short pulses, frequency chirps or single frequency illumination. The photons of the excitation light penetrate the tissue where they are scattered and eventually absorbed by the chromophores that are present in the illuminated volume. The absorbing chromophores are promoted to an excited state from which they relax to the more stable ground state. Assuming predominately vibrational relaxation pathways, the optical energy of the photons is converted to heat. Provided the heat deposition occurs sufficiently quickly to ensure thermal and stress confinement,[1] the heat deposition is accompanied by an increase in pressure. This initial pressure is proportional to the absorbed energy density, i.e. the pressure increase is greatest in regions of high optical absorption. The spatial distribution of the tissue chromophores is therefore encoded onto the pressure field, which relaxes by emitting a broadband acoustic wave in the ultrasound frequency spectrum. The acoustic field is then detected outside the organism, for example on the skin, using ultrasound transducers. By recording acoustic transients at multiple locations, the differences in the time-of-arrival of the PA waves are exploited to reconstruct the location of the source, i.e. the initial pressure distribution, using image reconstruction algorithms. The initial pressure distribution, $p_0(r)$, is defined as

$$p_0(r) = \Gamma \mu_a(r) \Phi(r)$$ (13.1)

where r is the spatial coordinate, $\Gamma = \beta c^2 / C_p$ is the dimensionless Grüneisen parameter, which represents the conversion efficiency of heat energy to pressure (β—the volume thermal expansivity, c—speed of sound, C_p—specific heat capacity), μ_a is the absorption coefficient, and $\Phi(r)$ is the fluence [3]. The local absorption coefficient and its wavelength dependence is typically expressed as

$$\mu_a(r,\lambda) = \sum_i c_i(r) \alpha_i(\lambda)$$ (13.2)

where λ is the wavelength, c_i is the local chromophore concentration and α_i is the wavelength-dependent specific absorption coefficient. While PA imaging can be

---

[1] Thermal confinement requires the heating pulse to be much shorter than thermal relaxation of the source. Since thermal diffusion in tissue is orders of magnitude slower than typical excitation pulse durations, thermal confinement is not a strongly limiting factor in PA imaging. Stress confinement requires the heating pulse duration to be shorter than the time it takes the photoacoustic wave to propagate across the heated source region. Let us assume that the excitation laser provides optical pulses of $t_p = 10$ ns duration and that photoacoustic waves are excited in a water-based medium, i.e. the speed of sound is $c_s = 1500$ ms^{-1} = 1.5 μm ns^{-1}. Within the duration of the excitation pulse, an acoustic wave will therefore travel a distance of $s = c_s t_p = 1.5$ μm ns^{-1} × 10 ns = 15 μm. In terms of photoacoustic imaging, this figure also approximates the maximum spatial resolution that can be achieved with this pulse duration.

described as providing absorption-based contrast, the initial pressure in turbid media is not a linear function of μ_a, and therefore chromophore concentration, because optical scattering by cells or cell components affects the fluence distribution. The $\Phi(r; \mu_a, \mu_s, g)$ in scattering tissue is a non-linear function of μ_a, the scattering coefficient, μ_s, and the scattering anisotropy, g. A simplified model of stochastic wave scattering in non-absorbing media is discussed in Chap. 2. The combined effect of absorption and scattering is the gradual decrease in fluence with depth, which can be described using the effective attenuation coefficient $\mu_{\text{eff}} = (3\ \mu_a\ (\mu_a + \mu_s'))^{-1/2}$ derived from diffusion theory where $\mu_s' = \mu_s(1 - g)$ is the reduced scattering coefficient. Its reciprocal value is the optical penetration depth, $l_{\text{eff}} = 1/\mu_{\text{eff}}$, which represents the depth at which the fluence has been reduced to $1/e$ of that incident on the tissue surface. An example for the wavelength dependence of the optical penetration depth is shown in Fig. 13.2 (black solid line) for typical chromophore concentrations and scattering coefficients of human tissue. The largest optical penetration depths are observed between 650 and 850 nm. This is explained be the absorption spectra of the endogenous chromophores and the wavelength dependence of the scattering coefficient. The absorption spectra of oxyhaemoglobin and deoxyhaemoglobin, lipid, water and melanin are shown in Fig. 13.2. Oxyhaemoglobin and deoxyhaemoglobin, for example, exhibit strong absorption at wavelengths shorter than 600 nm while the absorption and lipid becomes more dominant in the near-infrared wavelength region ($\lambda > 950$ nm).

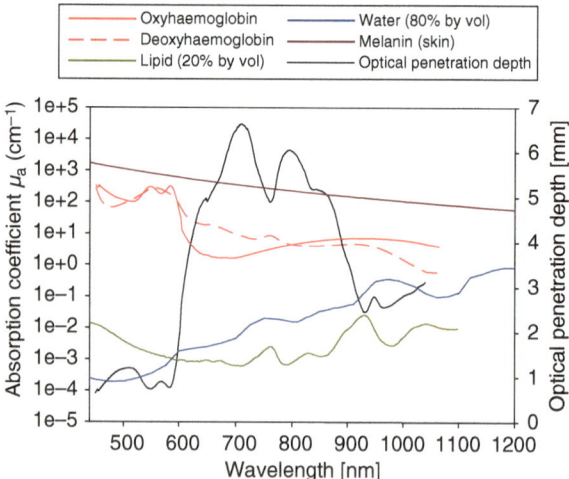

Fig. 13.2 Spectra of the absorption of the main endogenous tissue chromophores and the optical penetration depth in tissue. The absorption coefficient spectrum of oxyhaemoglobin and deoxyhaemoglobin was calculated assuming a concentration of 2.3 mM. For lipid and water, tissue volume fractions of 20% and 80%, respectively, were assumed. The absorption spectrum of melanin corresponds to that found in skin (http://omlc.org/spectra/melanin/mua.html). The optical penetration depth was calculated by assuming a total blood volume of 2%, a blood oxygenation of 70%, and a reduced scattering coefficient of 10 cm^{-1}

To excite PA waves in tissue, low energy optical pulses below the maximum permissible exposure are typically used and result in temperature increases that are typically below 100 mK, which results in comparatively low initial pressures in the tens to hundreds of kPa [1]. These waves, which are broadband in terms of frequency content, then propagate to the surface where they are detected by ultrasound sensors. The propagation of the waves through the tissue results in a reduction in amplitude due to geometric spreading and frequency dependent acoustic attenuation. The combination of the optical and acoustic attenuation can be estimated to reduce the signal amplitude by an order of magnitude per cm, which makes it challenging to obtain good signal-to-noise ratios. This can be mitigated through careful choice of excitation wavelengths and detectors but the strong attenuation both optically and acoustically place significant demands on the capabilities of ultrasound detectors, in particular in terms of acoustic sensitivity and frequency response.

13.3 Ultrasound Detection Technologies

The ultrasonic pressure field generated via the PA effect is broadband in frequency (kHz to hundreds of MHz) and low in amplitude. By mapping the PA field on the tissue surface using a detector array, data sets are obtained from which 3D images can be reconstructed. To obtain images with high spatial resolution, an ideal acoustic detector should possess the following attributes: (1) small active element size, (2) high acoustic sensitivity, (3) high acoustic bandwidth and (4) flat frequency response. In addition, optical transparency is advantageous as it allows backward mode imaging, i.e. PA signal generation and detection on the same side of the target.

Piezoelectric transducers, such as those used in conventional ultrasound scanners, are the by far most popular type of detector for PA imaging. However, while they are convenient due to their wide availability and flexible design, they have a number of disadvantages when applied to high resolution PA tomography. First, the acoustic sensitivity scales with active element size, i.e. small active element sizes result in low sensitivity. Second, the frequency response of piezoelectric detectors tends to be resonant, which results in high pass filtering effects and hence a loss of information. Third, piezoelectric detectors are typically opaque—a disadvantage for backward mode PA imaging.

Optical methods for ultrasound detection have been shown to be attractive alternatives that meet many of the above criteria. Methods such as Schlieren [4] and phase contrast imaging [5] rely on the detection of acoustically induced modulations of the refractive index in a transparent coupling medium. While these methods have been shown to provide high spatial resolution of tens of microns and potentially short image acquisition times (from a few seconds to real-time), their disadvantage lies in relatively low acoustic sensitivity (several kPa mm) due to the typically small elasto-optic coefficients of the coupling medium. While integrating line detectors provide higher acoustic sensitivity (0.5 kPa mm) [6, 7], image acquisition is slow due to the need for mechanical scanning of the sample. Microring resonators (MRR) are another type of optical ultrasound sensor that has been applied to PA imaging. They have been shown to provide large acoustic detection bandwidths (hundreds of MHz), high acoustic sensitivity (0.1 kPa for a detection bandwidth up to 350 MHz), and high axial (or vertical) resolution of a few micrometres [8, 9]. However, the lateral resolution is likely to be limited by the minimum diameter of the microring required to ensure total internal reflection (typically >60 μm). Also, while the fabrication of MRR arrays for parallelised detection is feasible, PA imaging has to date only been achieved using single, mechanically scanned MRRs [10–13], resulting in slow image acquisition. Non-contact surface displacement measurements have also been reported [14, 15] with a sensitivity of hundreds of Pa (20 MHz bandwidth). Again, the acquisition of images is slow due to the need for mechanical scanning.

Over the last decade, Fabry–Pérot interferometer (FPI) based optical ultrasound sensors have arguably set a standard for high resolution 3D imaging to cm depths [16]. Figure 13.2a shows the structure of the sensor. The FPI is formed by two dielectric mirrors separated by a transparent polymer spacer, and is supported by a polymer substrate. It is illuminated by a wavelength tuneable cw interrogation laser. The reflections from the mirrors, which have a phase difference determined by the optical thickness of the FPI, interfere and are directed to a photodetector. The interferometer transfer function, shown in Fig. 13.2c, describes the reflectivity of the FPI as a function of optical phase, ϕ. By tuning the interrogation wavelength to a point where the slope of the transfer function, i.e. the phase sensitivity, is at a maximum, small acoustically induced modulations in the phase difference result in a large change the reflected optical power. The FPI sensor design has a number of distinct advantages over piezoelectric detectors: (1) small active element sizes (tens of microns) for high lateral resolution and near omnidirectional response, (2) high acoustic sensitivity (0.2 kPa, 20 MHz bandwidth), (3) broad detection bandwidth (DC to potentially 100 MHz) with a flat frequency response and (4) transparency (600–1100 nm) (Fig. 13.2b) for backward mode operation, multiwavelength excitation, and compatibility with other optical imaging technologies, such as optical coherence tomography [17].

Capacitive micromachined ultrasound transducers (CMUT) have also been explored as an alternative to piezoelectric and optical ultrasound detectors [18–20] as they provide detector arrays with small active elements that can be made using established semiconductor fabrication technologies. These detectors have been shown to provide acoustic sensitivity of ~100 mPa $Hz^{-1/2}$ but tend to exhibit resonant frequency responses. In addition, their opacity requires more complex optical arrangements to couple the excitation pulses into the target.

13.4   PA Imaging Modalities

13.4.1 PA Tomography

PA tomography relies on the detection of PA waves using detectors at multiple locations around the target. Three detection geometries are typically used: spherical, cylindrical, and planar. A spherical arrangement of ultrasound detectors allows, at least in theory, a perfect reconstruction of the initial pressure since the propagating PA wave is recorded in all spatial directions. However, its practical implementation is less straightforward and half-spherical or circular line detector arrangements are often used instead [21]. The cylindrical detector geometry has been adopted in a large number of PA scanners. Some of the first PA imaging systems relied on circular scanning of focussed piezoelectric ultrasound detectors [22] and allowed the acquisition of 2D images. By translating the imaging plane, 3D image stacks can be obtained. This work led to the development of focussed, arc-shaped detector arrays [23, 24], which are now used routinely in commercial PA small animal imaging systems. A cylindrical detection geometry is also suitable for integrating line detectors, in particular those based on optical interferometers. This approach provides high acoustic sensitivity but also requires the rotation of the object or the detectors to enable 3D tomographic imaging [6, 7]. By contrast, planar detection geometries are the most versatile and practical. A drawback of the planar geometries for tomographic imaging is the generation of image reconstruction artefacts that are caused by the limited detection aperture (Fig. 13.3).

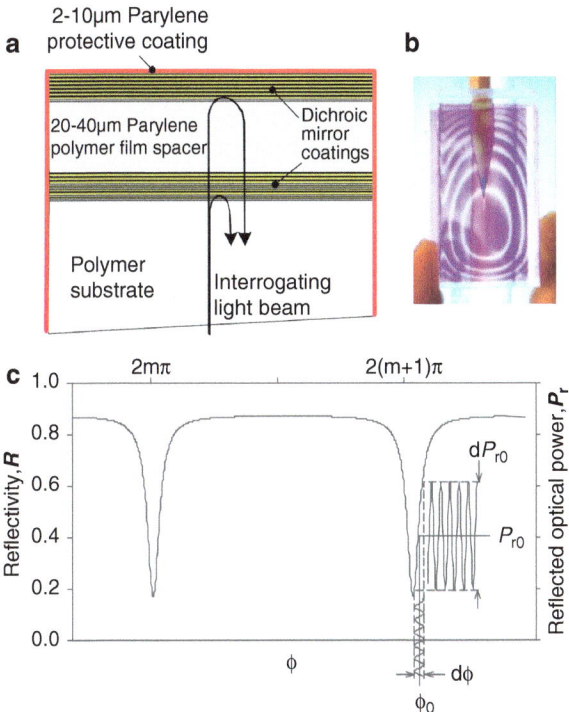

Fig. 13.3 (**a**) Structure of the FPI (Fabry–Pérot interferometer) ultrasound sensor. (**b**) Photograph of the sensor showing its transparency in the visible wavelength region. (**c**) FPI transfer function

13.4.2 Optical-Resolution and Acoustic-Resolution PA Microscopy

Over the last decade, two major microscopic PA modalities have emerged: (1) optical resolution and (2) acoustic resolution PA microscopy [25], typical experimental configurations of which are illustrated in Fig. 13.4. Optical resolution PA microscopy (OR-PAM), where the excitation light is focussed below the tissue surface to diffraction limited spot sizes (Fig. 13.4a), has been shown to provide the highest spatial resolution.

For linear PA excitation, the lateral resolution is determined by the beam waist at the optical focus and is typically of the order of a few microns [26] while submicron resolution up to 0.22 μm [27] has been achieved using a water immersion objective. Moreover, by exploiting non-linear effects observed during dual pulse excitation [28] and photobleaching [29], resolutions of 0.41 μm and 0.12 μm, respectively, have been achieved. Due to optical scattering, the maximum imaging depth is limited to approximately 1 mm and therefore comparable to that of optical microscopy. The vertical resolution is dependent upon the frequency response of the ultrasound transducer, the duration of the excitation pulse, and the frequency-dependent acoustic attenuation along the source–detector path. For example, using 5 ns pulses and a 125 MHz transducer, an axial resolution of 7.5 μm has been reported [30]. Nonlinear effects, such as the temperature dependence of the Grüneisen coefficient, have been used to achieve vertical resolutions of up to 2.3 μm [28].

In acoustic resolution PA microscopy (AR-PAM, Fig. 13.4b), by contrast, the excitation beam is weakly focussed into the tissue to generate PA waves, and a focussed ultrasound detector, typically made from piezoelectric materials, is used to acquire the PA signals. The lateral resolution is limited by acoustic diffraction at the

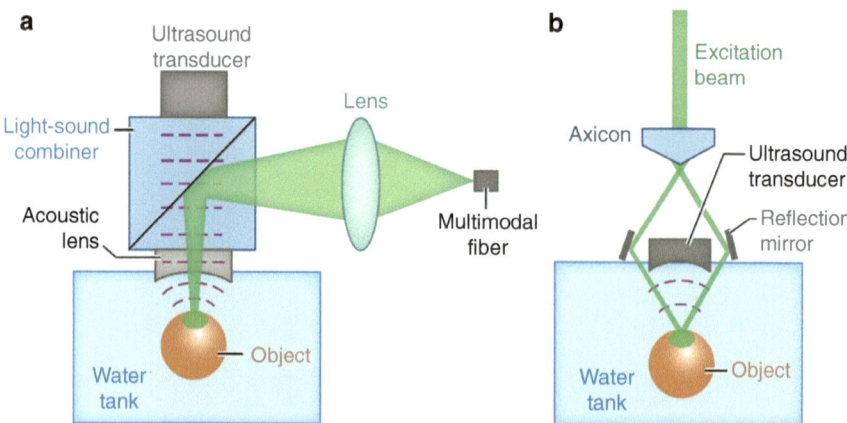

Fig. 13.4 Typical detection geometries for PA microscopy. (**a**) optical-resolution PA microscopy, where the excitation beam is focussed into the tissue, and (**b**) acoustic-resolution PA microscopy using a focussed ultrasound transducer. (reproduced with permission from Wang and Gao [25])

transducer focus. Away from the focal region, the lateral resolution degrades rapidly. As in OR-PAM, vertical resolution is determined by the detector frequency response and the excitation pulse duration. To acquire acoustic diffraction limited 3D images requires mechanical scanning of the co-aligned transducer and excitation beam. AR-PAM has been used to image, for example, the vasculature in the mouse brain through the intact skin and skull to yield lateral and axial resolutions of 70 μm and 27 μm, respectively, with an imaging depth of 3.6 mm [31]. In soft tissue, lateral and axial resolutions of 44 μm and 15 μm, respectively, with an imaging depth of 4.8 mm have also been achieved [32].

Typical OR-PAM and AR-PAM systems rely on large area piezoelectric transducers to provide sufficient acoustic sensitivity. The disadvantages of these transducers are (a) opacity, which requires some form of acoustic or optical beam splitting to couple the excitation light into the tissue, (b) a resonant frequency response, which reduces the information content of the signals and (c) acoustic sensitivity scales with active area, i.e. low sensitivity for small active element sizes. Other novel ultrasound detection approaches include optical methods, such as plasmonic detection [33] and the use of microring resonators [9, 34, 35].

13.5   Spatial Resolution

Figure 13.5 provides an overview of the spatial resolution and penetration depth provided by some of the main PA imaging modalities. The greatest penetration depth in tissue is achieved using tomographic scanners, which offer imaging depths on the order of cm. Their spatial resolution ranges from tens of microns for the Fabry–Pérot-based PA scanner (OD-PACT in Fig. 13.5, i.e. optical detection PA computed tomography) to hundreds of microns for acoustic-resolution PA macroscopy (AR-PAMac) and C-PACT. By contrast, PA microscopy modalities achieve sub-micron spatial resolutions but the trade-off is imaging depth, which ranges from

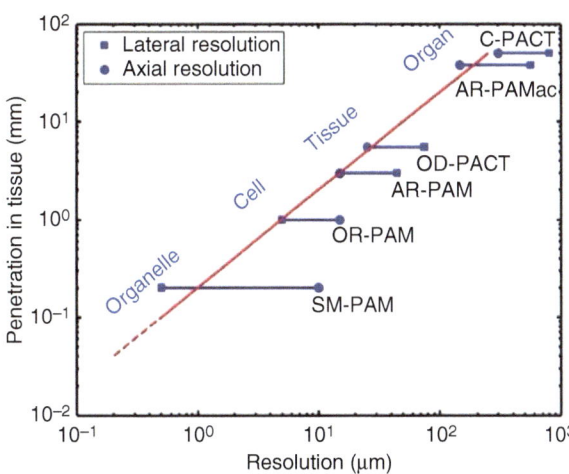

Fig. 13.5 Imaging depth vs spatial resolution provided by microscopic and tomographic PA imaging modalities (reproduced with permission from [36])

a few hundred microns for sub-micron PA microscopy (SM-PAM) to <3 mm for AR-PAM.

13.6   Image Reconstruction

The PA signals acquired using OR-PAM and AR-PAM, where the excitation beam and the acoustic detector are raster scanned similar to B- and C-scans in conventional ultrasound imaging (see Chap. 12), a 3D image is from the individual waveforms following signal processing, such as frequency filtering and Hilbert transformation [37].

The reconstruction of tomographic images is based on the backprojection of the recorded PA signals. This is illustrated in Fig. 13.6, which shows a turbid target with a single optically absorbing inclusion that is illuminated with excitation light from the top. The resulting PA pressure field is mapped as a function of time and location on the surface of the target, r, using an ultrasound detector array. The time-of-arrival of the PA waves, t, detected by each array element varies according to its distance from the source, R. Assuming that the array elements are point-like acoustic detectors, which are characterised by an omnidirectional response, the PA wave must have originated from a position that is located somewhere on a spherical surface with the detector at its centre and a radius R that is related to the time-of-arrival via the speed of sound, i.e. $R = ct$. By backprojecting the PA signals detected at each array element into a spatial domain, the "projection arcs" combine to reveal the

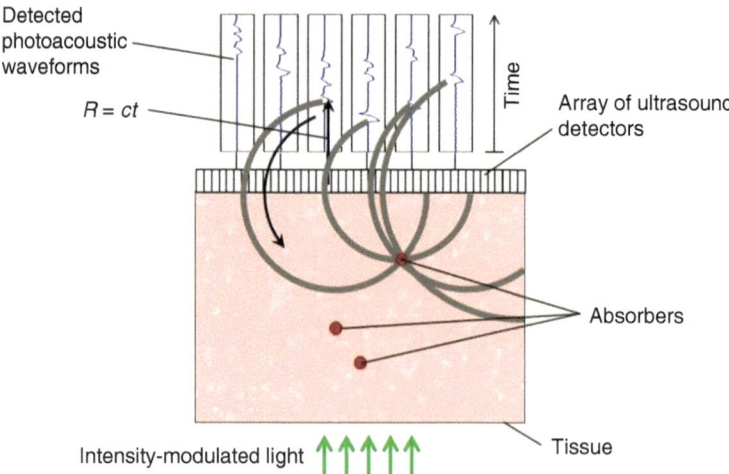

Fig. 13.6 Principle of PA image reconstruction using backprojection for a planar detection geometry. PA signals are recorded by each detector array element at position r and backprojected over spherical surfaces of radius $R = ct$ where c is the speed of sound into the image volume. (In this illustration, the output of each detector is depicted as a time-integrated pressure waveform for illustrative purposes as the backprojected quantity is the velocity potential. In practice, the detectors record a pressure waveform and the time integration is performed computationally)

location of the PA source. The principle of backprojection reconstruction can be implemented in different ways, such as delay-and-sum [38], Fourier-transform-based algorithms [39], and time-reversal pseudo-spectral methods [40, 41].

13.7 In Vivo Applications: Morphological, Functional and Molecular PA Imaging

Due to the strong absorption of excitation pulses in the visible to near-infrared wavelength region by haemoglobin, PA imaging is highly suited to the study of the changes in the vascular morphology during disease and therapy. Early studies have shown that PA tomography provides 3D images of the vasculature in superficial and deep tissue in preclinical applications, such as PA imaging of the skin [42], the brain [22, 43, 44] and subcutaneous tumours [2] (Fig. 13.7). The ability to monitor changes in the vasculature in the same organism repeatedly over time has been demonstrated in a longitudinal study in which the effects of a vascular disrupting agent were investigated [2]. The agent was administered systemically and resulted in the enlargement of the epithelial cells lining the blood vessels of the tumour. This effectively blocks the supply of nutrients and oxygen to the tumour core, where the cells undergo rapid apoptosis. The images obtained in this study are shown in Fig. 13.7 and illustrate that the PA contrast detected in the intact tumour vasculature before the treatment (Fig. 13.7b) is largely completely removed 24 and 48 h after the injection of the vascular disrupting agent (Fig. 13.7c, d). It is noticeable, however, that the PA contrast is strong in the image region that corresponds to the rim of the tumour. This is because this region of the tumour, including the blood vessels within it, received sufficient oxygen and nutrients from the surrounding tissue to survive.

A number of groups have pursued the clinical application of PA imaging of the vasculature in the area of mammography [21, 45–47]. The PA scanners typically rely on large area piezoelectric ultrasound detectors to achieve an optimal combination of acoustic sensitivity, and hence imaging depth, and spatial resolution. Figure v

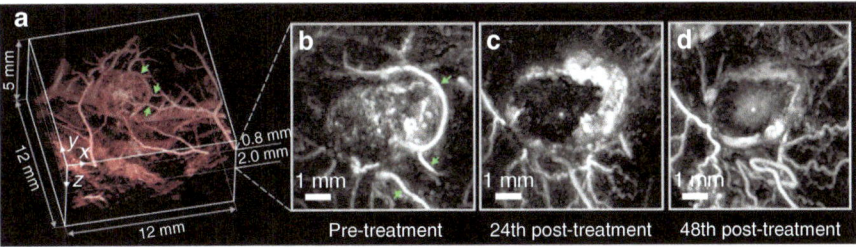

Fig. 13.7 Longitudinal photoacoustic imaging of the effect of a vascular disrupting agent (combratestatin OXi4503) on the blood vessel network of a tumour. (**a**) 3D volume rendered image of the tumour and the surrounding region prior to the administration of the agent. x-y MIPs through the centre of the tumour (**b**) before, (**c**) 24 h and (**d**) 48 h after treatment for $z = 0.8$–2.0 mm. The green arrows in (**a**) and (**b**) indicate common vascular features in the skin. All images were acquired at an excitation wavelength of 640 nm. (Reproduced with permission from [2])

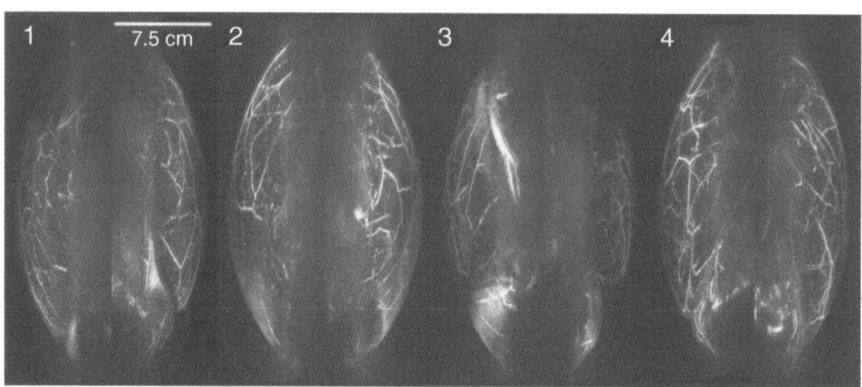

Fig. 13.8 Maximum intensity projections in the medial-lateral (ML) projection of bilateral PAM exams of four healthy volunteers with known mammographic breast density and brassiere cup size (back-to-back images: left breast on right, right breast on left as is normally presented clinically for X-ray mammograms): (1) heterogeneously dense, D cup; (2) scattered fibroglandular densities, DD cup; (3) scattered fibroglandular densities, C cup; and (4) scattered fibroglandular densities, DD cup. (Reproduced with permission from Kruger et al. [48])

shows PA images obtained with a dedicated PA mammography system developed by Optosonics Inc [48], which illustrate that detailed images of the vasculature in the breast are obtained with imaging depths of several cm.

By acquiring PA image data sets at difference excitation wavelengths and by using some form of spectral unmixing that exploits the wavelength dependence of oxyhaemoglobin and deoxyhaemoglobin, PA imaging has been shown to enable functional imaging. This approach relies on determining the relative concentrations of oxyhaemoglobin and deoxyhaemoglobin from the wavelength dependence of the PA image intensity from which the derived parameter of blood oxygen saturation, i.e. the ratio of the concentration of oxyhaemoglobin and the total haemoglobin concentration, can be calculated. This approach has been shown to provide reliable quantitative results when modalities based on optical resolution PA microscopy were used since the adverse effects of optical scattering on the accuracy of the spectral inversion are minimal. This is shown in Fig. 13.9, which depicts OR-PAM images of blood oxygenation acquired in the mouse brain using two excitation wavelengths. The colour-codes blood oxygenation image shown in Fig. 13.9a clearly visualises the location of arteries and veins while Fig. 13.9b shows the absolute blood oxygenation values along the vessel trees. While acoustic resolution PA microscopy and PA tomography have also been shown capable of acquiring high resolution images of the mouse brain (Fig. 13.9c, e), these modalities probe deeper tissue regions than OR-PAM for which experimentally validated and generally applicable methods for deep-tissue quantitative PA imaging do not yet exist. This often restricts functional PA imaging to qualitative measurements of changes in blood oxygenation as shown in Fig. 13.9d, f.

However, OR-PAM also offers alternative methods for measuring blood oxygenation by exploiting non-linear phenomena that arise from the high photon densities

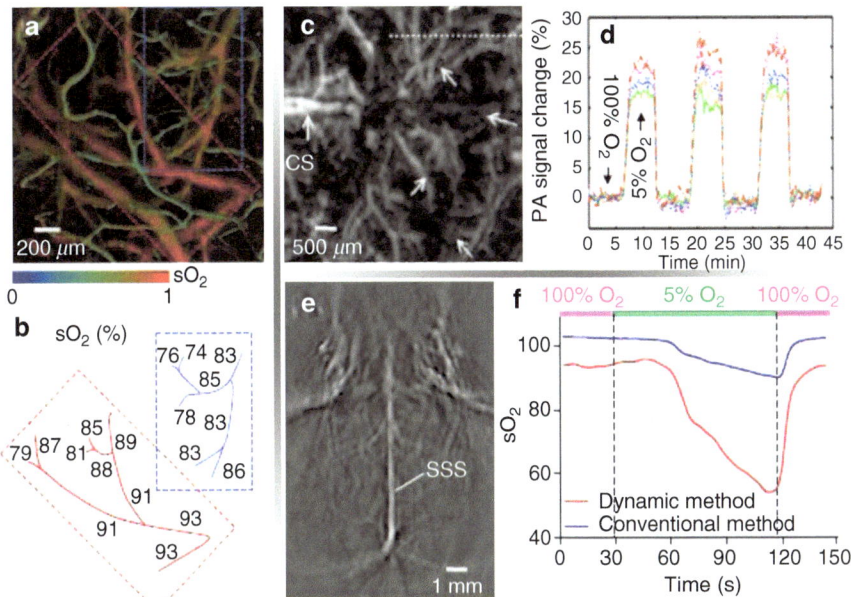

Fig. 13.9 Photoacoustic tomography (PAT) of mouse brain oxygenation. (**a**) OR-PAM of oxygen saturation (sO$_2$) in a mouse brain based on two-wavelength measurements at 570 and 578 nm [49]. (**b**) sO$_2$ values in percentage along an arterial tree and a venous tree marked by the dashed boxes in (**a**), showing decreased sO$_2$ with vessel branch orders. (**c**) AR-PAM of cortical vasculature in a living mouse [50]. The dotted white line indicates the line scanning range for oxygenation measurement. *CS* coronal suture. (**d**) Dynamic vessel responses acquired through a hypoxic challenge, shown in percent change of ratiometric PA signals at 561 and 570 nm. Each coloured trace corresponds to the respective cortical vessel crossed by the dotted line in (**c**). (**e**) PACT of cortical vasculature in a living mouse [51]. *SSS* superior sagittal sinus. (**f**) Dynamics of absolute sO$_2$ measured by PACT on the SSS in response to a hypoxic challenge. The measured sO$_2$ values based on the new oxygenation-state method are compared with the conventional two-wavelength method. (Reproduced with permission from Refs. [31, 49–51])

at the focus of the excitation beam. An example is a recent study by Yao et al. [52], which demonstrated quantitative blood oxygenation imaging using single-wavelength dual pulse excitation. By generating PA waves using subsequent excitation pulses of picosecond and nanosecond duration, differences in the excited state lifetime of oxyhaemoglobin and deoxyhaemoglobin lead to varying PA signal amplitudes as illustrated in Fig. 13.10c. This was then used to obtain maps of blood oxygenation in the mouse brain to depths of 0.8 mm (Fig. 13.10e–f). The advantage of this method is that it eliminates the effects of the wavelength-dependent optical attenuation, which has to be accounted for in conventional multiwavelength acquisition and spectral unmixing approaches to functional PA imaging.

The absorption-based contrast provided by PA imaging modalities has also been applied to molecular imaging, where exogenous or genetically expressed absorbing compounds are used to label specific tissue types or cells. A wide range of exogenous optical contrast agents has been reported in the literature [53]. One of the

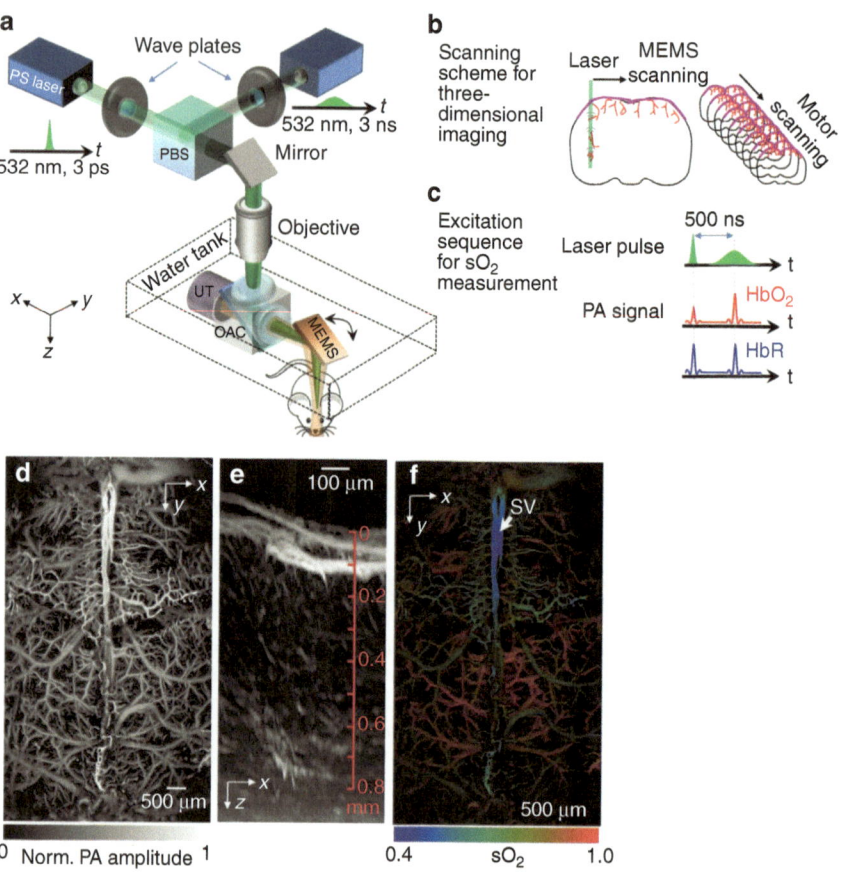

Fig. 13.10 Experimental set-up and principle of fast functional photoacoustic microscopy (PAM) of the mouse brain acquired using the single-wavelength pulse-width-based method. (**a**) Schematic of the PAM system. *OAC* optical-acoustic combiner, *PBS* polarising beam splitter, *UT* ultrasonic transducer. (**b**) Illustration of the experimental protocol for 3D OR-PAM scanning. (**c**) Sequence of PAM excitation and detection. The picosecond pulse incident on oxyhaemoglobin (HbO$_2$) results in a smaller PA signal amplitude than that produced by the following nanosecond pulse. The difference in signal amplitude for deoxyhaemoglobin (HHb) is negligible. (**d**) *x-y* maximum intensity projection image of the brain vasculature imaged through an intact skull. (**e**) *x-z* maximum intensity projection of the brain vasculature acquired over a 0.6 × 0.6 mm^2 region with depth scanning. (**f**) PAM image of blood oxygen saturation in the same mouse brain as shown in (**d**). *SV* skull vessel. (Reproduced with permission from Yao et al. [52])

earliest reports used systemically injected single-walled carbon nanotubes into tumour-bearing mice to demonstrate targeted molecular PA imaging [54]. A subsequent study by the same group addressed the limited molecular absorption of carbon nanotubes by chemically attaching dyes, such as indocyanine green (ICG) or fluorophores [55] (Fig. 13.11). This achieved an increase in optical absorption by two orders of magnitude since fluorophores typically exhibit high molar extinction.

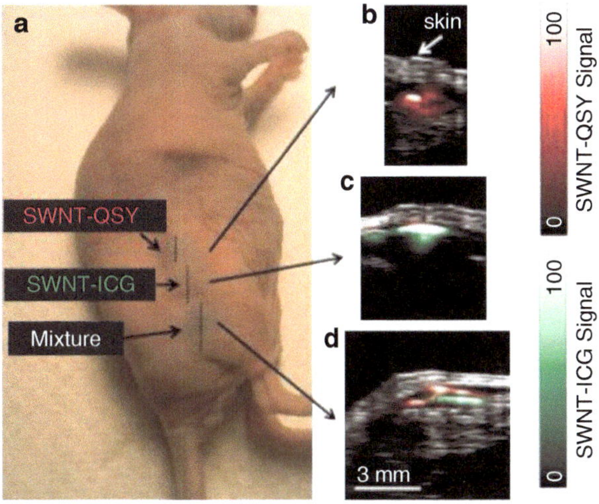

Fig. 13.11 In vivo PA imaging of dye-loaded single-walled carbon nanotubes. (**a**) A mouse was injected with 30 µL of 50 nM single-walled carbon nanotubes (SWNT) and dye QSY (upper inclusion), 30 µL of 50 nM SWNT and ICG (middle inclusion), and 30 µL of an equal mixture of 50 nM of SWNT-QSY and SWNT-ICG (lower inclusion). (**b**) Spectrally unmixed PA vertical slice through the upper inclusion, showing only SWNT-QSY signal (red). (**c**) Unmixed photoacoustic slice through the middle inclusion, showing mostly SWNT-ICG signal (green). (**d**) Unmixed PA slice through the lower inclusion, showing both SWNT-QSY and SWNT-ICG signals spread throughout the inclusion area. (Reproduced with permission from de la Zerda et al. [55]. Copyright 2012 American Chemical Society)

In addition, the close proximity of the fluorescent dye molecules to the carbon nanotube resulted in efficient quenching of the fluorescence, which results in predominately vibrational relaxation pathways to generate heat, and hence PA pressure. By contrast, the radiative relaxation observed in fluorophores results in a reduction in PA pressure compared to a non-fluorescent absorber of equal molar absorption.

Another example of particle-based molecular PA imaging is the use of gold nanocages to visualise extravasation in tumours, i.e. the passage of the compounds through the leaky tumour vasculature [56]. Systemically administered exogenous molecular absorbers, such as ICG, were used to, for example, to show the ability of 2D PA imaging to visualise the time course of their biodistribution in mice using multiwavelength image acquisition and spectral unmixing based on a fluence-corrected linear inversion [57]. An alternative, experimental methodology for the detection of fluorescent contrast agents was recently developed by using pump-probe excitation to acquire tomographic PA difference images [58–60]. The attraction of this approach lies in the generation of fluorophore-specific PA contrast while the endogenous contrast is eliminated (Fig. 13.12).

While the development of methods for the detection of contrast agents using PA imaging is important for its long term translation to clinical applications, the use of reporter genes that encode absorbing proteins and pigments are a powerful tool for basic research in the life sciences and preclinical applications. The reporter gene in

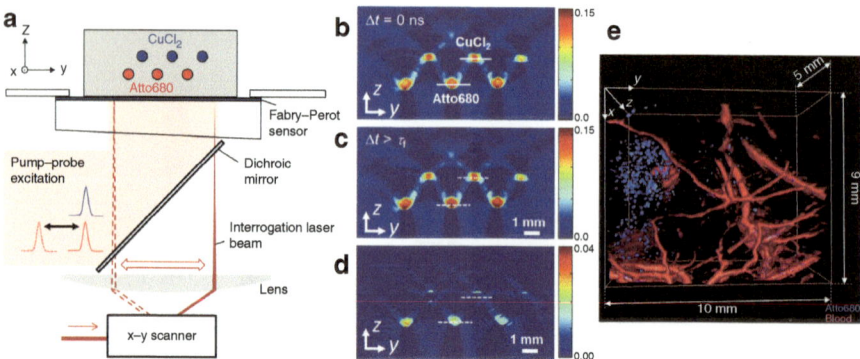

Fig. 13.12 PA difference imaging of fluorophores using pump-probe excitation. (**a**) Schematic of the experimental set-up, in which an all-optical Fabry–Perot PA scanner was used to image a tissue phantom consisting of tubes filled with solutions of a fluorophore (Atto680) and a non-fluorescent absorber (CuCl₂) to mimic blood vessels. (**b**, **c**) Cross-sectional PA image of the phantom for simultaneous and time-delayed pump and probe pulses, respectively. (**d**) Difference image of (**b**) and (**c**), which shows the location of the fluorophore-filled tubes while the background is eliminated. (**e**) In vivo image of a subcutaneously injected fluorophore solution in a mouse. (Taken with permission from Märk et al. [60])

conjunction with a suitable promoter is incorporated into the cells either transiently or permanently, for example via viral transduction. Once the gene is inside the cell, it is transcribed to result in the expression of either proteins that absorb light directly, and therefore provide the absorption-based PA contrast, or enzymes which convert endogenous cellular components into light-absorbing pigments. This method was first demonstrated by Razansky et al. who visualised fluorescent proteins in comparatively small and translucent organisms, such as fruit fly pupae and zebrafish [61], using multiwavelength imaging and linear spectral unmixing. PA reporter gene imaging of mammalian cells was later demonstrated using near-infrared fluorescent proteins [62], while deep tissue PA imaging was shown using the genetic expression of the enzyme tyrosinase, which results in the synthesis of light-absorbing eumelanin [63] (Fig. 13.13).

13.8 Quantitative PA Tomography

In the previous sections, the potential of PA imaging to enable functional measurements and to detect exogenous contrast agents or genetically expressed labels has been demonstrated. However, to recover chromophore concentrations or concentration ratios, such as blood oxygen saturation, from tomographic images acquired in deep tissue, methods are required that correctly account for the physical processes involved in PA excitation and image acquisition. This area of research is referred to as quantitative PA tomography (qPAT) and the following section will give a brief overview of its principles.

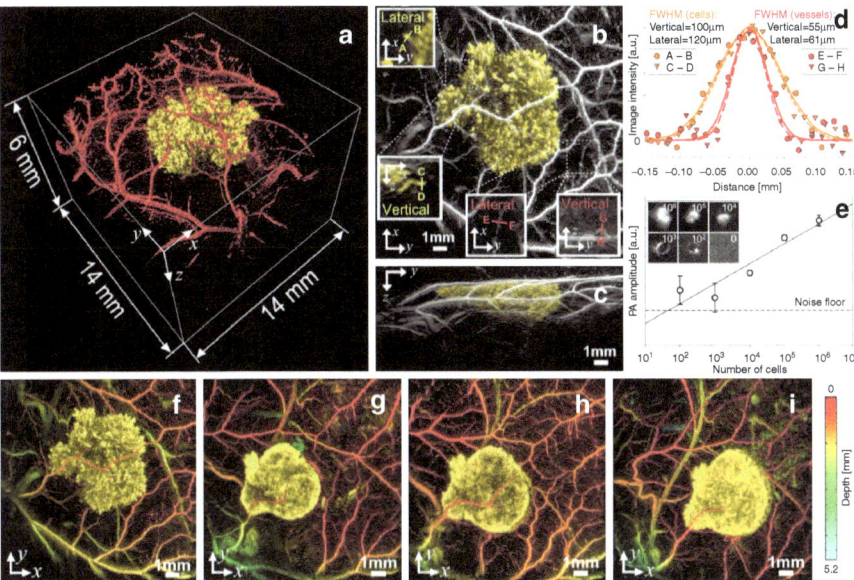

Fig. 13.13   Deep-tissue PA reporter gene imaging using genetically expressed tyrosinase, which results in the synthesis of eumelanin. (**a**) 3D fused-colour volume rendered PA image of tyrosinase expressing subcutaneous tumour cells acquired in a mouse. The tumour cells are shown in yellow and the vasculature of the surrounding tissue regions is shown in red. (**b**, **c**) x-y and y-z maximum intensity projections (MIP) of the data set shown in (**a**). Image intensity profiles of a cell outcrop and a small blood vessel are plotted in (**d**) to illustrate the spatial resolution that was obtained. (**e**) PA amplitude as a function of the number of cells. The detection limit was around 100 cells. (**f**–**i**) Longitudinal imaging: x-y MIPs of a tumour acquired on different days post-injection up to day 52. (Reproduced with permission from [63])

The forward problem of qPAT [3, 64] can be split into the optical and the acoustic forward problem (Fig. 13.14). The distribution of the tissue chromophores and scatterers determines the optical properties, i.e. the absorption coefficient, μ_a, and scattering coefficient, μ_s, and therefore the light transport in the tissue. The light transport can be described by the fluence distribution, Φ, from which the absorbed energy, $H = \mu_a\,\Phi$, is obtained. The absorbed energy is converted into an initial pressure distribution, $p_0 = \Gamma H$, where Γ is the Grüneisen parameter. The propagation of the PA waves to the surface is a function of the acoustic properties. PA signals are detected at multiple points on the surface to yield a time series, $p(t)$, and are affected by the detector response. Finally, images are obtained using reconstruction algorithms. The inverse problem in qPAT lies in recovering the spatial distribution of the chromophore concentrations from the measured PA time series or images, i.e. going up the flowchart in Fig. 13.14. This means that the inversion scheme has to account for the generation, the propagation, and the detection of PA waves. qPAT typically requires the acquisition of multiwavelength images to (a) avoid the non-uniqueness problem in determining μ_a and μ_s at a single excitation wavelength [65], and to (b) exploit the differences in the absorption spectra of the chromophores for

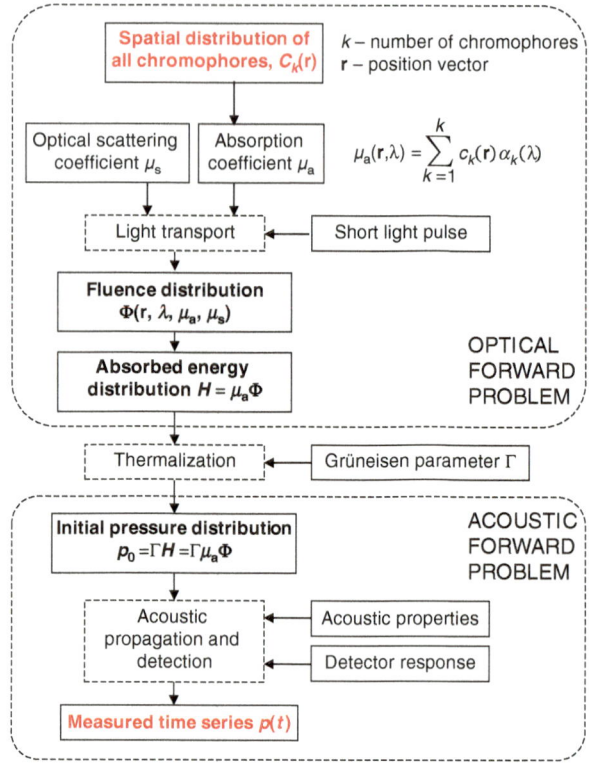

Fig. 13.14 The forward problem in quantitative PA imaging

quantitative imaging. Assuming PA images represent p_0, the inversion can be reduced to solving the optical inverse problem using one of two general approaches [3]. The first is a two-stage inversion, in which the spatial distribution of the absorption coefficient, $\mu_a(r, \lambda)$ is obtained from images of p_0 using an approximate fluence correction. Using the known molar absorption coefficient spectra, $\alpha_k(\lambda)$, the chromophore abundance can then be obtained via a linear matrix inversion. The second approach, termed model-based inversion, involves a forward model to predict PA signals or images directly as a function of chromophore concentrations. In a single-stage inversion, the model output is fitted to measured data using minimisation techniques to recover the spatial distribution of the chromophore concentrations, c_k.

Accounting for the fluence is at the core of the inverse problem in qPAT as identified in two recent reviews [3, 66]. If a linear relationship between the absorbed energy and the concentrations could be assumed, c_k could easily be recovered from measured absorbed energy, $H(\lambda)$, using the known molar absorption spectra, $\alpha_k(\lambda)$, and a linear matrix inversion. Unfortunately, H is also dependent upon the fluence ($H = \mu_a \Phi$). The fluence, Φ, is typically unknown and a nonlinear function of μ_a and μ_s. Its effects on PA images have been described as *spectral colouring* and *structural distortion*. *Spectral colouring* arises because photons may travel long and convoluted paths in scattering tissue before they are absorbed. The fluence at one point will therefore be affected by the optical properties of the surrounding region. *Structural distortion* refers to the corruption of the image of $\mu_a(\lambda)$ by the heterogeneous fluence [3]. For example, a superficial blood vessel may cast a "shadow" onto a deeper vessel. Given these effects, it is not reasonable to assume proportionality between the absorbed energy, H, and the absorption coefficient, μ_a, unless the effects of fluence are accounted for. This, however, is a distinctly non-trivial problem and one of the main reasons why qPAT has not yet been demonstrated in vivo.

Three-dimensional in vivo imaging of absolute chromophore concentrations represents a formidable challenge likely to require a long-term research effort. However, there is sufficient evidence to suggest that PAT of *absolute* blood oxygen saturation, sO_2, which is an important physiological parameter, can be achieved more readily. Blood sO_2, defined as the ratio of the concentrations of oxyhaemoglobin, c_{HbO_2}, to total haemoglobin, c_{THb}, has been shown to be a more robust parameter compared to absolute concentrations [67]. This is because errors in the absolute concentrations are divided out, making blood sO_2 dependent upon the shape rather than the amplitude of the PA spectrum and allowing the Grüneisen parameter to be neglected. The measurement of absolute blood sO_2 is therefore an achievable intermediate goal. The demonstration of non-invasive, longitudinal qPAT of blood sO_2 will have a major impact on a range of areas in life sciences and is vital for the widespread adoption of PAT as a biomedical research tool.

Optical resolution PA microscopy methods are mentioned here for completeness as they have enabled innovative approaches to the measurement of blood sO_2. They allow, for example, the exploitation of nonlinear absorption phenomena to measure blood sO_2 [68] or fast imaging of blood sO_2 changes in individual red blood cells [69]. Since the imaging depth of PA microscopy is restricted to superficial regions where ballistic photon propagation can be assumed (<1 mm), linear inversion schemes have been shown to be applicable. In addition, non-linear phenomena can easily be created by focussing the excitation light and thus creating high fluences. However, PA microscopy approaches are largely unsuitable for qPAT. First, nonlinear absorption effects are unlikely to be generated in deep tissue since the maximum permissible exposure limits the fluence at the skin. Second, linear inversion schemes are not valid due to effect of the fluence.

Conclusion

PA microscopy and tomography provide multiscale imaging capability ranging from sub-micron resolutions at superficial depths (<1 mm) using OR-PAM to cm imaging depths using PA tomography with spatial resolutions of few hundreds of micrometres. PA imaging is a hybrid technology that combines the strong absorption-based contrast and spectroscopic specificity of purely optical modalities with high spatial resolution afforded by ultrasound imaging. While the potential for enabling functional and molecular imaging has been demonstrated in a large number of basic science and preclinical studies, the translation of these methodologies to clinical applications nevertheless remains challenging. This is because deep-tissue 3D PA tomography of functional parameters or contrast agent accumulations requires robust methods for their recovery from multiwavelength PA images. Based on recent progress in the field of quantitative PA imaging and in the development of experimental methods for difference imaging of contrast agents, it is reasonable to assume that these goals will be reached in the intermediate future. PA imaging can then be expected to be translated to a broad range of potential applications in biology and medicine, such as imaging of angiogenesis, tumour microenvironments, drug response, brain function and tissue metabolism.

References

1. Beard P. Biomedical photoacoustic imaging. Interface Focus. 2011;1:602–31.
2. Laufer J, et al. In vivo preclinical photoacoustic imaging of tumor vasculature development and therapy. J Biomed Opt. 2012;17:56016.
3. Cox B, Laufer JG, Arridge SR, Beard PC. Quantitative spectroscopic photoacoustic imaging: a review. J Biomed Opt. 2012;17:61202.
4. Niederhauser JJ, Frauchiger D, Weber HP, Frenz M. Real-time optoacoustic imaging using a Schlieren transducer. Appl Phys Lett. 2002;81:571.
5. Nuster R, Slezak P, Paltauf G. High resolution three-dimensional photoacoutic tomography with CCD-camera based ultrasound detection. Biomed Opt Express. 2014;5:2635.
6. Nuster R, et al. Photoacoustic microtomography using optical interferometric detection. J Biomed Opt. 2010;15:21307.
7. Paltauf G, Nuster R. Artifact removal in photoacoustic section imaging by combining an integrating cylindrical detector with model-based reconstruction. J Biomed Opt. 2014;19:26014.
8. Zhang C, Ling T, Chen S, Guo LJ. Ultrabroad bandwidth and highly sensitive optical ultrasonic detector for photoacoustic imaging. ACS Photonics. 2014;1:1093–8.
9. Li H, Dong B, Zhang Z, Zhang HF, Sun C. A transparent broadband ultrasonic detector based on an optical micro-ring resonator for photoacoustic microscopy. Sci Rep. 2014;4:4496.
10. Hsieh B-Y, Chen S-L, Ling T, Guo LJ, Li P-C. All-optical scanhead for ultrasound and photoacoustic dual-modality imaging. Opt Express. 2012;20:1588.
11. Hsieh B-Y, Chen S-L, Ling T, Guo LJ, Li P-C. All-optical scanhead for ultrasound and photoacoustic imaging—imaging mode switching by dichroic filtering. Photo-Dermatology. 2014;2:39–46.
12. Ling T, Chen S-L, Guo LJ. High-sensitivity and wide-directivity ultrasound detection using high Q polymer microring resonators. Appl Phys Lett. 2011;98:204103.
13. Chen S-L, et al. Miniaturized all-optical photoacoustic microscopy based on microelectromechanical systems mirror scanning. Opt Lett. 2012;37:4263–5.
14. Rousseau G, Blouin A, Monchalin J. Non-contact photoacoustic tomography and ultrasonography for tissue imaging. Biomed Opt Express. 2012;3:3233–5.
15. Rousseau G, Monchalin J, Gauthier B, Blouin A. Non-contact biomedical photoacoustic and ultrasound imaging. J Biomed Opt. 2012;17:61217.
16. Zhang E, Laufer J, Beard P. Backward-mode multiwavelength photoacoustic scanner using a planar Fabry-Perot polymer film ultrasound sensor for high-resolution three-dimensional imaging of biological tissues. Appl Opt. 2008;47:561–77.
17. Zhang EZ, et al. Multimodal photoacoustic and optical coherence tomography scanner using an all optical detection scheme for 3D morphological skin imaging. Biomed Opt Express. 2011;2:2202–15.
18. Chee R, Sampaleanu A, Rishi D, Zemp R. Top orthogonal to bottom electrode (TOBE) 2-D CMUT arrays for 3-D photoacoustic imaging. IEEE Trans Ultrason Ferroelectr Freq Control. 2014;61:1393–5.

19. Vaithilingam S, et al. Three-dimensional photoacoustic imaging using a two-dimensional CMUT array. IEEE Trans Ultrason Ferroelectr Freq Control. 2009;56:2411–9.
20. Bhuyan A, et al. Integrated circuits for volumetric ultrasound imaging with 2-D CMUT arrays. IEEE Trans Biomed Circuits Syst. 2013;7:796–804.
21. Kruger RA, Lam RB, Reinecke DR, Del Rio SP, Doyle RP. Photoacoustic angiography of the breast. Med Phys. 2010;37:6096.
22. Wang X, et al. Noninvasive laser-induced photoacoustic tomography for structural and functional in vivo imaging of the brain. Nat Biotechnol. 2003;21:803–6.
23. Xia J, Wang LV. Small-animal whole-body photoacoustic tomography: a review. IEEE Trans Biomed Eng. 2014;61:1380–9.
24. Razansky D, Buehler A, Ntziachristos V. Volumetric real-time multispectral optoacoustic tomography of biomarkers. Nat Protoc. 2011;6:1121–9.
25. Wang LV, Gao L. Photoacoustic microscopy and computed tomography: from bench to bedside. Annu Rev Biomed Eng. 2014;16:155–85.
26. Yao J, Wang LV. Sensitivity of photoacoustic microscopy. Photo-Dermatology. 2014;2:87–101.
27. Zhang C, Maslov K, Wang LV. Subwavelength-resolution label-free photoacoustic microscopy of optical absorption in vivo. Opt Lett. 2010;35:3195–7.
28. Wang L, Zhang C, Wang LV. Grueneisen relaxation photoacoustic microscopy. Phys Rev Lett. 2014;113:1–5.
29. Yao J, Wang L, Li C, Zhang C, Wang LV. Photoimprint photoacoustic microscopy for three-dimensional label-free subdiffraction imaging. Phys Rev Lett. 2014;112:14302.
30. Zhang C, Maslov K, Yao J, Wang LV. In vivo photoacoustic microscopy with 7.6-µm axial resolution using a commercial 125-MHz ultrasonic transducer. J Biomed Opt. 2012;17:116016.
31. Yao J, Wang LV. Photoacoustic brain imaging: from microscopic to macroscopic scales. Neurophotonics. 2014;1:1877516.
32. Wang L, Maslov K, Xing W, Garcia-Uribe A, Wang L. Video-rate functional photoacoustic microscopy at depths. J Biomed Opt. 2012;17:106007.
33. Wang T, et al. All-optical photoacoustic microscopy based on plasmonic detection of broadband ultrasound. Appl Phys Lett. 2015;107:153702.
34. Dong B, Chen S, Zhang Z, Sun C, Zhang HF. Photoacoustic probe using a microring resonator ultrasonic sensor for endoscopic applications. Opt Lett. 2014;39:4372–5.
35. Chen S-L, Ling T, Guo LJ. Low-noise small-size microring ultrasonic detectors for high-resolution photoacoustic imaging. J Biomed Opt. 2011;16:56001.
36. Hu S, Wang LV. Neurovascular photoacoustic tomography. Front Neuroenerg. 2010;2:10.
37. Li G, Li L, Zhu L, Xia J, Wang LV. Multiview Hilbert transformation for full-view photoacoustic computed tomography using a linear array. J Biomed Opt. 2015;20:66010.
38. Pramanik M. Improving tangential resolution with a modified delay-and-sum reconstruction algorithm in photoacoustic and thermoacoustic tomography. J Opt Soc Am A. 2014;31:621.
39. Köstli KP, Beard PC. Two-dimensional photoacoustic imaging by use of Fourier-transform image reconstruction and a detector with an anisotropic response. Appl Opt. 2003;42:1899–908.
40. Treeby BE, Zhang EZ, Cox BT. Photoacoustic tomography in absorbing acoustic media using time reversal. Inverse Prob. 2010;26:115003.
41. Cox BT, Treeby BE. Artifact trapping during time reversal photoacoustic imaging for acoustically heterogeneous media. IEEE Trans Med Imaging. 2010;29:387–96.
42. Zhang EZ, Laufer JG, Pedley RB, Beard PC. In vivo high-resolution 3D photoacoustic imaging of superficial vascular anatomy. Phys Med Biol. 2009;54:1035–46.
43. Laufer J, Zhang E, Raivich G, Beard P. Three-dimensional noninvasive imaging of the vasculature in the mouse brain using a high resolution photoacoustic scanner. Appl Opt. 2009;48:D299–306.
44. Wang LH. Multiscale photoacoustic microscopy and computed tomography. Nat Photonics. 2009;3:503.
45. Kitai T, et al. Photoacoustic mammography: initial clinical results. Breast Cancer. 2014;21:146–53.
46. Xia W, Steenbergen W, Manohar S. Photoacoustic mammography: prospects and promises. Breast Cancer Manag. 2014;3:387–90.

47. Heijblom M, et al. Appearance of breast cysts in planar geometry photoacoustic mammography using 1064-nm excitation. J Biomed Opt. 2013;18:126009.
48. Kruger RA, et al. Dedicated 3D photoacoustic breast imaging. Med Phys. 2013;40:113301.
49. Hu S, Maslov K, Tsytsarev V, Wang LV. Functional transcranial brain imaging by optical-resolution photoacoustic microscopy. J Biomed Opt. 2009;14:40503.
50. Stein EW, Maslov K, Wang LV. Noninvasive, in vivo imaging of blood-oxygenation dynamics within the mouse brain using photoacoustic microscopy. J Biomed Opt. 2010;14:20502.
51. Xia J, Danielli A, Liu Y, Wang LLV, Maslov K. Calibration-free quantification of absolute oxygen saturation based on the dynamics of photoacoustic signals. Opt Lett. 2013;38:2800–3.
52. Yao J, et al. High-speed label-free functional photoacoustic microscopy of mouse brain in action. Nat Methods. 2015;12:407–10.
53. Weber J, Beard PC, Bohndiek SE. Contrast agents for molecular photoacoustic imaging. Nat Methods. 2016;13:639–50.
54. De la Zerda A, et al. Carbon nanotubes as photoacoustic molecular imaging agents in living mice. Nat Nanotechnol. 2008;3:557–62.
55. de la Zerda A, et al. Family of enhanced photoacoustic imaging agents for high-sensitivity and multiplexing studies in living mice. ACS Nano. 2012;6:4694–701.
56. Li M-L, et al. In-vivo photoacoustic microscopy of nanoshell extravasation from solid tumor vasculature. J Biomed Opt. 2009;14:10507.
57. Burton NC, et al. Multispectral opto-acoustic tomography (MSOT) of the brain and glioblastoma characterization. NeuroImage. 2013;65:522–8.
58. Märk J, et al. Photoacoustic imaging of fluorophores using pump-probe excitation. Biomed Opt Express. 2015;6:2522–35.
59. Märk J, Schmitt F, Laufer J. Photoacoustic imaging of the excited state lifetime of fluorophores. J Opt. 2016;18:54009.
60. Märk J, Wagener A, Zhang E, Laufer J. Photoacoustic pump-probe tomography of fluorophores in vivo using interleaved image acquisition for motion suppression. Sci Rep. 2017;7:1–9.
61. Razansky D, et al. Multispectral opto-acoustic tomography of deep-seated fluorescent proteins in vivo. Nat Photonics. 2009;3:412–7.
62. Krumholz A, Shcherbakova DM, Xia J, Wang LV, Verkhusha VV. Multicontrast photoacoustic in vivo imaging using near-infrared fluorescent proteins. Sci Rep. 2014;4:3939.
63. Jathoul AP, et al. Deep in vivo photoacoustic imaging of mammalian tissues using a tyrosinase-based genetic reporter. Nat Photonics. 2015;9:239–46.
64. Cox BT, Laufer JG, Beard PC. The challenges for quantitative photoacoustic imaging. Proc SPIE. 2009;7177:717713.
65. Cox B, Tarvainen T, Arridge S. Multiple illumination quantitative photoacoustic tomography using transport and diffusion models. Contemp Math. 2011;559:1–12.
66. Lutzweiler C, Razansky D. Optoacoustic imaging and tomography: reconstruction approaches and outstanding challenges in image performance and quantification. Sensors. 2013;13:7345–84.
67. Laufer J, Delpy D, Elwell C, Beard P. Quantitative spatially resolved measurement of tissue chromophore concentrations using photoacoustic spectroscopy: application to the measurement of blood oxygenation and haemoglobin concentration. Phys Med Biol. 2007;52:141–68.
68. Danielli A, Favazza CP, Maslov K, Wang LV. Picosecond absorption relaxation measured with nanosecond laser photoacoustics. Appl Phys Lett. 2010;97:163701.
69. Wang L, Maslov K, Wang LV. Single-cell label-free photoacoustic flowoxigraphy in vivo. Proc Natl Acad Sci U S A. 2013;110:5759–64.

Fundamentals of X-Ray Computed Tomography: Acquisition and Reconstruction

14

Marc Dewey and Marc Kachelrieß

Abstract

Computed tomography (CT) provides quantitative assessment of tissue properties by a unique linear relationship between signal and CT contrast agents. Clinically, CT is widely used in the acute setting but also for chronic conditions. High radiation dose and the potential for contrast-induced acute kidney injury are the two major challenges for CT. This chapter briefly summarizes the clinical use of CT and presents the physical and technical basis of CT data acquisition and image reconstruction.

14.1 Introduction/Aims

14.1.1 Historical Summary

Since the invention of CT by Godfrey Hounsfield [1, 2] and Allan Cormack [3, 4] in the 1960s and 1970s of the last century and with the advent of spiral CT in the 1980s [5], remarkable advances in speed, temporal resolution, and voxel size have enabled CT to become the most sensitive test for many diseases. Among these clinical scenarios is, for instance, the noninvasive detection of coronary artery stenosis [6–8]. Moreover, CT may even be safer than invasive angiography techniques for the management of certain patients [9]. Thus, there is great clinical need to further

M. Dewey
Department of Radiology, Charité - Universitätsmedizin Berlin, Berlin, Germany
e-mail: marc.dewey@charite.de

M. Kachelrieß (✉)
Medical Physics in Radiology, German Cancer Research Center (DKFZ),
Heidelberg, Germany
e-mail: marc.kachelriess@dkfz.de

© Springer International Publishing AG 2018   325
I. Sack, T. Schaeffter (eds.), *Quantification of Biophysical Parameters in Medical Imaging*, https://doi.org/10.1007/978-3-319-65924-4_14

improving the widespread availability of CT as it would allow added clinical value while reducing the burden on patients.

14.1.2 CT Examination Setting

A typical CT examination is shown in Fig. 14.1. The patient is accompanied here by a physician as a stress cardiac examination is shown (for details see Chap. 22): besides the contrast agent injector, also a stress agent infusion system is placed on the CT table. Apart from that, the situation is similar from the perspective of the patient for any other CT examination with contrast agent injection.

14.1.3 Major CT Issue: Radiation Exposure

One of the major issues with CT is the requirement of ionizing radiation. The first and comprehensive European Commission report on medical radiation dose shows that CT accounts for only 8.7% of all examinations but the CT-related

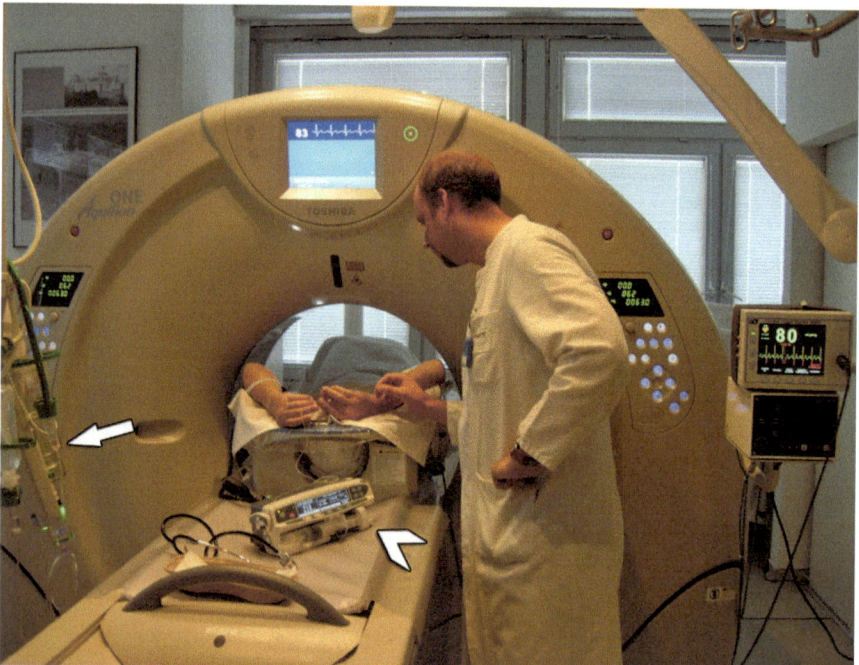

Fig. 14.1 Typical picture of a CT examination. The patient lies on the CT table and is positioned within the CT gantry. The contrast agent is injected via a contrast agent pump (*arrow*). In this specific examination, the patient is closely observed by a board-certified radiologist for vital signs as a stress agent (adenosine) is intravenously infused using an infusion system (*arrowhead*). As a result of that, the heart rate increased to 80 beats/min. CT examinations are typically short with an acquisition duration of just a few seconds during about 2–3 breath holds

radiation dose is far higher with about 57% of all medically induced radiation exposures. This demonstrates the great need to further reduce CT-related radiation dose [10].

14.1.4 Major Current and Future Clinical Potential of CT

The major current clinical potential of CT is in its high accuracy of depicting anatomical details inside the human body without interventional procedures. With this ability of submillimeter three-dimensional depiction of human anatomy, clinical approaches to trauma imaging, lung and liver imaging, as well as oncologic, cardiac, and vascular imaging have been revolutionized. Also, early and fast imaging of patients with suspected stroke is possible using anatomic depiction of hypoperfused and hemorrhagic areas in the brain. Adding quantitative information about the perfusion of infarcted or ischemic tissue is one of the key challenges for CT. Thus, the major future potential of CT is its unique ability to quantify tissue perfusion and tissue changes over time. This is due to an optimal linear relationship between the amount of CT tracers and the resulting signal on CT images [11]. Thus, CT is theoretically in the pole position for quantification of tissue changes as well as tissue perfusion.

The aim of this chapter is thus twofold: (1) present the current clinical use of CT and (2) explain the physics and technical basis of CT. A further outlook on biophysical parameters that can be derived by 3D as well as 4D CT perfusion of the myocardium is presented in Chap. 22.

14.2 Clinical Use of 3D CT

14.2.1 Clinical Standing of CT in Guidelines

3D CT is already included in many guidelines as the primary diagnostic test to consider in most acute presentations to the emergency rooms of hospitals but also for many chronic clinical scenarios. A primary example for this are any patients with stable typical or atypical angina where CT is actually recommended as the frontline test for ruling out obstructive coronary artery disease (CAD) according to the National Institute for Health and Care Excellence (NICE) guideline drafted in the UK in 2016 [12].

14.2.2 Example of Cardiac and Chest Imaging Guidelines

In addition to the above NICE guideline from 2016, CT is also recommended in patients with 15–50% pretest probability of obstructive CAD in the most recent European stable angina and myocardial revascularization guidelines from 2013 and 2014 [10, 11]. These recommendations were made because CT is an effective gatekeeper for invasive coronary angiography based on its high sensitivity and negative

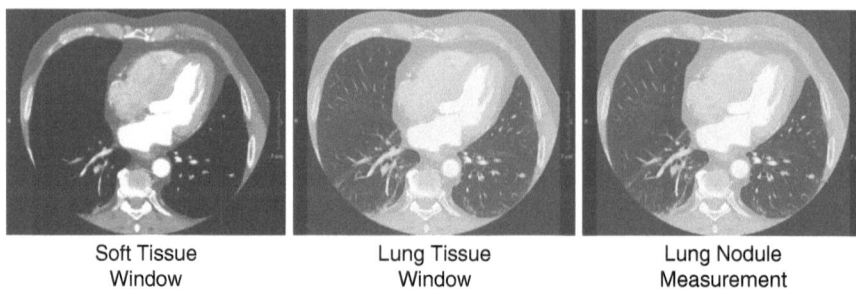

| Soft Tissue | Lung Tissue | Lung Nodule |
| Window | Window | Measurement |

Fig. 14.2 Typical result and visualization of chest CT examinations. Data are reconstructed using soft tissue as well as lung tissue kernels and are then displayed using soft and lung tissue window-level settings. In case of pulmonary nodules such as the 7 mm nodule shown here, any measurements of dimensions are performed on the lung kernel-based images that are viewed using lung window-level settings

predictive value [6, 7], as we have confirmed in a randomized trial of patients with atypical symptoms [9], and is best suited to reduce the over 2 million negative invasive coronary angiographies performed in Europe each year [13]. This is also supported by encouraging findings in the SCOT-Heart trial where patients were randomized to CT or standard of care and myocardial infarction were reduced on follow-up in the CT group [14]. Another example is the use of CT as the primary test for screening patients for the presence of lung nodules (Fig. 14.2), according to the United States Preventive Services Task Force (USPSTF) recommendation [15–19], based on findings of the very large randomized National Lung Screening Trial [20–24] in patients with a long-standing history of smoking and thus increased risk of lung cancer.

14.2.3 Clinical Challenges: Radiation Dose and Risk of Nephrotoxicity

One of the results of the rather frequent use of CT as a radiation exposure test, with higher exposure levels than other modalities, is that the population-based radiation exposure from CT is now higher than that from natural radiation sources [25]. This is a cause of concern. In any case the radiation exposure by CT needs to be as low as reasonably achievable. This can be achieved by better individualizing radiation output to the specific patient and related clinical scenario.

Moreover, another important cause of concern for clinical decision making about the use of CT is the requirement of contrast agents. Contrast-enhanced CT allows the differentiation of different types of tumors and is used for angiographic CT examinations, e.g., of the coronary arteries (Fig. 14.3). CT contrast agents are intravenously injected for most diagnostic questions but can lead to contrast-induced nephrotoxicity [26], which is the third most common cause of hospital-acquired acute kidney injury [27]. With about 5–6% contrast-induced nephropathy rates and about 70 million CT examinations each year in the United States of America and Europe alone [28, 29], an estimated 3–4 million people have contrast-induced nephropathy after a CT examination with contrast every year. Thus, close

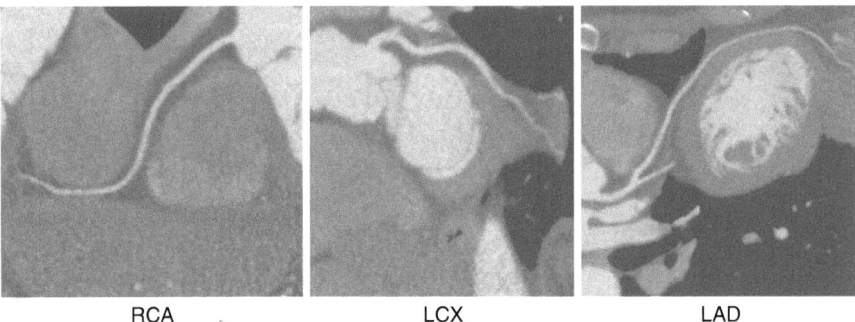

RCA   LCX   LAD

Fig. 14.3 Cardiac CT results. Shown are the curved multiplanar reformations along the course of the three coronary arteries (right coronary artery *RCA*; left circumflex coronary artery *LCX*; and left anterior descending coronary artery *LAD*) based on the volumetric 3D CT data that encompass between 80 and 120 million voxels. An examination is acquired within a fraction of a second (typically less than 200 ms with state-of-the-art CT). The curved reformations in this patient show no obstructions or vessel wall changes

monitoring and meticulous decision making about the appropriateness of imaging involving contrast agent use in general (also for MRI) is thus key to avoiding complications and providing high-value clinical care.

14.2.4 Clinical Potential: Quantitative Image Scale

A major clinical advantage of CT is the quantitative Hounsfield unit (HU) scale (Table 14.1) which is used for CT image analysis (Fig. 14.4) using different organ-based window-level settings (Table 14.2). This facilitates the quantitative analysis of tissue properties which may become crucial for diagnosis and prediction. Such quantification is likely to become most relevant when follow-up imaging is performed and compared with baseline assessment. During such follow-up imaging, it would be possible to quantify not only the extent but also the composition of coronary artery plaques by CT. Such clinical applications are not feasible by any of the other noninvasive imaging modalities in clinical practice with quantitative results.

Table 14.1 Tissue types and CT image scale results (in HU)

Organ/tissue	Min/average	Max
Air	−1000	
Lung	−900	−500
Fat	−100	−70
Water	0	
Kidney	20	40
Pancreas	20	50
Blood	30	60
Liver	40	70
Bone (inner)	70	350
Bone (cortical)	350	2000

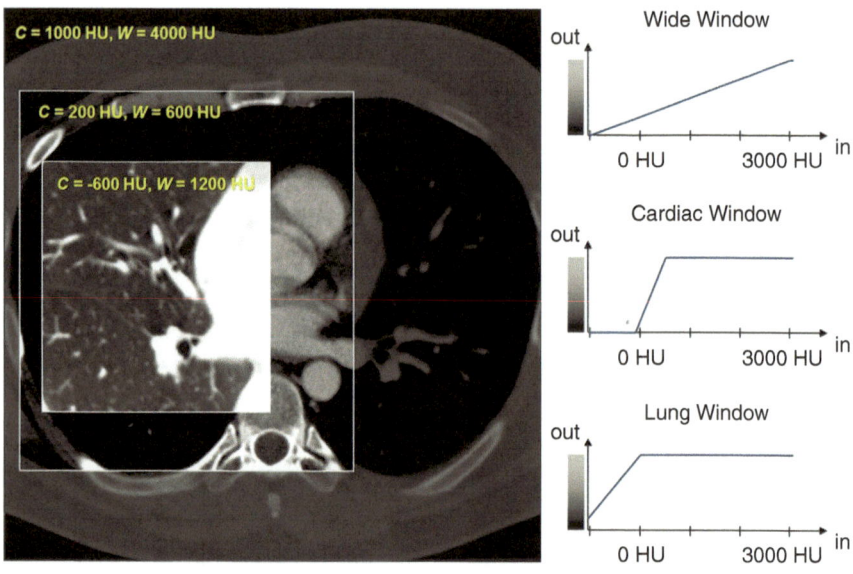

Fig. 14.4 Quantitative Hounsfield unit (HU) scale and window selection (window center C and window width W) for image analysis. Wide windows encompass all gray levels of the CT image and are thus rather low contrast but are used (with variation), e.g., for bone analysis. Cardiac windows are focused on depicting soft tissue and vascular structures and are thus suitable for vessel and plaque analysis. Lung windows have a negative center of the window and nicely demonstrate contrasting tissues with the lung parenchyma and are thus suitable for lung nodule detection. Published with kind permission of copyright Marc Kachelrieß/Marc Dewey 2017. All Rights Reserved

Table 14.2 Organ-based window-level settings in CT (in HU)

Organ/tissue	Window center C	Window width W
Pelvis	35	350
Abdomen	40	300
Liver	40	200
Lung	−600	1200
Heart	200	600
Bone	450	1500
Spine	40	350
Shoulder	400	2000
Extremities	300	1400
Mediastinum	40	400
Larynx	50	250
Inner EArt	700	4000
Cerebrum	35	80
Nasal sinuses	400	2000
Dental	400	2000
Angiography	80	700

14.3   Physics and Technical Basis of CT

14.3.1 Hardware

The CT system geometry is show in Fig. 14.5. The CT system typically consists of one tube-detector unit. There are special dual source CT systems on the market that comprise two tube-detector units and thereby achieve twice the temporal resolution. The main hardware components are, besides the gantry mechanics, the X-ray tube and the X-ray detector.

The X-ray tube design and cooling is shown in Fig. 14.6. Directly cooled X-ray tubes are the most advanced cooling design which, however, is more expensive to build. Direct cooling offers shorter cooling times which in turn increases versatility of CT scanners, for instance, in regard to dynamic 4D scanning and high-mA low-kV scanning. Collimators within the CT systems allow that only the X-ray beam required for generating images hits the patient and that low-energy photons are removed by prefilters in order to reduce radiation dose to the patients (Fig. 14.7). Bow-tie filters are commonly used in CT to shape the X-ray beam to the curvature of the human body by reducing the amount of photons in the periphery. Moreover, scatter grids mounted on the detector improve image quality by removing scattered radiation

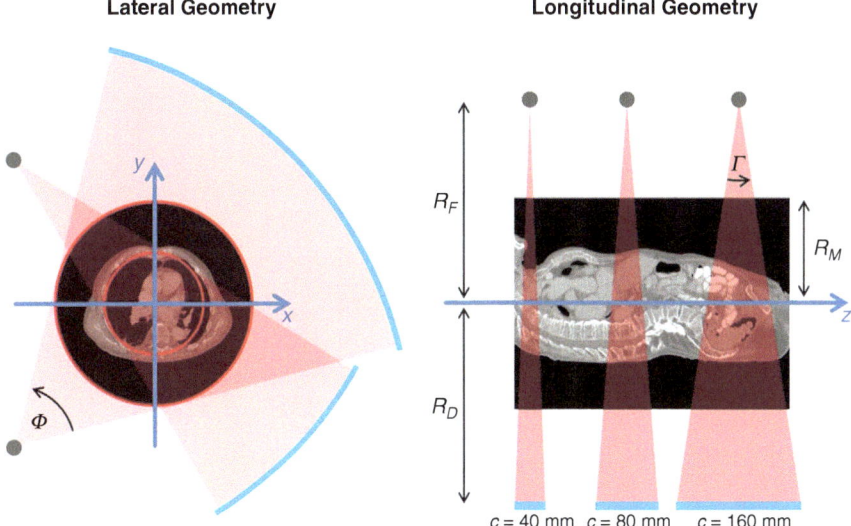

Fig. 14.5 CT system geometry. The lateral geometry shows that CT systems have either one or two tube-detector units shown as points and blue curvatures. Each detector unit consists of almost 1000 detector elements per CT detector row. The longitudinal geometry along the patients' Z-axis is variable depending on the CT type and can cover a collimation C of up to 160 mm based on the acquisition of up to 320 detector rows. R_F is the distance from the focus of the tube to the rotation center and is typically about 0.6 m. R_D the distance from the rotation center to the detector. The entire distance from the focus of the tube to the detector is typically about 1 m. Published with kind permission of copyright Marc Kachelrieß/Marc Dewey 2017. All Rights Reserved

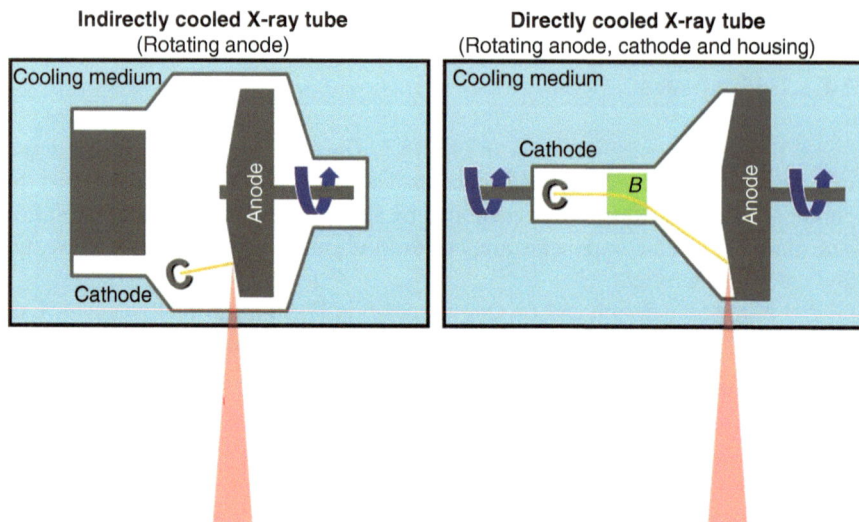

Fig. 14.6 Indirectly and directly cooled X-ray tube design. Conventional X-ray tubes are indirectly cooled with heat generated by the anode being transported to the cooling medium. Directly cooled X-ray tubes have an anode that is in direct contact with the cooling medium allowing shorter cooling times but requiring higher technical efforts as the housing also has to rotate. Published with kind permission of copyright Marc Kachelrieß/Marc Dewey 2017. All Rights Reserved

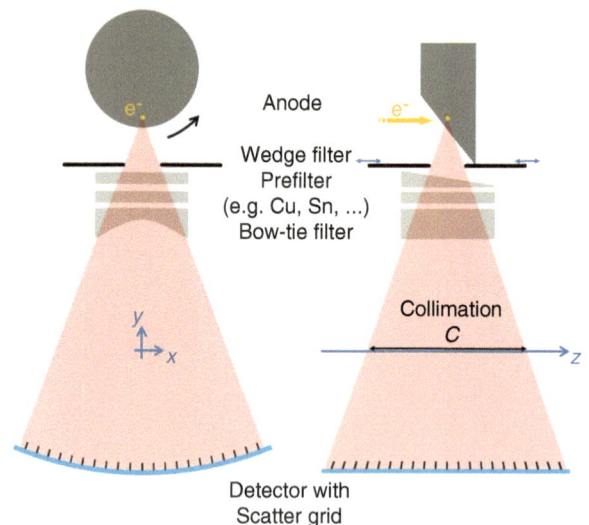

Fig. 14.7 Collimation of the X-ray beam in CT systems as shown in the X-Y plane and along the Z-axis. Wedge and prefilters improve the quality of the X-ray beams that hit the patient, e.g., by removing low-energy photons that will never reach the detector but only increase biological radiation dose. Bow-tie filters are important to shape the X-ray beam to the configuration and shape of the patient and reduce the amount of photons in the periphery where the diameters of the "round object" patients are typically smaller. Scatter grids remove scattered radiation and improve image quality. Note that not all vendors use all filter approaches shown here in their specific configurations. Published with kind permission of copyright Marc Kachelrieß/Marc Dewey 2017. All Rights Reserved

(Fig. 14.7). The size of the focal spot (focus) in the tube influences spatial resolution achievable by the specific CT with small focal spots allowing for higher spatial resolution at the cost of lower maximum tube currents compared to larger focal spots.

The detector is opposite to the X-ray tube and has a curved shape. High-end CT systems from the four main vendors (GE, Philips, Siemens, Toshiba/Canon) include more than 100 detector rows in the Z-axis (longitudinal geometry, Fig. 14.5) per detector unit. Each of the CT detector rows consists of nearly 1000 detector elements in the X-Y plane (lateral geometry, Fig. 14.5). With up to 320 detector rows, the Z-axis collimation is up to 160 mm which allows covering the entire heart in a single tube-detector unit rotation. The structure and design of CT detector elements, which are inclined against each other so that a circular arc is formed, are shown in Fig. 14.8. The detector components along the direction of the X-ray beams consist of the anti-scatter grid, the scintillator, a photodiode array, and the analog-to-digital converter (Fig. 14.8). The technology that will most likely dominate the configuration of CT detectors in the near future is that of direct converters which avoid analog-to-digital conversion and may enable reliable photon counting and signal measurement that is proportional to the energy of the specific photon (Fig. 14.9).

Typical rotation times of CT gantries are nowadays below 0.5 s and down to 0.25 s, and maximum Z-axis coverage is between 57.6 and 320 mm (Table 14.3). Exponentially increasing centrifugal forces on the CT gantry with decreasing rotation times (Table 14.4) and raw data transfer challenges are the main barrier to further speeding up rotation times.

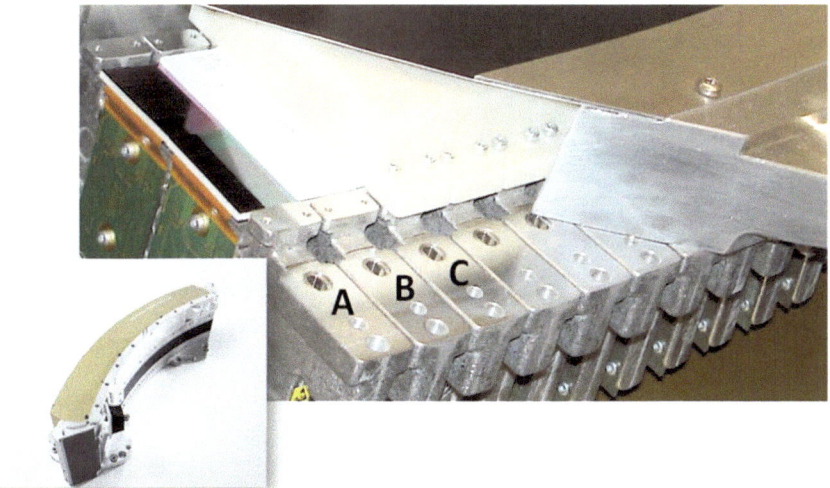

Fig. 14.8 Design of a CT detector. A modern CT detector consists of several detector modules, which are inclined against each other so that they form an approximate circular traverse (small inset on the left). For illustration the detector elements A–C are mounted in the large panel in different stages of building. A shows a detector module with the photodiode array (*black*), below which the analog-to-digital converter is located. B shows the scintillator (*white*) already on top of the photodiode. C shows a module where the scatter grid (*beige*) has already been mounted on top of the scintillator. For further information on the functionality of a detector, see Fig. 14.9. Printed with permission of Siemens Healthineers, Forchheim, Germany

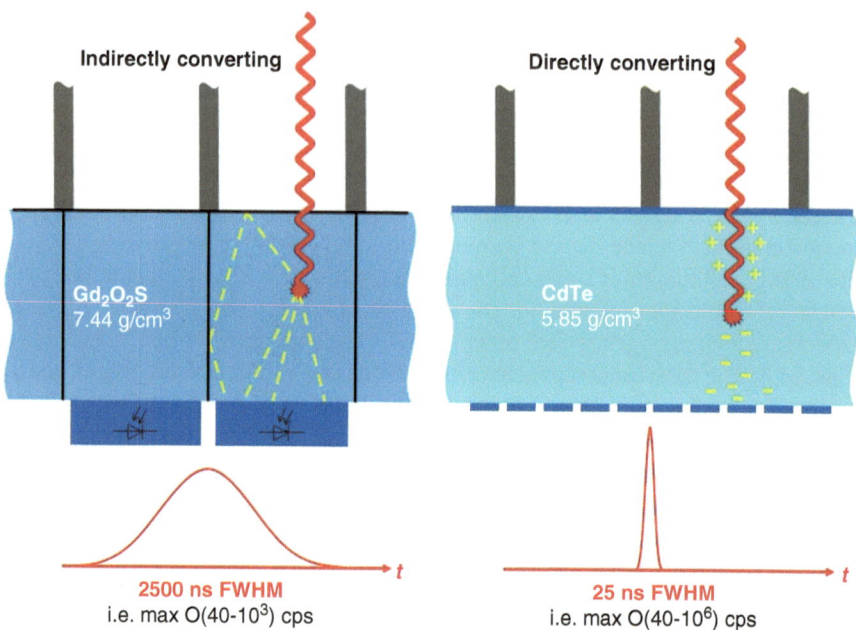

Fig. 14.9 Functionality of CT detectors. Currently, CT detectors are "indirectly converting" X-rays into visible light which is then transformed by photodiodes into photon current which can finally be measured and digitized for image reconstruction. Future concepts foresee "directly converting" detectors which allow to directly transform X-ray into electric impulses enabling single-photon counting. Published with kind permission of copyright Marc Kachelrieß/Marc Dewey 2017. All Rights Reserved

Table 14.3 Detector configuration and collimation of different high-end CT systems

CT system	Vendor	Configuration	Collimation (mm)	Fan angle (°)	Rotation time (s)
Aquilion ONE vision edition	Toshiba (Canon)	320 × 0.5 mm	160	15	0.275
Somatom force	Siemens	2 × 96 × 0.6	57.6	5.5	0.25
Revolution CT	GE	256 × 0.624 mm	160	15	0.28
Brilliance iCT	Philips	128 × 0.625 mm	80	7.7	0.27

Configuration means number of CT detector rows times the thickness

Table 14.4 CT detector rotation times and centrifugal forces

Rotation time (s)	Centrifugal force (g)
1.00	2.41
0.75	4.29
0.50	9.66
0.40	15.1
0.30	26.8
0.25	38.6
0.20	60.4

14.3.2 Acquisition

There are three main acquisition types of a CT system. The simplest one is a circle scan where the patient table remains stationary. A number of images, typically corresponding to the number of active detector rows, are simultaneously acquired during a half or a full rotation. The circle scan gains more and more importance because the z-coverage of the CT systems tends to increase. Thus, it becomes more likely that a complete anatomical area can be covered by a single circle scan.

If the z-coverage of a single circle is too small, one may append multiple circle scans, with a table shift in-between. This scan mode is known as the sequence scan, or as the step-and-shoot scan. The disadvantage of this scan type is the delay between two adjacent circles, which may be problematic when the patient moves or when the contrast agent distribution changes quickly.

The most important and most widely used trajectory is the spiral trajectory. Here, the CT system is continuously acquiring while, simultaneously, the patient table is continuously shifted through the gantry. In the patient's coordinate system, the X-ray focal spot follows a spiral, or helical, trajectory. The advantage of the spiral data acquisition is its high symmetry: it is the only trajectory that symmetrically covers long scans with all voxels being treated on an equal footing. Moreover, the ratio of the table increment per rotation and the collimation of the CT system, which is known as the pitch value, is a parameter that can be freely selected, typically between 0.1 and 1.5. Low values imply an overlapping data acquisition (more data than necessary are acquired) which may be advantageous for thicker patients or for certain contrast agent injection types, while high pitch values mean faster scans. One may be tempted to think that low pitch values mean higher dose and vice versa. But this is not the case: the CT systems adjust the tube current accordingly, i.e., proportional to the pitch value.

14.3.3 Reconstruction

For reconstruction, projection data need to cover at least 180° rotation of the lateral geometry (which implies that the scan range needs to cover 180° plus fan angle) and the number of projections (line integrals) needs to be sufficiently high (Fig. 14.10). During a rotation of the tube-detector unit, typically about 1000 projections are obtained. With up to 320 detector rows and 1000 detector elements per row, more than 300 million projection data are acquired in the raw data space per gantry rotation.

The principles of CT image reconstruction are shown in Fig. 14.10. The measured intensities of the raw data are normalized, logarithmized, and finally convolved (filtered) with the reconstruction kernel. Then, the data are backprojected into the image (or volume). This means that for each filtered data point, which corresponds to a measurement along a line in space, its value is added onto the pixels of the image just along that line. Once all data have been backprojected, the image is reconstructed. Filtered backprojection (FBP) is the cornerstone of analytic image reconstruction.

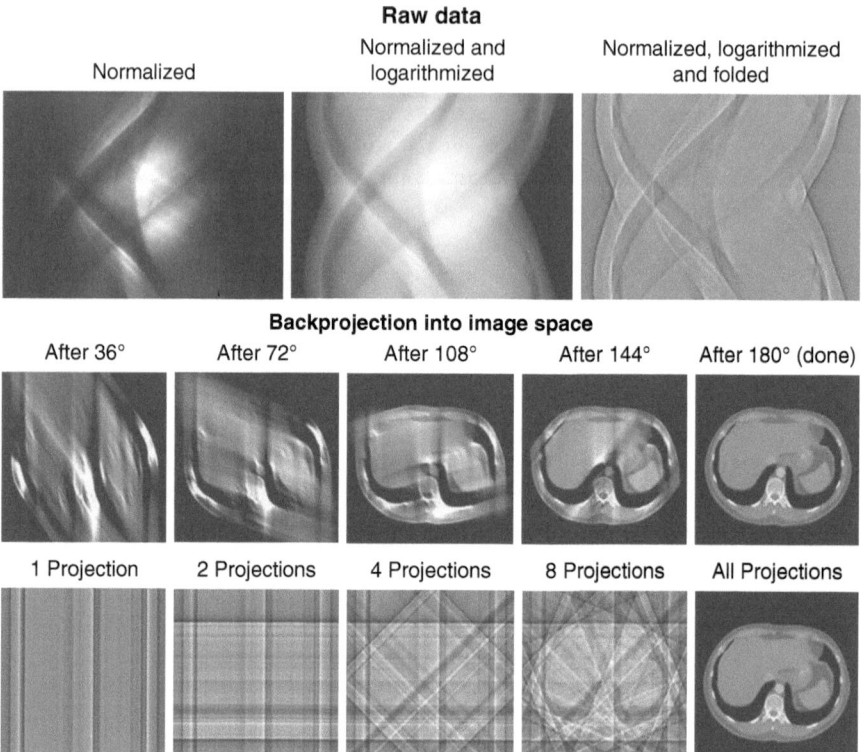

Fig. 14.10 Principles of CT image data reconstruction. Before backprojection into image space, the measured intensities are normalized and logarithmized and finally convolution happens with the reconstruction kernel in a line-wise fashion. To achieve high-quality images, two requirements need to be fulfilled: (1) data come from rotation of at least 180° and the number of projections is high enough. The window-level settings of the CT images in the last two rows are Center 0 HU and Window 1000 HU. Published with kind permission of copyright Marc Kachelrieß/Marc Dewey 2017. All Rights Reserved

During the last decade, iterative image reconstruction algorithms have become available. In contrast to the analytical algorithms, such as the abovementioned filtered backprojection algorithm, they iteratively estimate the image. This can be thought of reconstructing a first image (e.g., by using FBP), by imposing some prior knowledge on this image (e.g., by applying an edge-preserving filter to reduce noise but to preserve spatial resolution), by computing another raw data set from this first image using a forward projection algorithm, and by comparing these forward-projected raw data with the measured raw data. If there is any difference between these two raw data sets, this difference is reconstructed (e.g., by FBP) and added to the initial image. Then, one iteration is finished. One may repeat this procedure as often as necessary to obtain a good image. In some iterative reconstruction algorithms, the FBP step may be replaced by an unfiltered (or direct) backprojection (Fig. 14.11). This has the downside of requiring much more iterations and thus is typically not done in diagnostic CT. The aim of iterative reconstruction is to obtain

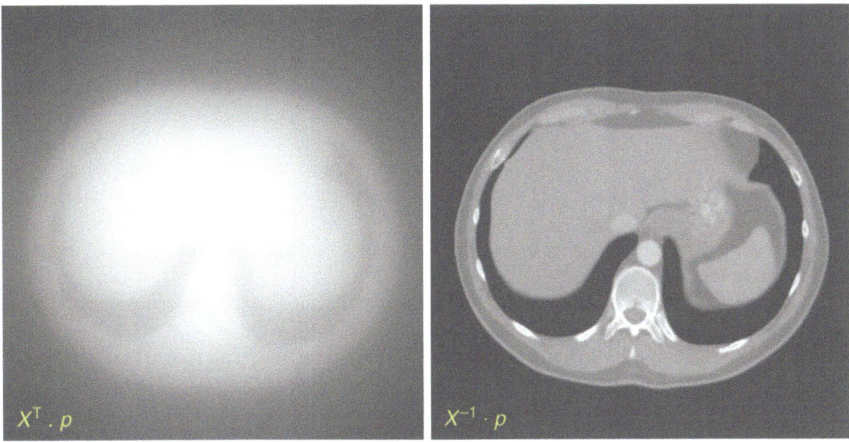

Fig. 14.11 Difference between unfiltered (*left*) and filtered backprojection (*right*). CT images of the lower lung and the upper abdomen based on unfiltered backprojection do not contain high frequencies as only positive raw data values are backprojected. Published with kind permission of copyright Marc Kachelrieß/Marc Dewey 2017. All Rights Reserved

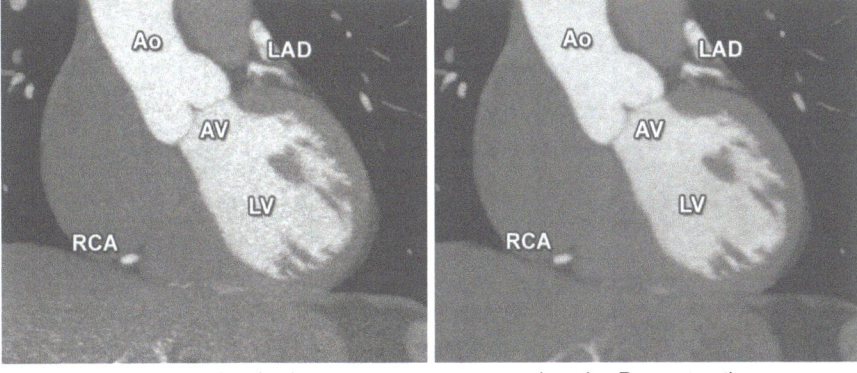

Fig. 14.12 Difference between filtered backprojection (*left*) and iterative reconstruction (*right*). Example of CT along the left ventricular outflow tract, aortic valve, and ascending aorta. Images based on iterative reconstruction contain relevantly less noise than standard filtered backprojection images. *Ao* aorta, *AV* aortic valve, *LV* left ventricle, *LAD* left anterior descending coronary artery, *RCA* right coronary artery. Published with kind permission of copyright Marc Kachelrieß/Marc Dewey 2017. All Rights Reserved

images that are sharper, less noisy, and less prone to artifacts than the FBP images. Resulting advantages in image quality by reduced noise and better perception of fine anatomical details by iterative reconstruction in comparison to standard filtered backprojection are shown in Fig. 14.12. For special applications one may use the standard iterative reconstruction techniques in combination with dedicated methods, such as in cardiac CT, for example [30].

Acknowledgment Professor Dewey would like to thank Drs. Feger, Rief, and Zimmermann for excellent support.

References

1. Hounsfield GN. Computerized transverse axial scanning (tomography). 1. Description of system. Br J Radiol. 1973;46:1016–22.
2. Ambrose J, Hounsfield G. Computerized transverse axial tomography. Br J Radiol. 1973;46:148–9.
3. Cormack AM. Nobel award address. Early two-dimensional reconstruction and recent topics stemming from it. Med Phys. 1980;7:277–82.
4. Cormack AM. Representation of a function by its line integrals, with some radiological applications. J Appl Phys. 1964;35:2908–13.
5. Kalender WA, Seissler W, Klotz E, Vock P. Spiral volumetric CT with single-breath-hold technique, continuous transport, and continuous scanner rotation. Radiology. 1990;176:181–3.
6. Schuetz GM, Schlattmann P, Dewey M. Avoiding overestimation of clinical performance by applying a 3x2 table with an intention-to-diagnose approach: an exemplary meta-analytical evaluation of coronary CT angiography studies. BMJ. 2012;345:e6717.
7. Schuetz GM, Zacharopoulou NM, Schlattmann P, Dewey M. Meta-analysis: noninvasive coronary angiography using computed tomography versus magnetic resonance imaging. Ann Intern Med. 2010;152:167–77.
8. Kachelriess M, Kalender WA. Electrocardiogram-correlated image reconstruction from subsecond spiral computed tomography scans of the heart. Med Phys. 1998;25:2417–31.
9. Dewey M, Rief M, Martus P, et al. Evaluation of computed tomography in patients with atypical angina or chest pain clinically referred for invasive coronary angiography: randomised controlled trial. BMJ. 2016;355:i5441.
10. EC. Medical radiation exposure of the European population; 2014.
11. Kalender W. CT: John Wiley & Sons; 2011.
12. NICE. CG 95 2016.
13. Moschovitis A, Cook S, Meier B. Percutaneous coronary interventions in Europe in 2006. EuroIntervention. 2010;6:189–94.
14. Williams MC, Hunter A, Shah AS, et al. Use of coronary computed tomographic angiography to guide management of patients with coronary disease. J Am Coll Cardiol. 2016;67:1759–68.
15. Field JK, Devaraj A, Duffy SW, Baldwin DR. CT screening for lung cancer: is the evidence strong enough? Lung Cancer. 2016;91:29–35.
16. Mauchley DC, Mitchell JD. Current estimate of costs of lung cancer screening in the United States. Thorac Surg Clin. 2015;25:205–15.
17. Mulshine JL, D'Amico TA. Issues with implementing a high-quality lung cancer screening program. CA Cancer J Clin. 2014;64:352–63.
18. Smieliauskas F, MacMahon H, Salgia R, Shih YC. Geographic variation in radiologist capacity and widespread implementation of lung cancer CT screening. J Med Screen. 2014;21:207–15.
19. Kamel M, Stiles B, Altorki NK. Clinical issues in the surgical management of screen-identified lung cancers. Oncology. 2015;29:944–9.
20. Aberle DR, DeMello S, Berg CD, et al. Results of the two incidence screenings in the national lung screening trial. N Engl J Med. 2013;369:920–31.
21. Black WC, Gareen IF, Soneji SS, et al. Cost-effectiveness of CT screening in the national lung screening trial. N Engl J Med. 2014;371:1793–802.
22. Aberle DR, Adams AM, National Lung Screening Trial Research Team, et al. Reduced lung-cancer mortality with low-dose computed tomographic screening. N Engl J Med. 2011;365:395–409.
23. Church TR, Black WC, National Lung Screening Trial Research Team, et al. Results of initial low-dose computed tomographic screening for lung cancer. N Engl J Med. 2013;368:1980–91.

24. Kovalchik SA, Tammemagi M, Berg CD, et al. Targeting of low-dose CT screening according to the risk of lung-cancer death. N Engl J Med. 2013;369:245–54.

25. Brenner DJ, Hall EJ. Computed tomography–an increasing source of radiation exposure. N Engl J Med. 2007;357:2277–84.

26. ACR. ACR manual on contrast media. Version 10.1; 2015.

27. Nash K, Hafeez A, Hou S. Hospital-acquired renal insufficiency. Am J Kidney Dis. 2002;39:930–6.

28. IMV. International marketing ventures. CT market outlook report. Des Plaines, IL; 2014.

29. European Commission (EC). Radiation protection No. 180. Medical radiation exposure of the European population. Part 1/2. Brussels, BE: directorate-general for energy. Directorate D— nuclear safety & fuel cycle. Unit D3—radiation protection; 2014.

30. Kachelriess M. Iterative reconstruction techniques: what do they mean for cardiac CT? Curr Cardiovasc Imaging Rep. 2013;6:268–81.

Quantification of Myocardial Effective Transverse Relaxation Time with Magnetic Resonance at 7.0 Tesla for a Better Understanding of Myocardial (Patho)physiology

15

Till Huelnhagen, Teresa Serradas-Duarte, Fabian Hezel, Katharina Paul, and Thoralf Niendorf

Abstract

Cardiovascular magnetic resonance imaging (CMR) has become an indispensable tool in the assessment of cardiac structure, morphology, and function. CMR also affords myocardial tissue characterization and probing of cardiac physiology, both being in the focus of ongoing research. These developments are fueled by the move to ultrahigh magnetic field strengths, which permits enhanced sensitivity and spatial resolution that help to overcome limitations of current clinical MR systems.

This chapter reviews the potential of using CMR as a means to assess physiology in the heart muscle by exploiting quantification of myocardial effective transverse relaxation times (T_2^*) for the better understanding of myocardial (patho)physiology. For this purpose the basic principles of T_2^* mapping, the biophysical mechanisms governing T_2^*, and Otherwise this implies that all preclinical applications of myocardial T_2^* mapping ever done are being presented which is not the case. Technological challenges and solutions for

T. Huelnhagen • T. Serradas-Duarte • F. Hezel • K. Paul
Berlin Ultrahigh Field Facility (B.U.F.F.), Max Delbrück Center for Molecular Medicine in the Helmholtz Association, Berlin, Germany

T. Niendorf (✉)
Berlin Ultrahigh Field Facility (B.U.F.F.), Max Delbrück Center for Molecular Medicine in the Helmholtz Association, Berlin, Germany

DZHK (German Centre for Cardiovascular Research), Partner Site, Berlin, Germany

MRI.TOOLS GmbH, Berlin, Germany
e-mail: thoralf.niendorf@mdc-berlin.de

© Springer International Publishing AG 2018
I. Sack, T. Schaeffter (eds.), *Quantification of Biophysical Parameters in Medical Imaging*, https://doi.org/10.1007/978-3-319-65924-4_15

343

T_2^*-sensitized CMR at ultrahigh magnetic field strengths are discussed followed by a survey of acquisition techniques and post processing approaches. Preliminary results derived from myocardial T_2^* mapping of healthy subjects and in patients at 7.0 T are presented. A concluding section provides an outlook including future developments and potential applications.

15.1 Introduction

15.1.1 Basics of T_2^*-Sensitized CMR

A growing number of reports refer to mapping the effective transverse relaxation time T_2^* in basic cardiovascular magnetic resonance (CMR) research and emerging clinical CMR applications. By making use of the blood oxygenation level-dependent (BOLD) effect [1], T_2^*-sensitized CMR has been suggested as a means of assessing myocardial tissue oxygenation and perfusion. T_2^* mapping has been shown to be capable of detecting myocardial ischemia caused by a stenotic coronary artery [2], to reveal myocardial perfusion deficits under pharmacological stress [3–8], to study endothelial function [9], or to assess breathing maneuver-dependent oxygenation changes in the myocardium [10–14]. T_2^* mapping is a proven clinical tool for myocardial iron quantification, an essential parameter for guiding therapy in patients with myocardial iron overload [15–19].

The fundamental principle behind T_2^* relaxation is the loss of phase coherence of an ensemble of spins contained within a volume of interest or voxel. Unlike T_1 relaxation which is based on spin–lattice interactions or T_2 relaxation which is caused by spin–spin interactions both being inherent tissue properties of tissues in a magnetic field, T_2^* relaxation includes contributions from external magnetic field perturbations [20]. These magnetic field inhomogeneities affect the effective transversal MR relaxation time T_2^* [21, 22]. T_2^* describes the loss of coherence and decay of the MR signal and is governed by [20]

$$\frac{1}{T_2^*} = \frac{1}{T_2} + \frac{1}{T_2'} \tag{15.1}$$

with T_2 being the transverse relaxation time and T_2' embodying magnetic susceptibility-related contributions [23].

The most common way of acquiring T_2^*-weighted images is gradient recalled echo (GRE) imaging. The magnitude signal intensity response $S_m(\theta)$ created by a basic GRE pulse sequence is

$$S_m(\theta) = S_0 \sin(\theta) \exp(-TE/T_2^*) \frac{\left[1 - \exp(-TR/T_1)\right]}{\left[1 - \cos(\theta)\exp(-TR/T_1)\right]} \tag{15.2}$$

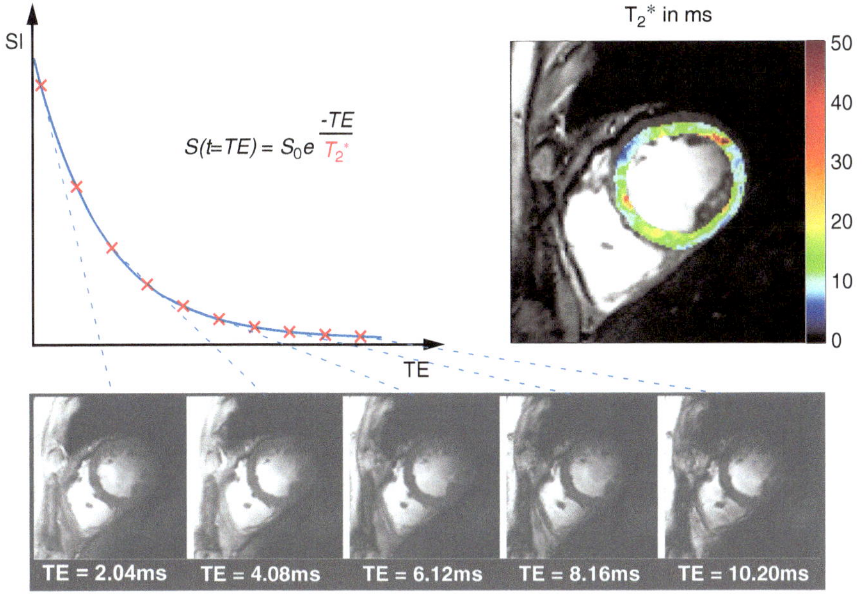

Fig. 15.1 T_2* decay and T_2*-weighted image contrast. (*Top left*) Plot of signal intensity over echo time. (*Bottom*) Example of a mid-ventricular short-axis view of the human heart at 7.0 T acquired with increasing T_2* weighting (*from left to right*). (*Top right*) Corresponding myocardial T_2* map superimposed to a CINE FLASH image

with ρ_0 representing the tissue spin density, *TR* the repetition time, *TE* the echo time defined by the time between MR signal excitation and MR signal readout [24], T_1 and T_2* tissue-specific longitudinal and effective transversal relaxation time constants, and θ the tip angle about which the magnetization is deflected by the excitation RF pulse. If *TR* and T_1 are being kept constant, Eq. (15.2) can be simplified to

$$S_m\left(\theta\right) \propto \exp\left(-TE / T_2^*\right) \tag{15.3}$$

Making use of this relationship, T_2*-driven decay of the MR signal intensity can be estimated by acquiring a series of images at different echo time *TE* followed by an exponential fit of the measured signal intensity versus the echo time *TE*. This is commonly achieved by using multi-echo gradient echo (MEGRE) pulse sequences, which take advantage of refocusing gradients to quickly acquire a series T_2*-sensitized images at several echo times as illustrated in Fig. 15.1. T_2*-sensitized MRI is most sensitive to field perturbations when *TE* is equal to T_2* [25]. Exponential fitting can be done for each voxel individually or for the mean signal within a region of interest. The former is more prone to noise but provides spatially resolved information in the form of relaxation maps (Fig. 15.1). Besides mono-exponential fitting, also multi-exponential fitting can be applied, if multiple signal compartments with different relaxation times are expected within an imaging voxel.

15.1.2 Biophysics of T_2^* and Relation to Physiology

T_2^* relaxation is blood oxygenation level dependent (BOLD) and provides a functional MR contrast [1, 22]. The effect is based on a change of the magnetic susceptibility of hemoglobin (Hb) depending on its oxygenation state. While oxygenated hemoglobin is diamagnetic and has minor effect on magnetic field homogeneity, deoxygenated hemoglobin is paramagnetic and causes magnetic field perturbations on a microscopic level resulting in spin dephasing and signal loss. T_2^*-weighted MRI is sensitive to changes in the amount of deoxygenated Hb (deoxy Hb) per tissue volume element (voxel). T_2^* decreases and a signal attenuation in T_2^*-weighted MR images occurs if the volume fraction of deoxy Hb increases. The phenomenon led to the development of functional MRI for mapping of human brain function but also inspired research into myocardial T_2^* mapping [7, 26].

BOLD imaging, T_2^*-sensitized imaging, and T_2^* mapping are widely assumed to provide a surrogate of oxygenation. Yet the factors influencing the transverse relaxation rate other than oxygenation are numerous including macroscopic magnetic field inhomogeneities, blood volume fraction, and hematocrit [14]. The magnetic field perturbations can have different origins. The most prominent B_0 effects are of macroscopic nature including strong susceptibility transitions at air tissue interfaces which are due to gradients in the magnetic susceptibility of the interfacing tissues and which can be moderated by dedicated magnetic field shimming techniques. B_0 perturbations can also be of microscopic nature. Considering a biologic tissue with a certain blood volume BVf, a hematocrit Hct, and a local blood oxygen saturation So_2, T_2^* can be modeled as

$$\frac{1}{T_2^*} = \frac{1}{T_2} + \gamma \left| \Delta B \right| = \frac{1}{T_2} + BVf \cdot \gamma \cdot \frac{4}{3} \cdot \pi \cdot \Delta\chi_0 \cdot Hct \cdot \left(1 - So_2\right) \cdot B_0 + \gamma \left| \Delta B_{\text{other}} \right| \quad (15.4)$$

with $\gamma|\Delta B_{\text{other}}|$ describing additional field inhomogeneities such as macroscopic field changes [27, 28] and $\Delta\chi_0 = 3.318$ ppm being the difference between the magnetic susceptibilities of fully oxygenated and fully deoxygenated hemoglobin (in SI units) [29]. Equation (15.4) can be utilized to noninvasively estimate tissue oxygenation using MRI when tissue blood volume fraction, hematocrit, and macroscopic B_0 contributions are known and echo times are greater than a characteristic time [28]. It is important to note that a reduction in the tissue blood volume fraction can result in a T_2^* increase which could be misinterpreted as an oxygenation increase and hence result in premature conclusions if the blood volume fraction is not considered [30]. If all the parameters are considered correctly, T_2^* can serve as a noninvasive means to probe physiology in vivo.

15.1.3 Why Myocardial T_2^* Mapping Benefits from Higher Magnetic Field Strengths

The linear relationship between magnetic field strength and microscopic susceptibility effects (Fig. 15.2) renders it conceptually appealing to perform myocardial T_2^* mapping at ultrahigh magnetic fields ($B_0 \geq 7.0$ T) [31]. The enhanced susceptibility

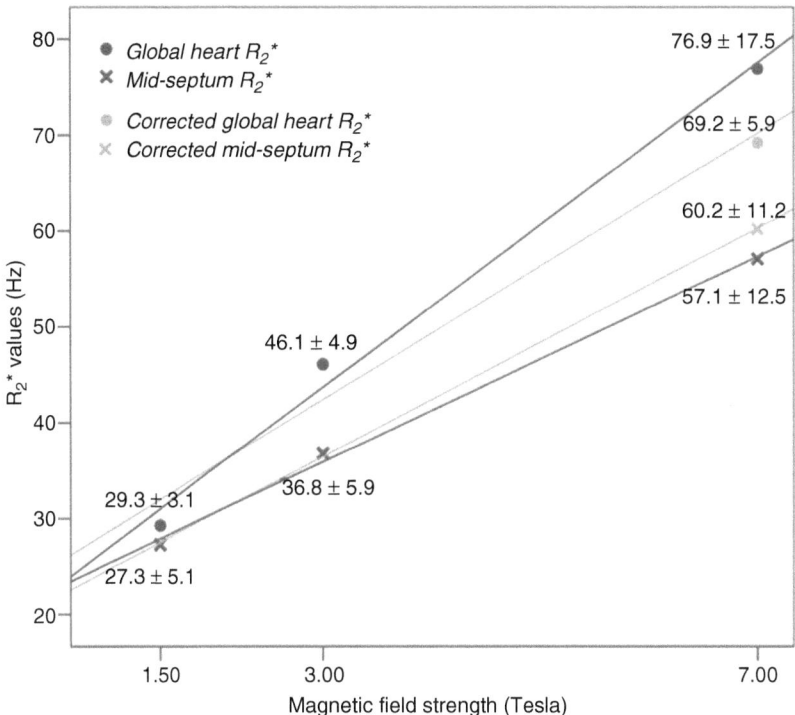

Fig. 15.2   Relation of ventricular septal (*crosses*) and global heart (*circles*) R_2^* and static magnetic field strength. The regression lines are indicated. Myocardial R_2^* increases linearly with magnetic field strength [31]

effects at 7.0 T may be useful to lower the detection level and to extend the dynamic range of the sensitivity for monitoring T_2^* changes. Transitioning to higher magnetic field strengths runs the boon that the in-phase inter-echo time governed by the fat-water phase shift is reduced from 4.8 ms (210 Hz) at 1.5 T to 1.02 ms (980 Hz) at 7.0 T. This enables rapid acquisition of multiple echoes with different T_2^* sensitization and facilitates high spatiotemporally resolved myocardial CINE T_2^* mapping of the human heart [32]. Taking advantage of this technique, T_2^* mapping at ultrahigh magnetic fields has been suggested as a means to probe myocardial physiology and to advance myocardial tissue characterization.

15.2   Challenges and Technical Solutions for Cardiac MRI at Ultrahigh Magnetic Fields

Magnetic resonance signal excitation and readout rely on the transmission and reception of radio-frequency (RF) waves at the resonance frequency of the nucleus under investigation. With the move to higher frequencies, the RF wavelength λ becomes shorter. At 7.0 T the resonance frequency of the proton is equal to about 298 MHz which corresponds to a wavelength in the myocardium of $\lambda_\text{myocardium} \approx 12$ cm.

This wavelength is relatively short compared to the size of the upper torso which poses a severe challenge for a uniform signal excitation. The heart being a deep-lying organ surrounded by the lung within the comparatively large volume of the thorax is a target region that is particularly susceptible to nonuniformities in the RF transmission field (B_1^+). These detrimental transmission field phenomena can cause shading, massive signal drop-off, or even signal void in the images and hence bear the potential to offset the benefits of UHF-CMR due to non-diagnostic image quality. These constraints are not a surprise as it might appear at the first glance, since transmission field nonuniformities, though somewhat reduced, remain significant in CMR at 3.0 T [33]. Further to the challenges imposed by B_1^+ nonuniformities, magnetohydrodynamic effects severely disturb the electrocardiogram (ECG) commonly used for cardiac triggering at clinical field strengths. This challenge evoked the need for practical solutions that support synchronization of MR data acquisition with cardiac activity at ultrahigh magnetic field strengths [34–36].

To address the practical obstacles of UHF-CMR, technical innovations in RF antenna design have been implemented [37–41]. Novel pulse sequences for transmission field mapping and shaping as well as innovative RF pulse designs along with multichannel RF transmission were developed to overcome the detrimental B_1^+ phenomena at 7.0 T with the goal to enable cardiac imaging [42–47]. Novel triggering techniques that are immune to electromagnetic fields have been established as an alternative to ECG. In this light this section surveys enabling technical innovations tailored for UHF-CMR.

15.2.1 Hardware for UHF-CMR

15.2.1.1 Enabling Radio-Frequency Antenna Technology

To overcome the obstacle of compromised transmission field uniformity in UHF-CMR, a dedicated effort has been invested in technical innovations in RF antenna design. Developments include (1) local transceiver (TX/RX) arrays and (2) multichannel transmission arrays in conjunction with multichannel local receive arrays.

A trend toward higher number of transmit and receive elements—up to 32—can be observed with the intention to increase the degrees of freedom for transmission field shaping [43], to enhance anatomic coverage, and to allow faster acquisition by exploiting local RF antenna sensitivity profiles [37–41, 48, 49]. Rigid, flexible, and modular RF antenna configurations have been reported. Recent developments exploited building blocks including stripline elements [42, 50–53], electric dipoles [41, 53–58], dielectric resonant antennas [59], and loop elements [37–40, 48, 49].

Loop element-based 7.0 T transceiver configurations optimized for CMR were reported for a 4-channel TX/RX [37] (Fig. 15.3a), an 8-channel TX/RX [38] (Fig. 15.3b), and a two-dimensional 16-channel TX/RX design [39] (Fig. 15.3c). A modular 32-channel TX/RX array [40] (Fig. 15.3d) extended the two-dimensional element layout.

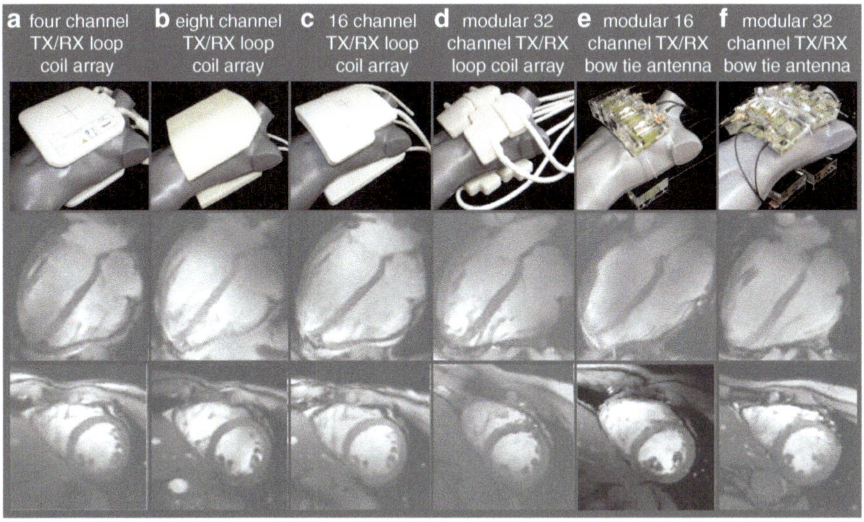

Fig. 15.3 Examples of multichannel transceiver arrays tailored for cardiac MR at 7.0 T. (*Left*) Photographs of cardiac optimized 7.0 T transceiver coil arrays including (*top to bottom*) a 4-channel, an 8-channel, a 16-channel, and a 32-channel loop array configuration together with an 8-channel and 16-channel bow tie antenna array. For all configurations, the RF elements are used for transmission and reception. (*Middle and right*) Four-chamber and short-axis views of the heart derived from 2D CINE FLASH acquisitions using the RF coil arrays shown on the left and spatial resolution (1.4 × 1.4. × 4.0) mm^3

A pioneering eight-element transverse electromagnetic field (TEM) transceiver array design was proposed where each stripline element was independently connected to a dedicated RF power amplifier [50]. Other stripline configurations exploit flexible designs [42, 51] or automated tuning with piezoelectric actuators [60].

Electric dipoles come with a linearly polarized current pattern, where RF energy is directed perpendicular to the dipole along the Poynting vector to the subject. The resulting symmetrical, uniform excitation field with good depth penetration [55] is beneficial for cardiac MR at 7.0 T and provided the momentum for explorations into electric dipole configurations [41, 54, 55, 57, 58]. Straight dipole elements [55] are detrimental for multichannel transceiver coil arrays due to size constraints. This limitation inspired the design of a short building block containing a bow tie-shaped λ/2 dipole antenna immersed in D$_2$O that shortens the effective antenna length. With this progress electric dipole configurations comprising 8 or 16 bow tie antenna building blocks tailored for UHF-CMR were implemented [41] (Fig. 15.3e, f).

15.2.1.2   Ancillary Devices for Cardiac Gating/Triggering

At ultrahigh magnetic fields, magnetohydrodynamic (MHD) effects severely disturb the electrocardiogram (ECG) [36, 61, 62] commonly used for cardiac triggering at clinical field strengths [63–65]. The MHD effect is pronounced during cardiac phases of systolic aortic flow resulting in distortions of the ECG's S-T segment

ECG triggering/gating

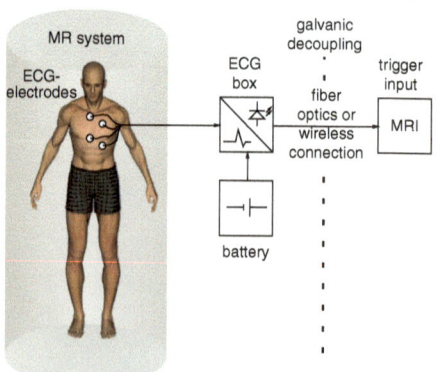 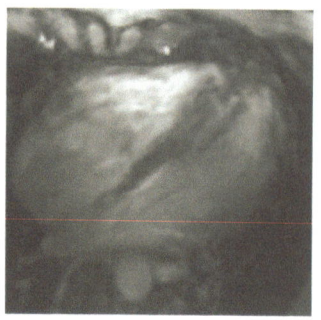

Acoustic triggering/gating

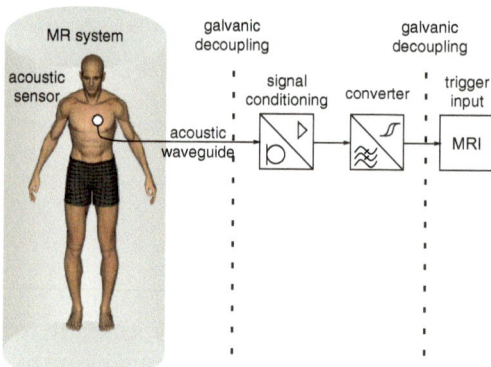 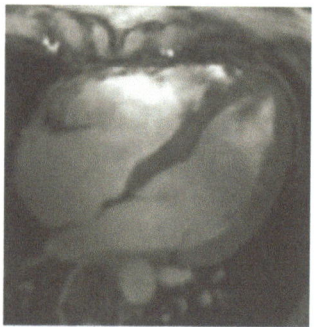

Fig. 15.4 Comparison of ECG and acoustic triggering or gating showing a basic scheme of the main technical principals (*left*) and four-chamber views of the heart acquired at 7.0 T (*right*) using ECG (*top*) and acoustic (*bottom*) gating. The ECG-gated image shows severe artifacts due to incorrect cardiac synchronization, while the acoustically triggered image reveals decent image quality with good blood myocardium contrast and clear delineation of subtle structures

already apparent in ECG traces acquired at 1.5 T [66]. Artifacts in the ECG trace might be misinterpreted as R-waves so that image quality can be impaired due to mis-detected cardiac activity, a limitation pronounced at ultrahigh fields [35, 42, 67]. Realizing this constraint, an MR stethoscope was proposed (Fig. 15.4) as an alternative to conventional ECG to support cardiac gating and triggering within strong magnetic fields [34–36]. An MR stethoscope employs acoustic instead of electrical signals. It builds on the first heart tone of the phonocardiogram which resembles the onset of the cardiac cycle. The phonocardiogram was reported to be immune to interferences with electromagnetic fields. It has been demonstrated that the phonocardiogram offers reliable trigger information in UHR-CMR as illustrated in Fig. 15.4 [34–36].

15.2.2 Imaging Methodology for T_2^* Mapping

15.2.2.1   RF Pulse Sequences

T_2^*-sensitized imaging and mapping can be achieved through gradient echo imaging with changing echo times. To decrease scan time and minimize motion artifacts, multi-echo techniques can be employed acquiring multiple echoes with each excitation instead of only one echo per repetition time *TR* (Fig. 15.5a, top). For myocardial T_2^* mapping, acquisitions are commonly performed in end-expiratory breath-hold conditions to avoid respiratory motion and to reduce macroscopic B_0 field fluctuations.

The ability to acquire multiple echoes within a single *TR* renders MEGRE a valuable candidate for T_2^* mapping. The echo times should be adapted to cover the T_2^* decay properly and to support the fitting and mapping algorithm. As the contributing fat and water signal are oscillating at different frequencies, mapping algorithms must either account for or compensate the varying signal intensity from fat and water. Acquiring T_2^*-weighted images at times when fat and water are equally contributing (in-phase) is the simplest approach to achieve this goal. At 7.0 T, fat and water are in phase for echo times being a multiple of 1.02 ms due to the chemical shift between fat and water of approximately 980 Hz. Acquisition of echoes at every in-phase time point within a single TR is challenging due to gradient amplitude and rise time limitations, especially when large acquisition matrix sizes are used for imaging. As an alternative, the acquisition of echoes can be interleaved and distributed to multiple excitations (Fig. 15.5a, middle). This approach permits low inter-echo spacing even at high spatial resolution but results in longer acquisition times since more than one TR is required to acquire a full T_2^* decay series. While T_2^* mapping at clinical field strengths is limited to single cardiac phase acquisitions, CINE T_2^* mapping covering the whole cardiac cycle is feasible at UHF [32]. This is mainly due to two reasons. First, thanks to transversal relaxation time shortening at ultra-high magnetic fields, *TE* can be limited to a range of *TE* = 0 ms to *TE* = 20 ms to properly sample the T_2^* decay. This approach is beneficial for reducing the duration of the gradient echo trains versus lower magnetic field strengths. Second, the reduced in-phase echo spacing permits acquisition of a sufficient number of echoes needed to cover the signal decay and to provide an appropriate number of data points for signal fitting. Interleaving of echo times can be used to ease gradient constraints while distributing the acquisition to multiple breath-holds limiting breath-hold duration for each acquisition (Fig. 15.5a, bottom). All described acquisition strategies are capable of producing T_2^* maps of similar fidelity as illustrated in Fig. 15.5b for a homogenous MR phantom resembling the relaxation properties of human myocardium. To reduce the effect of macroscopic magnetic field contributions on spin dephasing and T_2^*, a small voxel size is preferable. Of course this preference has to be balanced with SNR requirements for accurate mapping which can be challenging particularly at lower magnetic field strengths. Figure 15.5b compares the effect of slice thickness on T_2^*. While maps acquired with slice thicknesses of 8 or 6 mm show intravoxel dephasing and T_2^* decrease pronounced at the phantom interfaces, this effect is reduced if the slice thickness is reduced to 4 or 2.5 mm resulting in a

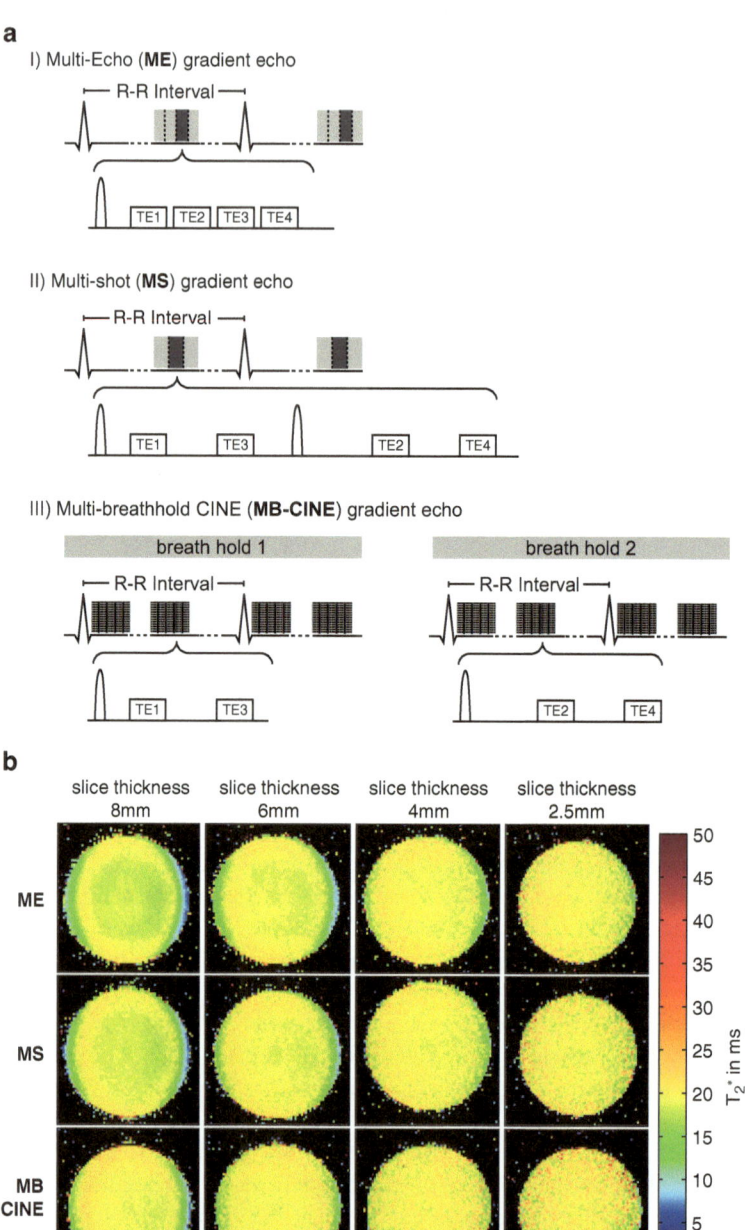

Fig. 15.5 Acquisition schemes used for T_2^*-weighted imaging/mapping and corresponding T_2^* maps. (**a**) (1) Multi-echo gradient echo (MEGRE) acquisition, (2) multi shot (MS) interleaved multi gradient echo acquisition, (3) multi breath-hold CINE (MB-CINE) interleaved multi gradient echo acquisition. (**b**) Comparison of T_2^* maps derived from a homogenous phantom resembling myocardial tissue acquired using the three different approaches and slice thicknesses from 8 to 2.5 mm. All acquisition strategies provide similar T_2^* maps. Through-plane dephasing is reduced for lower slice thickness. Adapted from [32]

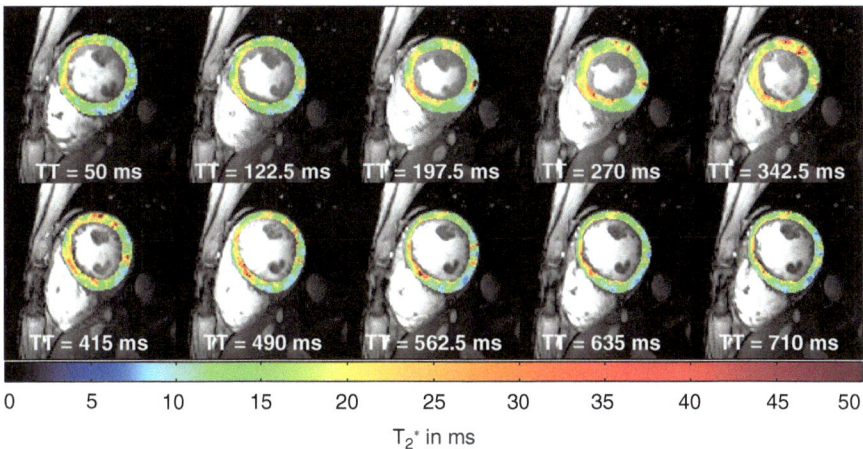

Fig. 15.6 Example of cardiac phase resolved myocardial T_2^* mapping of a short-axis view of a healthy volunteer (10 out of 20 phases shown). Spatial resolution (1.1 × 1.1 × 4.0) mm³. Temporal resolution = 37 ms. T_2^* variations can be observed across the cardiac cycle. TT indicates the time since the trigger provided by the first heart tone

more uniform T_2^* map. Employing the described multi-breath-hold CINE technique at 7.0 T, CINE T_2^* mapping with more than 20 cardiac phases is feasible which allows monitoring of myocardial T_2^* across the cardiac cycle (Fig. 15.6).

In contrast to gradient echo imaging, rapid acquisition with relaxation enhancement (RARE) imaging is largely immune to B_0 inhomogeneities, provides images free of distortion due to the use of RF refocused echoes, and inherently suppresses blood signal. The applicability of cardiac RARE at 3.0 T [68] provided momentum for RARE-based myocardial T_2^* mapping at 7.0 T [69]. T_2^* weighting in RARE is accomplished by inserting an evolution time τ after the excitation RF pulse whereby an additional phase is accrued reflecting the T_2^* effect [70] (Fig. 15.7a). This approach runs the benefit that T_2^* can be adjusted from zero upward to maximize functional or tissue contrast [68]. Inserting the evolution time τ results in violation of the Carr–Purcell–Meiboom–Gill (CPMG) condition [71, 72] and requires measures to avoid destructive interferences between odd and even echo groups that constitute the signal in RARE imaging. Displaced RARE has been proposed [70] and avoids image artifacts by discarding one of both echo groups but comes with a loss in SNR of factor two. Alternatively a split-echo variant [73] can be employed where both echo groups are used, reconstructed separately, and superimposed to restore the full signal intensity. Figure 15.7b shows a series of T_2^*-weighted images derived from RARE using evolution times τ ranging from 0–10 ms. For comparison MEGRE images covering the same echo time range are presented in Fig. 15.7b. The geometric integrity of the RARE images is maintained over the range of T_2^* weighting. Myocardial T_2^* mapping was feasible for the RARE-based technique as demonstrated in Fig. 15.7c. In comparison, MEGRE imaging exhibited less myocardium to blood contrast since signal contributions from the blood pool were not suppressed. As a consequence, the delineation of the myocardium in the corresponding

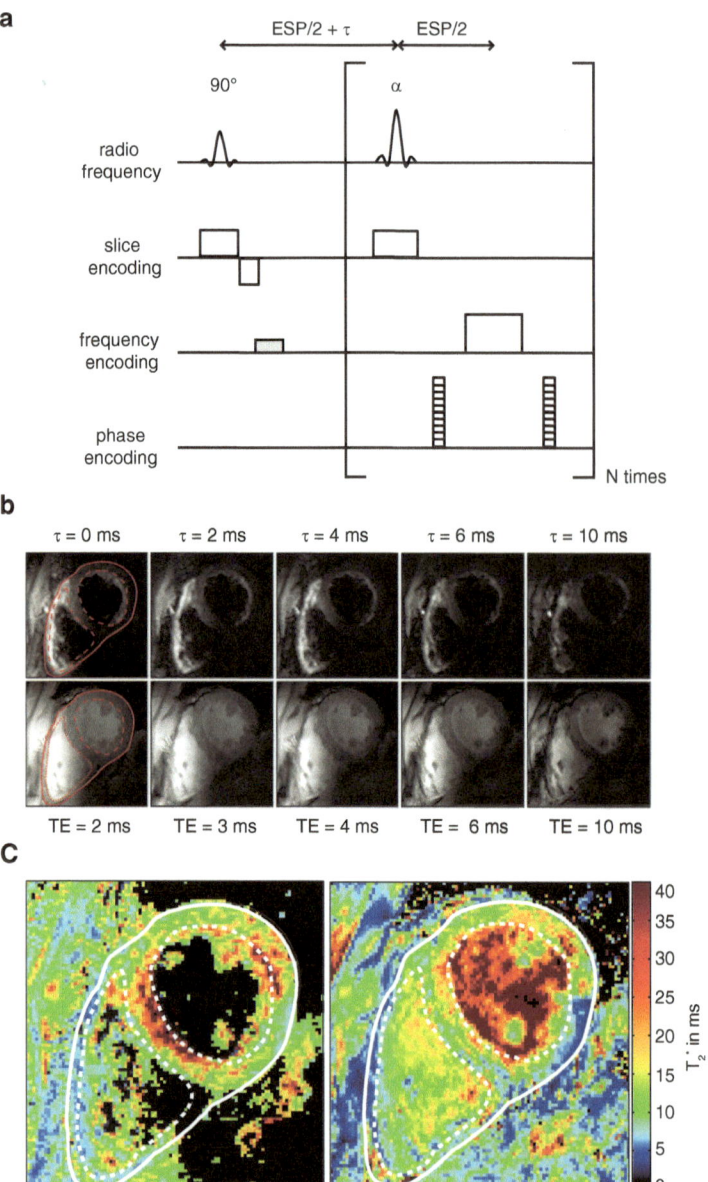

Fig. 15.7 RARE-based myocardial T_2^* mapping at 7.0 T. (**a**) Basic pulse sequence diagram of T_2^*-weighted split-echo RARE. T_2^* weighting is introduced into RARE by adding an evolution time τ after the excitation RF pulse. The dephasing gradient in frequency encoding direction is imbalanced (marked in *gray*) to avoid destructive interferences between odd and even echo groups. (**b**) T_2^*-weighted split-echo RARE images of a short-axis view employing evolution times ranging from $\tau = 0$ to $\tau = 10$ ms (*top*). MEGRE images with *TE* ranging from 2 to 10 ms (*bottom*). For improved visualization, the images do not exhibit identical windowing over the range of increasing susceptibility weighting. The myocardium was delineated and the contour is shown in the images with minimal τ/TE. (**c**) T_2^* maps derived from data shown in (**b**) are depicted for the RARE (*left*) and the GRE (*right*) approach. The contours defined in the images with minimal τ/TE were copied to the T_2^* maps for better delineation of the myocardium

T_2^* map (Fig. 15.7c) is more challenging compared to the RARE-based T_2^* map. The T_2^* map derived from RARE imaging shows an averaged effective transversal relaxation time which accords well with previously reported results based on MEGRE [32]. These preliminary results demonstrate that the concerns of RF power deposition and RF nonuniformity of RARE imaging can be offset at 7.0 T to enable myocardial T_2^* mapping using split-echo RARE. This approach holds the potential to provide a valuable alternative for T_2^* mapping free of geometric distortion and viable blood myocardium contrast.

15.2.2.2   Assessment and Adjustment of Main Magnetic Field Homogeneity: B_0 Mapping and Shimming

T_2^*-sensitized images are commonly obtained through gradient echo (GRE) techniques, where the complex MR signal at a location r is given by

$$S(r,t) = \hat{S}(t) \cdot e^{-i\phi(r,t)}, \quad \hat{S}(t) \propto S_0 \cdot e^{-\frac{t}{T_2^*}} \tag{15.5}$$

where $\hat{S}(t)$ is the magnitude signal, i is the imaginary unit, and ϕ is the phase signal, which has both time-independent and time-dependent components. The phase ϕ can be written as a function of spatial location and time [74]:

$$\phi(r,t) = \phi_0 - \gamma \cdot \Delta B(r) \cdot t \tag{15.6}$$

with γ the gyromagnetic constant of the nucleus (for $^1H\gamma = 2.675 \times 10^8 \mathrm{rad/s/T}$) and $\Delta B(r)$ the local magnetic field deviations (compare Eq. (2.64) in Chap. 2, Eq. (2) in Chaps. 8 and 9). While the constant component ϕ_0 is a receiver phase offset arising from several artefactual factors, the time-dependent component is dominated by the deviation from the static magnetic field and linearly evolving over the time [75]. Thus, assuming there are no other external sources of dephasing (e.g. motion or flow), the phase ϕ serves as a direct measure of deviations from main B_0.

It should also be noted that the net signal and net phase within a voxel are the sum of an ensemble of spins, each spin having its own phase. As a consequence, the signal reaches its maximum when all spins are in phase. Because of local magnetic field differences, spins within a voxel dephase over time resulting in a loss of phase coherence and an overall loss of signal. T_2^*-weighted images are particularly affected by this effect since T_2^* is given by [27, 74]

$$\frac{1}{T_2^*} = \frac{1}{T_2} + \frac{1}{T_2'} \cong \frac{1}{T_2} + \gamma \left|\Delta B\right| \tag{15.7}$$

The approximation in Eq. (15.7) is valid if a linear B_0 gradient within a voxel is assumed, which is justified for typical cardiac voxel sizes at 7.0 T using an in-plane spatial resolution of about 1 mm and a slice thickness of 2–4 mm. This dependency highlights the need to monitor and possibly compensate B_0 inhomogeneities when applying GRE-based techniques.

By acquiring a series of GRE images at multiple echo times (*TE*) and by employing a phase difference method to eliminate the initial phase offset, the local magnetic field variations can be calculated at each voxel by making use of Eq. (15.6):

$$\Delta B(r) = \frac{\phi(r, TE_2) - \phi(r, TE_1)}{\gamma(TE_2 - TE_1)} \quad (15.8)$$

where $\Delta B(r)$ is given in Tesla. It is also very common to find local magnetic field variations represented in Hz by means of off-center frequency maps: $\Delta B_{Hz}(r) = (\phi(r, TE_2) - \phi(r, TE_1))/(2\pi(TE_2 - TE_1))$.

The sources of magnetic field perturbations may be either macroscopic (scale of about the voxel size) or microscopic (well below the size of a voxel).While the macroscopic field inhomogeneities are commonly associated with poor magnet B_0 homogeneity and strong susceptibility transitions producing image distortion and miscalculation of the real tissue T_2^*, the microscopic magnetic field variations can, e.g., be caused by physiologic changes and may thus provide information about tissue function at a mesoscopic level above that of the cell.

B_0 shimming in MRI describes the process of applying additional magnetic field gradients to improve macroscopic B_0 homogeneity. Active shimming employs generation of compensatory currents driving dedicated gradient coils, thereby creating compensatory magnetic fields up to the fifth-order spherical harmonics [76]. Various shimming modes are implemented on commercial MR scanners, and their selection has a major impact on the ability to homogenize B_0. Shimming options include fixed shim current settings (*tune-up*) stored in a configuration file or shimming modes based on an individual B_0 map (which for cardiac imaging may be acquired with or without cardiac triggering). The possibility of adjusting a shim volume to a target anatomy is also an important B_0 shimming feature. Ideally the target volume should cover a small region of interest like the heart, to allow good field homogeneity even when a limited order of shim coils is available. Figure 15.8 shows a comparison between the applications of a volume shim (Fig. 15.8a), which is focused only on the heart, and a global shim (Fig. 15.8b), which includes the entire field of view for a healthy subject at 7.0 T. Volume-selective shimming was found to lead to a significant increase in macroscopic B_0 homogeneity versus global B_0 shimming. The position of the adjustment volume also plays an important role for the shimming quality; e.g., adjusting the volume tightly to the heart using long- and short-axis views appeared to provide better B_0 homogeneity when compared to planning the volume based on conventional orthogonal (coronal, axial, sagittal) views of the heart. Depending on the system manufacturer, the available geometries of the shim volume may vary. Possible options include cuboids, cylinders, or ellipsoids.

Due to increased susceptibility effects, magnetic field inhomogeneities are pronounced at higher magnetic field strengths [31]. However, employing dedicated shimming approaches, a mean peak-to-peak off-resonance frequency across the heart of 80 Hz [32] was reported at 7.0 T (Fig. 15.8d) and is comparable to the 71 Hz field dispersion across the heart reported for 1.5 T [77]. Yet, peak-to-peak off-resonance is a suboptimal metric for the assessment of magnetic field (non)

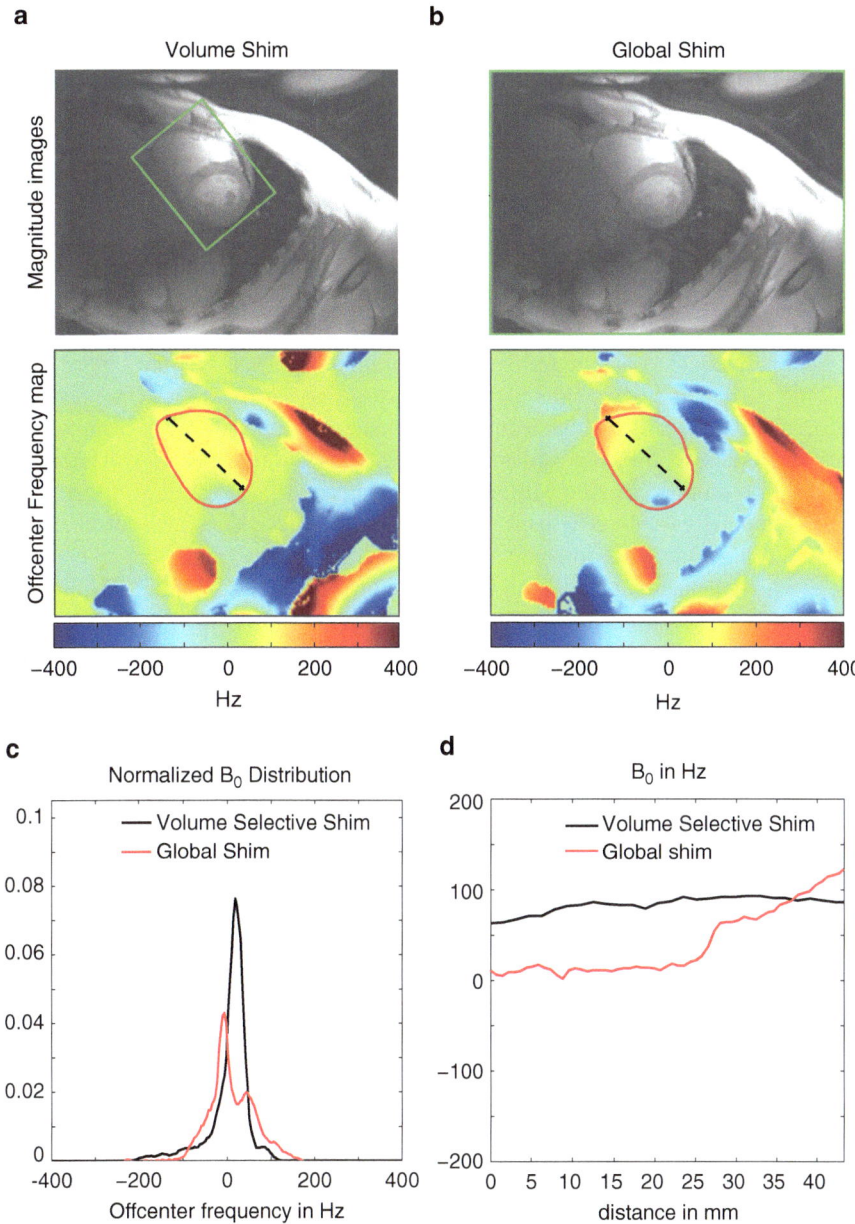

Fig. 15.8 Comparison between volume (**a**) and global (**b**) shim in the heart of a healthy volunteer at 7.0 T. (*Top*) Placement of the adjustment volume in a magnitude image. (*Middle*) B_0 maps with the region of interest (*red*) and a profile across the heart through the septum (*dashed black line*) overlaid. (*Bottom*) Plot of the histogram detailing the distribution of B_0 in the ROI outlined in the B_0 maps (**c**) and plots along the profile in the B_0 map (**d**) of volume selective (*black*) and global (*red*) shim. A clear improvement of field inhomogeneity can be observed after volume selective shimming indicated by a narrowing of the histogram and a more flat B_0 profile

uniformity. The full width half maximum (FWHM) of the off-resonance frequency (Fig. 15.8c) provides a viable alternative.

Obtaining a macroscopic B_0 homogeneity across the heart at 7.0 T which is competitive with that obtained at lower magnetic field strengths provides encouragement to pursue susceptibility-based myocardial T_2^* mapping at ultrahigh fields. Notwithstanding this encouragement temporally resolved B_0 assessment across the cardiac cycle, and detailing its implications on T_2^* is required for a meaningful interpretation of dynamic T_2^*-weighted acquisitions. Shah et al. reported temporal variation of the main magnetic field B_0 to be negligible across the cardiac cycle at 1.5 T [78], but B_0 inhomogeneities are increased at UHF [31]. Moreover, T_2^*-weighted contrast is determined by magnetic field gradients rather than absolute magnetic field strength; thus, it is essential to investigate the change of these gradients over the cardiac cycle. For this purpose, a study conducted at 7.0 T assessed dynamic intravoxel macroscopic B_0 gradients together with high temporal and spatial resolution T_2^* maps in the heart of healthy volunteers over the cardiac cycle [79, 80]. T_2^*-weighted series of short-axis views were acquired using a MEGRE CINE approach (Fig. 15.9a, top) for T_2^* mapping. Temporally resolved B_0 maps of short-axis views of the heart were calculated and filtered using a Gaussian low-pass to reduce high-frequency (microscopic) contributions while maintaining macroscopic B_0 variations (Fig. 15.9a, middle). The in-plane B_0 gradient maps were calculated for each voxel as the norm of the gradients in x (G_x) and y (G_y) directions. Analysis was focused on mid-ventricular septal segments including segments 8 and 9 (according to the AHA convention, [81]) which are less prone to susceptibility artifacts than other myocardial segments [31] and hence are commonly assessed in clinical routine. Through-plane B_0 gradients (G_z) were approximated by the slope of a line profile placed in the septum of B_0 maps of a four-chamber view of the heart (perpendicular to the in-plane short-axis view) at each phase of the cardiac cycle.

The septal in-plane gradients were found to be significantly larger compared to through-plane gradients within a voxel (with a mean in-plane field dispersion of 2.5 ± 0.2 Hz/mm against a mean through-plane field dispersion of 0.4 ± 0.1 Hz/mm) [82]. This might be unexpected for an in-plane resolution almost four times higher than the through-plane slice thickness. Yet, these results are plausible, considering the pronounced anisotropic anatomy of the heart which is changing only a little along the long axis of the heart but significantly in the short-axis plane of the heart.

In-plane and through-plane macroscopic gradient information were combined in a total intravoxel B_0 gradient map (Fig. 15.9a, bottom) by calculating the square root sum of square at each voxel (G_{xyz}):

$$G_{xyz} = \sqrt{G_x^2 + G_y^2 + G_z^2} \tag{15.9}$$

For simplification, the through-plane B_0 gradients were approximated by the slope of a line profile of a single end-diastolic four-chamber view, which was found to be reasonable since total gradients were dominated by their in-plane contribution and results were similar compared to using temporally resolved through-plane gradients. In order to evaluate how these B_0 gradients affect T_2^* measurements, the B_0 gradient-induced ΔT_2^* was estimated based on Eq. (15.7):

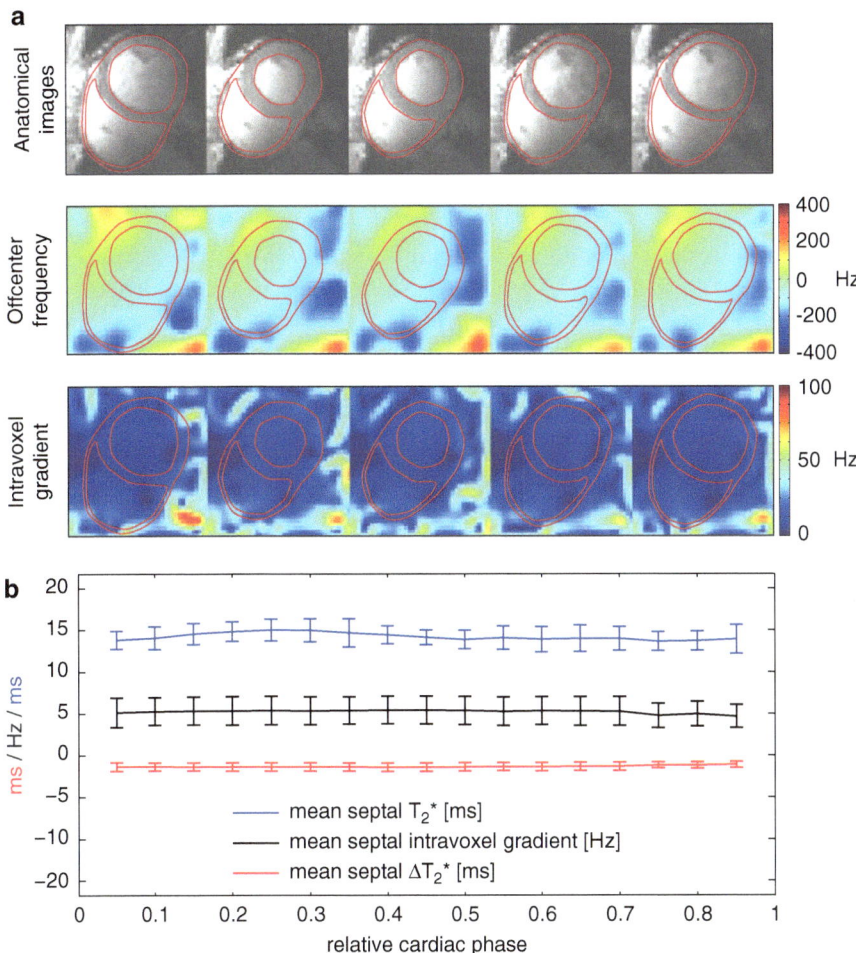

Fig. 15.9 Spatial and temporal variation of macroscopic intravoxel B_0 gradients in the in vivo human heart. (**a**) Magnitude images of a short-axis view (*top*), in-plane macroscopic B_0 maps (*middle*), and intravoxel macroscopic B_0 gradient maps (*bottom*) over the cardiac cycle. (**b**) Mean septal T_2* (*blue*), intravoxel B_0 gradient (*black*), and estimated ΔT_2* (*red*) over the cardiac cycle, averaged among a group of healthy volunteers. Temporal macroscopic magnetic field changes over the cardiac cycle are minor regarding their effects on T_2*

$$\Delta \frac{1}{T_2^*} = \frac{1}{{}^2T_2^*} - \frac{1}{{}^1T_2^*} = \left(\frac{1}{{}^2T_2} - \frac{1}{{}^1T_2} \right) + \gamma \left({}^2|\Delta B| - {}^1|\Delta B| \right) \qquad (15.10)$$

where ${}^2T_2^*$ is the measured T_2^* including macroscopic susceptibility effects, ${}^1T_2^*$ is the hypothetical real T_2^* free from these B_0 effects, and $\left({}^2|\Delta B| - {}^1|\Delta B| \right)^1$ is the

---

[1] Where ${}^1|\Delta B|$ is the hypothetic magnetic field gradient without macroscopic contributions and

intravoxel macroscopic magnetic field gradient. If T_2 is assumed to be constant for each cardiac phase $\left(\Delta \dfrac{1}{T_2} = 0 \right)$, Eq. (15.10) can be simplified to

$$\Delta T_2^* = {}^2 T_2^* \cdot \left(1 - \frac{1}{1 - {}^2 T_2^* \cdot \gamma \left({}^2 |\Delta B| - {}^1 |\Delta B| \right)} \right) \tag{15.11}$$

where ΔT_2^* is the portion of ${}^2 T_2^*$ induced by ${}^2 |\Delta B| - {}^1 |\Delta B|$. The calculated septal mean T_2^* and B_0 gradient plots were used to estimate mean gradient-induced ΔT_2^* in the septum over the cardiac cycle. Figure 15.9b shows the plot of mean septal T_2^*, intravoxel B_0 gradients, and estimated gradient-induced ΔT_2^* over the cardiac cycle averaged over a group of healthy subjects. The mean septal T_2^* per cardiac phase was found to vary over the cardiac cycle in a range of approximately 23% of the total mean T_2^* for all phases. Yet, the temporal range of mean ΔT_2^* induced by the calculated intravoxel macroscopic B_0 gradients represented only a 5% change of total mean T_2^*. The remaining 18% were suggested to reflect microscopic B_0 gradient changes (potentially caused by physiological events) rather than macroscopic field inhomogeneities [80, 82]. In conclusion, it was shown that, after applying dedicated shimming, macroscopic B_0 gradients in the intraventricular septum at 7.0 T are minor regarding their effects on T_2^*. This provides support for temporally resolved susceptibility sensitized CMR at ultrahigh magnetic field strengths.

15.2.2.3   Data Post Processing

The relaxation time T_2^* can be assessed by fitting a series of gradient echo images with increasing T_2^* weighting, i.e., increasing echo times describing the T_2^* decay with an exponential function (see Eq. (15.3)). One way of calculating such a fit avoiding nonlinear fitting procedures is to take the natural logarithm of the acquired signal intensities and apply a least squares linear fit to the resulting data. This procedure is fast and produces the best solution in a least squares sense. Further to this, nonlinear fitting approaches which can be applied directly to the measured data are available which use nonlinear models in conjunction with optimization algorithms like Levenberg–Marquardt or Simplex [83]. No matter what kind of fitting procedure is applied, care should be taken, not to include voxels with intensity being at the noise level and hence potentially deteriorating the fit quality. Such voxels should consequently be excluded from the fit, which is referred to as truncation.

While most commercial MR systems support exponential fitting algorithms as part of the (black box) system software, using customized fitting routines is very much beneficial for cardiac research. First, it is often unclear what model or fitting approach is used by commercial software and how good the fit quality was, i.e., how well the fit describes the measured data. Measures like the coefficient of determination R^2 or the standard deviation of the T_2^* fit [84] should be used to evaluate the reliability of the results. Taking fitting results for granted without considering fit quality may lead to wrong results and eventually wrong conclusions. Second, individual fitting procedures offer the freedom to select the most appropriate fit model,

---

[2]$|\Delta B|$ the magnetic field gradient including macroscopic inhomogeneities

optimization approach, truncation threshold, etc. for the given application. Depending on the kind of application, it may make sense to use a bi-exponential model instead of a mono-exponential approach.

Measurement of relaxation times offers the advantage over qualitative signal intensity images of providing quantitative, comparable results. However, effects like B_0 inhomogeneities or noise can impair assessment of T_2^* and deteriorate fitting results. Dedicated B_0 shimming approaches help to reduce the impact of macroscopic magnetic field inhomogeneities. A reduction in voxel size can further reduce the impact of B_0 gradients on T_2^* (Fig. 15.5b). Yet, reducing voxel size is always accompanied by a loss in SNR, which in turn can result in poor fit quality. While this reduction in SNR might be counteracted by signal averaging and increasing acquisition times in static acquisition situations (e.g., MRI of the brain), it is not feasible in cardiac applications, where acquisitions need to be synchronized with the cardiac cycle and are often performed under breath-hold conditions constraining the viable acquisition window to few seconds. This issue is even further pronounced in patients suffering from cardiac diseases and for acquisitions at high spatial and temporal resolution such as dynamic T_2^* mapping.

Image de-noising provides a viable alternative to address this constraint and to allow the use of small voxel sizes while enabling acceptable fit quality. Powerful de-noising approaches like nonlocal means filtering [85] are available that can greatly improve SNR with minimal loss of information in contrast to simple de-noising approaches such as Gaussian filtering which are accompanied by loss of detail.

While direct de-noising of the T_2^* maps can lead to wrong results, filtering of the magnitude images prior to fitting represents a robust way of improving fit quality and has been shown to increase fitting accuracy and precision [80, 86, 87]. Figure 15.10 shows an example how T_2^* mapping can benefit from noise filtering. In this example a 30% reduction of T_2^* fits standard deviation was achieved. If fitting results are affected by low SNR, de-noising approaches should be considered.

15.3   In Vivo Insights from Myocardial T_2^* Mapping in Humans at Ultrahigh Fields

Making use of the technological and methodological developments outlined above, MRI at ultrahigh magnetic fields permits for the first time the in vivo assessment of temporal myocardial T_2^* changes across the cardiac cycle. The following section gives first insights of using this technique in healthy volunteers and patients suffering from cardiovascular diseases and provides interpretation of the results.

15.3.1 Cardiac Phase Resolved Myocardial T_2^* Mapping in Healthy Subjects and Its Physiological Interpretation

The first pioneering study investigating the temporal changes of myocardial T_2^* in healthy subjects at 7.0 T was published in 2016 [80]. The authors investigated the time course of myocardial T_2^* across the cardiac cycle along with basic morphological parameters of the heart such as ventricular septal wall thickness and inner left ventricular radius as potential confounders of T_2^*. The results showed that T_2^* in the

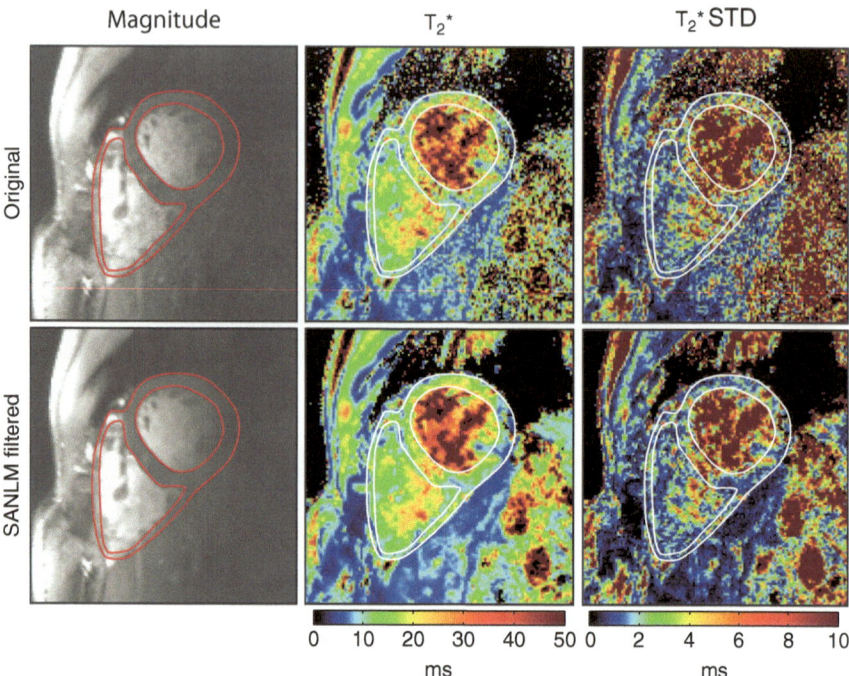

Magnitude T_2^* T_2^* STD

Original

SANLM filtered

0 10 20 30 40 50 0   2   4   6   8   10
        ms   ms

Fig. 15.10 Impact of spatially adaptive nonlocal means (SANLM) noise filtering on T_2^* maps of a mid-ventricular short-axis view of the heart. (*Left*) Original and SANLM filtered signal magnitude images of the first echo ($TE = 2.04$ ms) of a series of multi-echo gradient echo images. (*Center*) Corresponding T_2^* maps in ms. (*Right*) T_2^* standard deviation maps illustrating the precision of the T_2^* maps. By applying the noise filter, an average decrease in T_2^* standard deviation of about 30% in the left ventricular myocardium was achieved

ventricular septum changes periodically across the cardiac cycle, increasing in systole and decreasing in diastole with a mean systole to diastole ratio of about 1.1 (Fig. 15.11a). Despite the numerous influential factors such as blood volume, hematocrit, etc., myocardial T_2^* is still commonly regarded as a surrogate for tissue oxygenation. Interpreting T_2^* to reflect tissue oxygenation, the observed systolic T_2^* increase would imply an increase in left myocardial oxygenation during systole, which is contrary to physiological knowledge. Instead, changes in myocardial blood volume fraction induced by variations in blood pressure and resulting myocardial wall stress are assumed to be responsible for the observed T_2^* changes [80]. The contraction of the myocardium compresses the intramyocardial vasculature such that inflow of arterial blood ceases while deoxygenated blood is squeezed out of the myocardium toward the venous coronary sinus [88–90] (Fig. 15.11b). The major decrease in blood volume fraction of the myocardium reduces the amount of deoxygenated blood per tissue volume, thereby increasing—instead of lowering—T_2^* during systole.

Besides this finding, a significant correlation of ventricular septal wall thickness and T_2^* was reported. The hypothesis, that the observed T_2^* changes could be induced by changes of the macroscopic B_0 field variations induced by morphologic changes

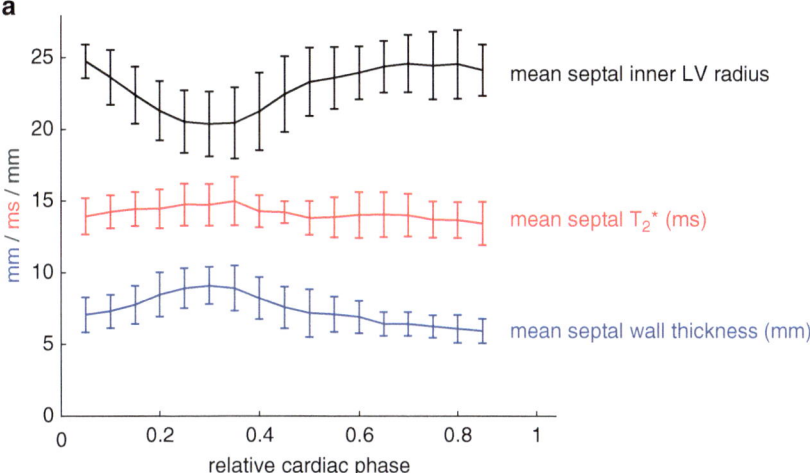

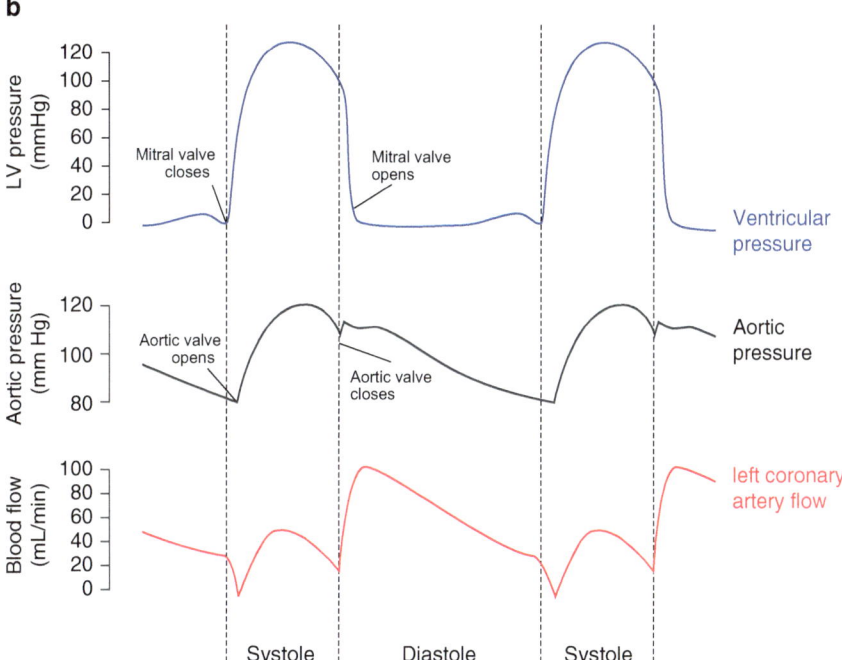

Fig. 15.11 Results of temporally resolve myocardial T_2* mapping in healthy volunteers at 7.0 T. (**a**) Course of mean septal wall thickness, mean LV inner radius, and mean septal T_2* plotted over the cardiac cycle averaged for all volunteers. Error bars indicate SD. Myocardial T_2* changes periodically across the cardiac cycle increasing in systole and decreasing in diastole. (**b**) The periodic T_2* changes can be explained by cyclic variations of myocardial blood volume fraction related to differences in blood pressure and myocardial wall stress. The massive pressure increase in the left ventricle in the beginning of systole results in high myocardial wall stress compressing the myocardial vessels leading to a reduced blood supply to the myocardium, while the blood contained in the tissue is squeezed out. The resulting reduced myocardial blood volume fraction explains the systolic T_2* increase, which cannot be explained by increased oxygenation

occurring between systole and diastole, was investigated but not confirmed. Both in silico magnetostatic simulations and in vivo temporally resolved B_0 mapping showed negligible impact on the macroscopic B_0 field in the ventricular septum and hence T_2^* [80].

15.3.2 Myocardial T_2^* Mapping in Patients with Cardiovascular Diseases

Besides the application of myocardial T_2^* mapping at ultrahigh magnetic fields in healthy volunteers, first investigations were carried out to explore the potential of the technique to distinguish between healthy and pathologic myocardium. These early UHF-CMR studies focused on T_2^* mapping in patients with hypertrophic cardiomyopathy (HCM). HCM is the most common inherited heart defect affecting about 0.5% in the general population. The disease is characterized by an increase in myocardial wall thickness related to myocyte hypertrophy, microstructural changes like myocardial disarray, fibrosis, and microvascular dysfunction. Based on these structural and physiologic changes, a difference in myocardial T_2^* was expected for HCM patients compared to age- and gender-matched healthy controls. This was investigated using high spatiotemporal resolution T_2^* mapping at 7.0 T (Fig. 15.12). It was shown that septal T_2^* is significantly increased in HCM with mean septal T_2^* being (17.5 ± 1.4) ms in a cohort of HCM patients compared to (13.7 ± 1.1) ms in a healthy age- and body mass index-matched group of healthy controls. Patient and control group could clearly be distinguished by septal T_2^* septal wall thickness (Fig. 15.12b). While temporal variations of myocardial T_2^* have been attributed to changes in myocardial blood volume fraction related to left ventricular blood pressure and resulting wall stress rather than changes in tissue oxygenation as described in the previous section [80], two main factors are assumed to cause the observed overall T_2^* increase in HCM. Improved tissue oxygenation in HCM is unlikely. Instead, T_2 has been reported to be elevated in HCM [91] related to inflammation and edema which would also increase T_2^* as outlined in Eq. (15.1). Also, reduced myocardial perfusion and ischemia are common in HCM [92], effectively reducing the tissue blood volume fraction resulting in a T_2^* increase as suggested by Eq. (15.4). These conditions are also associated with a higher risk for a poor outcome in those patients [93]. With this in mind, it is fair to conclude that myocardial T_2^* mapping could be beneficial for a better understanding of cardiac (patho)physiology in vivo with the ultimate goal to support risk stratification in HCM.

15.4   Conclusion and Outlook

The progress in myocardial T_2^* mapping at ultrahigh magnetic fields creates excitement. Yet, the clinical benefit remains to be carefully investigated. This includes putting further weight behind the solution of unsolved problems and unmet needs

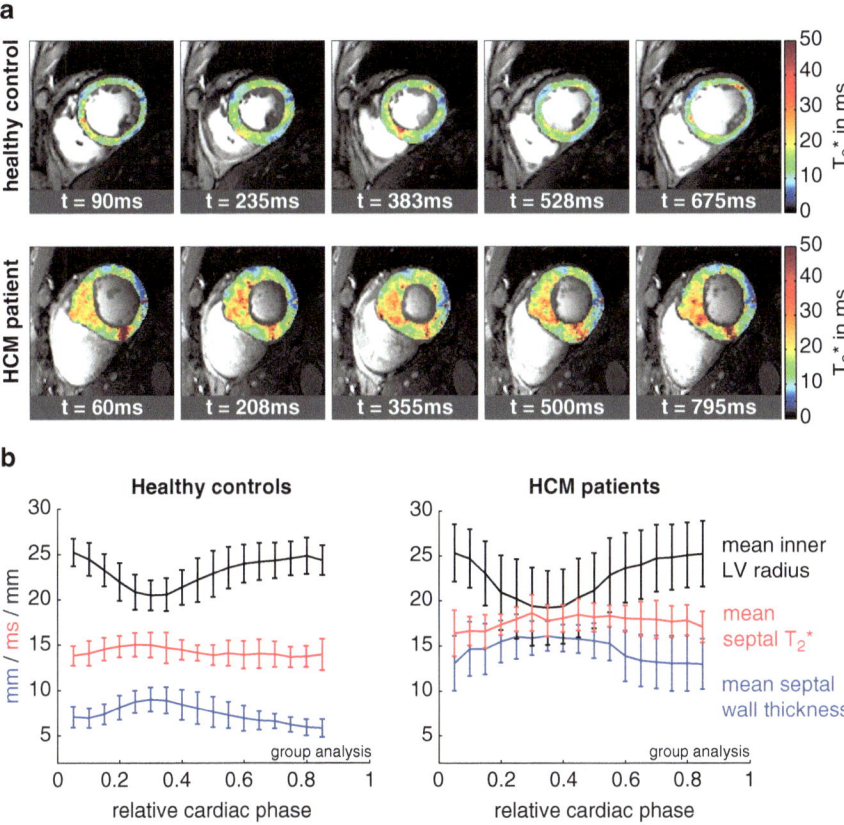

Fig. 15.12 Cardiac phase resolved myocardial T_2^* mapping in healthy volunteers and HCM patients. (**a**) Temporally resolved myocardial T_2^* maps of a short-axis view of a healthy control (*top*) and an HCM patient (*bottom*), (5 out of 20 phases shown) of a healthy volunteer (10 out of 20 phases shown). Spatial resolution $(1.1 \times 1.1 \times 4.0)$ mm^3. T_2^* variations can be observed across the cardiac cycle. (**b**) Course of mean septal wall thickness, inner LV radius, and mean septal T_2^* plotted over the cardiac cycle averaged for groups of healthy controls (*left*) and HCM patients (*right*). Relative cardiac phase 0 indicates the beginning of the cardiac cycle. T_2^* changes periodically over the cardiac cycle increasing in systole and decreasing in diastole in both healthy controls and HCM patients but is significantly higher in HCM patients

standing in the way en route to broader clinical studies. For example, the relatively long breath-hold times required for the acquisition of high spatiotemporal T_2^* maps constitute a challenge particularly in cardiac patients. Free breathing acquisition techniques could offset this constraint, permit broader application, and allow for full 3D heart coverage. This would also help to further investigate the effect of through-plane motion. Techniques like simultaneous multi-slice excitation can be used to reduce scan times, while multichannel transmit systems can be employed to balance excitation field homogeneity and RF power deposition constraints [46, 47].

Based on the multifaceted contributions of physiological parameters on T_2^*, research will not stop at just mapping myocardial T_2^*. Tailored acquisition schemes, data post processing, analysis, and interpretation will allow exploiting the breadth of information encoded into T_2^*. For example, it is conceivable that high-fidelity spatial resolution T_2^* mapping facilitated by ultrahigh magnetic field strengths might be beneficial to gain a better insight into the myocardial microstructure in vivo with the ultimate goal to visualize myocardial fibers or to examine helical angulation of myocardial fibers using T_2^* mapping, since the susceptibility effects depend on the tilt angle between blood-filled capillaries and the external magnetic field [94]. Myocardial fiber tracking using T_2^* mapping holds the promise to be less sensitive to bulk motion than diffusion-weighted MR of the myocardium [95, 96]. The results reported in this book chapter also suggest that the increased susceptibility contrast available at 7.0 T could be exploited to quantitatively study iron accumulations in in the heart with high sensitivity and temporal and spatial resolution superior to what can be achieved at 1.5 and 3.0 T. This requires the determination of norm values for healthy myocardial T_2^* at 7.0 T as a mandatory precursor to broader clinical studies.

The ability to probe for changes in myocardial tissue oxygenation using T_2^*-sensitized imaging/mapping offers the potential to address some of the spatial and temporal resolution constraints of conventional first-pass perfusion imaging and holds the promise to obviate the need for exogenous contrast agents. Since microscopic susceptibility increases with field strength, thus making oxygenation sensitivity due to ischemic (patho)physiology more pronounced, T_2^* mapping at 7.0 T might be beneficial to address some of the BOLD sensitivity constraints reported for the assessment of regional myocardial oxygenation changes in the presence of coronary artery stenosis [97] or for the characterization of vasodilator-induced changes of myocardial oxygenation at 1.5 and at 3.0 T [8].

Meanwhile, the pace of discovery is heartening and a powerful motivator to transfer the lessons learned from T_2^* mapping research at 7.0 T into the clinical scenario. These efforts are fueled by the quest for advancing the capabilities of quantitative MRI, a story worth following since the implications feed into a broad spectrum of MR physics, biomedical engineering, radiology, cardiology, internal medicine, and other related fields of basic research and clinical science. The requirements of T_2^* mapping at 7.0 T are likely to pave the way for further advances in MR technology and MR system design. With appropriate multitransmit systems that offer more than 16 TX channels each providing at least 4 kW peak power, an optimistically inclined scientist might envisage the implementation of high-density transceiver arrays with 64 and more elements with the ultimate goal to break ground for a many element upper torso or even a body coil array. This vision continues to motivate new research on integrated multichannel transmission systems [98], on novel RF pulse design, and on RF coil design together with explorations into ideal current patterns yielding optimal signal-to-noise ratio for UHF-CMR [99]. Perhaps another development is the move toward myocardial T_2^* mapping using reduced field of views zoomed into the target anatomy enabled by spatially selective excitation techniques which put the capabilities of parallel transmission technology to good use.

References

1. Ogawa S, Menon RS, Tank DW, Kim SG, Merkle H, Ellermann JM, Ugurbil K. Functional brain mapping by blood oxygenation level-dependent contrast magnetic resonance imaging. A comparison of signal characteristics with a biophysical model. Biophys J. 1993;64(3):803–12. https://doi.org/10.1016/S0006-3495(93)81441-3.
2. Wacker CM, Hartlep AW, Pfleger S, Schad LR, Ertl G, Bauer WR. Susceptibility-sensitive magnetic resonance imaging detects human myocardium supplied by a stenotic coronary artery without a contrast agent. J Am Coll Cardiol. 2003;41(5):834–40.
3. Friedrich M, Niendorf T, Schulz-Menger J. Blood oxygen level-dependent magnetic resonance imaging in patients with stress-induced angina. ACC Curr J Rev. 2004;13(4):30.
4. Jahnke C, Gebker R, Manka R, Schnackenburg B, Fleck E, Paetsch I. Navigator-gated 3D blood oxygen level–dependent CMR at 3.0-T for detection of stress-induced myocardial ischemic reactions. JACC Cardiovasc Imaging. 2010;3(4):375–84.
5. Karamitsos TD, Leccisotti L, Arnold JR, Recio-Mayoral A, Bhamra-Ariza P, Howells RK, Searle N, Robson MD, Rimoldi OE, Camici PG, Neubauer S, Selvanayagam JB. Relationship between regional myocardial oxygenation and perfusion in patients with coronary artery disease insights from cardiovascular magnetic resonance and positron emission tomography. Circ Cardiovasc Imaging. 2010;3(1):32–40.
6. Tsaftaris SA, Tang R, Zhou X, Li D, Dharmakumar R. Ischemic extent as a biomarker for characterizing severity of coronary artery stenosis with blood oxygen-sensitive MRI. J Magn Reson Imaging. 2012;35(6):1338–48.
7. Wacker CM, Bock M, Hartlep AW, Beck G, van Kaick G, Ertl G, Bauer WR, Schad LR. Changes in myocardial oxygenation and perfusion under pharmacological stress with dipyridamole: assessment using T*2 and T1 measurements. Magn Reson Med. 1999;41(4):686–95.
8. Vohringer M, Flewitt JA, Green JD, Dharmakumar R, Wang J Jr, Tyberg JV, Friedrich MG. Oxygenation-sensitive CMR for assessing vasodilator-induced changes of myocardial oxygenation. J Cardiovasc Magn Reson. 2010;12:20. https://doi.org/10.1186/1532-429X-12-20.
9. Utz W, Jordan J, Niendorf T, Stoffels M, Luft FC, Dietz R, Friedrich MG. Blood oxygen level-dependent MRI of tissue oxygenation: relation to endothelium-dependent and endothelium-independent blood flow changes. Arterioscler Thromb Vasc Biol. 2005;25(7):1408–13. https://doi.org/10.1161/01.ATV.0000170131.13683.d7.
10. Guensch DP, Fischer K, Flewitt JA, Friedrich MG. Impact of intermittent apnea on myocardial tissue oxygenation–a study using oxygenation-sensitive cardiovascular magnetic resonance. PLoS One. 2013;8(1):e53282. https://doi.org/10.1371/journal.pone.0053282.
11. Guensch DP, Fischer K, Flewitt JA, Friedrich MG. Myocardial oxygenation is maintained during hypoxia when combined with apnea–a cardiovascular MR study. Phys Rep. 2013;1(5):e00098.
12. Guensch DP, Fischer K, Flewitt JA, Yu J, Lukic R, Friedrich JA, Friedrich MG. Breathing manoeuvre-dependent changes in myocardial oxygenation in healthy humans. Eur Heart J Cardiovasc Imaging. 2014;15(4):409–14. https://doi.org/10.1093/ehjci/jet171.
13. Fischer K, Guensch DP, Shie N, Lebel J, Friedrich MG. Breathing maneuvers as a vasoactive stimulus for detecting inducible myocardial ischemia - an experimental cardiovascular magnetic resonance study. PLoS One. 2016;11(10):e0164524. https://doi.org/10.1371/journal.pone.0164524.
14. Guensch DP, Nadeshalingam G, Fischer K, Stalder AF, Friedrich MG. The impact of hematocrit on oxygenation-sensitive cardiovascular magnetic resonance. J Cardiovasc Magn Reson. 2016;18(1):42. https://doi.org/10.1186/s12968-016-0262-1.
15. He T. Cardiovascular magnetic resonance T2* for tissue iron assessment in the heart. Quant Imaging Med Surg. 2014;4(5):407–12. https://doi.org/10.3978/j.issn.2223-4292.2014.10.05.
16. Anderson L, Holden S, Davis B, Prescott E, Charrier C, Bunce N, Firmin D, Wonke B, Porter J, Walker J. Cardiovascular T 2-star(T 2*) magnetic resonance for the early diagnosis of myocardial iron overload. Eur Heart J. 2001;22(23):2171–9.

17. Mavrogeni S. Evaluation of myocardial iron overload using magnetic resonance imaging. Blood Transfus. 2009;7(3):183.
18. Carpenter J-P, He T, Kirk P, Roughton M, Anderson LJ, de Noronha SV, Sheppard MN, Porter JB, Walker JM, Wood JC, Galanello R, Forni G, Catani G, Matta G, Fucharoen S, Fleming A, House M, Black G, Firmin D, St Pierre T, Penell D. On T2* magnetic resonance and cardiac iron. Circulation. 2011;123(14):1519–28.
19. Friedrich MG, Karamitsos TD. Oxygenation-sensitive cardiovascular magnetic resonance. J Cardiovasc Magn Reson. 2013;15:43. https://doi.org/10.1186/1532-429X-15-43.
20. Haacke EM, Brown RW, Thompson MR, Venkatesan R (1999) Magnetization, relaxation and the bloch equation. In: Magnetic resonance imaging - physical principles and sequence design. 1 edn. Wiley, New York, pp 51–64
21. Zhao JM, Clingman CS, Närväinen MJ, Kauppinen RA, van Zijl P. Oxygenation and hematocrit dependence of transverse relaxation rates of blood at 3T. Magn Reson Med. 2007;58(3):592–7.
22. Ogawa S, Lee TM, Kay AR, Tank DW. Brain magnetic resonance imaging with contrast dependent on blood oxygenation. Proc Natl Acad Sci U S A. 1990;87(24):9868–72.
23. Brown RW, Cheng Y-CN, Haacke EM, Thompson MR, Venkatesan R. Magnetic resonance imaging: physical principles and sequence design. Hoboken: Wiley; 2014.
24. Bernstein MA, King KF, Zhou XJ. Handbook of MRI pulse sequences. Amsterdam: Elsevier; 2004.
25. Ogawa S, Tank DW, Menon R, Ellermann JM, Kim SG, Merkle H, Ugurbil K. Intrinsic signal changes accompanying sensory stimulation: functional brain mapping with magnetic resonance imaging. Proc Natl Acad Sci. 1992;89(13):5951–5.
26. Bauer WR, Nadler W, Bock M, Schad LR, Wacker C, Hartlep A, Ertl G. Theory of the BOLD effect in the capillary region: an analytical approach for the determination of T2 in the capillary network of myocardium. Magn Reson Med. 1999;41(1):51–62.
27. Yablonskiy DA, Haacke EM. Theory of NMR signal behavior in magnetically inhomogeneous tissues: the static dephasing regime. Magn Reson Med. 1994;32(6):749–63.
28. Christen T, Lemasson B, Pannetier N, Farion R, Segebarth C, Rémy C, Barbier EL. Evaluation of a quantitative blood oxygenation level-dependent (qBOLD) approach to map local blood oxygen saturation. NMR Biomed. 2011;24(4):393–403.
29. Spees WM, Yablonskiy DA, Oswood MC, Ackerman JJ. Water proton MR properties of human blood at 1.5 tesla: magnetic susceptibility, T1, T2, T2*, and non-Lorentzian signal behavior. Magn Reson Med. 2001;45(4):533–42.
30. Niendorf T, Pohlmann A, Arakelyan K, Flemming B, Cantow K, Hentschel J, Grosenick D, Ladwig M, Reimann H, Klix S. How bold is blood oxygenation level-dependent (BOLD) magnetic resonance imaging of the kidney? Opportunities, challenges and future directions. Acta Physiol. 2015;213(1):19–38.
31. Meloni A, Hezel F, Positano V, Keilberg P, Pepe A, Lombardi M, Niendorf T. Detailing magnetic field strength dependence and segmental artifact distribution of myocardial effective transverse relaxation rate at 1.5, 3.0, and 7.0 T. Magn Reson Med. 2014;71(6):2224–30. https://doi.org/10.1002/mrm.24856.
32. Hezel F, Thalhammer C, Waiczies S, Schulz-Menger J, Niendorf T. High spatial resolution and temporally resolved T2* mapping of normal human myocardium at 7.0 tesla: an ultrahigh field magnetic resonance feasibility study. PLoS One. 2012;7(12):e52324. https://doi.org/10.1371/journal.pone.0052324.
33. Mueller A, Kouwenhoven M, Naehle CP, Gieseke J, Strach K, Willinek WA, Schild HH, Thomas D. Dual-source radiofrequency transmission with patient-adaptive local radiofrequency shimming for 3.0-T cardiac MR imaging: initial experience. Radiology. 2012;263(1):77–85. https://doi.org/10.1148/radiol.11110347.
34. Frauenrath T, Niendorf T, Kob M. Acoustic method for synchronization of magnetic resonance imaging (MRI). Acta Acust Acust. 2008;94:148–55.
35. Frauenrath T, Hezel F, Heinrichs U, Kozerke S, Utting JF, Kob M, Butenweg C, Boesiger P, Niendorf T. Feasibility of cardiac gating free of interference with electro-magnetic fields at 1.5 tesla, 3.0 tesla and 7.0 tesla using an MR-stethoscope. Investig Radiol. 2009;44(9):539–47. https://doi.org/10.1097/RLI.0b013e3181b4c15e.

36. Frauenrath T, Hezel F, Renz W, d'Orth Tde G, Dieringer M, von Knobelsdorff-Brenkenhoff F, Prothmann M, Schulz Menger J, Niendorf T. Acoustic cardiac triggering: a practical solution for synchronization and gating of cardiovascular magnetic resonance at 7 tesla. J Cardiovasc Magn Reson. 2010;12:67. https://doi.org/10.1186/1532-429X-12-67.

37. Dieringer MA, Renz W, Lindel T, Seifert F, Frauenrath T, von Knobelsdorff-Brenkenhoff F, Waiczies H, Hoffmann W, Rieger J, Pfeiffer H, Ittermann B, Schulz-Menger J, Niendorf T. Design and application of a four-channel transmit/receive surface coil for functional cardiac imaging at 7T. J Magn Reson Imaging. 2011;33(3):736–41. https://doi.org/10.1002/jmri.22451.

38. Graessl A, Winter L, Thalhammer C, Renz W, Kellman P, Martin C, von Knobelsdorff-Brenkenhoff F, Tkachenko V, Schulz-Menger J, Niendorf T. Design, evaluation and application of an eight channel transmit/receive coil array for cardiac MRI at 7.0 T. Eur J Radiol. 2013;82(5):752–9. https://doi.org/10.1016/j.ejrad.2011.08.002.

39. Thalhammer C, Renz W, Winter L, Hezel F, Rieger J, Pfeiffer H, Graessl A, Seifert F, Hoffmann W, von Knobelsdorff-Brenkenhoff F, Tkachenko V, Schulz-Menger J, Kellman P, Niendorf T. Two-dimensional sixteen channel transmit/receive coil array for cardiac MRI at 7.0 T: design, evaluation, and application. J Magn Reson Imaging. 2012;36(4):847–57. https://doi.org/10.1002/jmri.23724.

40. Graessl A, Renz W, Hezel F, Dieringer MA, Winter L, Oezerdem C, Rieger J, Kellman P, Santoro D, Lindel TD, Frauenrath T, Pfeiffer H, Niendorf T. Modular 32-channel transceiver coil array for cardiac MRI at 7.0T. Magn Reson Med. 2014;72(1):276–90. https://doi.org/10.1002/mrm.24903.

41. Oezerdem C, Winter L, Graessl A, Paul K, Els A, Weinberger O, Rieger J, Kuehne A, Dieringer M, Hezel F, Voit D, Frahm J, Niendorf T. 16-channel bow tie antenna transceiver array for cardiac MR at 7.0 tesla. Magn Reson Med. 2016;75(6):2553–65. https://doi.org/10.1002/mrm.25840.

42. Snyder CJ, DelaBarre L, Metzger GJ, van de Moortele PF, Akgun C, Ugurbil K, Vaughan JT. Initial results of cardiac imaging at 7 Tesla. Magn Reson Med. 2009;61(3):517–24. https://doi.org/10.1002/mrm.21895.

43. Adriany G, Ritter J, Vaughan JT, Ugurbil K, Moortele PF. Experimental verification of enhanced B1 Shim performance with a Z-encoding RF coil array at 7 Tesla. In: Proc. Intl. Soc. Mag. Reson. Med., Stockholm, Sweden; 2010. p. 3831.

44. Santoro D, von Samson-Himmelstjerna F, Carinci F, Hezel F, Dieringer MA, Niendorf T. Cardiac Triggered B1+ mapping using Bloch Siegert in the heart at 3T. In: Proc Intl. Soc. Mag. Reson. Med.; 2012. p. 3840.

45. Lindel T, Greiser A, Waxmann P, Dieterle M, Seifert F, U. Fontius, Renz W, Dieringer M, Frauenrath T, Schulz Menger J, Niendorf T, Ittermann B. Cardiac CINE MRI at 7 T using a transmit array. In: Proc. Intl. Soc. Mag. Reson. Med.; 2012. p. 1227.

46. Schmitter S, Wu X, Uğurbil K, de Moortele V. Design of parallel transmission radiofrequency pulses robust against respiration in cardiac MRI at 7 tesla. Magn Reson Med. 2015;74(5):1291–305.

47. Schmitter S, Moeller S, Wu X, Auerbach EJ, Metzger GJ, Van de Moortele PF, Ugurbil K. Simultaneous multislice imaging in dynamic cardiac MRI at 7T using parallel transmission. Magn Reson Med. 2016;77(3):1010–20. https://doi.org/10.1002/mrm.26180.

48. Versluis MJ, Tsekos N, Smith NB, Webb AG. Simple RF design for human functional and morphological cardiac imaging at 7tesla. J Magn Reson. 2009;200(1):161–6. https://doi.org/10.1016/j.jmr.2009.06.014.

49. Winter L, Kellman P, Renz W, Grassl A, Hezel F, Thalhammer C, von Knobelsdorff-Brenkenhoff F, Tkachenko V, Schulz-Menger J, Niendorf T. Comparison of three multichannel transmit/receive radiofrequency coil configurations for anatomic and functional cardiac MRI at 7.0T: implications for clinical imaging. Eur Radiol. 2012;22(10):2211–20. https://doi.org/10.1007/s00330-012-2487-1.

50. Vaughan JT, Snyder CJ, DelaBarre LJ, Bolan PJ, Tian J, Bolinger L, Adriany G, Andersen P, Strupp J, Ugurbil K. Whole-body imaging at 7T: preliminary results. Magn Reson Med. 2009;61(1):244–8. https://doi.org/10.1002/mrm.21751.

51. Maderwald S, Orzada S, Schaefer LC, Bitz AK, Brote I, Kraff O, Theysohn JM, Ladd ME, Ladd SC, Quick HH 7T human in vivo cardiac imaging with an 8-channel transmit/receive array. In: Proc. Intl. Soc. Mag. Reson. Med., Honolulu, Hawaii, USA; 2009. p. 822.

52. van den Bergen B, Klomp DW, Raaijmakers AJ, de Castro CA, Boer VO, Kroeze H, Luijten PR, Lagendijk JJ, van den Berg CA. Uniform prostate imaging and spectroscopy at 7 T: comparison between a microstrip array and an endorectal coil. NMR Biomed. 2011;24(4):358–65. https://doi.org/10.1002/nbm.1599.

53. Ipek O, Raaijmakers AJ, Klomp DW, Lagendijk JJ, Luijten PR, van den Berg CA. Characterization of transceive surface element designs for 7 tesla magnetic resonance imaging of the prostate: radiative antenna and microstrip. Phys Med Biol. 2012;57(2):343–55. https://doi.org/10.1088/0031-9155/57/2/343.

54. Winter L, Özerdem C, Hoffmann W, Santoro D, Müller A, Waiczies H, Seemann R, Graessl A, Wust P, Niendorf T. Design and evaluation of a hybrid radiofrequency applicator for magnetic resonance imaging and RF induced hyperthermia: electromagnetic field simulations up to 14.0 tesla and proof-of-concept at 7.0 tesla. PLoS One. 2013;8(4):e61661. https://doi.org/10.1371/journal.pone.0061661.

55. Raaijmakers AJ, Ipek O, Klomp DW, Possanzini C, Harvey PR, Lagendijk JJ, van den Berg CA. Design of a radiative surface coil array element at 7 T: the single-side adapted dipole antenna. Magn Reson Med. 2011;66(5):1488–97. https://doi.org/10.1002/mrm.22886.

56. Ipek O, Raaijmakers AJ, Lagendijk JJ, Luijten PR, van den Berg CA. Intersubject local SAR variation for 7T prostate MR imaging with an eight-channel single-side adapted dipole antenna array. Magn Reson Med. 2014;71(4):1559–67. https://doi.org/10.1002/mrm.24794.

57. Erturk MA, Raaijmakers AJ, Adriany G, Ugurbil K, Metzger GJ. A 16-channel combined loop-dipole transceiver array for 7 tesla body MRI. Magn Reson Med. 2016;77(2):884–94. https://doi.org/10.1002/mrm.26153.

58. Raaijmakers AJE, Italiaander M, Voogt IJ, Luijten PR, Hoogduin JM, Klomp DWJ, van den Berg CAT. The fractionated dipole antenna: a new antenna for body imaging at 7 tesla. Magn Reson Med. 2016;75(3):1366–74. https://doi.org/10.1002/mrm.25596.

59. Aussenhofer SA, Webb AG. An eight-channel transmit/receive array of TE_{01} mode high permittivity ceramic resonators for human imaging at 7T. J Magn Reson. 2014;243:122–9. https://doi.org/10.1016/j.jmr.2014.04.001.

60. Keith GA, Rodgers CT, Hess AT, Snyder CJ, Vaughan JT, Robson MD. Automated tuning of an eight-channel cardiac transceive array at 7 tesla using piezoelectric actuators. Magn Reson Med. 2015;73(6):2390–7.

61. Togawa T, Okai O, Oshima M. Observation of blood flow E.M.F. in externally applied strong magnetic field by surface electrodes. Med Biol Eng. 1967;5(2):169–70.

62. Stuber M, Botnar RM, Fischer SE, Lamerichs R, Smink J, Harvey P, Manning WJ. Preliminary report on in vivo coronary MRA at 3 tesla in humans. Magn Reson Med. 2002;48(3):425–9. https://doi.org/10.1002/mrm.10240.

63. Lanzer P, Barta C, Botvinick EH, Wiesendanger HU, Modin G, Higgins CB. ECG-synchronized cardiac MR imaging: method and evaluation. Radiology. 1985;155(3):681–6.

64. Fischer SE, Wickline SA, Lorenz CH. Novel real-time R-wave detection algorithm based on the vectorcardiogram for accurate gated magnetic resonance acquisitions. Magn Reson Med. 1999;42(2):361–70. https://doi.org/10.1002/(SICI)1522-2594(199908)42:2<361::AID-MRM18>3.0.CO;2-9.

65. Chia JM, Fischer SE, Wickline SA, Lorenz CH. Performance of QRS detection for cardiac magnetic resonance imaging with a novel vectorcardiographic triggering method. J Magn Reson Imaging. 2000;12(5):678–88. https://doi.org/10.1002/1522-2586(200011)12:5<678::aid-jmri4>3.0.co;2-5.

66. Becker M, Frauenrath T, Hezel F, Krombach GA, Kremer U, Koppers B, Butenweg C, Goemmel A, Utting JF, Schulz-Menger J, Niendorf T. Comparison of left ventricular function assessment using phonocardiogram- and electrocardiogram-triggered 2D SSFP CINE MR imaging at 1.5 T and 3.0 T. Eur Radiol. 2010;20(6):1344–55. https://doi.org/10.1007/s00330-009-1676-z.

67. Brandts A, Westenberg JJ, Versluis MJ, Kroft LJ, Smith NB, Webb AG, de Roos A. Quantitative assessment of left ventricular function in humans at 7 T. Magn Reson Med. 2010;64(5):1471–7. https://doi.org/10.1002/mrm.22529.

68. Heinrichs U, Utting JF, Frauenrath T, Hezel F, Krombach GA, Hodenius MA, Kozerke S, Niendorf T. Myocardial T₂* mapping free of distortion using susceptibility-weighted fast spin-echo imaging: a feasibility study at 1.5 T and 3.0 T. Magn Reson Med. 2009;62(3):822–8. https://doi.org/10.1002/mrm.22054.

69. Fuchs K, Hezel F, Winter L, Oezerdem C, Graessl A, Dieringer M, Kraus O, Niendorf T. Feasibility of cardiac fast spin echo imaging at 7.0 T using a two-dimensional 16 channel array of bowtie transceivers. In: Proc. Intl. Soc. Mag. Reson. Med. 21, Salt Lake City, USA; 2013. p. 1413.

70. Norris DG, Boernert P, Reese T, Leibfritz D. On the application of ultra-fast rare experiments. Magn Reson Med. 1992;27(1):142–64.

71. Carr HY, Purcell EM. Effects of diffusion on free precession in nuclear magnetic resonance experiments. Phys Rev. 1954;94(3):630–8.

72. Meiboom S, Gill D. Modified spin-echo method for measuring nuclear relaxation times. Rev Sci Instrum. 1958;29(8):688–91. https://doi.org/10.1063/1.1716296.

73. Schick F. SPLICE: sub-second diffusion-sensitive MR imaging using a modified fast spin-echo acquisition mode. Magn Reson Med. 1997;38(4):638–44.

74. Haacke E, Brown R, Thompson M, Venkatesan R, Bernstein M, King K, Zhou X. Magnetic resonance imaging - physical principles and sequence design. New York: Wiley; 1999.

75. Robinson S, Schodl H, Trattnig S. A method for unwrapping highly wrapped multi-echo phase images at very high field: UMPIRE. Magn Reson Med. 2013;72:80–92. https://doi.org/10.1002/mrm.24897.

76. Jaffer FA, Wen H, Balaban RS, Wolff SD. A method to improve the B0 homogeneity of the heart in vivo. Magn Reson Med. 1996;36(3):375–83.

77. Reeder SB, Faranesh AZ, Boxerman JL, McVeigh ER. In vivo measurement of T2* and field inhomogeneity maps in the human heart at 1.5 T. Magn Reson Med. 1998;39(6):988–98.

78. Shah S, Kellman P, Greiser A, Weale PJ, Zuehlsdorff S, Jerecic R. Rapid fieldmap estimation for cardiac shimming. In: Intl. Soc. Mag. Reson. Med. 17, Hawai, USA; 2009. p. 565.

79. Serradas Duarte T, Huelnhagen T, Niendorf T. Assessment of myocardial B0 over the cardiac cycle at 7.0T: implications for susceptibility-based cardiac MR techniques. In: Proc. Intl. Soc. Mag. Reson. Med. 24, Singapore; 2016. p. 2541.

80. Huelnhagen T, Hezel F, Serradas Duarte T, Pohlmann A, Oezerdem C, Flemming B, Seeliger E, Prothmann M, Schulz-Menger J, Niendorf T. Myocardial effective transverse relaxation time T2* correlates with left ventricular wall thickness: a 7.0 T MRI study. Magn Reson Med. 2017;77:2381–9. https://doi.org/10.1002/mrm.26312.

81. Cerqueira MD, Weissman NJ, Dilsizian V, Jacobs AK, Kaul S, Laskey WK, Pennell DJ, Rumberger JA, Ryan T, Verani MS. Standardized myocardial segmentation and nomenclature for tomographic imaging of the heart. A statement for healthcare professionals from the Cardiac Imaging Committee of the Council on Clinical Cardiology of the American Heart Association. Circulation. 2002;105(4):539–42.

82. Serradas Duarte T. Detailing myocardial B0 across the cardiac cycle at UHF: B0 assessment and implications for susceptibility-based CMR techniques. Lisbon: Universidade Nova de Lisboa; 2016.

83. Kelley CT. Iterative methods for optimization, vol. 18. Siam: SIAM Frontiers in Applied Mathematics; 1999.

84. Sandino CM, Kellman P, Arai AE, Hansen MS, Xue H. Myocardial T2* mapping: influence of noise on accuracy and precision. J Cardiovasc Magn Reson. 2015;17(1):7. https://doi.org/10.1186/s12968-015-0115-3.

85. Manjon JV, Coupe P, Marti-Bonmati L, Collins DL, Robles M. Adaptive non-local means denoising of MR images with spatially varying noise levels. J Magn Reson Imaging. 2010;31(1):192–203. https://doi.org/10.1002/jmri.22003.

86. Feng Y, He T, Feng M, Carpenter JP, Greiser A, Xin X, Chen W, Pennell DJ, Yang GZ, Firmin DN. Improved pixel-by-pixel MRI R2* relaxometry by nonlocal means. Magn Reson Med. 2014;72(1):260–8. https://doi.org/10.1002/mrm.24914.

87. Huelnhagen T, Pohlmann A, Niendorf T. Improving T2* mapping accuracy by spatially adaptive non local means noise filtering. In: Proc. Intl. Soc. Mag. Reson. Med., vol. 23, Toronto, Canada; 2015. p. 3753.

88. Guyton AC, Hall JE (2000) Muscle blood flow and cardiac output during exercise; the coronary circulation and ischemic heart disease. In: Guyton and Hall textbook of medical physiology. 10 edn. Saunders: Philadelphia, pp 223-234

89. Schmidt RF, Lang F, Heckmann M (2010) Herzmechanik. In: Physiologie des Menschen: Mit Pathophysiologie. 31 edn. Springer, Heidelberg, pp 539-564

90. Schmidt RF, Lang F, Heckmann M (2010) Herzstoffwechsel und Kroronardurchblutung. In: Physiologie des Menschen: Mit Pathophysiologie. 31 edn. Springer, Heidelberg, pp 565-571

91. Abdel-Aty H, Cocker M, Strohm O, Filipchuk N, Friedrich MG. Abnormalities in T2-weighted cardiovascular magnetic resonance images of hypertrophic cardiomyopathy: regional distribution and relation to late gadolinium enhancement and severity of hypertrophy. J Magn Reson Imaging. 2008;28(1):242–5. https://doi.org/10.1002/jmri.21381.

92. Johansson B, Mörner S, Waldenström A, Stål P. Myocardial capillary supply is limited in hypertrophic cardiomyopathy: a morphological analysis. Int J Cardiol. 2008;126(2):252–7.

93. Cecchi F, Olivotto I, Gistri R, Lorenzoni R, Chiriatti G, Camici PG. Coronary microvascular dysfunction and prognosis in hypertrophic cardiomyopathy. N Engl J Med. 2003;349(11):1027–35.

94. Reichenbach JR, Haacke EM. High-resolution BOLD venographic imaging: a window into brain function. NMR Biomed. 2001;14(7–8):453–67.

95. Reese TG, Weisskoff RM, Smith RN, Rosen BR, Dinsmore RE, Wedeen VJ. Imaging myocardial fiber architecture in vivo with magnetic resonance. Magn Reson Med. 1995;34(6):786–91.

96. MT W, Tseng WY, MY S, Liu CP, Chiou KR, Wedeen VJ, Reese TG, Yang CF. Diffusion tensor magnetic resonance imaging mapping the fiber architecture remodeling in human myocardium after infarction: correlation with viability and wall motion. Circulation. 2006;114(10):1036–45. https://doi.org/10.1161/CIRCULATIONAHA.105.545863.

97. Dharmakumar R, Arumana JM, Tang R, Harris K, Zhang Z, Li D. Assessment of regional myocardial oxygenation changes in the presence of coronary artery stenosis with balanced SSFP imaging at 3.0 T: theory and experimental evaluation in canines. J Magn Reson Imaging. 2008;27(5):1037–45. https://doi.org/10.1002/jmri.21345.

98. Poulo L, Alon L, Deniz C, Haefner R, Sodickson D, Stoeckel B, Zhu Y. A 32-channel parallel exciter/amplifier transmit system for 7T imaging. In: ISMRM (ed) Proc. Intl. Soc. Mag. Reson. Med., vol. 19; 2011. p. 1867.

99. Lattanzi R, Sodickson DK. Ideal current patterns yielding optimal signal-to-noise ratio and specific absorption rate in magnetic resonance imaging: computational methods and physical insights. Magn Reson Med. 2012;68(1):286–304. https://doi.org/10.1002/mrm.23198.

Extracellular Matrix-Specific Molecular MR Imaging Probes for the Assessment of Aortic Aneurysms

Julia Brangsch, Carolin Reimann, and Marcus R. Makowski

Abstract

All tissues in the human body are composed of cells that are embedded in the extracellular matrix. The extracellular matrix has, besides their structural role, several important functions. These functions include important regulatory mechanisms for signal transduction and matrix cell interactions. If pathological processes, e.g., in atherosclerosis or aortic aneurysms, occur, the extracellular matrix changes in response. This includes alterations in the structural and functional components of the extracellular matrix.

While traditional imaging technologies, such as X-ray or computed tomography (CT), are mainly aimed at imaging morphological changes, molecular magnetic resonance (MR) imaging is a technique that enables the visualization and quantification of pathological changes on a molecular scale. Different techniques can be used for molecular MR imaging. The most commonly employed techniques include the use of specific molecular magnetic resonance probes. These probes are, in most cases, either based on iron oxide particles or gadolinium chelates for signal generation.

Aortic abdominal aneurysms represent an irreversible dilation of the aortic wall which could cause severe consequences, including wall rupture with a mortality rate >90%. Due to the absence of symptoms during the development of aortic aneurysms, early diagnosis remains challenging.

J. Brangsch • C. Reimann
Department of Radiology, Charité-Universitätsmedizin Berlin, Berlin, Germany

M.R. Makowski (✉)
Department of Radiology, Charité-Universitätsmedizin Berlin, Berlin, Germany

Division of Imaging Sciences, King's College London, London, UK
e-mail: marcus.makowski@charite.de

© Springer International Publishing AG 2018
I. Sack, T. Schaeffter (eds.), *Quantification of Biophysical Parameters in Medical Imaging*, https://doi.org/10.1007/978-3-319-65924-4_16

In the following chapter, we will outline major developments regarding extracellular matrix-specific molecular magnetic resonance imaging for the assessment of aortic aneurysms.

16.1 Overview

All tissues in the human body are composed of cells which are embedded in the extracellular matrix. The extracellular matrix (ECM) has, besides their structural role, several important functions. These functions include important regulatory mechanisms for signal transduction and matrix cell interactions. If pathological processes, e.g., in atherosclerosis or aortic aneurysms, occur, the extracellular matrix changes in response. This includes alterations in the structural and functional components of the extracellular matrix. More reading can be found in Chap. 6.

Abdominal aortic aneurysms represent a cardiovascular disease which is associated with a dilatation of the aorta. In the case of abdominal aortic aneurysms, a dilation beyond 3 cm or more than 50% compared to the normal aorta is considered an aneurysm. One of the most dangerous complications of aortic aneurysms is aortic rupture, which is associated with a high mortality rate [1, 2]. The incidence of aortic aneurysms has been rising over the last years and decades, mainly due to an increased aging of the population [3]. Besides age, risk factors include genetic predisposition, male sex, smoking, hypertension, and increase in low density lipoprotein (LDL) [4, 5]. Even though aortic aneurysms have a relatively high incidence in the general population, their pathophysiology is not fully elucidated yet. This is mainly due to the fact that it is difficult to obtain tissue specimens of aortic aneurysms. Different causes can lead to the development of aortic aneurysms; these include infections, trauma, and connective tissue disorders [6–8].

While traditional imaging technologies, such as X-ray or computed tomography (CT), are mainly aimed at imaging morphological changes, such as the aortic diameter, molecular magnetic resonance (MR) imaging is an imaging technique that enables the visualization and quantification of pathological changes on a molecular scale. Different techniques can be used for molecular MR imaging. The most commonly employed techniques include the use of specific molecular probes. These probes are, in most cases, either based on iron oxide particles or gadolinium chelates for signal generation. These molecular probes have potential to improve the in vivo evaluation of aortic aneurysms, as they enable imaging which looks beyond morphological changes.

In the following chapter, we will outline, from our point of view, the most important developments regarding ECM-specific molecular MR imaging probes for the assessment of aortic aneurysms.

16.2 Extracellular Matrix-Specific Molecular MR Imaging Probes

16.2.1 Experimental Extracellular Matrix-Specific Molecular MR Imaging Probes

Compared to imaging modalities such as CT (computed tomography) and PET (positron emission tomography), MRI (MR imaging) has the advantage that it is an imaging modality which works without requiring ionizing radiation. In MRI, radiofrequency pulses which interact with a static magnetic field enable the generation of images with a high soft tissue contrast [9]. For molecular MRI, different types of molecular probes can be used. These probes are either highly specific for a molecular target or they accumulate passively in certain regions of the extracellular matrix. Gadolinium-based MR probes are called T1 molecular probes. This type of probes is mainly detected and quantified by T1-weighted sequences or T1 mapping sequences. MR mapping techniques enable a more accurate in vivo quantification of the molecular probe. On the other hand, iron oxide based probes are T2/T2* molecular probes which can be used for molecular imaging. This type of probe is usually detected using T2-/T2*-weighted sequences, including T2/T2* mapping. Both probes have certain advantages and disadvantages for the visualization of molecular targets. One advantage of T1 molecular probes is their positive contrast effect, which enables the visualization of molecular targets with a bright or positive signal on T1-weighted images. An advantage of T2 molecular probes is that they can be detected in vivo with a higher sensitivity compared to T1 molecular probes [10, 11].

16.2.2 Characterization of Collagen in the Extracellular Matrix

As illustrated in Fig. 6.1 of Chap. 6, collagen is, besides elastin, a highly abundant protein in the extracellular matrix. It contributes significantly to the stability of the arterial wall [12, 13]. Several subtypes of collagen (types 1–3) are known. In the aorta mainly types 1 and 3 are found. If collagen is degraded by matrix metalloproteinases, it leads to a weakening of the arterial wall. This can lead to the development of an aortic aneurysm.

Different types of molecular MR probes have been used for the characterization of collagen in the context of aortic aneurysms. One important approach comprises fluorescent micellar nanoparticles which can bind to collagen. Using this probe, it was shown that it is feasible to differentiate between stable and unstable aortic aneurysms in an experimental mouse model [14].

16.2.3 Characterization of Elastin in the Extracellular Matrix

Besides collagen, elastin is a major structural protein of the extracellular matrix [15]. It is highly important for the stabilization of the arterial wall enabling it to withstand the constant pulsatile pressure from the lumen. It is mainly secreted by fibroblasts, smooth muscle cells, and, to a smaller extent, macrophages, which can be found in the extracellular matrix of the arterial wall. If elastin is degraded by matrix metalloproteinases, the arterial wall is weakened. This can lead to a dilatation and the development of an aortic aneurysm [16]. As already described in a previous publication (see Fig. 6.3 of Chap. 6 [17]), an increase in the nonlinear elastic modulus (stiffening) can also be expected as a result of the elastin digestion [17].

Recently, an elastin-specific molecular MR probe enabling the characterization of the arterial wall, regarding its elastin content, has been introduced. This probe enables the visualization of the specific site of rupture in the aortic wall prior to the development of an aortic aneurysm (Fig. 16.1) [18–20]. A different group investigated the potential of the elastin-specific molecular MR probe in the context of a Marfan's disease model. They could demonstrate that an overall decrease in elastin content in the

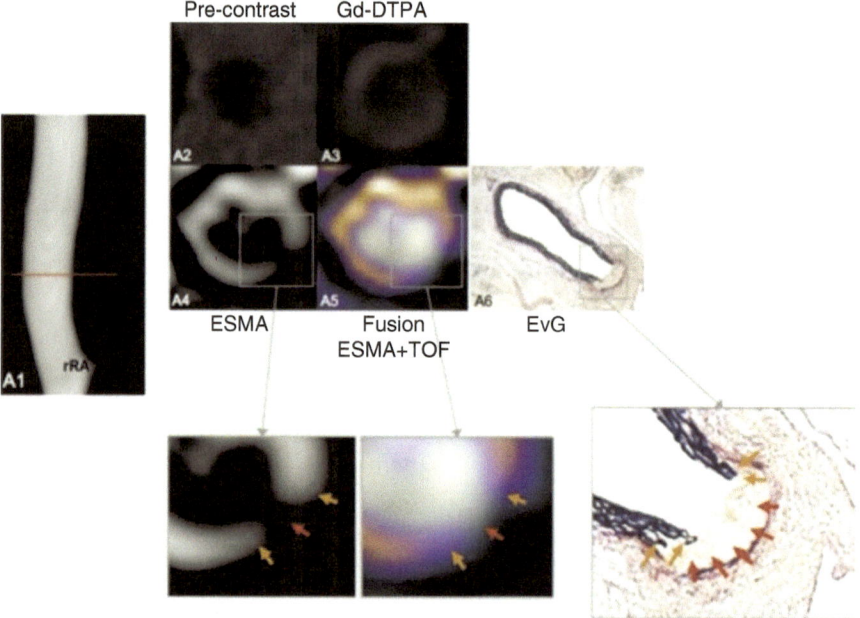

Fig. 16.1 In vivo characterization of abdominal aortic aneurysms in an experimental mouse model by molecular MR imaging. A: In vivo characterization of the relative elastin composition of the aorta in an aneurysmal mouse model using an elastin-specific MR probe. A1: Normal MR angiogram showing a nondilated regular aortic lumen. A2, A3: Precontrast scan and scan with control agent (Gd-DTPA) demonstrating no unspecific uptake. A4, A5: Elastin-specific magnetic resonance imaging demonstrating the aortic wall with a focal rupture of the elastic laminae (magnifications). A6: Corresponding histology demonstrating the rupture of the elastic laminae and a small hematoma. *rRA* right renal artery. Adapted from Botnar et al. [20]

arterial wall occurs in this model. These changes could be monitored and quantified in vivo. Such decrease in elastin could represent a risk factor for the weakening of the arterial wall and the subsequent development of an aortic aneurysm [21].

16.2.4 Characterization of Matrix Metalloproteinases in the Extracellular Matrix

The increased expression of matrix metalloproteinases (MMPs) leads, by degrading collagen and elastin, to the destabilization of the arterial wall. These proteins are expressed during proinflammatory processes in all layers of the arterial wall [22, 23]. They are able to degrade almost all extracellular matrix proteins, including elastin and collagen. Their activity is regulated by specific inhibitor proteins so-called TIMPs (specific tissue inhibitors of metalloproteinases) [24]. If the expression of TIMPs is reduced, an increased activity of MMPs can lead to degradation processes and ultimately to the thinning and destabilization of the arterial wall.

Probes specific for MMPs do not allow a direct visualization of the ECM. However, by visualizing matrix metalloproteinases, their effect on the ECM can be visualized and quantified in vivo. One research group developed a MMP-specific probe based on a Gd-DOTA chelate and showed that a specific in vivo visualization of MMPs is feasible using molecular MRI. Such a probe could be useful, e.g., to evaluate the risk of a potential aortic rupture [23].

16.2.5 Macrophages and the Extracellular Matrix

If inflammation occurs, macrophages are one of the most prominent cell types. While macrophages are in general considered to be proinflammatory cells, there are two distinct subtypes of macrophages. On the one hand, there are M1 macrophages, which actively contribute to the degradation of ECM proteins by expressing proinflammatory cytokines. On the other hand, M2 macrophages can be found during the process of inflammation. M2 macrophages promote the repair of the ECM and increase the proliferation of surrounding cells, including smooth muscle cells [25]. Overall, the ratio between M1 and M2 macrophages plays an important part in determining the further development and outcome of inflammatory processes [26, 27].

For the visualization of macrophages, different MRI techniques can be applied. In most cases, the imaging approach is based on iron oxide particles. Different types of iron oxide particles, e.g., ultra-small superparamagnetic iron oxide (USPIO) particles, are applied for the visualization of macrophages. The principle behind this approach is that, if iron oxide particles are administered intravenously, they circulate in the bloodstream. Proinflammatory macrophages phagocyte the circulating iron oxide particles. Subsequently these proinflammatory macrophages migrate to the area of inflammation. Due to their high intracellular iron oxide content, they can be visualized by T2/T2* MR imaging sequences [27]. Using such an approach, not only the development of proinflammatory processes can be visualized but also the response to, e.g., an anti-inflammatory therapy.

16.2.6 Clinical Imaging of the Extracellular Matrix

For the translation of experimental/preclinical results into clinical applications, it is highly important to validate preclinical results in clinical studies. Regarding molecular MRI of aortic aneurysms, the focus was put on imaging with iron oxide nanoparticles. Different-sized nanoparticles have been used in this context. In a clinical study with stable patients with symptomatic abdominal aortic aneurysms (29 patients), the potential of iron oxide particles for the characterization of local inflammatory processes in aortic aneurysms was tested (Fig. 16.2) [28]. In this study it was shown that patients with a specific uptake pattern for the iron oxide particles showed a faster growth rate of the abdominal aortic aneurysm. It is important to mention that for these results the diameter of the aortic aneurysm didn't

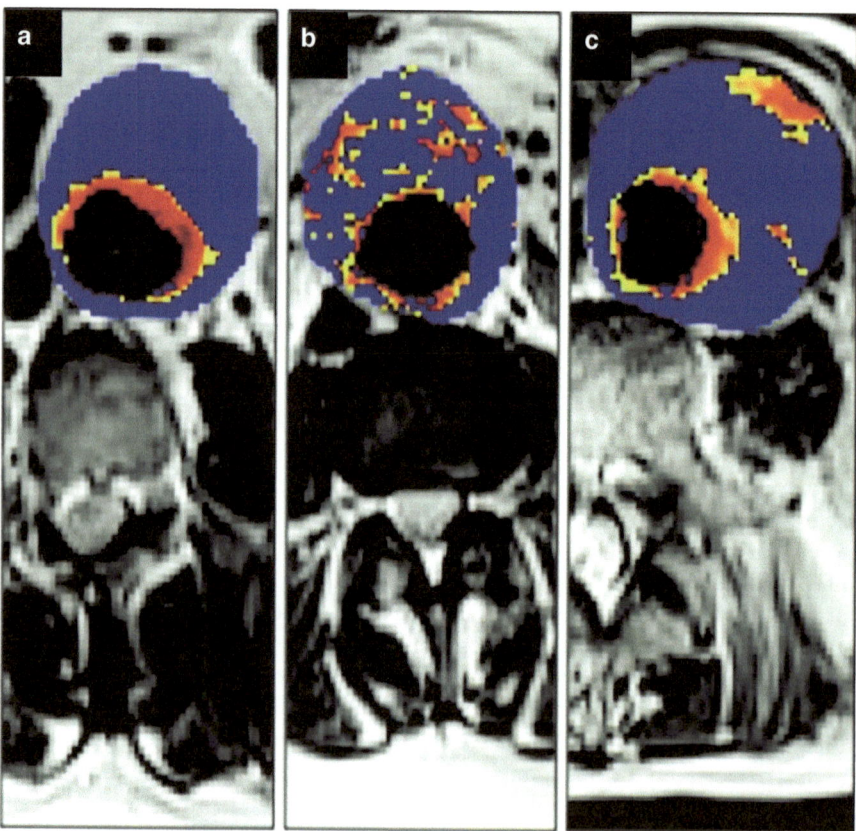

Fig. 16.2 In vivo imaging of aortic aneurysms in patients using an iron oxide particle (USPIO) for the characterization of different groups of aortic aneurysms. Different specific uptake patterns were described: (**a**) Uptake of the USPIO was mainly found directly adjacent to the lumen. (**b**) Diffuse focal uptake in the intraluminal thrombus was measured. (**c**) Discrete focal uptake in the aortic wall was demonstrated. Based on these different uptake patterns, the prediction of the aneurysmal growth rate could be performed. Adapted from Richards et al. [28]

differ between patient groups. In a different study, a correlation between the uptake of the iron oxide particles and the T2/T2* relaxation time was confirmed [29]. Currently it is planned to investigate the potential of iron oxide particles for the characterization of abdominal aortic aneurysms in a large patient study with more than 300 patients [30].

Conclusion

The extracellular matrix plays an important role in the onset and development of different cardiovascular diseases. These include atherosclerosis and aortic aneurysms. In the case of abdominal aortic aneurysms, different studies have shown that a weakening of the extracellular matrix, e.g., by a degradation of elastin or collagen, can lead to an increased risk for the development of an aortic dilation and ultimately of an aneurysm. Different cell types, including macrophages, interact with the extracellular matrix. This is done, e.g., by the expression of certain enzymes, including matrix metalloproteinases which can degrade extracellular matrix proteins.

Using different gadolinium-based or iron oxide-based molecular MR probes, these processes can be visualized in vivo. In the first step, these probes are investigated in experimental/preclinical studies. If these are performed successfully, probes are eventually translated into a clinical study. In the future, large-scale multicenter studies, testing the clinical potential of these probes, will be important. Ultimately these probes could represent novel in vivo biomarkers for the prediction of, e.g., the onset of disease, and improve the prediction of the response to therapy.

Acknowledgments The authors MRM, JB, and CR are grateful for the financial support from the Deutsche Forschungsgemeinschaft (DFG, 5943/31/41/91).

References

1. Nevitt MP, Ballard DJ, Hallett JW Jr. Prognosis of abdominal aortic aneurysms. A population-based study. N Engl J Med. 1989;321:1009–14.
2. Thompson RW, Geraghty PJ, Lee JK. Abdominal aortic aneurysms: basic mechanisms and clinical implications. Curr Probl Surg. 2002;39:110–230.
3. Hallett JW Jr. Management of abdominal aortic aneurysms. Mayo Clin Proc. 2000;75:395–9.
4. Alcorn HG, Wolfson SK Jr, Sutton-Tyrrell K, Kuller LH, O'Leary D. Risk factors for abdominal aortic aneurysms in older adults enrolled in The Cardiovascular Health Study. Arterioscler Thromb Vasc Biol. 1996;16:963–70.
5. Johansen K, Koepsell T. Familial tendency for abdominal aortic aneurysms. JAMA. 1986;256:1934–6.
6. Sakalihasan N, Limet R, Defawe OD. Abdominal aortic aneurysm. Lancet (Lond, Engl). 2005;365:1577–89.
7. Matsumura K, Hirano T, Takeda K, Matsuda A, Nakagawa T, Yamaguchi N, Yuasa H, Kusakawa M, Nakano T. Incidence of aneurysms in Takayasu's arteritis. Angiology. 1991;42:308–15.
8. Towbin JA, Casey B, Belmont J. The molecular basis of vascular disorders. Am J Hum Genet. 1999;64:678–84.

9. Lin JB, Phillips EH, Riggins TE, Sangha GS, Chakraborty S, Lee JY, Lycke RJ, Hernandez CL, Soepriatna AH, Thorne BR, Yrineo AA, Goergen CJ. Imaging of small animal peripheral artery disease models: recent advancements and translational potential. Int J Mol Sci. 2015;16:11131–77.

10. Botnar RM, Ebersberger H, Noerenberg D, Jansen CH, Wiethoff AJ, Schuster A, Kasner M, Walter TC, Knobloch G, Hoppe P, Diederichs G, Hamm B, Makowski MR. Molecular imaging in cardiovascular diseases. RoFo Fortschr Geb Rontgenstr Nuklearmed. 2015;36:92–101.

11. Johnson GA, Benveniste H, Black RD, Hedlund LW, Maronpot RR, Smith BR. Histology by magnetic resonance microscopy. Magn Reson Q. 1993;9:1–30.

12. Menashi S, Campa JS, Greenhalgh RM, Powell JT. Collagen in abdominal aortic aneurysm: typing, content, and degradation. J Vasc Surg. 1987;6:578–82.

13. Burton AC. Relation of structure to function of the tissues of the wall of blood vessels. Physiol Rev. 1954;34:619–42.

14. Klink A, Heynens J, Herranz B, Lobatto ME, Arias T, Sanders HM, Strijkers GJ, Merkx M, Nicolay K, Fuster V, Tedgui A, Mallat Z, Mulder WJ, Fayad ZA. In vivo characterization of a new abdominal aortic aneurysm mouse model with conventional and molecular magnetic resonance imaging. J Am Coll Cardiol. 2011;58:2522–30.

15. Krettek A, Sukhova GK, Libby P. Elastogenesis in human arterial disease: a role for macrophages in disordered elastin synthesis. Arterioscler Thromb Vasc Biol. 2003;23:582–7.

16. Sakalihasan N, Delvenne P, Nusgens BV, Limet R, Lapiere CM. Activated forms of MMP2 and MMP9 in abdominal aortic aneurysms. J Vasc Surg. 1996;24:127–33.

17. Roach MR, Burton AC. The reason for the shape of the distensibility curves of arteries. Can J Biochem Physiol. 1957;35:681–90.

18. Makowski MR, Preissel A, von Bary C, Warley A, Schachoff S, Keithan A, Cesati RR, Onthank DC, Schwaiger M, Robinson SP, Botnar RM. Three-dimensional imaging of the aortic vessel wall using an elastin-specific magnetic resonance contrast agent. Investig Radiol. 2012;47:438–44.

19. von Bary C, Makowski M, Preissel A, Keithahn A, Warley A, Spuentrup E, Buecker A, Lazewatsky J, Cesati R, Onthank D, Schickl N, Schachoff S, Hausleiter J, Schomig A, Schwaiger M, Robinson S, Botnar R. MRI of coronary wall remodeling in a swine model of coronary injury using an elastin-binding contrast agent. Circ Cardiovasc Imaging. 2011;4:147–55.

20. Botnar RM, Wiethoff AJ, Ebersberger U, Lacerda S, Blume U, Warley A, Jansen CH, Onthank DC, Cesati RR, Razavi R, Marber MS, Hamm B, Schaeffter T, Robinson SP, Makowski MR. In vivo assessment of aortic aneurysm wall integrity using elastin-specific molecular magnetic resonance imaging. Circ Cardiovasc Imaging. 2014;7:679–89.

21. Okamura H, Pisani LJ, Dalal AR, Emrich F, Dake BA, Arakawa M, Onthank DC, Cesati RR, Robinson SP, Milanesi M. Assessment of elastin deficit in a Marfan mouse aneurysm model using an elastin-specific magnetic resonance imaging contrast agent. Circ Cardiovasc Imaging. 2014;7:690–6.

22. Birkedal-Hansen H. Proteolytic remodeling of extracellular matrix. Curr Opin Cell Biol. 1995;7:728–35.

23. Bazeli R, Coutard M, Duport BD, Lancelot E, Corot C, Laissy JP, Letourneur D, Michel JB, Serfaty JM. In vivo evaluation of a new magnetic resonance imaging contrast agent (P947) to target matrix metalloproteinases in expanding experimental abdominal aortic aneurysms. Investig Radiol. 2010;45:662–8.

24. Thompson RW, Parks WC. Role of matrix metalloproteinases in abdominal aortic aneurysms. Ann N Y Acad Sci. 1996;800:157–74.

25. Mills CD. M1 and M2 macrophages: oracles of health and disease. Crit Rev Immunol. 2012;32:463–88.

26. Hellenthal FA, Buurman WA, Wodzig WK, Schurink GW. Biomarkers of abdominal aortic aneurysm progression. Part 2: inflammation. Nat Rev Cardiol. 2009;6:543–52.

27. Turner GH, Olzinski AR, Bernard RE, Aravindhan K, Boyle RJ, Newman MJ, Gardner SD, Willette RN, Gough PJ, Jucker BM. Assessment of macrophage infiltration in a murine model of abdominal aortic aneurysm. J Magn Reson Imaging. 2009;30:455–60.

28. Richards JM, Semple SI, MacGillivray TJ, Gray C, Langrish JP, Williams M, Dweck M, Wallace W, McKillop G, Chalmers RT, Garden OJ, Newby DE. Abdominal aortic aneurysm growth predicted by uptake of ultrasmall superparamagnetic particles of iron oxide: a pilot study. Circ Cardiovasc Imaging. 2011;4:274–81.

29. Sadat U, Taviani V, Patterson AJ, Young VE, Graves MJ, Teng Z, Tang TY, Gillard JH. Ultrasmall superparamagnetic iron oxide-enhanced magnetic resonance imaging of abdominal aortic aneurysms – a feasibility study. Eur J Vasc Endovasc Surg. 2011;41:167–74.

30. Mcbride OMB, Berry C, Burns P, Chalmers RTA, Doyle B, Forsythe R, Garden OJ, Goodman K, Graham C, Hoskins P, Holdsworth R, MacGillivray TJ, McKillop G, Murray G, Oatey K, Robson JMJ, Roditi G, Semple S, Stuart W, van Beek EJR, Vesey A, Newby DE. MRI using ultrasmall superparamagnetic particles of iron oxide in patients under surveillance for abdominal aortic aneurysms to predict rupture or surgical repair: MRI for abdominal aortic aneurysms to predict rupture or surgery – the MA(3)RS study. Open Heart. 2015;2:e000190.

Michael Scheel

Abstract

Diffusion-weighted imaging (DWI) is an accepted and widely used MRI technique today. DWI image contrast is based on water molecule displacement at a micrometer scale. DWI, therefore, does not excel at high anatomic accuracy but provides a physically meaningful parameter for tissue characterization. Pathological tissue changes often go along with changes in diffusion properties, which make DWI so useful clinically.

In contrast to water molecules diffusing freely in an open system, diffusion in biological tissues is restricted and influenced in many ways. The main diffusion barriers that are probed with DWI are cell membranes. Depending on the density and orientation of cell membranes in a given tissue, the diffusion of water molecules is influenced in both magnitude and orientation.

DWI is a relatively fast and robust imaging technique, which makes its integration into a clinical imaging protocol relatively easy. It is currently an indispensable imaging tool in the diagnostic workup of cerebral stroke imaging and is also becoming an important cornerstone in cancer imaging.

It has been widely demonstrated that DWI and its advanced derivatives (diffusion tensor imaging, diffusion kurtosis imaging, etc.) all provide meaningful tissue characterization parameters. Higher image resolution, reduced image distortion artifacts, and standardized DWI protocols will further enhance the benefits of DWI in many clinical applications and improve diagnostic accuracy and patient care.

M. Scheel
Department of Radiology, Charité – Universitätsmedizin Berlin, Berlin, Germany
e-mail: Michael.Scheel@charite.de

© Springer International Publishing AG 2018
I. Sack, T. Schaeffter (eds.), *Quantification of Biophysical Parameters in Medical Imaging*, https://doi.org/10.1007/978-3-319-65924-4_17

17.1   Introduction

Diffusion-weighted imaging (DWI) is an accepted and widely used MRI technique today, most importantly in the diagnostic workup of cerebral stroke imaging. However, DWI has also proved to be a useful technique in many other clinical applications and is currently becoming an important cornerstone in cancer imaging.

DWI can be performed today on almost all clinical MRI scanners. The main advantages over many other advanced MR techniques is that (1) it is not dependent on contrast agent application like dynamic contrast-enhanced perfusion imaging, (2) it does not need specialized hardware like MR elastography, and (3) it is a relatively robust and easy technique to apply, which requires little user intervention, unlike, for example, spectroscopy. This makes its integration into a clinical imaging protocol relatively easy.

The information that can be gained from DWI is complementary to clinical routine MRI techniques, i.e., T1- and T2-weighted imaging. T1- and T2-relaxation times are governed by proton interactions with the surrounding molecular environment and are, as parameters, meaningful only in the context of MRI. DWI, however, measures a biophysical tissue property that is meaningful both in and outside the context of MRI. DWI shares this capability with other techniques, such as dynamic contrast-enhanced imaging or arterial spin labeling (measuring perfusion), MR spectroscopy (measuring metabolite concentration), and MR elastography (measuring mechanical tissue properties, e.g., stiffness).

Compared to standard clinical routine imaging, the spatial resolution of most DWI images is low (at the scale of around $2 \times 2 \times 2$ mm^3). However, the DWI image contrast is based on water molecule displacement at a micrometer scale. DWI therefore does not excel at high anatomic accuracy but provides a physically meaningful parameter for tissue characterization. Pathological tissue changes often go along with changes in diffusion properties, which DWI can detect, and so it is very useful clinically.

This chapter will give a general introduction into DWI and its clinical application. It will explain the basics of the diffusion process, DWI signal acquisition and postprocessing, including examples on how DWI is used in clinical routine imaging and in clinically oriented imaging research.

17.2   Technical and Biological Aspects of Diffusion-Weighted MRI

17.2.1 Sensitizing MRI Sequences to Diffusion

Most MRI sequences can be sensitized to diffusion. For example, a standard spin-echo sequence can be diffusion sensitized by using two additional gradients symmetrically placed around the 180° refocusing pulse (see Fig. 2.6c in Chap. 2).

As known from any MRI sequence, the insertion of a gradient will introduce a phase shift in spins, and the phase shift amount is directly dependent on the magnetic field strength of that gradient at each spatial location. In diffusion-sensitized spin-echo sequences, the first diffusion-sensitizing gradient will introduce a phase shift depending on the gradient strength at each position. This phase shift is then flipped by the 180° refocusing pulse. The second diffusion-sensitizing gradient is identical in length and strength to the first, and will then reverse all phase shifts that were initially introduced. However, this only holds true for spins that stay stationary. All spins undergoing a displacement within the time between both gradients will experience a different field strength during the second gradient compared to the first, which results in an incomplete rephasing and consequently in a signal loss. Two aspects are important to note here: (1) that the average displacement of *all* spins in an imaging voxel governs the signal and (2) that in each measurement only the net displacement along the direction of the applied diffusion-weighting gradient can be quantified. To obtain diffusion strength in different directions, multiple measurements with different gradient directions must be performed.

As a consequence, image intensity in a DWI image is low in regions where diffusion was high along the direction of the diffusion-sensitizing gradient. Another important fact that should be remembered when interpreting DWI images is that most DWI sequence schemes used today are T2 weighted (long TR typically around 6000–10,000 ms and long TE typically around 60–120 ms). Consequently, any lesion or tissue with long T2-relaxation times will have a high signal in T2 and depending on the diffusion weighting of the sequence will also show a relatively high signal in DWI. This has been termed the "T2 shine-through" effect [1].

As outlined in Sect. 1.7.3, quantification of diffusion in any image voxel is based on the Stejskal-Tanner equation. This equation describes the relationship between diffusion coefficient and MRI signal intensity. The DWI signal intensity (S) is the solution of the integral equation given in Eq. (2.68) of Chap. 2 [2]:

$$S = S_0 \cdot e^{-bD} \text{ with } b = \gamma^2 g^2 \tau^2 \left(\tau_\Delta - \tau / 3 \right) \qquad (17.1)$$

(S = DWI signal, S_0 = signal of non-diffusion-weighted T2 image, D = diffusion coefficient, γ = gyromagnetic ratio of the protons, g = gradient strength, δ = gradient duration, τ_Δ = time between the diffusion gradients)

The diffusion process in biological tissue is almost never a "free diffusion" but is governed by the complex interaction of water molecules with cellular or extracellular structures. Thus, the diffusion coefficient in tissues is therefore termed the "apparent diffusion coefficient" (ADC). The diffusion time, i.e., the time τ_Δ between both gradients, is proportional to the distances that are probed with a specific DWI sequence. When the diffusion time is too short, water molecules are hardly experiencing any diffusion barriers, compromising the sensitivity with respect to tissue characterization. Longer diffusion times give more sensitivity to changes or differences in diffusion barriers. At the same time, longer diffusion times lead to a prolonged echo time of the MR sequence, causing additional signal loss due to T2-signal decay. Careful adjustment of DWI protocols is therefore required to find a compromise between diffusion sensitivity and signal quality. The sensitivity to diffusion is adjusted in DWI protocols by means of the b-value, which combines all the relevant diffusion imaging parameters such as the amplitude of the encoding gradients and their duration and temporal distance in the sequence (see Eq. (17.1)).

It is important to note that signal loss due to motion occurs with any type of motion, not only with diffusion. At very low b-values (up to 100 s/mm^2), perfusion effects play an important role [3, 4]. Already in the early stages of DWI, Denis Le Bihan introduced this approach to assess motion from perfusion, a concept termed intravoxel incoherent motion [4]. Also any other motion, e.g., by subject motion or pulsating vessels or other sources of motion, will influence diffusion measurements. Vice versa, the encoding gradients in flow MRI or MR elastography are also sensitive to diffusion. A recently introduced method exploits the effect of water diffusion on the MRI magnitude signal in MR elastography for measuring both diffusion (at low b-values, i.e., in the range of 100–1000 s/mm^2 in a preclinical MRI scanner) and elasticity in the mouse at the same time [5, 6]. For the sake of simplicity, the following discussion will neglect the influence of other types of motion, other than diffusion, to the DWI signal.

17.2.2 Basics of the Diffusion Process

Molecules in a liquid possess nonzero kinetic energy due to thermal exchange with their environment. Their inherent kinetic energy causes them to move and collide with each other, giving rise to random walk-like motion trajectories that are restricted by barriers, such as tissue boundaries or vessel walls. For regular diffusion processes such as free water diffusion, this random molecule motion is described best by a Gaussian displacement distribution, $P(x,y,z,t)$, which quantifies the probability to find a molecule at position (x,y,z) at time t (see Eq. (2.5) in Chap. 2) [7].

The displacement distribution is governed by the properties of the liquid, the temperature and the diffusion time. For example, the diffusion coefficient D for water at 37°C is approximately 3.0×10^{-3} mm^2/s [7]. Given that the standard deviation of a Gaussian distribution is $\sigma = \sqrt{2 \cdot D \cdot T}$, we find for water at 37°C that approx. 68% of all molecules are not farther than 17 μm away from their origin after a diffusion time of 50 ms. At the same time 95% of all molecules will be within 34 μm, i.e., two times the standard deviation, from their original position.

In contrast to water molecules diffusing freely in an open system, diffusion in biological tissues is restricted and influenced in many ways [8]. First of all, the liquid that is being investigated is not free water but widely immobilized by proteoglycans in the interstitial spaces (see Chap. 6) with diffusion properties different from pure water. Secondly, intracellular organelles and the cytoskeleton need to be taken into account as diffusion barriers, although they have been demonstrated to influence DWI only at a minor scale [9]. Last but not least, the main diffusion barriers influencing the diffusion process are cell membranes at the diffusion time scale that is usually captured by DWI (20–80 ms). Depending on the density and orientation of cell membranes in a given tissue, the diffusion of water molecules is influenced in both magnitude and orientation.

As a result, the diffusion process becomes restricted (e.g., brain ADC around 0.8×10^{-3} mm^2/s) and depending on the tissue directionally dependent, giving rise to a tensorial property of D. Analysis of anisotropic diffusion requires measurement

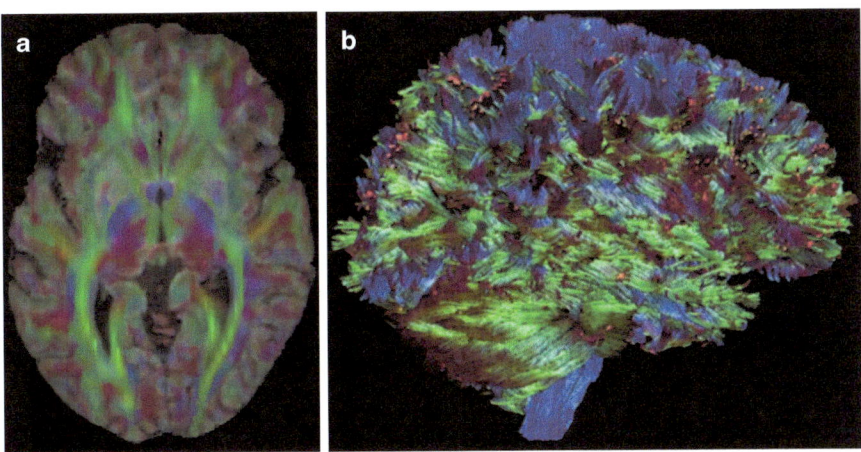

Fig. 17.1 Examples of diffusion tensor imaging-based postprocessing. Figure (**a**) shows the color-coded principle direction (*green* = anterior-posterior, *red* = left-right, *blue* = head-feet) of fibers in each voxel. Figure (**b**) is an example of a whole-brain tractography

of the tensor elements in *D*, which is no longer a scalar, but a rank-two tensor (see Table 2.2, Chap. 2). This is used, for example, in diffusion tensor imaging (DTI), to describe the spatial diffusion profile in every voxel [10]. DTI is mainly used in neuroscience research including human brain fiber tracking or microstructure assessment in neurodegenerative brain diseases, such as Alzheimer's disease, amyotrophic lateral sclerosis, normal pressure hydrocephalus, or schizophrenia [11–14] (Fig. 17.1).

17.3 Application of DWI in Clinical Imaging

17.3.1 Stroke Imaging

Nowadays, DWI is an essential imaging tool and part of almost every head MRI protocol in clinical routine. DWI has evolved from its invention in the late 1980s into a broad clinically applicable imaging technique in an exceptionally short time. Today, stroke imaging is by far the most important clinical application for DWI [15]. It is an important tool in cerebral ischemia diagnosis and has tremendous impact on patient management and treatment.

Compared to computed tomography (CT), the usage of MRI has substantially improved ischemic lesion detection. In CT scans ischemic brain regions are hard to identify, especially in the early stages, and for small (<0.5 cm) ischemic lesions since they initially demonstrate only a very subtle image intensity reduction. In MRI scans, ischemic brain lesions appear hyperintense in T2-weighted images and can be detected more easily [16]. Meanwhile, ischemic lesions in the acute phase and the chronic phase appear T2-hyperintense. It is of utmost importance to delineate an acute from an old lesion for adequate patient management and therapeutic

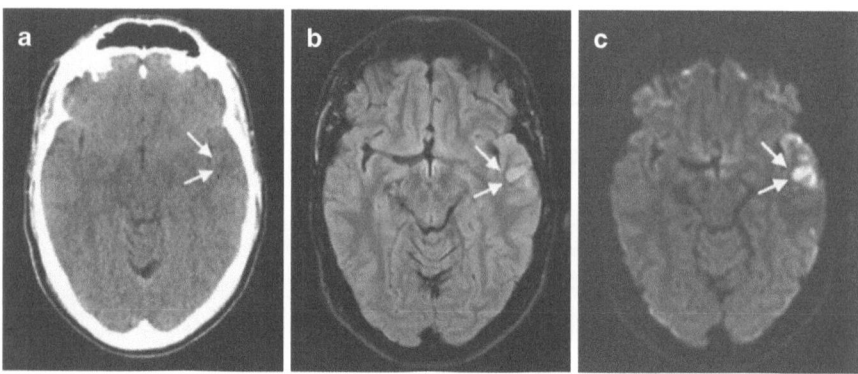

Fig. 17.2 Small cortical ischemic lesion that is not detectable on CT images (**a**) and hard to appreciate on T2-FLAIR images (**b**) and readily detected on DWI (**c**)

decision-making. Acute ischemic lesions are characterized by a relatively strong diffusion restriction and consequently show a hyperintensity in DWI images while "old" ischemic lesions exhibit normal or reduced signal DWI intensities. DWI was demonstrated to improve the accuracy of acute stroke lesion detection [17], most likely because the DWI changes in the early phase of stroke are more pronounced compared to changes in T2-weighted images.

While T2-weighted MRI in general has increased the detection accuracy of stroke lesions compared to CT, DWI has substantially increased the specificity of T2-hyperintense lesions in the acute stroke setting [17]. It is commonly accepted that in the chronic phase of ischemic lesions, i.e., after 2 weeks, the breakdown of cellular diffusion barriers causes ADC value elevation, but the pathophysiological basis of the substantial diffusion restriction during the acute phase is still a matter of debate (Fig. 17.2).

The temporal changes of DWI and T2-weighted imaging during different stages of ischemia have been extensively studied in the past. The interplay of DWI hyperintensity, ADC values, and pathophysiological changes is complex. ADC values are decreased minutes after stroke onset, and DWI signal is consequently hyperintense. ADC values stay low (~30%) during the acute stage (week 1) and slowly return to normal values during the subacute phase (week 2). However, the DWI signal stays hyperintense for a relatively long time. This is due to the increased intensity in T2-weighted images ("T2 shine-through effect").

In the chronic phase (>2 weeks) of an ischemic lesion, ADC values are elevated due to high T2-signal and low DWI signal intensities [18] (Fig. 17.3).

Ischemic brain tissue is characterized by a shift of water from the extracellular space into the intracellular space of neurons. Due to reduced energy supply in ischemic brain regions, the membrane pumps that are responsible for maintaining ionic gradients fail and subsequently an osmotically driven influx of water leads to intracellular swelling. This theory states that massive neuronal cell swelling leads to reduced extracellular space, which in turn increases the tortuosity of extracellular pathways, and consequently a diffusion restriction in the extracellular space occurs.

Recently, a different biophysical mechanism was proposed to explain the diffusion restriction in acute stroke named neurite beading [19]. Neuronal axons react to

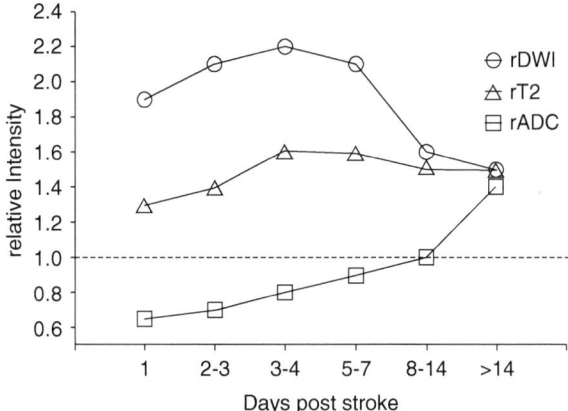

Fig. 17.3 Temporary course of T2-signal intensity, DWI signal intensity, and ADC in stroke lesions relative to healthy tissue (Adapted from [18], with permission)

cellular stress with focal constrictions and enlargement along the axon. In healthy neuronal tissue, diffusion perpendicular to the main axis of axons is mostly restricted due to the cell membrane. In contrast, diffusion parallel to the main axis of axons is almost unrestricted. In a beaded axonal structure, the main diffusion restriction would be based on restriction along the main axis of each neurite. It was demonstrated that the change from a tubular shape of an axon into a beaded shape is sufficient enough to explain the DWI changes [19].

However, the pathophysiological basis for diffusivity changes in ischemic tissue remains a matter of debate and future research in this field is warranted.

17.3.2 Cancer Imaging

In addition to its role in the diagnostic workup for cerebral ischemia, DWI is increasingly being used in cancer imaging. In general, tissue diffusion in solid tumors is restricted. Cell membranes are the main contributor in tissue diffusion properties [8]; therefore tissue ADC is directly related to the membrane density and consequently to the cellularity of a region. Malignant tumors generally show lower ADC values compared to benign tumors. Even though overlap exists, these values are remarkably similar across different organs, e.g., breast, liver, and prostate [20]. DWI has also proved to be useful in monitoring treatment response. Effective chemotherapy or radiotherapy results in tumor necrosis and hence in elevated diffusion that can be monitored with DWI. It should be noted, however, that DWI has a rather low resolution and thus is always complementary to classic T1w and T2-weighted images which have a much higher image resolution. This chapter will focus on examples of cancer DWI in breast, liver, and prostate.

Several studies demonstrated that ADC values of malignant breast tumors (~0.8–1.2 mm²/s) are lower than that of benign tumors (1.4–2.0 mm²/s) or normal tissue [21]. The optimum imaging parameters for a reliable cutoff between these two tumor categories are still a matter of debate [22]. A relatively close correlation was also demonstrated between ADC values and tumor grading, reflecting the

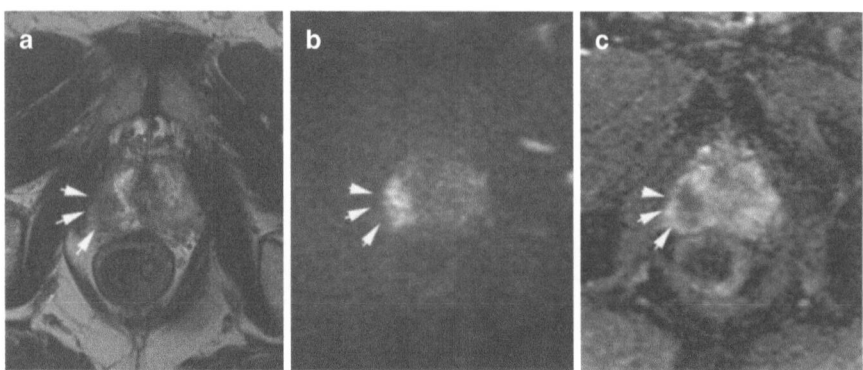

Fig. 17.4 Prostate cancer as a T2-hypointense lesion (**a**) with high DWI signal intensity (**b**) and consequently low ADC values (**c**)

aggressiveness of a tumor [23]. Interestingly, an ADC association to treatment response was also demonstrated by an approximately 50% increase in ADC values in patients who responded to treatment, while nonresponding patients showed ADC increases of only about 20% [24, 25].

Malignant liver tumors show ADC values (1.5×10^{-3} mm^2/s) that are significantly lower than that of benign tumors (~2.5×10^{-3} mm^2/s) [26]. However, some of the benign liver lesions with a rather high cellularity (e.g., focal nodular hyperplasia) are difficult to differentiate using ADC measurement alone. In the clinical setting, liver lesions are mainly characterized based on T2-weighted imaging and liver-specific MRI contrast agents. It was demonstrated that DWI-based characterization provides additional information as an adjunct to both those techniques. When DWI is combined with those techniques, a more precise characterization of primary liver malignancies and metastases can be achieved [27–29].

The diagnostic value of DWI in the diagnostic workup of prostate cancer has been shown extensively [30]. Therefore, DWI is now one of the imaging cornerstones in the standard image-based grading scheme for prostate cancer evaluation. Prostate cancer, especially the rather aggressive subtypes, is characterized by high cellularity [31, 32] and consequently lower ADC values (malignant 1.3–1.4×10^{-3} mm^2/s vs. benign 1.8–2.0×10^{-3} mm^2/s). In the monitoring of prostate cancer recurrence, DWI has been demonstrated to be more sensitive than T2-weighted images alone [33] (Fig. 17.4).

Despite great advances in tumor characterization in the last few years, the limited resolution of DWI makes it difficult to detect smaller malignant lesions. Future improvements in image resolution will be of great diagnostic and clinical benefit.

17.4   Medical Imaging as Quantitative Science

In this context it must be noted that most of the clinical radiological assessments performed today are not based on quantitative assessment, but rather qualitative signal changes. Pathologic tissue is characterized by signal changes that are relatively easy to observe by comparison to the surrounding "normal" tissue. However, the demand for true quantitative imaging is increasing. Diagnostic tests (biochemical, genetic, imaging, etc.) are characterizing patients and their specific diseases on an increasingly detailed level. In a true quantitative imaging approach, one would aim to compare an imaging parameter estimate, for example, the tissue mean diffusivity, to reference values. These reference values could be from a group of the healthy population or from a control group of patients. A simple analogy to quantitative imaging would be the results of a blood test, where "pathologic" means outside the normal distribution of healthy population values.

It is unlikely that a single laboratory or imaging test will ever decide upon the diagnosis of a specific patient alone. Clinically, laboratory and imaging results will always have to be integrated to give the physician certainty about diagnosis or which treatment option to proceed with. However, before clinical decisions are made based on imaging markers, both their accuracy and precision have to be ensured. Accuracy is the correspondence of a measure with its true value and precision being the reliability of that test. For valid and clinically useful interpretation, these measurement methods need to be standardized across equipment and institutions and also compared to healthy control populations. More advances in standardization need to be achieved before "imaging as a quantitative science" can become a reality in the clinical setting [34]. After more than 30 years of developments in DWI, this notion is still highly relevant for the translation of diffusion-based quantitative imaging markers into standard clinical examinations.

Conclusion

DWI and related methods provide a meaningful imaging contrast complementary to T1- and T2-weighted sequences. Pathological tissue changes are often associated with changes in the diffusion properties of water molecules. Although the spatial image resolution is comparatively low, DWI makes it possible to detect, quantify, and visualize these changes in water diffusivity which normally relate to a micrometer scale.

DWI is firmly established in stroke imaging, and it is currently becoming an important cornerstone in cancer imaging. More advanced derivatives of diffusion imaging could support clinical decisions in the future, especially in neurodegenerative diseases. However, further technological advances are needed, such as higher image resolution, faster acquisition, and reduction of distortion artifacts. Standardization of DWI protocols across vendors and institutions will further enhance the benefits of DWI in many clinical applications and improve diagnostic accuracy and patient care.

References

1. Burdette JH, Elster AD, Ricci PE. Acute cerebral infarction: quantification of spin-density and T2 shine-through phenomena on diffusion-weighted MR images. Radiology. 1999;212:333–9. https://doi.org/10.1148/radiology.212.2.r99au36333.

2. Stejskal EO, Tanner JE. Spin diffusion measurements: spin echoes in the presence of a time-dependent field gradient. J Chem Phys. 1965;42:288–92. https://doi.org/10.1063/1.1695690.
3. Le Bihan D, Breton E, Lallemand D, Aubin ML, Vignaud J, Laval-Jeantet M. Separation of diffusion and perfusion in intravoxel incoherent motion MR imaging. Radiology. 1988;168:497–505. https://doi.org/10.1148/radiology.168.2.3393671.
4. Le Bihan D, Breton E, Lallemand D, Grenier P, Cabanis E, Laval-Jeantet M. MR imaging of intravoxel incoherent motions: application to diffusion and perfusion in neurologic disorders. Radiology. 1986;161:401–7. https://doi.org/10.1148/radiology.161.2.3763909.
5. Yin Z, Kearney SP, Magin RL, Klatt D. Concurrent 3D acquisition of diffusion tensor imaging and magnetic resonance elastography displacement data (DTI-MRE): theory and in vivo application. Magn Reson Med. 2017;77:273–84. https://doi.org/10.1002/mrm.26121.
6. Yin Z, Magin RL, Klatt D. Simultaneous MR elastography and diffusion acquisitions: diffusion-MRE (dMRE). Magn Reson Med. 2014;71:1682–8. https://doi.org/10.1002/mrm.25180.
7. Hagmann P, Jonasson L, Maeder P, Thiran J-P, Wedeen VJ, Meuli R. Understanding diffusion MR imaging techniques: from scalar diffusion-weighted imaging to diffusion tensor imaging and beyond. Radiogr Rev Publ Radiol Soc N Am Inc. 2006;26(Suppl 1):S205–23. https://doi.org/10.1148/rg.26si065510.
8. Beaulieu C. The basis of anisotropic water diffusion in the nervous system – a technical review. NMR Biomed. 2002;15:435–55. https://doi.org/10.1002/nbm.782.
9. Beaulieu C. The biological basis of diffusion anisotropy. In: Johansen-Berg H, Behrens TEJ, editors. Diffusion MRI: from quantitative measurement to in-vivo neuroanatomy. Boston, MA: Academic; 2009. p. 105–27.
10. Basser PJ, Mattiello J, LeBihan D. MR diffusion tensor spectroscopy and imaging. Biophys J. 1994;66:259–67. https://doi.org/10.1016/S0006-3495(94)80775-1.
11. Agosta F, Galantucci S, Filippi M. Advanced magnetic resonance imaging of neurodegenerative diseases. Neurol Sci. 2017;38:41–51. https://doi.org/10.1007/s10072-016-2764-x.
12. Prokscha T, Guo J, Hirsch S, Braun J, Sack I, Meyer T, Scheel M. Diffusion tensor imaging in amyotrophic lateral sclerosis-increased sensitivity with optimized region-of-interest delineation. Clin Neuroradiol. 2013. https://doi.org/10.1007/s00062-013-0221-2.
13. Scheel M, Diekhoff T, Sprung C, Hoffmann K-T. Diffusion tensor imaging in hydrocephalus – findings before and after shunt surgery. Acta Neurochir. 2012;154:1699–706. https://doi.org/10.1007/s00701-012-1377-2.
14. Scheel M, Prokscha T, Bayerl M, Gallinat J, Montag C. Myelination deficits in schizophrenia: evidence from diffusion tensor imaging. Brain Struct Funct. 2013;218:151–6. https://doi.org/10.1007/s00429-012-0389-2.
15. Roberts TPL, Rowley HA. Diffusion weighted magnetic resonance imaging in stroke. Eur J Radiol. 2003;45:185–94.
16. Moreau F, Asdaghi N, Modi J, Goyal M, Coutts SB. Magnetic resonance imaging versus computed tomography in transient ischemic attack and minor stroke: the more you see the more you know. Cerebrovasc Dis Extra. 2013;3:130–6. https://doi.org/10.1159/000355024.
17. Latchaw RE, Alberts MJ, Lev MH, Connors JJ, Harbaugh RE, Higashida RT, Hobson R, Kidwell CS, Koroshetz WJ, Mathews V, Villablanca P, Warach S, Walters B, American Heart Association Council on Cardiovascular Radiology and Intervention, Stroke Council, and the Interdisciplinary Council on Peripheral Vascular Disease. Recommendations for imaging of acute ischemic stroke: a scientific statement from the American Heart Association. Stroke. 2009;40:3646–78. https://doi.org/10.1161/STROKEAHA.108.192616.
18. Lansberg MG, Thijs VN, O'Brien MW, Ali JO, de Crespigny AJ, Tong DC, Moseley ME, Albers GW. Evolution of apparent diffusion coefficient, diffusion-weighted, and T2-weighted signal intensity of acute stroke. AJNR Am J Neuroradiol. 2001;22:637–44.
19. Budde MD, Frank JA. Neurite beading is sufficient to decrease the apparent diffusion coefficient after ischemic stroke. Proc Natl Acad Sci U S A. 2010;107:14472–7. https://doi.org/10.1073/pnas.1004841107.
20. Malayeri AA, El Khouli RH, Zaheer A, Jacobs MA, Corona-Villalobos CP, Kamel IR, Macura KJ. Principles and applications of diffusion-weighted imaging in cancer detection, staging,

and treatment follow-up. Radiogr Rev Publ Radiol Soc N Am Inc. 2011;31:1773–91. https://doi.org/10.1148/rg.316115515.

21. Sinha S, Lucas-Quesada FA, Sinha U, DeBruhl N, Bassett LW. In vivo diffusion-weighted MRI of the breast: potential for lesion characterization. J Magn Reson Imaging JMRI. 2002;15:693–704. https://doi.org/10.1002/jmri.10116.

22. Pereira FPA, Martins G, Figueiredo E, Domingues MNA, Domingues RC, da Fonseca LMB, Gasparetto EL. Assessment of breast lesions with diffusion-weighted MRI: comparing the use of different b values. AJR Am J Roentgenol. 2009;193:1030–5. https://doi.org/10.2214/AJR.09.2522.

23. Costantini M, Belli P, Rinaldi P, Bufi E, Giardina G, Franceschini G, Petrone G, Bonomo L. Diffusion-weighted imaging in breast cancer: relationship between apparent diffusion coefficient and tumour aggressiveness. Clin Radiol. 2010;65:1005–12. https://doi.org/10.1016/j.crad.2010.07.008.

24. Park SH, Moon WK, Cho N, Song IC, Chang JM, Park I-A, Han W, Noh D-Y. Diffusion-weighted MR imaging: pretreatment prediction of response to neoadjuvant chemotherapy in patients with breast cancer. Radiology. 2010;257:56–63. https://doi.org/10.1148/radiol.10092021.

25. Sharma U, Danishad KKA, Seenu V, Jagannathan NR. Longitudinal study of the assessment by MRI and diffusion-weighted imaging of tumor response in patients with locally advanced breast cancer undergoing neoadjuvant chemotherapy. NMR Biomed. 2009;22:104–13. https://doi.org/10.1002/nbm.1245.

26. Miller FH, Hammond N, Siddiqi AJ, Shroff S, Khatri G, Wang Y, Merrick LB, Nikolaidis P. Utility of diffusion-weighted MRI in distinguishing benign and malignant hepatic lesions. J Magn Reson Imaging JMRI. 2010;32:138–47. https://doi.org/10.1002/jmri.22235.

27. Haradome H, Grazioli L, Morone M, Gambarini S, Kwee TC, Takahara T, Colagrande S. T2-weighted and diffusion-weighted MRI for discriminating benign from malignant focal liver lesions: diagnostic abilities of single versus combined interpretations. J Magn Reson Imaging JMRI. 2012;35:1388–96. https://doi.org/10.1002/jmri.23573.

28. Koh DM, Brown G, Riddell AM, Scurr E, Collins DJ, Allen SD, Chau I, Cunningham D, desouza NM, Leach MO, Husband JE. Detection of colorectal hepatic metastases using MnDPDP MR imaging and diffusion-weighted imaging (DWI) alone and in combination. Eur Radiol. 2008;18:903–10. https://doi.org/10.1007/s00330-007-0847-z.

29. Nishie A, Tajima T, Ishigami K, Ushijima Y, Okamoto D, Hirakawa M, Nishihara Y, Taketomi A, Hatakenaka M, Irie H, Yoshimitsu K, Honda H. Detection of hepatocellular carcinoma (HCC) using super paramagnetic iron oxide (SPIO)-enhanced MRI: added value of diffusion-weighted imaging (DWI). J Magn Reson Imaging JMRI. 2010;31:373–82. https://doi.org/10.1002/jmri.22059.

30. Weinreb JC, Barentsz JO, Choyke PL, Cornud F, Haider MA, Macura KJ, Margolis D, Schnall MD, Shtern F, Tempany CM, Thoeny HC, Verma S. PI-RADS prostate imaging – reporting and data system: 2015, version 2. Eur Urol. 2016;69:16–40. https://doi.org/10.1016/j.eururo.2015.08.052.

31. desouza NM, Reinsberg SA, Scurr ED, Brewster JM, Payne GS. Magnetic resonance imaging in prostate cancer: the value of apparent diffusion coefficients for identifying malignant nodules. Br J Radiol. 2007;80:90–5. https://doi.org/10.1259/bjr/24232319.

32. Sato C, Naganawa S, Nakamura T, Kumada H, Miura S, Takizawa O, Ishigaki T. Differentiation of noncancerous tissue and cancer lesions by apparent diffusion coefficient values in transition and peripheral zones of the prostate. J Magn Reson Imaging JMRI. 2005;21:258–62. https://doi.org/10.1002/jmri.20251.

33. Kim CK, Park BK, Lee HM. Prediction of locally recurrent prostate cancer after radiation therapy: incremental value of 3T diffusion-weighted MRI. J Magn Reson Imaging JMRI. 2009;29:391–7. https://doi.org/10.1002/jmri.21645.

34. Sullivan DC. Imaging as a quantitative science. Radiology. 2008;248:328–32. https://doi.org/10.1148/radiol.2482080242.

Quantification of Functional Heterogeneities in Tumors by PET Imaging

18

Winfried Brenner, Florian Wedel, and Janet F. Eary

Abstract

Among the many attributes of molecular imaging methods is generation of image data that can be subjected to a number of different analytical approaches to characterize tumor biology, report treatment effect, and predict risk for poor outcome. Using both semi-quantitative and quantitative image data, these methods can be applied to calculate tumor spatial heterogeneity in biologically specific imaging agent uptake and utilization. Tumor imaging heterogeneity characteristics represent the intra- and intertumoral differences in tumor genetics and biology and can be used to understand tumor behavior. Several image heterogeneity analysis methods have been validated in clinical image datasets and show strong correlations with patient outcome. In the near future, these types of measures will become a part of clinical practice in cancer image interpretation and tumor molecular characterization.

18.1 Introduction to Tumor Heterogeneity

A common feature of malignant tumors is biological heterogeneity based on genetic variations between tumor cells [1]. These genetic variations translate into different molecular and cellular characteristics of single tumor cells and cell clones within a primary tumor, between primary tumors and metastases, and between metastases within a patient. Under light microscopy, malignant tumors

W. Brenner (✉) • F. Wedel
Department of Nuclear Medicine, Charité – Universitaetsmedizin Berlin, Berlin, Germany
e-mail: winfried.brenner@charite.de

J.F. Eary
US National Cancer Institute Cancer Imaging Program, Bethesda, MD, USA

© Springer International Publishing AG 2018
I. Sack, T. Schaeffter (eds.), *Quantification of Biophysical Parameters in Medical Imaging*, https://doi.org/10.1007/978-3-319-65924-4_18

show cellular atypia, increased mitosis, crowding, invasion, and necrosis. As characterization of living tissues has become more refined, genetic, proteomic, and molecular expressions for the wide range in tumor biologic behavior have been emerging.

Investigators in search of biomarkers that indicate mutations and other disease characteristics for new targeted therapy approaches are evaluating specific tumors for their unique characteristics. In colon cancer and other tumors, large databases are being examined to discover genes that are more frequently mutated and whose products can be expected to become therapy targets [2]. Furthermore, genetic alterations often result in downstream changes in intracellular pathways which can change cellular function and characteristics and can become the target of new therapeutic approaches. Groups of tumors with common mutations might allow subtyping of tumors into treatment and risk groups of tumor aggressive behavior [3, 4]. Genetic analyses also have discovered the basis for heterogeneity in tumor biological characteristics [5].

These findings on tumor heterogeneity present a number of challenges to the targeted therapy discovery process and to identification of patients with tumors that would benefit by novel treatment combinations. Recent reviews describe the challenges that tumor biologic heterogeneity present [6–8] and recommend that tumor characterization be geared toward understanding the processes that drive differences between similar tumors in patients, between the primary tumor and metastases, and changes in tumor phenotype expression over time [9, 10].

The tumor phenotype represents the entirety of its presentation and behavior in the patient and usually defines how the tumor is treated by surgical excision, radio- and chemotherapy, and targeted therapy combinations. Personalized medicine is a goal in modern cancer therapy that aims for optimal treatment. Although it is dependent on tumor characteristics in an individual, it is often still based on limited information from tumor staging which includes imaging, histopathology, and immunohistochemistry, while genetic profiling and whole tumor genome sequencing are not routinely performed in individual patients in current clinical practice. Initial tumor staging, i.e., information on the extent of tumor spread, is usually based on imaging. CT, MRI, and positron emission tomography (PET), as well as simple image-derived parameters such as tumor size and contrast enhancement in CT and MRI, and tumor uptake reported as standardized uptake value (SUV) of a radioactive tracer within the tumor in PET.

Standard medical images however contain more usable information on tumors than size and tracer uptake. The field of *radiomics*, i.e., the extraction of a multitude of quantitative features from a digital image "may allow non-invasive molecular and genetic profiling of tumors as a further step toward personalized medicine" [11]. New imaging strategies for tumor texture analysis provide a possible advantage in characterizing tumor phenotype heterogeneity, as they can be used to image the entire body, encompassing all tumor sites (this cannot be accomplished by histopathology due to practical reasons), and can be performed non-invasively over time for diagnosis, throughout treatment for response monitoring, and during follow-up.

18.2   PET Imaging, Common PET Tumor Tracers, and Quantification of Tumor Tracer Uptake

PET is considered as a primarily functional and quantitative imaging tool. Depending on the administered tracer, molecular and/or cellular targets such as receptors, transporters, etc., as well as metabolic pathways can be quantitatively assessed.

18.2.1 Tracers

This section gives just a brief overview on PET tracers for clinical tumor characterization, more details on tracer chemistry in PET are provided in Chap. 11 *Innovative PET and SPECT Tracers*. F-18-2-fluoro-2-deoxy-glucose (FDG) is the most used tracer in PET in general and in oncology. This imaging agent is taken up by tumor cells via glucose transporters, mainly glucose transporter-1 (Glut-1), and then undergoes phosphorylation by the enzyme hexokinase into FDG-6-phosphate (FDG-6P) in the same glycolysis pathway as glucose. At this stage of cellular metabolism, further progress through the metabolic pathway is stopped because the next enzyme glucose-6-phosphate isomerase is highly substrate-specific and does not accept FDG-6P as a substrate. FGD-6P is a negatively charged anion and can neither leave the cell by diffusion nor by glucose transporters. FDG-6P dephosphorylation is a very slow process in tumors and all organs except the liver. Thus, FDG-6P is trapped within the cell. The amount of trapped FDG-6P depends on the expression and number of glucose transporters and hexokinase enzymatic activity [12]. FDG PET imaging therefore serves as a biomarker for tissue glucose metabolism which is typically increased in almost all malignant tumors.

Other commonly applied PET tracers in oncology are F-18-fluoro-ethyl-tyrosine (FET) as a substrate for amino-acid transporters which is used for brain tumor imaging [13]; F-18-misonidazole (FMISO) as a marker of intracellular hypoxia [14, 15]; Ga-68-DOTA-TOC or -TATE as a ligand for somatostatin cell surface receptors characteristically overexpressed by neuroendocrine tumors [16–20]; Ga-68-PSMA as a ligand for membrane-bound glycoprotein prostate membrane-specific antigen which is typically overexpressed by prostate cancer cells [21–23]; and oxygen-15 labeled water which as a freely diffusible molecule can be used as a marker for tumor blood flow (see also Chap. 21 *Radionuclide Imaging of Cerebral Blood Flow*) [24].

A combination of different PET tracers increases the information on characteristics of a given tumor: in patients with neuroendocrine tumors, the differential uptake of FDG and Ga-68-DOTA-TATE provides significant information on tumor aggressive features and prognosis (Fig. 18.1).

18.2.2 Quantification of Tumor Tracer Uptake

PET imaging is a quantitative imaging method which allows calculation of tracer uptake in a semi-quantitative as well as a true quantitative approach [26]. Using

Fig. 18.1 Imaging results of a patient with a well-differentiated pancreatic neuroendocrine tumor (NET) with a high receptor expression on Ga-68-DOTA-TATE, a relatively low FDG uptake (SUVmax 3.6), and a very high tumor blood flow on O-15 water PET. This is a typical tumor pattern in well-differentiated low-grade NET which is characterized by high somatostatin receptor expression and a very high tumor perfusion (on contrast-enhanced CT, arterial-phase images are considered mandatory for tumor detection [25]) but a relatively low FDG uptake. In high-grade, undifferentiated NET, the pattern changes to low or even missing somatostatin receptor expression but high FDG uptake, while tumor perfusion remains high

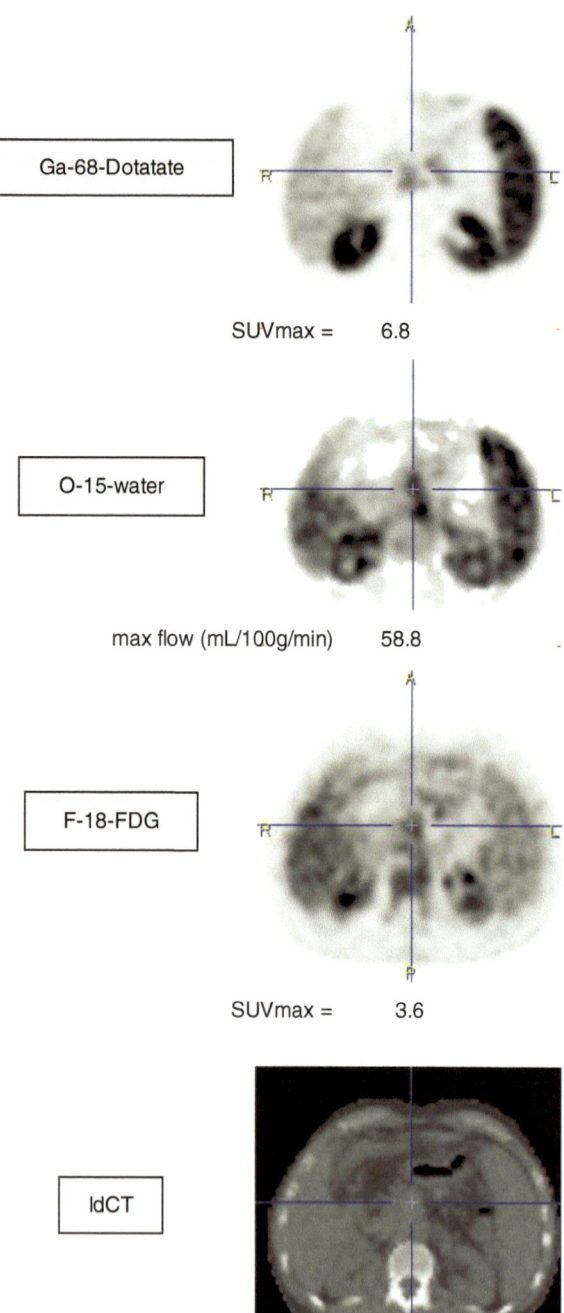

FDG for example, the trapped amount of FDG-6P can be measured semi-quantitatively as the tissue standardized uptake value (SUV) or quantitatively in μmol/min/g tissue by dynamic imaging and kinetic modeling such as Patlak or non-linear regression analysis [27].

18.2.2.1 Non-linear Regression Analysis

Based on the original three-compartment model of Sokoloff and coworkers [28, 29], a two-tissue, four-parameter compartment model for FDG tissue metabolism that consists of the plasma space and an unbound and a bound (phosphorylated) tissue compartment became the standard modeling approach for FDG [30]. In quantitating the physiological process of FDG tissue metabolism, the rate constants $K1$ to $k4$ describe the transport of FDG between the compartments: K_1 and k_2 represent the forward and reverse transport of FDG from plasma to the cell compartment, k_3 represents phosphorylation by hexokinase, and k_4 is dephosphorylation. In FDG models, k_4 is often set to zero ($k_4 = 0$) because of the very slow dephosphorylation process in tissues other than the liver. Tissue FDG net uptake K_{NLR} (mL/min/mL) can be calculated by non-linear regression as

$$K_{NLR} = K_1 \times k_3 / \left(k_2 + k_3 \right)$$

18.2.2.2 Patlak Analysis

As an alternative to non-linear regression, tissue FDG uptake in images can be estimated by Patlak graphical analysis [31, 32]. Results from this method correlate well with non-linear regression results although single rate constants are not calculated and k_4 is set to zero [30]. Patlak analysis involves linear regression and can be performed more easily and robustly than non-linear regression methods for tissue tracer uptake quantitation.

However, in the daily routine of a PET center, even simplified methods such as Patlak analysis are not used on a regular basis since they require both blood sampling and dynamic imaging for determining the tissue input function for the quantitative analysis. A sequence of dynamic PET scans over the tumor starting with the administration of the tracer is used to measure the tumor time-activity curves of the tracer. In parallel, arterial or venous blood samples are drawn during each time frame of the imaging sequence to measure the concentration of the tracer in the plasma. This blood time-activity curve serves as an input function in tracer kinetic modeling (see also R. Buchert for more details) in non-linear regression analysis as well as in Patlak graphical analysis.

18.2.2.3 Standardized Uptake Value (SUV)

The most widely used parameter for quantitation of PET studies in clinical practice is the SUV [33]. This semi-quantitative measure represents activity within a tissue region-of-interest (ROI) corrected for the injected activity and for patient weight or lean body mass as a substitute for the distribution volume of the patient [30]. There is a multitude of publications on this parameter in tumor treatment response assessment, and there are also widely accepted recommendations for using this parameter, the PERCIST criteria (PET Response Criteria in Solid Tumors) [34, 35]. The practical advantages of this approach are the need for static PET images only with no need for blood sampling or dynamic imaging. The calculation of tumor SUV in a PET image requires a transmission scan to generate true tissue activity in attenuation-corrected images which are currently standard in PET/CT clinical images:

$$SUV = \frac{A}{ID / m}$$

where A is the mean tissue activity (Mbq/mL or Mbq/g) within the ROI, ID is the injected activity (Mbq), and m is the patient body weight (kg).

18.3   Measures of Heterogeneity

PET tracer uptake can be estimated reliably in tumor tissue, and the tumor SUV is a widely used and accepted biomarker for tumor characterization in clinical trials for initial diagnosis and staging, treatment response evaluation (PERCIST), restaging, and follow-up. Modern high-resolution PET scanners with a spatial resolution in the range of 5 mm allow extraction of more information out of the tumor images than just the amount of tracer uptake in tissue regions of interest. The quantitative spatial and image analysis data present in today's clinical cancer imaging techniques provide datasets for characterizing tumor heterogeneity evaluating a variety of different parameters.

The focus in current image analysis literature is on tumor texture analysis and tumor spatial heterogeneity in FDG PET/CT, based on the measurement of spatial variations of voxel greyscale intensities within a tumor. This is termed intratumor heterogeneity. Differences between lesions within a patient, e.g., between primary tumor and metastases or between metastases, are termed intertumor heterogeneity [36].

18.3.1 Intratumor Heterogeneity

Intratumor heterogeneity of FDG uptake can be frequently observed in larger masses (>2 cm) of almost all tumor types and can be easily spotted by visual analysis (Fig. 18.2). Measurement of spatial variations in uptake and the generation of textural quantitative parameters, however, can be quite challenging. The first report on measuring tumor tissue heterogeneity in 3D FDG PET data was published in 2003 for a group of sarcoma patients [37]. In this paper, the authors described a statistical measure of the heterogeneity of the tissue characteristic "FDG uptake" that is based on the deviation of the distribution of the measured tissue uptake from a unimodal elliptically contoured spatial pattern, and presented an algorithm for computation of the measure based on volumetric ROI PET data. Details of the definition and mathematical computation of this new measure can be found in this paper by O'Sullivan,

Fig. 18.2 Intratumor heterogeneity of FDG uptake in a large mixed adeno-neuroendocrine carcinoma (MANEC) of the stomach, tumor size 63 × 78 × 47 mm on CT, Ki-67 index 90% indicating a highly mitotic and aggressive tumor. The heterogeneous FDG uptake in the tumor can be easily observed by visual inspection on fused axial PET/CT (*above*) and PET images (*below*)

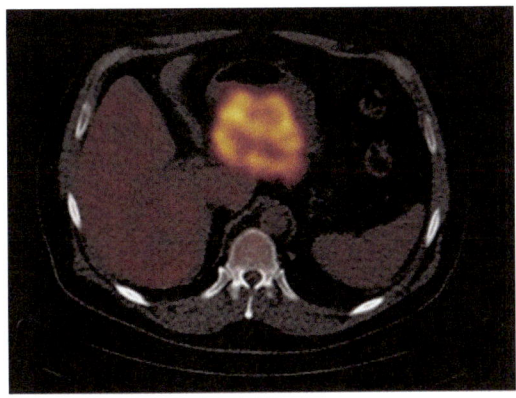

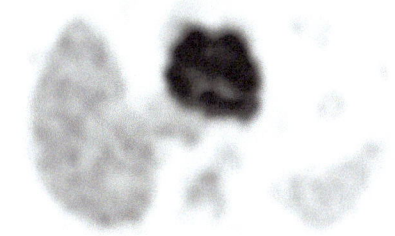

Eary, and coworkers [37] which is freely available on the Internet. Extracting data on FDG spatial uptake heterogeneity in a PET image, the authors showed in 76 patients that the degree of FDG spatial heterogeneity of their sarcoma was an independent predictor for patient survival even outperforming tumor SUV. Thus, this tumor characteristic was found to be a major risk factor for survival [38]. In subsequent papers, O'Sullivan, Eary, and coworkers presented updated and improved modeling approaches for calculating heterogeneity of FDG uptake in sarcoma tumor PET images. These further investigations found that analyses that utilized more complex tumor boundary shapes and edge detection algorithms improved the correlation of the tumor spatial heterogeneity results with patient outcome [39–42].

New strategies for analyzing intratumor heterogeneity of PET images have been published in the last decade that are based on a number of mathematical methods that describe the relationships between the counts per voxel and their position within an image. Statistics- and frequency-based methods as well as intensity histogram analyses have been most commonly applied. These calculate local tissue image features (depending on the respective PET tracer) of each voxel in the image and deduce new image parameters from the distributions of these local features. Further image processing and analysis methods to quantify the heterogeneity of voxel intensities are based on tumor image texture, fractal, and shape analyses. As texture analysis has been widely used for contrast-enhanced CT and MR imaging [43], more recent studies have also focused on extracting features from both PET and CT or MRI of hybrid PET/CT and PET/MR imaging [44–47]. For a more comprehensive summary of the various parameters and methods for quantifying tumor tissue heterogeneity, we refer the reader to the reviews of Chicklore et al. [43], Hatt and coauthors [48], or J.P. O'Connor [36]. In an interesting publication on intratumor heterogeneity for prediction of therapy response in esophageal cancer, Tixier et al. compared more than 40 quantitative parameters extracted from FDG PET images including maximum SUV, peak SUV, mean SUV, and a total of 38 textural features such as entropy, size, and magnitude of local and global heterogeneous and homogeneous tumor regions for comparison with patient treatment response. The authors could show "that tumor textural analysis can provide non-responder, partial responder, and complete-responder patient identification with higher sensitivity (76–92%) than any tumor SUV measurement" [49].

18.3.2 Asphericity

Inspired by the early image analysis research of Eary and O'Sullivan, we concentrated on creating an automated and easy-to-perform textural measure for tumor characterization which can be applied in daily PET routine image analysis. This novel measure of spatial irregularity of the FDG uptake in the primary tumor termed "asphericity" (ASP) was designed to characterize the deviation of the tumor's shape from sphere symmetry and to serve as a prognostic marker in tumors [50].

A common characteristic feature of malignant tumors is that aggressive tumors grow faster and more asymmetrical and with a more irregular shape than less proliferative tumors which are closer to a sphere's shape, mostly because of a lack of sufficient neoangiogenesis and variations of cellularity and cellular genetic heterogeneity. Exceptions may include low-grade gliomas with a less compact growth pattern than high-grade gliomas [51]. With respect to these well-known macroscopic features of tumors, we designed our new parameter as a measure for the deviation of the tumor's shape from sphere symmetry.

We defined asphericity as

$$ASP = 100^{*}\left(\sqrt[3]{H} - 1\right) \text{ with } H = \frac{1}{36\pi}\frac{S^3}{V^2}$$

where S and V are the surface and the metabolic tumor volume. ASP for a perfect sphere is zero and increases for any other lesion type with a more irregular and, thus, bigger surface.

Currently, there is no feasible way to calculate ASP on a fully automated basis because the segmentation for the initial discrimination of the tumor uptake from the uptake of nearby tissue is often difficult if not impossible for automated algorithms (Fig. 18.3). After automatic tumor delineation, e.g., by a threshold-based algorithm, the volumes-of-interest (VOI) often have to be adapted and corrected manually, e.g., to exclude inflamed but benign tissue within the borders of the VOI. Details of this process can be found in a publication by Hofheinz and coworkers [53].

ASP as a prognostic marker for patient outcome was developed as an easy-to-acquire numeric representation of tumor biology and malignancy in FDG PET/

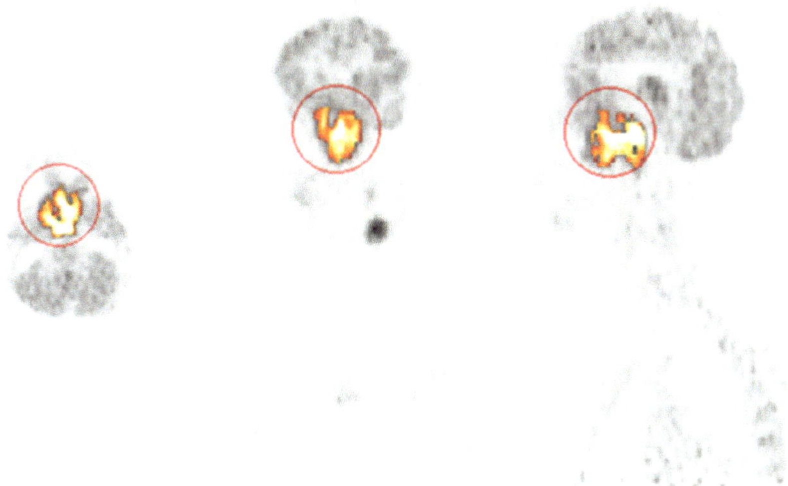

Fig. 18.3 Successful combined automatic and manual segmentation and discrimination of a nasopharynx carcinoma using ROVER software (http://www.abx.de/rover [52]) shown on representative axial (*left*), coronal (*center*), and sagittal (*right*) FDG PET images for further analysis of asphericity (ASP). Discrimination of FDG tumor uptake from background activity: in this case, the adjacent physiological uptake of FDG in the brain can be quite difficult by fully automatic segmentation

CT. It is evident that such a textural parameter would be greatly affected by therapies which aim to change metabolic functions of a tumor, like chemotherapy and radiation therapy. Therefore, studies so far are mostly concentrated on pre-therapeutic patients with different cancer entities (Fig. 18.4). In patients with local recurrences, ASP showed a rather limited prognostic value, as did any other investigated textural marker besides the presence of the recurrence itself [50].

In a first study on 52 patients with head and neck cancer, ASP could be shown as an independent prognostic marker for overall survival (OS) and progression free survival (PFS) [50]. The probability of a 2-year PFS decreased from 65% in patients with low ASP ($\leq$24) to 20% in patients with a higher ASP (>24) and 0% in patients with an additional large metabolic tumor volume ($\geq$11.5 mL). For OS, multivariate Cox regression analysis revealed high ASP (HR, 6.4; $p = 0.006$) as the only significant predictor for decreased OS, but not SUV and metabolic tumor volume [50]. These initial results were confirmed in a second study analyzing data from a separate institution (University Hospital Carl Gustav Carus, Dresden, Germany) with a different PET scanner and with a separate cohort of 37 patients with pre-therapeutic head and neck cancer. Again, ASP proved to be an independent prognostic factor for PFS and OS [54].

To investigate ASP as a general parameter for characterization of tumor biology in FDG PET, this analysis was applied in a cohort of 60 patients with non-small cell lung cancer (NSCLC) treated with curative intent [55]. In this study, ASP proved to be an independent prognostic factor for PFS (HR = 3.4, $p = 0.001$) and OS (HR OS = 2.97, $p = 0.03$), and it outperformed metabolic tumor volume as well as any other conventional metabolic parameter analyzed. Only the clinical parameter

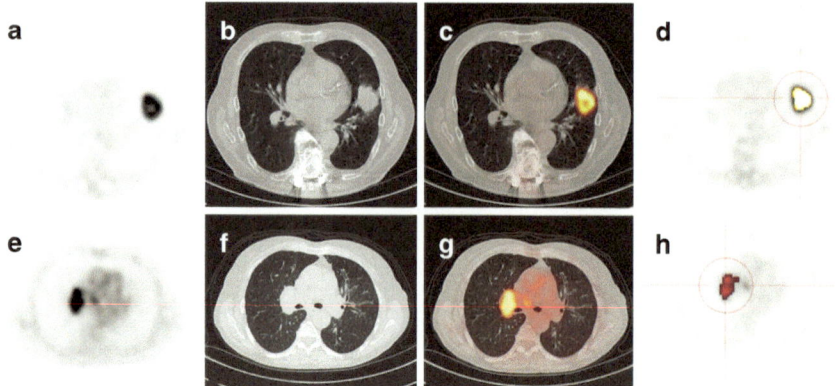

Fig. 18.4 Tumor lesion segmentation and VOI extraction (ROVER software) for further analysis of asphericity (ASP) in patients with non-small-cell lung cancer (axial PET (A/E), CT (B/F), fused PET/CT (C/G), and ROVER segmentation (D/H) images)

"primary surgical treatment" performed comparably well (HR PFS = 2.09, $p = 0.05$ – HR OS = 3.78, $p = 0.01$) [55].

These findings on prediction of survival are based on a strong correlation of ASP in FDG PET with histologic and genetic markers as shown recently in 83 patients with NSCLC prior to treatment [56]. In this study, ASP was significantly associated with TNM stage, the Ki-67 proliferation index, and a strong EGFR expression in the tumor. The correlation of survival with other image-derived markers such as SUV and metabolic tumor volume was non-significant [56].

In summary, ASP seems to have potential as an independent and comprehensive prognostic tumor texture marker mirroring tumor biological processes and genetic aberrations in tumors that can be used to identify patients with a high risk of an unfavorable course of disease. Of note, ASP can be used as a prognostic tumor texture marker on SPECT images as well as shown in 20 patients with gastro-entero-pancreatic neuroendocrine neoplasms undergoing SPECT/CT imaging with In-111 octreotide [57]. Thus, ASP can work as a spatial texture parameter of tissue heterogeneity independently of the physiology of the applied tracer (metabolic tracer, receptor ligand) and the imaging technology used (SPECT, PET).

18.3.3 Flow-Metabolism Mismatch

Another approach of our group for extracting spatial tissue information for tumor characterization and risk assessment is calculation of blood flow-metabolism mismatch. This is an image data comparison using FDG SUV for calculating tumor metabolism and oxygen-15 (O-15) labeled water for tumor blood flow assessment. O-15 water images are acquired by dynamic PET imaging, and the data is analyzed with kinetic modeling techniques.

In his seminal work on tumor glucose metabolism in the 1930s, Otto Warburg showed that anaerobic metabolism of glucose is a fundamental property of all tumors and that there is a relationship between the degree of anaerobic metabolism and tumor growth rate. A comprehensive summary on this topic with respect to modern imaging approaches can be found in Miles and Williams [58].

Based on Warburg's work, the group of Mankoff and coworkers used PET quantitative imaging for blood flow measurements with O-15 water and metabolism with FDG in tumor tissue. They could demonstrate that tumor blood flow and metabolism, although tightly coupled in most normal tissues, are often not well matched in tumors and that a flow-metabolism mismatch, i.e., high metabolism relative to blood flow, was associated with poor treatment response and poor survival in patients with locally advanced breast cancer [59–61].

In our pilot study on tumor blood flow-metabolism mismatch in patients with advanced cervical cancer, we aimed at evaluating the rational of this approach in another tumor entity. We developed a new automated voxel-based method for quantitative characterization of the spatial patterns of O-15 water flow represented by the rate constant K1, FDG metabolism represented by median SUV, and its mismatch [24]. First, the median SUV and the median K1 in the tumor were computed for each patient. Then a tumor voxel was classified to represent a flow-metabolism mismatch if its SUV was larger than the median SUV and at the same time its K1 was smaller than the median K1. The absolute mismatch volume was obtained by multiplying the number of mismatch voxels with the voxel volume [24]. This novel parameter for the tumor absolute mismatch volume and the spatial extent of the mismatch can be easily calculated by an automated quantification algorithm.

In the future, the clinical value of this parameter for prediction of response to treatment and disease-free and overall survival needs to be tested in prospective multi-center studies as well as in other tumor histologies.

Another practical aspect of future research will be the replacement of O-15 water as a marker for tumor blood flow. As O-15 and O-15 water are only available in a few research centers in the world supplied with a dedicated cylotron for production of O-15, O-15 water PET will not likely become a routine diagnostic tool. Moreover, O-15 water PET requires dynamic imaging and kinetic modeling for calculating tumor blood flow which are both laborious in clinical routine. Dynamic contrast-enhanced magnetic resonance imaging (DCE-MRI) and dynamic contrast-enhanced CT can measure regional tissue perfusion; however, this parameter differs from tissue blood flow measured by O-15 water PET. Using modern hybrid scanners, PET/CT and PET/MRI, CT- or MRI-based parametric substitutes for O-15 water blood flow will be a future topic of research for assessing the presence and extent of a tumor flow-metabolism mismatch.

18.4   General Aspects and Clinical Implications of Imaging Tumor Heterogeneity

More than two decades ago, there was a huge effort on decoding tumor genetic profiles, and the ultimate goal was whole-genome sequencing, first on a general scale as in the Human Genome Project and later especially under commercial aspects on a personal individual scale. Although these projects were successful and the human genome sequence has been fully known since 2003, all this information is only slowly being translated into an understanding of physiological and patho-physiological processes of diseases and into new treatment approaches. The same is to be expected with imaging and radiomics. It is easy to extract a series of numbers and parameters from digital images as shown by Tixier and coworkers [49], but what do these data mean, and how can we use them in medicine to make radiomics a success story [62]? Three major accomplishments are necessary for a future successful application of heterogeneity and tumor texture parameters in analogy to the PERCIST criteria representing the amount of tumor tracer uptake: first, the imaging community should concentrate on a limited number of promising, pathophysiologically understood and easy-to-calculate parameters which, secondly, have to be clearly defined by a standardized mathematical algorithm to be, thirdly, evaluated and validated in prospective multi-center clinical trials. Using the potential of modern molecular imaging as mainly represented by PET and MRI and ideally by hybrid PET/MRI, composite parameters combining the information from both imaging modalities will make important contributions to cancer patient care.

The newly established graduate school *BIOQIC—BIOphysical Quantitative Imaging Towards Clinical Diagnosis* (http://bioqic.de), funded by Deutsche Forschungsgemeinschaft, will support and help defining useful and clinically relevant image-derived parameters which we will fully understand from both a mathematical and a pathophysiological point of view and which we can successfully integrate into tumor imaging and daily medical practice.

Conclusion

Image-derived tumor heterogeneity characteristics and tumor texture parameters represent the intra- and intertumoral differences in tumor genetics and biology and can be used to characterize tumor phenotype, report treatment effects, and predict risk for outcome in an individual patient. Extracted from PET and MRI digital images, tumor texture parameters as well as novel composite parameters combining the information from both imaging modalities will become a part of routine clinical practice in cancer image interpretation and tumor molecular characterization in the near future. For accomplishing this goal and, thus, making *radiomics* a success story as a further step toward personalized cancer patient management, pathophysiologically well-understood and clinically relevant image-derived parameters have to be clearly defined by standardized mathematical algorithms and validated in prospective multi-center clinical trials.

References

1. Meacham CE, Morrison SJ. Tumour heterogeneity and cancer cell plasticity. Nature. 2013;501:328–37.

2. Dienstmann R, Vermeulen L, Guinney J, Kopetz S, Tejpar S, Tabernero J. Consensus molecular subtypes and the evolution of precision medicine in colorectal cancer. Nat Rev Cancer. 2017;17:79–92.
3. De Palma M, Hanahan D. The biology of personalized cancer medicine: facing individual complexities underlying hallmark capabilities. Mol Oncol. 2012;6:111–27.
4. Donovan MJ, Cordon-Cardo C. Implementation of a precision pathology program focused on oncology-based prognostic and predictive outcomes. Mol Diagn Ther. 2016;21(2):115–23.
5. Surrey LF, Luo M, Chang F, Li MM. The genomic era of clinical oncology: integrated genomic analysis for precision cancer care. Cytogenet Genome Res. 2016;150(3–4):162–75.
6. Horn H, Staiger AM, Ott G. New targeted therapies for malignant lymphoma based on molecular heterogeneity. Expert Rev Hematol. 2017;10:39–51.
7. Punt CJ, Koopman M, Vermeulen L. From tumour heterogeneity to advances in precision treatment of colorectal cancer. Nat Rev Clin Oncol. 2016;14(4):235–46.
8. Serie DJ, Joseph RW, Cheville JC, Ho TH, Parasramka M, Hilton T, Thompson RH, Leibovich BC, Parker AS, Eckel-Passow JE. Clear cell type A and B molecular subtypes in metastatic clear cell renal cell carcinoma: tumor heterogeneity and aggressiveness. Eur Urol. 2017;71:979–85.
9. Horak P, Frohling S, Glimm H. Integrating next-generation sequencing into clinical oncology: strategies, promises and pitfalls. ESMO Open. 2016;1:E000094.
10. Simone G. Stochastic phenotypic interconversion in tumors can generate heterogeneity. Eur Biophys J. 2016;46(2):189–94.
11. Lambin P, Rios-Velazquez E, Leijenaar R, Carvalho S, Van Stiphout RG, Granton P, Zegers CM, Gillies R, Boellard R, Dekker A, Aerts HJ. Radiomics: extracting more information from medical images using advanced feature analysis. Eur J Cancer. 2012;48:441–6.
12. Von Forstner C, Egberts JH, Ammerpohl O, Niedzielska D, Buchert R, Mikecz P, Schumacher U, Peldschus K, Adam G, Pilarsky C, Grutzmann R, Kalthoff H, Henze E, Brenner W. Gene expression patterns and tumor uptake of 18F-FDG, 18F-FLT, and 18F-FEC in PET/MRI of an orthotopic mouse xenotransplantation model of pancreatic cancer. J Nucl Med. 2008;49:1362–70.
13. Langen KJ, Hamacher K, Weckesser M, Floeth F, Stoffels G, Bauer D, Coenen HH, Pauleit D. O-(2-[18F]fluoroethyl)-L-tyrosine: uptake mechanisms and clinical applications. Nucl Med Biol. 2006;33:287–94.
14. Grunbaum Z, Freauff SJ, Krohn KA, Wilbur DS, Magee S, Rasey JS. Synthesis and characterization of congeners of misonidazole for imaging hypoxia. J Nucl Med. 1987;28:68–75.
15. Rajendran JG, Krohn KA. Imaging hypoxia and angiogenesis in tumors. Radiol Clin N Am. 2005;43:169–87.
16. Hofmann M, Maecke H, Borner R, Weckesser E, Schoffski P, Oei L, Schumacher J, Henze M, Heppeler A, Meyer J, Knapp H. Biokinetics and imaging with the somatostatin receptor PET radioligand (68)Ga-DOTATOC: preliminary data. Eur J Nucl Med. 2001;28:1751–7.
17. Kowalski J, Henze M, Schuhmacher J, Macke HR, Hofmann M, Haberkorn U. Evaluation of positron emission tomography imaging using [68Ga]-DOTA-D Phe(1)-Tyr(3)-Octreotide in comparison to [111In]-DTPAOC SPECT. First results in patients with neuroendocrine tumors. Mol Imaging Biol. 2003;5:42–8.
18. Prasad V, Brenner W, Modlin IM. How smart is peptide receptor radionuclide therapy of neuroendocrine tumors especially in the salvage setting? The clinician's perspective. Eur J Nucl Med Mol Imaging. 2014;41:202–4.
19. Prasad V, Steffen IG, Pavel M, Denecke T, Tischer E, Apostolopoulou K, Pascher A, Arsenic R, Brenner W. Somatostatin receptor PET/CT in restaging of typical and atypical lung carcinoids. EJNMMI Res. 2015;5:53.
20. Prasad V, Tiling N, Denecke T, Brenner W, Plockinger U. Potential role of (68)Ga-DOTATOC PET/CT in screening for pancreatic neuroendocrine tumour in patients with von Hippel-Lindau disease. Eur J Nucl Med Mol Imaging. 2016;43:2014–20.
21. Afshar-Oromieh A, Haberkorn U, Eder M, Eisenhut M, Zechmann CM. [68Ga]Gallium-labelled PSMA ligand as superior PET tracer for the diagnosis of prostate cancer: comparison with 18F-FECH. Eur J Nucl Med Mol Imaging. 2012;39:1085–6.

22. Kratochwil C, Afshar-Oromieh A, Kopka K, Haberkorn U, Giesel FL. Current status of prostate-specific membrane antigen targeting in nuclear medicine: clinical translation of chelator containing prostate-specific membrane antigen ligands into diagnostics and therapy for prostate cancer. Semin Nucl Med. 2016;46:405–18.
23. Prasad V, Steffen IG, Diederichs G, Makowski MR, Wust P, Brenner W. Biodistribution of [(68)Ga]PSMA-HBED-CC in patients with prostate cancer: characterization of uptake in normal organs and tumour lesions. Mol Imaging Biol. 2016;18:428–36.
24. Apostolova I, Hofheinz F, Buchert R, Steffen IG, Michel R, Rosner C, Prasad V, Kohler C, Derlin T, Brenner W, Marnitz S. Combined measurement of tumor perfusion and glucose metabolism for improved tumor characterization in advanced cervical carcinoma. A PET/CT pilot study using [15O]water and [18F]fluorodeoxyglucose. Strahlenther Onkol. 2014;190:575–81.
25. Schreiter NF, Maurer M, Pape UF, Hamm B, Brenner W, Froeling V. Detection of neuroendocrine tumours in the small intestines using contrast-enhanced multiphase Ga-68 DOTATOC PET/CT: the potential role of arterial hyperperfusion. Radiol Oncol. 2014;48:120–6.
26. Brenner W, Vernon C, Muzi M, Mankoff DA, Link JM, Conrad EU, Eary JF. Comparison of different quantitative approaches to 18F-fluoride PET scans. J Nucl Med. 2004;45:1493–500.
27. Weber WA, Schwaiger M, Avril N. Quantitative assessment of tumor metabolism using FDG-PET imaging. Nucl Med Biol. 2000;27:683–7.
28. Phelps ME, Huang SC, Hoffman EJ, Selin C, Sokoloff L, Kuhl DE. Tomographic measurement of local cerebral glucose metabolic rate in humans with (F-18)2-fluoro-2-deoxy-D-glucose: validation of method. Ann Neurol. 1979;6:371–88.
29. Reivich M, Kuhl D, Wolf A, Greenberg J, Phelps M, Ido T, Casella V, Fowler J, Gallagher B, Hoffman E, Alavi A, Sokoloff L. Measurement of local cerebral glucose metabolism in man with 18F-2-fluoro-2-deoxy-D-glucose. Acta Neurol Scand Suppl. 1977;64:190–1.
30. Graham MM, Peterson LM, Hayward RM. Comparison of simplified quantitative analyses of FDG uptake. Nucl Med Biol. 2000;27:647–55.
31. Gjedde A. Calculation of cerebral glucose phosphorylation from brain uptake of glucose analogs in vivo: a re-examination. Brain Res. 1982;257:237–74.
32. Patlak CS, Blasberg RG, Fenstermacher JD. Graphical evaluation of blood-to-brain transfer constants from multiple-time uptake data. J Cereb Blood Flow Metab. 1983;3:1–7.
33. Lucignani G, Paganelli G, Bombardieri E. The use of standardized uptake values for assessing FDG uptake with pet in oncology: a clinical perspective. Nucl Med Commun. 2004;25:651–6.
34. O JH, Lodge MA, Wahl RL. Practical PERCIST: a simplified guide to PET response criteria in solid tumors 1.0. Radiology. 2016;280:576–84.
35. Wahl RL, Jacene H, Kasamon Y, Lodge MA. From RECIST to PERCIST: evolving considerations for PET response criteria in solid tumors. J Nucl Med. 2009;50(Suppl 1):122s–50s.
36. O'connor JP. Cancer heterogeneity and imaging. Semin Cell Dev Biol. 2016;64:48–57.
37. O'sullivan F, Roy S, Eary J. A statistical measure of tissue heterogeneity with application to 3D PET sarcoma data. Biostatistics. 2003;4:433–48.
38. Eary JF, O'sullivan F, O'sullivan J, Conrad EU. Spatial heterogeneity in sarcoma 18F-FDG uptake as a predictor of patient outcome. J Nucl Med. 2008;49:1973–9.
39. O'sullivan F, Roy S, O'sullivan J, Vernon C, Eary J. Incorporation of tumor shape into an assessment of spatial heterogeneity for human sarcomas imaged with FDG-PET. Biostatistics. 2005;6:293–301.
40. O'sullivan F, Wolsztynski E, O'sullivan J, Richards T, Conrad EU, Eary JF. A statistical modeling approach to the analysis of spatial patterns of FDG-PET uptake in human sarcoma. IEEE Trans Med Imaging. 2011;30:2059–71.
41. Vernon CB, Eary JF, Rubin BP, Conrad EU 3rd, Schuetze S. FDG PET imaging guided re-evaluation of histopathologic response in a patient with high-grade sarcoma. Skelet Radiol. 2003;32:139–42.
42. Yan J, Jones RL, Lewis DH, Eary JF. Impact of (18)F-FDG PET/CT imaging in therapeutic decisions for malignant solitary fibrous tumor of the pelvis. Clin Nucl Med. 2013;38:453–5.

43. Chicklore S, Goh V, Siddique M, Roy A, Marsden PK, Cook GJ. Quantifying tumour heterogeneity in 18F-FDG PET/CT imaging by texture analysis. Eur J Nucl Med Mol Imaging. 2013;40:133–40.
44. Gao X, Chu C, Li Y, Lu P, Wang W, Liu W, Yu L. The method and efficacy of support vector machine classifiers based on texture features and multi-resolution histogram from (18)F-FDG PET-CT images for the evaluation of mediastinal lymph nodes in patients with lung cancer. Eur J Radiol. 2015;84:312–7.
45. Lartizien C, Rogez M, Niaf E, Ricard F. Computer-aided staging of lymphoma patients with FDG PET/CT imaging based on textural information. IEEE J Biomed Health Inform. 2014;18:946–55.
46. Vallieres M, Freeman CR, Skamene SR, El Naqa I. A radiomics model from joint FDG-PET and MRI texture features for the prediction of lung metastases in soft-tissue sarcomas of the extremities. Phys Med Biol. 2015;60:5471–96.
47. Xu R, Kido S, Suga K, Hirano Y, Tachibana R, Muramatsu K, Chagawa K, Tanaka S. Texture analysis on (18)F-FDG PET/CT images to differentiate malignant and benign bone and soft-tissue lesions. Ann Nucl Med. 2014;28:926–35.
48. Hatt M, Tixier F, Pierce L, Kinahan PE, Le Rest CC, Visvikis D. Characterization of PET/CT images using texture analysis: the past, the present... any future? Eur J Nucl Med Mol Imaging. 2017;44:151–65.
49. Tixier F, Le Rest CC, Hatt M, Albarghach N, Pradier O, Metges JP, Corcos L, Visvikis D. Intratumor heterogeneity characterized by textural features on baseline 18F-FDG PET images predicts response to concomitant radiochemotherapy in esophageal cancer. J Nucl Med. 2011;52:369–78.
50. Apostolova I, Steffen IG, Wedel F, Lougovski A, Marnitz S, Derlin T, Amthauer H, Buchert R, Hofheinz F, Brenner W. Asphericity of pretherapeutic tumour FDG uptake provides independent prognostic value in head-and-neck cancer. Eur Radiol. 2014;24:2077–87.
51. Van Den Bent MJ, Snijders TJ, Bromberg JE. Current treatment of low grade gliomas. Memo. 2012;5:223–7.
52. Torigian DA, Lopez RF, Alapati S, Bodapati G, Hofheinz F, Van Den Hoff J, Saboury B, Alavi A. Feasibility and performance of novel software to quantify metabolically active volumes and 3D partial volume corrected SUV and metabolic volumetric products of spinal bone marrow metastases on 18F-FDG-PET/CT. Hell J Nucl Med. 2011;14:8–14.
53. Hofheinz F, Potzsch C, Oehme L, Beuthien-Baumann B, Steinbach J, Kotzerke J, Van Den Hoff J. Automatic volume delineation in oncological PET. Evaluation of a dedicated software tool and comparison with manual delineation in clinical data sets. Nuklearmedizin. 2012;51:9–16.
54. Hofheinz F, Lougovski A, Zophel K, Hentschel M, Steffen IG, Apostolova I, Wedel F, Buchert R, Baumann M, Brenner W, Kotzerke J, Van Den Hoff J. Increased evidence for the prognostic value of primary tumor asphericity in pretherapeutic FDG PET for risk stratification in patients with head and neck cancer. Eur J Nucl Med Mol Imaging. 2015;42:429–37.
55. Apostolova I, Rogasch J, Buchert R, Wertzel H, Achenbach HJ, Schreiber J, Riedel S, Furth C, Lougovski A, Schramm G, Hofheinz F, Amthauer H, Steffen IG. Quantitative assessment of the asphericity of pretherapeutic FDG uptake as an independent predictor of outcome in NSCLC. BMC Cancer. 2014;14:896.
56. Apostolova I, Ego K, Steffen IG, Buchert R, Wertzel H, Achenbach HJ, Riedel S, Schreiber J, Schultz M, Furth C, Derlin T, Amthauer H, Hofheinz F, Kalinski T. The asphericity of the metabolic tumour volume in NSCLC: correlation with histopathology and molecular markers. Eur J Nucl Med Mol Imaging. 2016;43:2360–73.
57. Wetz C, Apostolova I, Steffen IG, Hofheinz F, Furth C, Kupitz D, Ruf J, Venerito M, Klose S, Amthauer H. Predictive value of asphericity in pretherapeutic [111In]DTPA-Octreotide SPECT/CT for response to peptide receptor radionuclide therapy with [177Lu] DOTATATE. Mol Imaging Biol. 2016;19(3):437–45.
58. Miles KA, Williams RE. Warburg revisited: imaging tumour blood flow and metabolism. Cancer Imaging. 2008;8:81–6.

59. Dunnwald LK, Gralow JR, Ellis GK, Livingston RB, Linden HM, Specht JM, Doot RK, Lawton TJ, Barlow WE, Kurland BF, Schubert EK, Mankoff DA. Tumor metabolism and blood flow changes by positron emission tomography: relation to survival in patients treated with neoadjuvant chemotherapy for locally advanced breast cancer. J Clin Oncol. 2008;26:4449–57.
60. Mankoff DA, Dunnwald LK, Partridge SC, Specht JM. Blood flow-metabolism mismatch: good for the tumor, bad for the patient. Clin Cancer Res. 2009;15:5294–6.
61. Tseng J, Dunnwald LK, Schubert EK, Link JM, Minoshima S, Muzi M, Mankoff DA. 18F-FDG kinetics in locally advanced breast cancer: correlation with tumor blood flow and changes in response to neoadjuvant chemotherapy. J Nucl Med. 2004;45:1829–37.
62. Orlhac F, Theze B, Soussan M, Boisgard R, Buvat I. Multiscale texture analysis: from 18F-FDG PET images to histologic images. J Nucl Med. 2016;57:1823–8.

Thomas Fischer, Anke Thomas, and Dirk-André Clevert

Abstract

State-of-the-art techniques of ultrasound elastography can contribute to the characterization of focal breast lesions and detection of malignant tumors in various organs. Another area of current interest is the differentiation of focal liver lesions in terms of vascularization patterns measured by perfusion ultrasound. This chapter provides an overview of the different techniques and their diagnostic role in clinical routine based on a review of the current literature. The most important techniques are compression or vibration elastography, shear wave elastography (SWE), and contrast-enhanced ultrasound (CEUS). Currently available scientific evidence suggests that elastography provides important supplementary information for the differentiation of breast lesions under routine clinical conditions. The information is immediately available and improves specificity. Strain ratio (SR) is especially useful in women with a high pretest likelihood of breast cancer. Prostate cancer also shows characteristic differences in terms of elastographic properties compared with surrounding tissue. Here, elastography can improve targeted biopsy for the workup of suspicious focal lesions and is superior to routine prostate biopsy guided by B-mode ultrasound. CEUS has high diagnostic accuracy and is comparable to computed tomography (CT) and magnetic resonance imaging (MRI) in terms of tumor characterization. Having a low rate of

T. Fischer (✉)
Department of Radiology, Institut für Radiologie und Ultraschallforschungslabor, Charité – Universitätsmedizin Berlin, Berlin, Germany
e-mail: thom.fischer@charite.de

A. Thomas
Ultraschallforschungslabor, Charité – Universitätsmedizin Berlin, Berlin, Germany

D.-A. Clevert
Institut für Klinische Radiologie, Klinikum der Universität München-Grosshadern, München, Germany

© Springer International Publishing AG 2018   411
I. Sack, T. Schaeffter (eds.), *Quantification of Biophysical Parameters in Medical Imaging*, https://doi.org/10.1007/978-3-319-65924-4_19

adverse effects, CEUS can be used in patients with impaired renal function or contraindications to CT or MRI contrast agents. Quantifiable elastography and CEUS have recently started to expand the role of classic B-mode ultrasound in oncology. Quantification of tumor stiffness and perfusion can improve the differential diagnosis. These two ultrasound techniques are beginning to enter the clinic and offer a fascinating potential for further advances including improved standardization of ultrasound diagnosis.

19.1 Part A: Application of Ultrasound Elastography

19.1.1 Introduction

Tissues have inherent elasticity, which changes with normal aging or when disease such as inflammation or a tumor is present. A variety of ultrasound techniques, jointly known as sonoelastography, have been developed since the early 1990s to assess the elasticity of biological tissues [1–3] (see Chap. 12). The external force required to induce tissue deformation depends on the tissue's shear modulus [4, 5] and appears to be altered in tumor tissue (see Chaps. 2 and 5 for constitutive equations and biophysical background). Clinical studies in different organ systems have shown that determination of tissue elasticity provides important supplementary diagnostic information. The organs investigated include the parotid [6], thyroid [7, 8], liver [9, 10], prostate [11], and cervix [12]. Special attention has been paid to the sonoelastographic characterization of tumorous lesions in the breast [13–15] and prostate. For breast imaging, it has been shown that using tissue elasticity as an additional criterion improves specificity [13–17] and lowers the number of false-positive findings [14]. Breast sonoelastography thus has the potential to reduce the need for biopsy in the future. An important clinical advantage of sonoelastography is that it is generally available and quick to perform at little extra cost. The high spatial resolution of ultrasound allows determination of the elastic properties of small structures (less than 5 mm). On the other hand, sonoelastography has some technical disadvantages including the limited penetration depth of only 5–6 cm (for transient-based techniques), sensitivity to the axial displacement component only, and the examiner dependence. More details on the comparison between state-of-the art sonoelastographic modalities can be found in Chap. 12. The following sections present clinical applications of sonoelastography in the diagnostic evaluation of malignant tumors of the breast, thyroid, and prostate.

19.1.2 Breast Cancer

Breast lesions may be very small (on the order of 5 mm), and their detection therefore requires an elastographic imaging technique with high spatial resolution. Research has focused on techniques using absence of elasticity for the evaluation of focal breast lesions. This criterion contributes to the characterization of breast lesions detected by ultrasound. The use of tissue elasticity as an additional diagnostic parameter has been shown to increase specificity and to improve the separation of benign and malignant focal lesions classified as BI-RADS (Breast Imaging Reporting and Data System) category 3 or 4 [18]. The technique thus reduces the number of false-positive findings and could spare many women unnecessary breast biopsies in the future. When the technique was first introduced, differentiation of benign and malignant breast lesions relied on subjective and/or semi-quantitative approaches (Fig. 19.1a, b).

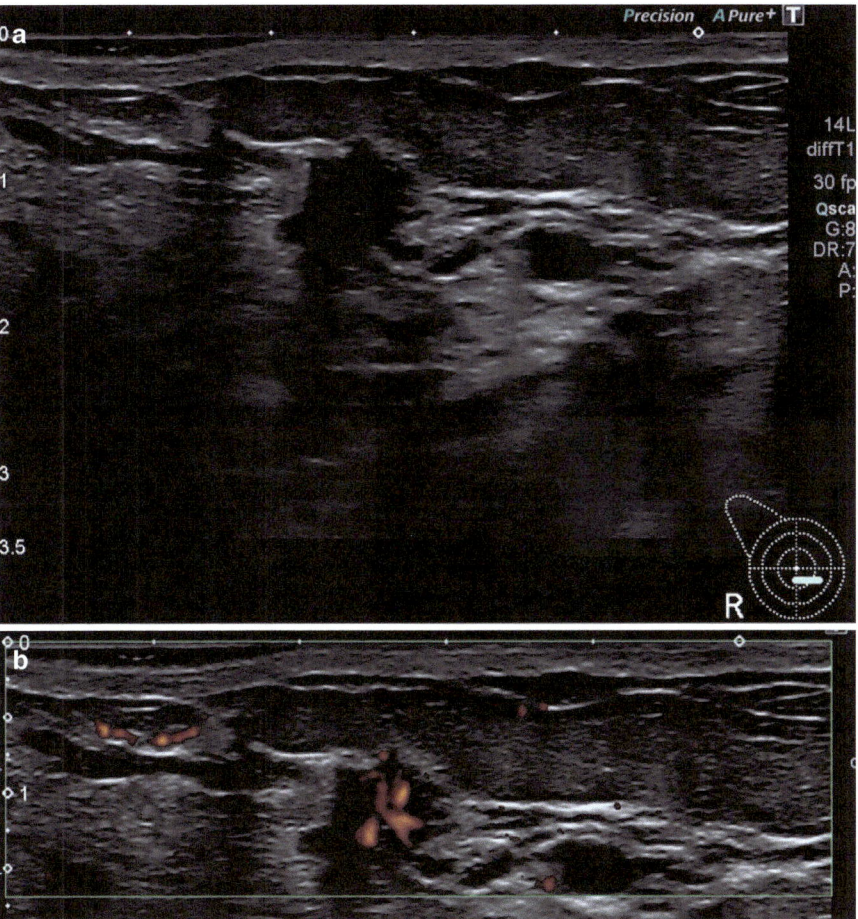

Fig. 19.1 A 48-year-old patient presenting with a suspicious lesion palpated in the right breast. The ultrasound B-mode image shows an irregular, microlobulated, spiculated focal lesion measuring 1.2 cm with marked ductal dilation in the vicinity of the lesion (**a**). The B-mode appearance is consistent with a BI-RADS 5 lesion. Ductal dilation in the vicinity of the lesion may indicate a DCIS component. There is ample perfusion of the lesion and a vessel entering the focal lesion perpendicularly (**b**). TDI shows complete absence of color pixels in the lesion, consistent with incompressibility (**c**). Both techniques confirm breast cancer. The fat-to-lesion strain ratio (FLR) is only slightly elevated at 1.62 (**d**). Shear wave elastography (SWE) shows faster transmission of signals through the breast lesion (**e**), which is reflected in an increase in the SWE ratio (**f**)

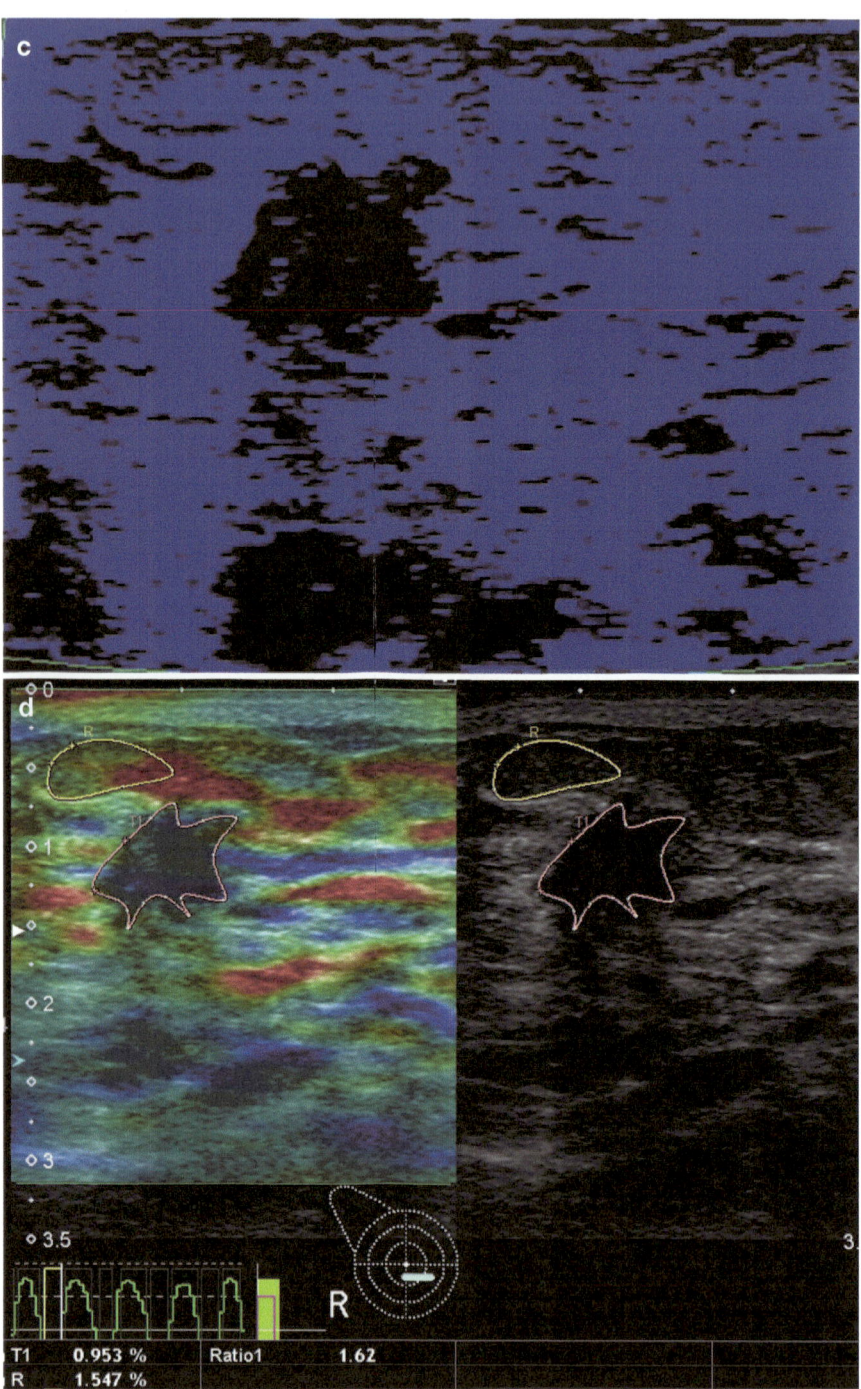

Fig. 19.1 (continued)

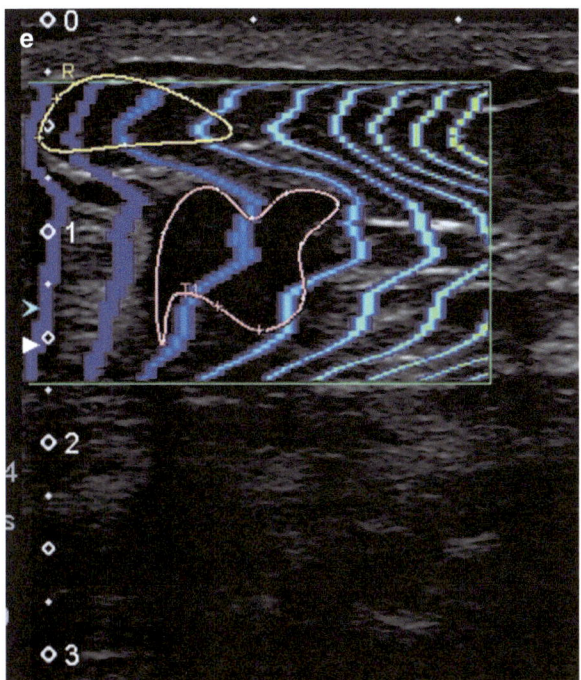

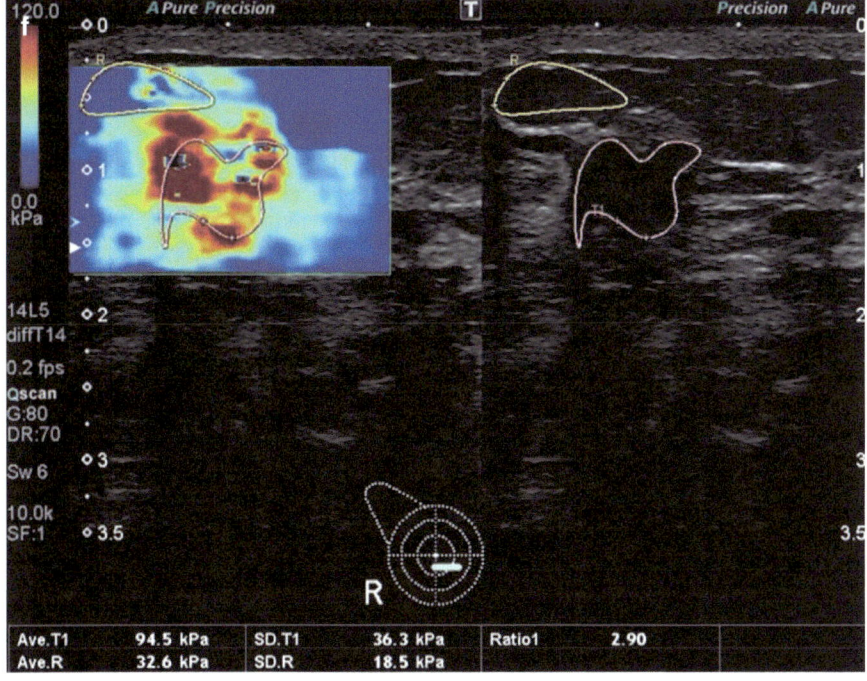

Fig. 19.1 (continued)

In real-time elastography, the examiner uses the ultrasound transducer to compress the tissue from outside the body and then to measure the resulting tissue movement or deformation. The elasticity information obtained in this way is displayed as a color map superimposed on the grayscale image in real time. In real-time elastography, many investigators use the blue color spectrum to represent low strain, which is equivalent to a high intrinsic elastic modulus (hard tissue), while green and red colors are used to indicate intermediate to high strain, i.e., low elastic modulus or soft tissue. Of note, most shear wave-based systems have a reversed scale, that is, red for stiff tissues and blue for soft (normal) tissues. State-of-the-art ultrasound systems allow free selection and reversal of the available color scales. Standardization would be desirable. In women with low breast tissue density, elastography has been found to markedly improve detection and characterization of focal lesions. In women with involuted glandular tissue, real-time elastography has been shown to increase specificity from 69 to 80%. This is important since the diagnostic accuracy of B-mode ultrasound decreases with involution of breast parenchyma. A multicenter study of 779 women has confirmed these results [14].

In the further development of sonoelastography of the breast, standardization was improved by the introduction of a *fat-to-lesion strain ratio* (*FLR*), which defines the relationship between the elasticity of fatty tissue and that of the breast lesion (Fig. 19.1d). The FLR is calculated from a region of interest (ROI) encircling the entire breast lesion and a second ROI placed in the surrounding fatty tissue and is compared individually and intraindividually [15, 19]. The most recent studies have shown that FLR calculation improves the characterization of breast lesions and allows differentiation of benign and malignant focal breast lesions. In a European patient population, an FLR cutoff value for discrimination of benign and malignant lesions ranging between 2.3 and 2.5 has been identified using different US systems, which differs from the cutoff of 3.1 defined in a population of Chinese women. These variations may be attributable to ethnic variations in normal glandular breast density, and they preclude the definition of a single standardized FLR. Nevertheless, this ratio is a simple and reproducible parameter for the characterization of known breast lesions. An FLR below the cutoff is highly indicative of a benign breast lesion, while an FLR above the cutoff is suspicious for a malignant breast tumor. A suspected malignant lesion requires confirmation by biopsy. Ultrasound elastography allows no differentiation of recurrent breast cancer from scar tissue. Scar tissue developing after surgery and radiotherapy has little intrinsic elasticity and thus has an FLR in the same range as malignant breast lesions. This is why magnetic resonance imaging (MRI) or biopsy with ultrasound guidance will continue to be necessary for ruling out cancer recurrence.

Tissue Doppler imaging (TDI) can be understood as a special pressure-independent version of strain elastography and also allows real-time analysis. Tissue reflection with TDI is very low; however, the signal has very high amplitude compared with the fast signals obtained by classic color Doppler ultrasound, where red blood cells serve as reflectors. When operated in the dual mode, information on tissue distortion is superimposed on B-mode views using red and blue as with conventional color Doppler imaging. Malignant breast lesions are characterized by the absence of color pixels, while benign lesions are filled with color pixels and typically appear markedly smaller in TDI (Fig. 19.1c). Therefore, TDI allows significant differentiation of benign and malignant focal breast lesions ($p < 0.001$) (Fig. 19.1c). This technique is particularly easy to use and the gain in diagnostic information is immediately apparent [20]. Although it can theoretically be implemented into all ultrasound systems, TDI has not yet become established as a routine clinical procedure.

A first large meta-analysis of sonoelastography was conducted in 2012 and included 5511 breast lesions [21]. In this analysis, specificity increased from 70 to 88% with the use of elastography. The use of sonoelastography can reduce the need for breast biopsy particularly in screening populations with a low breast cancer risk. In the screening situation, however, elastography should not be used as the first ultrasound method but should ideally be used when B-mode ultrasound findings suggest a breast lesion. In contrast, when examining women with a high risk of breast cancer, the technique with the highest correct classification rate should be used. For breast examinations, this is FLR calculation, which has higher sensitivity compared with subjective assessment [22].

SWE and transient elastography (TE) rely on a different physical principle and require a special transducer. In TE, the transducer generates the classic ultrasound waves and additional low-frequency shear waves in the 50 Hz frequency range. The speed at which shear waves or transverse waves propagate in the tissue is measured to derive the tissue's elasticity modulus. This technique has gained much attention in recent years for the grading of liver fibrosis [23]. Recently, Stock et al. [24] investigated the acoustic radiation force impulse (ARFI) technique for the quantification of renal transplant fibrosis and found a correlation between elastography and histologic fibrosis grading. Only a few clinical studies have evaluated the potential of this technique for the differentiation of focal breast lesions. Two studies, Evans et al. [25] and Berg et al. [26], found SWE to increase specificity and thus improve the characterization of breast lesions (Fig. 19.1e, f). Other techniques, such as magnetic resonance elastography (MRE), hold promise for further improving imaging characterization of breast lesions. Inherent limitations of MRE such as long examination times and reduced spatial resolution can be overcome by state-of-the-art single-shot acquisition techniques and multifrequency vibration. Overall, it is expected that breast MRE will in the future be used as a short supplementary examination in patients with a clinical indication for conventional breast MRI. Another promising method for determining mechanical properties of breast tissue is tomosynthesis elastography, which uses a tomosynthesis technique to scan tissue layers before and after static distortion and image registration for the subsequent computation of distortion maps [27].

Regardless of the medical imaging modality used, elastography is a valid tool for detecting pathological differences in the cohesiveness of breast tissue. Mechanical stimulation can be used to derive diagnostic information on tissue properties otherwise requiring invasive procedures.

19.1.3 Thyroid Cancer

The anatomic location makes elastography of the thyroid more difficult than examination of the breast, where the surrounding fat can be used as reference for comparison.

Therefore, absolute elasticity properties of individual thyroid nodules need to be compared, and definition of standardized reference values is difficult. A comprehensive meta-analysis of the characterization of focal thyroid lesions by elastography is still lacking. While single-center studies suggest a benefit of elastography in characterizing thyroid nodules, elastography cannot replace cytology for a definitive diagnosis [28].

Nonpalpable thyroid nodules are a common incidental finding in asymptomatic individuals. Ultrasound is the first-line imaging modality in the workup of small incidentally detected thyroid nodules. While small nodules <1 cm in size are typically managed by follow-up, the further procedure in individuals with nodules ≥1 cm in size depends on the initial ultrasound findings and may include laboratory tests, fine-needle aspiration cytology (FNAC), and scintigraphy. Besides the general appearance at B-mode or color duplex ultrasound, a number of individual features are assessed to identify malignant thyroid nodules including echotexture/hypoechogenicity, presence of microcalcifications, absence or poorly defined margin and central hypervascularization, as well as conspicuous lymph nodes [28]. Diagnostic accuracy relies on the examiner's experience [28]. Some of the features that are typical of malignancy may also be present in benign thyroid nodules. As a result, high-resolution US has low specificity, and elastography has been investigated to determine its potential for improving the specificity of thyroid US. However, currently available data were obtained in small, selected patient populations with an indication for FNAC [29]. The first studies of thyroid sonoelastography were performed using the technique of strain elastography and manual tissue compression with the ultrasound transducer [29, 30]. Different scoring systems such as the Ueno score have been proposed (ranging from 1 for mostly soft tissue to 4/5 for completely hard tissue); however, thyroid nodule categorization using these scores has been found to have only moderate interobserver validity of <68% in unselected patient populations [29–31]. These results were improved with the introduction of semi-quantitative elasticity indices and definition of cutoff values, which resulted in reported sensitivities of 74–98% and specificities of 72–100% [32, 33]. There is agreement among investigators that strain elastography of the thyroid is a supplementary technique for improving the specificity of high-resolution B-mode ultrasound and that the sensitivity of combined B-mode ultrasound and color-coded Doppler ultrasound (CD-US) appears to be superior to elastography [32, 33].

Another approach to standardization is to exploit carotid artery pulsation for inducing pressure-dependent deformation of the thyroid, and a study has shown that this approach reduces examiner dependence [34]. Furthermore, attempts have been made, using ARFI [35] and SWE, to develop an examiner-independent technique for the measurement of the propagation velocity in m/s or of pressure in kPa. Again, cutoff values were determined to discriminate benign and malignant lesions. Initial optimistic results with specificities of 93–95% [35] were not confirmed in later studies, where specificities of 71–78% were found [1, 36]. This is below the specificity of strain elastography and suggests that examiner dependence might not be an issue. Multicenter studies should be performed in unselected patient populations with subsequent histological confirmation of findings. So far, the superior specificity of elastography mainly helps in identifying patients who should undergo FNAC. Having relatively low sensitivity, elastography cannot be recommended as the only test for the follow-up of presumably benign thyroid nodules, and FNAC continues to be required for diagnostic confirmation.

19.1.4 Prostate Cancer

Men with an elevated prostate-specific antigen (PSA) level or abnormal prostate findings in the digital rectal examination (DRE) undergo workup by transrectal ultrasound (TRUS) in combination with systematic biopsy for histologic confirmation. In a subgroup of these patients, TRUS-guided biopsy fails to detect cancer despite increasing PSA levels, and multiple biopsies may be necessary before a diagnosis can be made [37, 38]. Since negative TRUS-guided biopsy does not rule out prostate cancer, healthy men may be repeatedly exposed to the possible risks (infection, bleeding) of this invasive procedure. Moreover, the detection rate markedly decreases with each repetition of prostate biopsy [37]. Many suggestions have been made to improve the cancer detection rate of TRUS. Since it is known that prostate cancer is associated with changes in metabolism and perfusion [39, 40], techniques such as color Doppler US and contrast-enhanced ultrasound (CEUS) at high frequency as well as elastography have been proposed for prostate cancer detection without achieving decisive progress [41]. Data on TRUS elastography are highly variable with reported sensitivities for prostate cancer detection ranging from 25 to 92% [42, 43]. A breakthrough was finally achieved by combining multiparametric 3 T MRI without the use of an endorectal coil for localizing suspicious lesions within the prostate with subsequent use of these data for real-time MRI/US fusion biopsy. Initial results with MRI/US fusion biopsy in subgroups of patients showed detection rates that were comparable to that of the time-consuming and expensive method of MRI-guided biopsy [44]. Fusion biopsy is also performed using a multiparametric approach combining color Doppler, CEUS, and elastography. The advantage of this technique is in the assessment of focal lesions in a given plane like the MRI, which also takes the high detection rate of prostate cancer by MRI into account. Both CEUS and elastography have shown high specificity in multiparametric US. Approaches for using elastography aim at identifying suspicious lesions for subsequent targeted biopsy with routine TRUS-based techniques. Of particular interest is SWE, which yields absolute values for focal lesions compared with the unaffected side. Initial publications on this technique have proposed cutoff values on the order of 35 kPa [45]. Future studies must show whether the limited penetration depth of this technique can be improved further and whether these initial results can be confirmed by multicenter trials. However, elastography has the potential to provide supplementary information that could be used for routine TRUS-guided biopsy in patients with abnormal B-mode findings.

19.1.5 Summary of Part A

Based on currently available data, routine clinical elastography could provide important additional diagnostic information for the differentiation of breast tumors, for identifying patients with thyroid nodules (>1 cm) who should undergo FNAC, and for men with suspected prostate cancer who have abnormal B-mode ultrasound findings and are scheduled for TRUS-guided biopsy. The use of sonoelastography improves specificity and directly provides additional diagnostic information. Having high detail resolution, sonoelastography allows reliable evaluation of lesions once they have reached a size of 5 mm. Besides real-time elastography based on strain imaging, SWE will gain wider acceptance in the future as it allows quantification of parameters. With ultrasound having a minor role in the classification of tumors, as discussed here for different cancers, the expected role of elastography is also limited; however, this should not prevent researchers from considering all tumors of a specialty when evaluating the potential of a new imaging modality. In addition, larger studies should investigate whether elastography yields adequate results for various diagnostic queries even in the hands of less experienced examiners. Papillary thyroid cancer appears to be harder than follicular and medullary thyroid cancer. Invasive ductal carcinoma is harder than invasive lobular breast tumors. This is where elastography has the potential for identifying tumor subgroups. Nevertheless, sound statistical data or evidence from large multicenter studies is still lacking. Not all malignant tumors are hard and not all benign tumors are soft, which is a fundamental limitation of elastography. On the other hand, sonoelastography requires little extra time and the cost is very low. These advantages make sonoelastography an attractive option and could contribute to its wider use in different diagnostic settings.

19.2 Part B: Contrast-Enhanced Ultrasound

19.2.1 Introduction

The visualization of tissue properties and tissue perfusion is an important component of the diagnostic evaluation of tumors and kidneys and in trauma patients by any imaging modality. Conventional vascular ultrasound (US) techniques such as color duplex ultrasonography (CDUS) do not depict vessels with a diameter of less than about 30 mm. Furthermore, this method is susceptible to error due to examiner dependence and the effect of systemic disease such as atherosclerosis which often results in artifacts and posterior acoustic shadowing. The advent of nonspecific ultrasound contrast media (USCM) has markedly improved the detection of very slow blood flow in small vessels. The potential of target-specific USCM for demonstrating neoangiogenesis in cancer is a new approach. Possible candidates for such a contrast agent are microbubbles to which a vascular endothelial growth factor receptor 2 (VEGFR2)-binding peptide or antigen is coupled, which selectively mark areas of tumor neoangiogenesis. Unspecific USCM might be superior in the diagnosis of abnormal tumor perfusion compared with conventional US since tumor perfusion is associated with characteristic changes of the arterial inflow and the late washout phase of the contrast agent.

19.2.2 Contrast-Enhanced Ultrasound of the Liver

In 2001, phospholipid-stabilized microbubbles of a poorly water-soluble gas (e.g., sulfur hexafluoride, SF6, SonoVue®) became commercially available as a second-generation ultrasound contrast agent [46]. This microbubble preparation is very stable, providing prolonged contrast and enhancing the ultrasound signal in blood vessels including the capillary system by several orders of magnitude (by a factor of approx. 103). The SonoVue microbubbles have a mean diameter of 2.5 µm and are smaller than red blood cells (7 µm), allowing them to distribute freely in the blood vessels and capillaries [47–50].

When exposed to low ultrasound energy, the microbubbles generate only linear backscatter. With increasing energy, once certain range is reached, the microbubbles begin to oscillate at eigenfrequency with a characteristic resonance spectrum. Following injection into a peripheral vein, the microbubbles will reach the organs and distribute in their capillary beds, resulting in homogeneous opacification of normally perfused organs or parts [51]. State-of-the-art ultrasound devices can be operated in a special mode to sample and process the specific nonlinear reflection from the microbubbles for selective visualization with very high temporal resolution of parenchymal perfusion (typically as color-coded information) [47].

Unlike conventional CT and MRI contrast agents, the ultrasound microbubbles do not diffuse into the interstitial space, and they are not eliminated by the kidneys but are exhaled via the lungs within a few minutes [48, 52]. Ultrasound contrast agents are considered to be very safe because they are biologically inert, are not nephrotoxic, and do not interact with the thyroid, and the incidence of allergic reactions following microbubble administration is well below that of conventional CT contrast agents [53].

One of the strengths of CEUS is the high temporal resolution of perfusion visualization compared with other imaging modalities. The safety profile of ultrasound contrast agents allows repeated administration in serial follow-up examinations at short intervals [52].

Ultrasound is usually the first-line imaging modalities for diagnostic evaluation of patients with metastatic liver lesions. Focal liver lesions are common, with a reported prevalence of approx. 5% [54]. Liver ultrasound is performed at a frequency range of 2–9 MHz (Figs. 19.2 and 19.3).

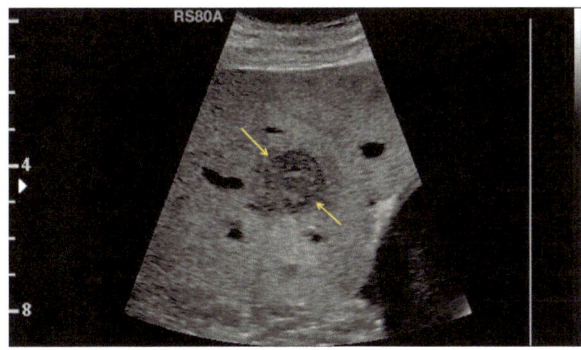

Fig. 19.2 In this patient with suspected pharyngeal cancer, abdominal staging by standard B-mode ultrasound reveals a hypoechoic liver lesion (*yellow arrows*)

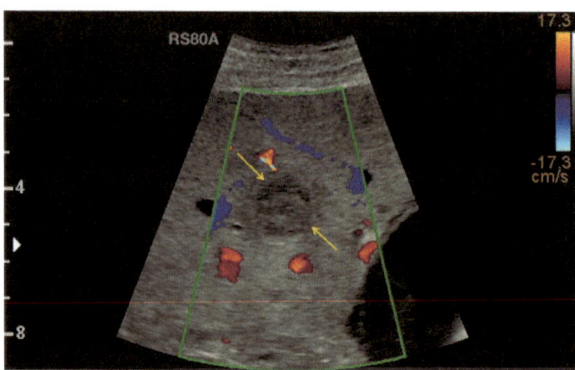

Fig. 19.3 The lesion (*yellow arrows*) does not have increased vascularization on color flow imaging (see patient description in Fig. 19.2)

The liver is the main target of metastatic disease with 25–50% of all cancer patients having liver metastases at the time of diagnosis [55]. The treatment options (surgical resection versus interventional treatment) depend on the size, number, and localization of liver metastases [56, 57]. This is why reliable detection and characterization of liver lesions is crucial for estimating the prognosis and the choice of treatment [58, 59]. In a meta-analysis, Kinkel et al. found the detection rate for liver metastases of gastrointestinal malignancies to be only 55% for B-mode ultrasound versus 72%, 76%, and 90% for contrast-enhanced CT, MRI, and PET, respectively [60, 61].

The advent of CEUS in 2001 fundamentally changed the diagnostic accuracy of ultrasound. In a German multicenter study, CEUS correctly characterized approx. 90% of focal liver lesions [58].

The EFSUMB Guidelines distinguish three phases of contrast enhancement in ultrasound: an arterial phase lasting until approx. 30 s after injection, a portal venous phase from 30 to 120 s, and a late phase after 120 s [62–64].

Benign liver lesions such as focal nodular hyperplasia (FNH) and hemangioma are characterized by isoenhancement to hyperenhancement in the late phase. The most important criterion distinguishing malignant from benign liver lesions is washout of the contrast agent in the late phase. Depending on the primary cancer, washout of a liver metastasis may begin in the late arterial phase and is nearly always seen in the portal venous phase (Figs. 19.4, 19.5, and 19.6).

In addition, CEUS can improve the monitoring of interventional treatment such as radiofrequency ablation (RFA) or transarterial chemoembolization (TACE) and intraoperative interventions [65, 66].

19.2.3 Summary of Part B

Ultrasound is usually the first-line imaging modalities for diagnostic evaluation of patients with metastatic liver lesions. The advent of CEUS fundamentally changed the diagnostic accuracy of ultrasound. CEUS can improve the monitoring of

Fig. 19.4 Following administration of the ultrasound contrast agent, there is strong marginal enhancement of the lesion in the arterial phase (*white arrows*, see patient description in Fig. 19.2)

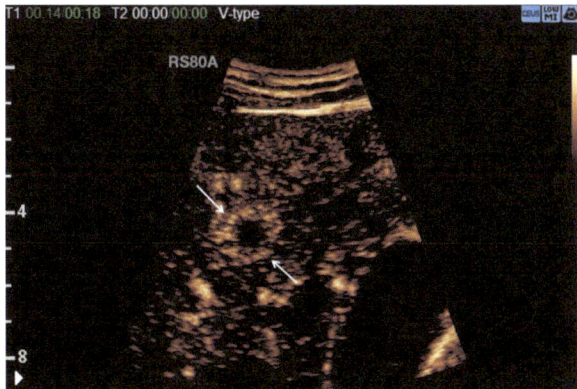

Fig. 19.5 In the portal venous phase, the lesion (*white arrows*) is demarcated from surrounding liver parenchyma by beginning washout (see patient description in Fig. 19.2)

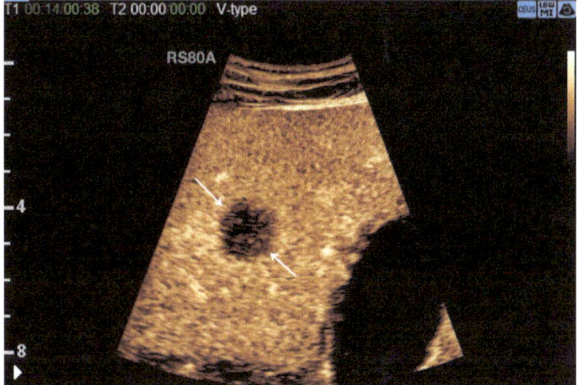

Fig. 19.6 In the late phase, there is increasing washout of the lesion (*white arrows*), confirming the diagnosis of a liver metastasis based in morphologic imaging. The subsequent liver biopsy confirmed liver metastasis from a poorly differentiated nonkeratinizing squamous cell carcinoma (see patient description in Fig. 19.2)

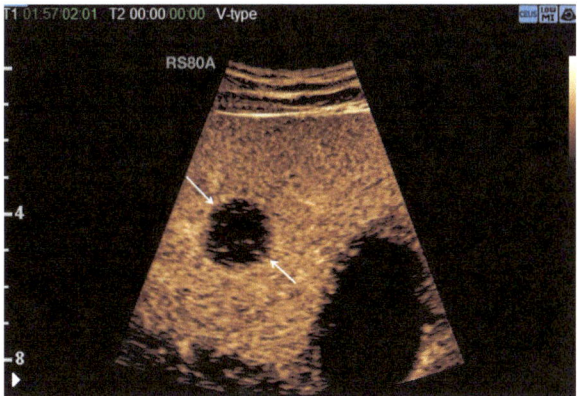

interventional treatment and the follow-up of tumor patients under drug treatment and monitor the effect of interventions. Ultrasound contrast agents are considered to be very safe because they are biologically inert, are not nephrotoxic, and do not interact with the thyroid, and the incidence of allergic reactions following microbubble administration is well below that of conventional CT contrast agents.

Conclusion

Meanwhile elastography and perfusion measurements have been established as clinical routine methods of ultrasound examinations. Quantifications of elasticity and perfusion provide objective parameters for tumor stiffness and specific perfusion. Ultrasound contrast agents and elastography are considered to be very safe because they are biologically inert. Many cancer entities show characteristic differences in terms of elastographic properties compared with surrounding tissue. Here, elastography can improve targeted biopsy for the workup of suspicious focal lesions. Elastography has been demonstrated to be superior to routine biopsy guided by B-mode ultrasound. CEUS has high diagnostic accuracy and is comparable to CT and MRI in terms of tumor characterization. Having a low rate of adverse effects, CEUS can be used in patients with impaired renal function or contraindications to CT or MRI contrast agents. Quantifiable elastography and CEUS have recently started to expand the role of classic B-mode ultrasound in oncology. Quantification of tumor stiffness and perfusion can improve the differential diagnosis. These two ultrasound techniques are beginning to enter the clinic and offer a fascinating potential for further advances including improved standardization of ultrasound diagnosis.

References

1. Céspedes I, Ophir J, Ponnekanti H, et al. Elastography: elasticity imaging using ultrasound with application to muscle and breast in vivo. Ultrason Imaging. 1993;15:73–88.
2. Garra BS, Cespedes EI, Ophir J, et al. Elastography of breast lesions: Initial clinical results. Radiology. 1997;202:79–86.
3. Krouskop TA, Wheeler TM, Kallel F, et al. Elastic moduli of breast and prostate tissues under compression. Ultrason Imaging. 1998;20:260–74.
4. Konofagou E, Ophir J. A new elastographic method for estimation and imaging of lateral displacements, lateral strains, corrected axial strains and Poisson's ratios in tissues. Ultrasound Med Biol. 1998;24:1183–99.
5. Frey H. Realtime-elastographie. Ein neues sonographisches verfahren für die darstellung der gewebeelastizität. Radiologe. 2003;43(10):850.
6. Klintworth N, Mantsopoulos K, Zenk J, et al. Sonoelastography of parotis gland tumours: initial experience and identification of characteristics patterns. Eur Radiol. 2012;22:947–56. https://doi.org/10.1007/s00330-011-2344-7.
7. Rubaltelli L, Corradin S, Dorigo A, et al. Differential diagnosis of benign and malignant thyroid nodules at elastosonography. Ultraschall Med. 2009;30:175–9. https://doi.org/10.105 5/s-2008-1027442.
8. Hong Y, Liu X, Li Z, et al. Real-time ultrasound elastography in the differential diagnosis in benign and malignant thyroid nodules. J Ultrasound Med. 2009;28:861–7.
9. Kanamoto M, Shimada M, Ikegami T, et al. Real time elastography for noninvasive diagnosis in liver fibrosis. J Hepato-Biliary-Pancreat Surg. 2009;16:463–7. https://doi.org/10.1007/s00534-009-0075-9.
10. Friedrich-Rust M, Nierhoff J, Lupsor M, et al. Performance of acoustic radiation force impulse imaging for the staging of liver fibrosis: a pooled meta-analysis. J Viral Hepat. 2012;19:e212–9. https://doi.org/10.1111/j.1365-2893.2011.01537.x.
11. Aigner F, Pallwein L, Junker D, et al. Value of real-time elastography targeted biopsy for prostate cancer detection in men with prostate specific antigen 1.25 ng/ml or greater and 4.00 ng/ml or less. J Urol. 2010;184:913–7. https://doi.org/10.1016/j.juro.2010.05.026.

12. Thomas A, Kümmel S, Gemeinhardt O, et al. Real-time sonoelastography of the cervix: tissue elasticity of the normal and abnormal cervix. Acad Radiol. 2007;14:193–200.

13. Thomas A, Kümmel S, Fritzsche F, et al. Real-time sonoelastography performed in addition to B-mode ultrasound and mammography: improved differentiation of breast lesions? Acad Radiol. 2006;13:1496–504.

14. Wojcinski S, Farrokh A, Weber S, et al. Multicenter study of ultrasound real-time tissue elastography in 779 cases for the assessment of breast lesions: improved diagnostic performance by combining the BI-RADS®-US classification system with sonoelastography. Ultraschall Med. 2010;31:484–91. https://doi.org/10.1055/s-0029-1245282.

15. Thomas A, Degenhardt F, Farrokh A, et al. Significant differentiation of focal breast lesions: calculation of strain ratio in breast sonoelastography. Acad Radiol. 2010;17:558–63. https://doi.org/10.1016/j.acra.2009.12.006.

16. Itoh A, Ueno E, Tohno E, et al. Breast disease: clinical application of US elastography for diagnosis. Radiology. 2006;239:341–50.

17. Thomas A, Fischer T, Ohlinger R, et al. An advanced method of ultrasound - real-time elastography: first experience on 106 patients with breast lesions. Ultrasound Obstet Gynecol. 2006;28:335–40.

18. D'Orsi CJ, Sickles EA, Mendelson EB, et al. ACR BI-RADS® atlas, breast imaging reporting and data system. Reston: American College of Radiology; 2013.

19. Fischer T, Peisker U, Fiedor S, et al. Significant differentiation of focal breast lesions: raw data-based calculation of strain ratio. Ultraschall Med. 2012;33:372–9.

20. Thomas A, Warm M, Diekmann F, et al. Tissue doppler and strain imaging for evaluating tissue elasticity of breast lesions. Acad Radiol. 2007;14:522–9.

21. Sadigh G, Carlos RC, Neal CH, et al. Ultrasonographic differentiation of malignant from benign breast lesions: a meta-analytic comparison of elasticity and BIRADS scoring. Breast Cancer Res Treat. 2012;133:23–35. https://doi.org/10.1007/s10549-011-1857-8.

22. Sadigh G, Carlos RC, Neal CH, et al. Accuracy of quantitative ultrasound elastography for differentiation of malignant and benign breast abnormalities: a meta-analysis. Breast Cancer Res Treat. 2012;134:923–31. https://doi.org/10.1007/s10549-012-2020-x.

23. Friedrich-Rust M, Schwarz A, Ong M, et al. Real-time tissue elastography versus FibroScan for noninvasive assessment of liver fibrosis in chronic liver disease. Ultraschall Med. 2009;30:478–84. https://doi.org/10.1055/s-0028-1109488.

24. Stock KF, Klein BS, Vo Cong MT, et al. ARFI-based tissue elasticity quantification in comparison to histology for the diagnosis of renal transplant fibrosis. Clin Hemorheol Microcirc. 2010;46:139–48. https://doi.org/10.3233/CH-2010-1340.

25. Evans A, Whelehan P, Thomson K, et al. Differentiating benign from malignant solid breast masses: value of shear wave elastography according to lesion stiffness combined with greyscale ultrasound according to BI-RADS classification. Br J Cancer. 2012;107:224–9. https://doi.org/10.1038/bjc.2012.253.

26. Berg WA, Cosgrove DO, Doré CJ, et al. Shear-wave elastography improves the specificity of breast US: the BE1 multinational study of 939 masses. Radiology. 2012;262:435–49. https://doi.org/10.1148/radiol.11110640.

27. Engelken FJ, Sack I, Klatt D, et al. Evaluation of tomosynthesis elastography in a breast-mimicking phantom. Eur J Radiol. 2012;81:2169–73. https://doi.org/10.1016/j.ejrad.2011.06.033.

28. Carneiro-Pla D. Ultrasound elastography in the evaluation of thyroid nodules for thyroid cancer. Curr Opin Oncol. 2013;25:1–5. https://doi.org/10.1097/CCO.0b013e32835a87c8.

29. Bhatia KS, Rasalkar DP, Lee YP, et al. Cystic change in thyroid nodules: a confounding factor for real-time qualitative thyroid ultrasound elastography. Clin Radiol. 2011;66:799–807. https://doi.org/10.1016/j.crad.2011.03.011.

30. Shuzhen C. Comparison analysis between conventional ultrasonography and ultrasound elastography of thyroid nodules. Eur J Radiol. 2012;81:1806–11. https://doi.org/10.1016/j.ejrad.2011.02.070.

31. Kim JK, Baek JH, Lee JH, et al. Ultrasound elastography for thyroid nodules: a reliable study? Ultrasound Med Biol. 2012;38:1508–13. https://doi.org/10.1016/j.ultrasmedbio.2012.05.017.

32. Xing P, Wu L, Zhang C, et al. Differentiation of benign from malignant thyroid lesions: calculation of the strain ratio on thyroid sonoelastography. J Ultrasound Med. 2011;30:663–9.
33. Ragazzoni F, Deandrea M, Mormile A, et al. High diagnostic accuracy and interobserver reliability of real-time elastography in the evaluation of thyroid nodules. Ultrasound Med Biol. 2012;38:1154–62. https://doi.org/10.1016/j.ultrasmedbio.2012.02.025.
34. Lim DJ, Luo S, Kim MH, et al. Interobserver agreement and intraobserver reproducibility in thyroid ultrasound elastography. Am J Roentgenol. 2012;198:896–901. https://doi.org/10.2214/AJR.11.7009.
35. Friedrich-Rust M, Romenski O, Meyer G, et al. Acoustic radiation force impulse-imaging for the evaluation of the thyroid gland: a limited patient feasibility study. Ultrasonics. 2012;52:69–74. https://doi.org/10.1016/j.ultras.2011.06.012.
36. Sebag F, Vaillant-Lombard J, Berbis J, et al. Shear wave elastography: a new ultrasound imaging mode for the differential diagnosis of benign and malignant thyroid nodules. J Clin Endocrinol Metab. 2010;95:5281–8. https://doi.org/10.1210/jc.2010-0766.
37. Djavan B, Ravery V, Zlotta A, et al. Prospective evaluation of prostate cancer detected on biopsies 1, 2, 3 and 4: when should we stop? J Urol. 2001;166:1679–83.
38. Presti JC Jr. Repeat prostate biopsy–when, where, and how. Urol Oncol. 2009;27:312–4. https://doi.org/10.1016/j.urolonc.2008.10.029.
39. Frauscher F, Pallwein L, Klauser A, et al. Ultrasound contrast agents and prostate cancer. Radiologe. 2005;45:544–51.
40. Fischer T, Paschen CF, Slowinski T, et al. Differentiation of parotid gland tumors with contrast-enhanced ultrasound. RöFo. 2010;182:155–62. https://doi.org/10.1055/s-0028-1109788.
41. Pallwein L, Mitterberger M, Gradl J, et al. Value of contrast-enhanced ultrasound and elastography in imaging of prostate cancer. Curr Opin Urol. 2007;17:39–47.
42. Yan Z, Jie T, Yan-Mi L, et al. Role of transrectal real-time tissue elastography in the diagnosis of prostate cancer. Zhongguo Yi Xue Ke Xue Yuan Xue Bao. 2011;33:175–9. https://doi.org/10.3881/j.issn.1000-503X.2011.02.015.
43. Nelson ED, Slotoroff CB, Gomella LG, et al. Targeted biopsy of the prostate: the impact of color Doppler imaging and elastography on prostate cancer detection and Gleason score. Urology. 2007;70:1136–40.
44. Maxeiner A, Stephan C, Durmus T, et al. Added value of multiparametric ultrasonography in magnetic resonance imaging and ultrasonography fusion-guided biopsy of the prostate in patients with suspicion for prostate cancer. Urology. 2015;86(1):108–14. https://doi.org/10.1016/j.urology.2015.01.055.
45. Correas JM, Tissier AM, Khairoune A, et al. Prostate cancer: diagnostic performance of real-time shear-wave elastography. Radiology. 2015;275(1):280–9. https://doi.org/10.1148/radiol.14140567.
46. Schneider M. SonoVue, a new ultrasound contrast agent. Eur Radiol. 1999;9:S347–8.
47. Greis C. Ultrasound contrast agents as markers of vascularity and microcirculation. Clin Hemorheol Microcirc. 2009;43(1):1–9. https://doi.org/10.3233/CH-2009-1216.
48. Clevert DA, Jung EM. Interventional sonography of the liver and kidneys. Radiologe. 2013;53:962–73. https://doi.org/10.1007/s00117-012-2459-0.
49. Jung EM, Clevert DA. Possibilities of sonographic image fusion: current developments. Radiologe. 2015;55:937–48. https://doi.org/10.1007/s00117-015-0025-2.
50. Schwarz F, Sommer WH, Reiser M, et al. Contrast enhanced sonography for blunt force abdominal trauma. Radiologe. 2011;51:475–82. https://doi.org/10.1007/s00117-010-2103-9.
51. Valentino M, Serra C, Pavlica P, et al. Contrast-enhanced ultrasound for blunt abdominal trauma. Semin Ultrasound CT MR. 2007;28:130–40.
52. Claudon M, Cosgrove D, Albrecht T, et al. Guidelines and good clinical practice recommendations for contrast enhanced ultrasound (CEUS) - update 2008. Ultraschall Med. 2008;29:28–44. https://doi.org/10.1055/s-2007-963785.
53. Piscaglia F, Bolondi L, Italian Society for Ultrasound in Medicine and Biology (SIUMB) Study Group on Ultrasound Contrast Agents. The safety of Sonovue® in abdominal applications: retrospective analysis of 23188 investigations. Ultrasound Med Biol. 2006;32(9):1369–75.

54. Strobel DB. Diagnostik bei fokalen Leberläsionen. Dtsch Arztebl Int. 2006;103:789–93.
55. Oldenburg A, Hohmann J, Foert E, et al. Detection of hepatic metastases with low MI real time contrast enhanced sonography and SonoVue. Ultraschall Med. 2005;26:277–84.
56. Harvey CJ, Blomley MJ, Eckersley RJ, et al. Developments in ultrasound contrast media. Eur Radiol. 2001;11:675–89.
57. Clevert DA, D'Anastasi M, Jung EM. Contrast-enhanced ultrasound and microcirculation: efficiency through dynamics–current developments. Clin Hemorheol Microcirc. 2013;53(1-2):171–86. https://doi.org/10.3233/CH-2012-1584.
58. Regge D, Campanella D, Anselmetti GC, et al. Diagnostic accuracy of portal-phase CT and MRI with mangafodipir trisodium in detecting liver metastases from colorectal carcinoma. Clin Radiol. 2006;61(4):338–47.
59. Jung EM, Wiggermann P, Stroszczynski C, et al. Ultrasound diagnostics of diffuse liver diseases. Radiologe. 2012;52(8):706–16. https://doi.org/10.1007/s00117-012-2307-2.
60. Kinkel K, Lu Y, Both M, et al. Detection of hepatic metastases from cancers of the gastrointestinal tract by using noninvasive imaging methods (US, CT, MR imaging, PET): a meta-analysis. Radiology. 2002;224(3):748–56.
61. Weskott HP. Detection and characterization of liver metastases. Radiologe. 2011;51:469–74. https://doi.org/10.1007/s00117-010-2100-z.
62. Strobel D, Seitz K, Blank W, et al. Contrast-enhanced ultrasound for the characterization of focal liver lesions – diagnostic accuracy in clinical practice1 (DEGUM multicenter trial). Ultraschall Med. 2008;29:499–505. https://doi.org/10.1055/s-2008-1027806.
63. Claudon M, Dietrich CF, Choi BI, et al. Guidelines and good clinical practice recommendations for contrast enhanced ultrasound (CEUS) in the liver–update 2012: a WFUMB-EFSUMB initiative in cooperation with representatives of AFSUMB, AIUM, ASUM, FLAUS and ICUS. Ultraschall Med. 2013;34:11–29. https://doi.org/10.1055/s-0032-1325499.
64. Clevert DA, Helck A, Paprottka PM, et al. Latest developments in ultrasound of the liver. Radiologe. 2011;51:661–70. https://doi.org/10.1007/s00117-010-2124-4.
65. Clevert DA, Paprottka PM, Helck A, et al. Image fusion in the management of thermal tumor ablation of the liver. Clin Hemorheol Microcirc. 2012;52:205–16. https://doi.org/10.3233/CH-2012-1598.
66. Clevert DA, Helck A, Paprottka PM, et al. Ultrasound-guided image fusion with computed tomography and magnetic resonance imaging. Clinical utility for imaging and interventional diagnostics of hepatic lesions. Radiologe. 2012;52:63–9. https://doi.org/10.1007/s00117-011-2252-5.

Sensitivity of Tissue Shear Stiffness to Pressure and Perfusion in Health and Disease

20

Jing Guo, Florian Dittmann, and Jürgen Braun

Abstract

This chapter discusses the sensitivity of the complex shear modulus to changes in pressure and perfusion under both physiological and pathological conditions. Biological tissue is considered an effective medium where fluid and solid phases are incorporated on a microscopic scale. The complex relationship between pressure/perfusion and effective tissue stiffness measured by elastography has so far been investigated in abdominal organs and the brain.

In the abdomen, postprandial variations in hepatic and splenic stiffness were observed in both healthy subjects and patients with liver fibrosis. In patients with portal hypertension, elevated stiffness was observed in both the liver and spleen, which instantly decreased after portal decompression. Reduced renal stiffness in patients with chronic kidney disease was found to be related to impaired renal perfusion. In the brain, different stiffness values were obtained in different gray matter regions due to their distinct perfusion characteristics. Preliminary experiments with hypercapnia or jugular compression resulted in elevated brain stiffness most likely due to increased intracranial pressure.

As shear stiffness has shown sensitivity to pressure and perfusion variations associated with changes in physiological and pathological conditions, it could be used as a parameter for noninvasive assessment of pressure and perfusion. However, as confounding factors, pressure and perfusion could mask other

J. Guo (✉) • F. Dittmann
Department of Radiology, Charité - Universitätsmedizin Berlin, Berlin, Germany
e-mail: Jing.Guo@charite.de

J. Braun
Department of Medical Informatics, Charité - Universitätsmedizin Berlin, Berlin, Germany

© Springer International Publishing AG 2018
I. Sack, T. Schaeffter (eds.), *Quantification of Biophysical Parameters in Medical Imaging*, https://doi.org/10.1007/978-3-319-65924-4_20

429

pathologies and bias the interpretation of stiffness, which reduces diagnostic accuracy. To avoid misinterpretation, potential effects of perfusion and pressure on shear stiffness measured by elastography must be disentangled and controlled by standardized measurement protocols.

20.1   Introduction

The cardiovascular system is precisely regulated to ensure an appropriate and reliable supply of oxygenated blood to different body tissues (Fig. 20.1). Blood flow in tissues depends on the pressure gradient existing across the vascular bed as well as vessel

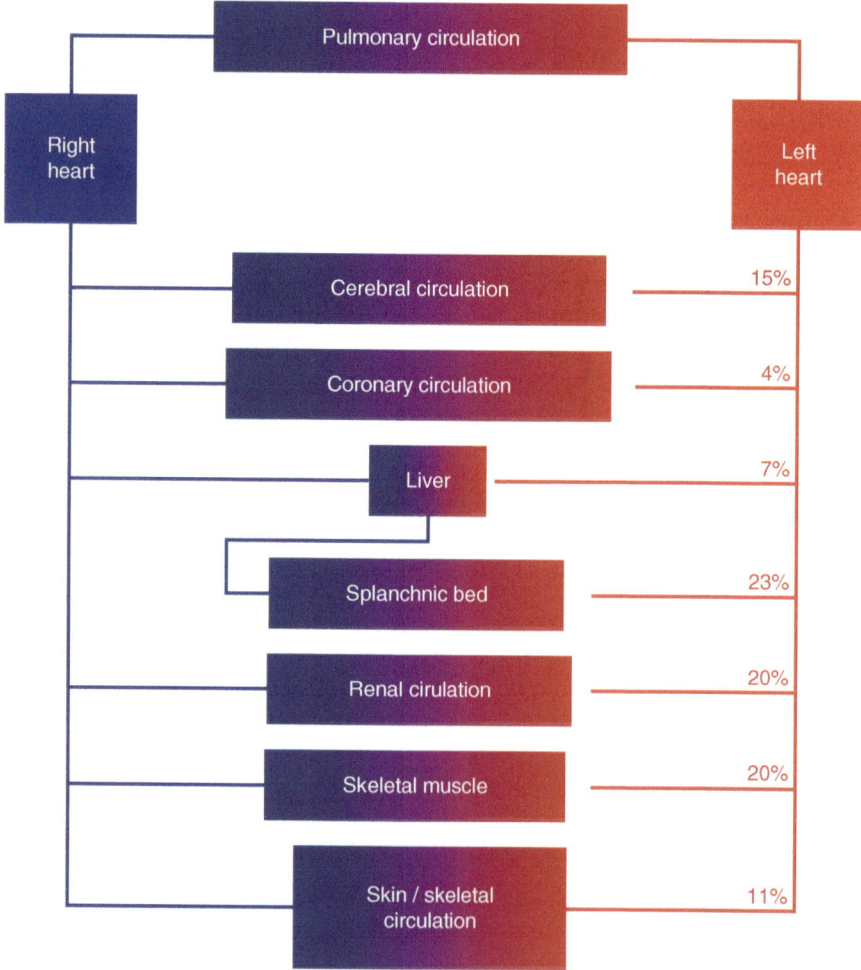

Fig. 20.1 The distribution of cardiac output to different organs. Individual organ blood flow is given as percentage of cardiac output

resistance. Perfusion is defined as blood flow at the capillary level. Blood flow is regulated by changing the internal diameter of the arterioles, which represent the major resistant element in the cardiovascular system. Vascular resistance in turn is controlled by hormones and autonomous nerves as well as by intrinsic mechanisms which are specific to individual organs. Common intrinsic mechanisms include pressure flow autoregulation, active hyperemia, and reactive hyperemia. Other mechanisms, e.g., in the brain, include acidosis. More detailed information on the autoregulation of blood flow can be found in standard textbooks of medical physiology [1, 2].

Hemodynamics are related to biomechanical properties such as blood viscosity and vascular resistance as well as compliance and geometry of the tissue matrix. Since the vascular system is embedded in the tissue matrix, one needs to consider biological tissue an effective medium with interactions between liquid and solid phases.

In Chaps. 3 and 4, the concept of poroelasticity was introduced to interpret the compression and shear properties of biphasic tissues. It was shown that analysis of the effective compression modulus in those media provides a quantitative marker of effective tissue pressure. In this chapter, we will focus, from an experimental perspective, on the effective shear modulus in effective media, where fluid and solid phases are integrated on a microscopic scale. We will discuss the sensitivity of the complex shear modulus to changes in pressure and perfusion under both physiological and pathological conditions. In the literature, the complex shear modulus is often reported by either the shear modulus μ (in kPa) or the shear wave speed c (in m/s) at a given frequency. In pure elastic materials, both are related to each other by $c = \sqrt{\mu / \rho}$ (see Eq. (2.23) in Chap. 2). If the material is viscoelastic, the wave speed is affected by the viscosity, e.g., the shear wave speed becomes $c = \mathrm{Re}\sqrt{G^* / \rho}$, where G^* is the complex shear modulus. By $G^* = G' + iG''$, the complex shear modulus is composed of storage modulus (G', sometimes named shear elasticity) and loss modulus (G'', sometimes named shear viscosity). Often in the literature, a lumped magnitude modulus $|G^*|$ is reported as "stiffness" or "shear elasticity," which is technically incorrect in particular when also used for shear wave speed c. However, the term stiffness makes intuitive sense to most people and to clinicians and will be used in this chapter to describe the tendency of a tissue to increase or decrease in its shear elasticity upon various flow-related changes.

The following sections are organized by organs, which have different functional and metabolic requirements giving rise to unique blood flow regulation mechanisms and eventually resulting in characteristic shear stiffness responses.

20.2 The Influence of Pressure and Perfusion on Abdominal Tissue Stiffness

Vascular flow and perfusion of abdominal organs are governed by different circulation systems. In this section, we will focus on three abdominal organs: the liver, spleen, and kidney. Both hepatic and splenic circulations belong to the splanchnic circulation, which includes the vascular system of the gastrointestinal tract, liver, spleen, and pancreas (Fig. 20.2). The splanchnic vessels transport the absorbed nutrients to the liver, and splanchnic vascular reserve can be mobilized during stress to

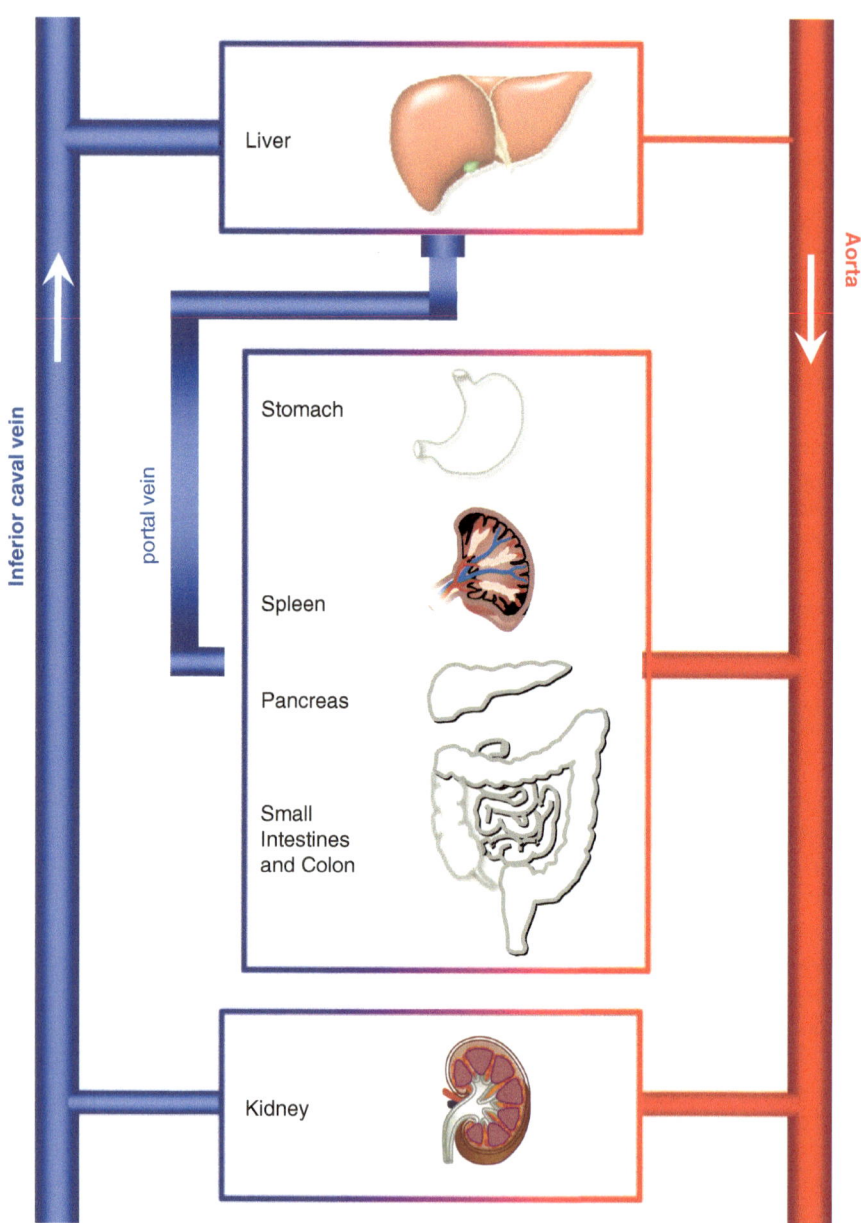

Fig. 20.2 Schematic representation of the arterial supply and venous return of abdominal organs

maintain overall cardiovascular homeostasis. Characteristic determining factors of splanchnic blood circulation are (a) active hyperemia elicited by food ingestion, (b) mutual blood flow control between regions of this organ system, and, in contrast to the brain, (c) a high sensitivity of splanchnic arterioles to extrinsic control mechanisms.

Unlike splenic circulation, which is regulated by the splenic artery and splenic vein, the liver possesses a dual blood supply system. The hepatic artery carries arterial flow, which maintains vital liver functions, and the hepatic portal vein receives blood from the stomach, intestines, pancreas, and spleen. The hepatic portal vein directs nutrient-rich blood from the gastrointestinal tract and spleen to the liver. With this unique vascular structure, the normal liver receives 70% portal venous flow and 30% hepatic arterial flow. To prevent sudden changes of flow in liver sinusoids, the amount of blood flow in the hepatic artery and the portal vein is inversely related: when blood flow through the portal vein increases, arterioles from the hepatic artery constrict and vice versa.

The kidneys serve as a natural blood filter. Renal circulation branches from the abdominal aorta through the renal artery and returns blood to the inferior vena cava via the renal vein (Fig. 20.2). The kidney removes water-soluble wastes which are diverted to the bladder. The main functions of the kidney include regulating electrolytes in blood and maintaining pH homeostasis. Overall, normal human kidneys generate about 150 L of filtrate per day. The two kidneys together constitute approximately 0.5% of total body mass but receive approximately 20% of cardiac output at rest. To manage such a heavy workload, the kidneys have a dense vascular network with low resistance and a large number of filter units, which are arranged in parallel and intensely packed. Similar to the brain, the kidneys have intrinsic ability to maintain a constant blood flow despite changes in systemic blood pressure. The renal autoregulation is important to maintain a stable glomerular filtration rate (GFR).

Circulation in the abdominal organs can be monitored by measuring blood flow and perfusion using medical imaging modalities such as Doppler ultrasound, contrast-enhanced ultrasound, or magnetic resonance imaging (MRI). In clinical routine, pressure in the abdominal organs is often measured invasively using pressure probes mounted to catheters. Here, elastography could offer a way to noninvasively monitor stiffness changes in response to an altered tissue pressure. Given that the effective shear stiffness of tissue is sensitive to perfusion and blood flow, physiological and pathological variations that cause changes in hemodynamic conditions could lead to the mechanical response of the tissue in terms of alerting the shear stiffness. We will discuss the pressure- and/or perfusion-related changes in shear stiffness in the following paragraphs.

20.2.1 Physiological Variations in Pressure and Perfusion

It is known from Doppler ultrasonography and MRI measurements in healthy individuals that mesenteric venous flow increases in the postprandial state and that an increase in hepatic portal blood flow is associated with a decrease in intrahepatic vascular resistance [3, 4]. A postprandial increase in portal venous perfusion and a decrease in arterial perfusion were recently observed by Schalkx et al. using spin labeling MRI [5]. In the spleen, no postprandial effects were found in splenic venous flow, as stated in [6, 7]. However, a postprandial reduction in splenic volume was reported [8]. Postprandial changes in shear stiffness were observed by elastography as a mechanical response to hemodynamic variations. A recent time-harmonic ultrasound elastography (THE; see Chap. 12) study on healthy volunteers revealed a significant increase in hepatic stiffness as a result of water or meal ingestion [9]. This finding was reproduced, and the experiment was expanded to other abdominal organs in subsequent studies. A 2D THE study in healthy volunteers reported that ingestion of 1 liter of water led to increased hepatic stiffness and decreased splenic stiffness [10]. In this study, the liver and spleen were firstly examined in a lower-frequency band centered at 40 Hz. The spleen was additionally investigated in a higher-frequency band centered at 120 Hz. In both lower- and higher-frequency bands, significant softening of the spleen was observed, as shown in Fig. 20.3.

A magnetic resonance elastography (MRE) study on healthy volunteers provided similar results. Significant stiffening of the liver as well as softening of the spleen after 1 liter water ingestion was reported in the frequency range from 50 to 60 Hz [11]. Additionally, the authors also observed a reduced pancreatic stiffness as a result of water intake. It was hypothesized in [10, 11] that different mechanical responses observed in abdominal organs could be related to differences in tissue

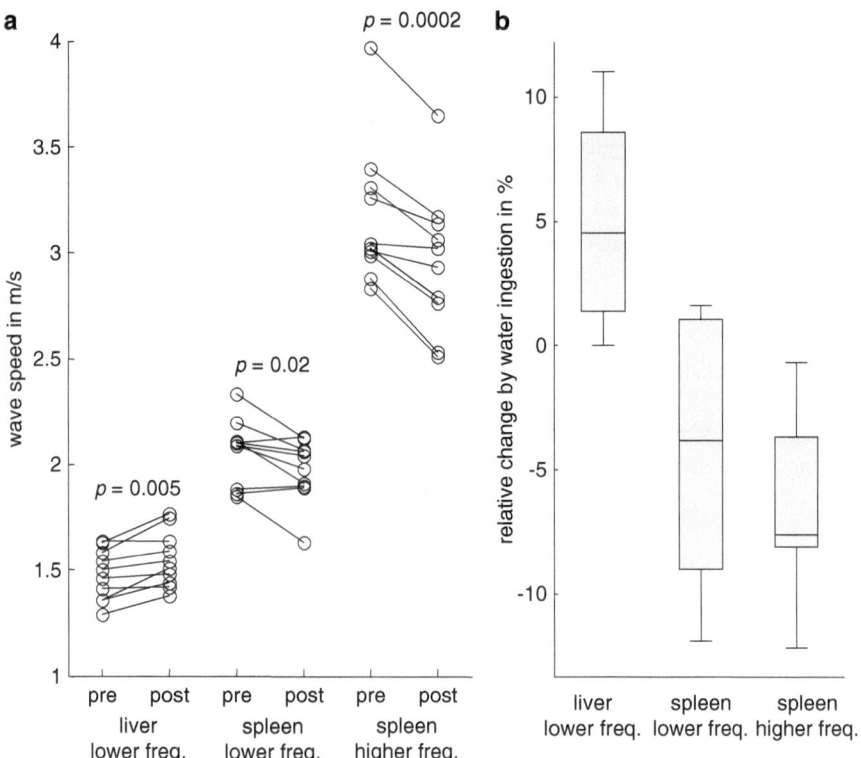

Fig. 20.3 (a) Wave speed values measured in the liver and spleen before and after intake of 1 L of water. (b) Relative change in hepatic and splenic stiffness after water ingestion. Figures are taken from [10] with journal permission

compliance. More discussion on the direction of the effect is provided below. In addition to healthy volunteers, postprandial variation in tissue stiffness was also observed in patients. In an MRE study of patients with liver cirrhosis, elevated liver stiffness was observed after administering a liquid test meal, which was found to result in a transient increase in portal pressure [12]. In the same study, healthy volunteers also showed a significant increase in liver stiffness, but the increase was much smaller than in patients with liver cirrhosis.

The postprandial increase in hepatic stiffness was also revealed by transient elastography (TE) in patients with cirrhosis and portal hypertension [13]. It was further reported in this study that the changes in hepatic stiffness were correlated with hepatic artery blood flow. In another TE study, portal blood flow rather than hepatic artery blood flow was assessed in patients with minimal or moderate fibrosis. A significant correlation between the increase in liver stiffness and the elevation of portal blood flow after water and meal consumption was observed [14]. As pointed out in [13, 14], the observation that, in patients with liver cirrhosis, the postprandial change in liver stiffness was only correlated with hepatic artery blood flow and not with portal blood flow might be due to liver arterialization, a process that only occurs in the presence of cirrhosis [14]. However, precise measurements at well-defined stimulation frequencies as used in THE demonstrated that venous blood flow influences shear stiffness [15]. This study used the Valsalva maneuver plus abdominal muscle contraction to increase abdominal pressure. The Valsalva maneuver, which is the forceful attempt of exhalation against a closed airway, is known to result in collapse of the inferior vena cava and intrahepatic veins [16]. In [15] THE was paired with color Doppler flow measurements demonstrating that a decrease in hepatic venous flow is correlated with reduction of hepatic stiffness.

In conclusion, shear stiffness of the abdominal organs obtained by elastography is sensitive to blood flow and perfusion, which can be altered by physiological variations such as water/meal ingestion and breathing maneuvers. As hemodynamic variation is a confounding factor in measured shear stiffness, elastography examinations should be performed during standardized fasting (nutrition state and fluid balance) and breathing conditions to improve the reproducibility and diagnostic accuracy of the method.

20.2.2 Pathological Variations in Pressure and Perfusion

20.2.2.1 Portal Hypertension

In patients with liver cirrhosis, accumulation of collagen leads to obliteration of portal venules, sinusoids, and central veins. These vascular changes affect hepatic hemodynamics in terms of reducing portal flow and elevating arterial flow as a compensatory mechanism to reperfuse remaining sinusoids [17]. Restricted portal flow in the cirrhotic liver leads to portal hypertension, esophageal varices, and eventually hepatic failure. In comparison to healthy controls, liver perfusion measured by dynamic CT showed a reduced portal/total liver perfusion and increased arterial perfusion in patients with portal hypertension [18]. In another study including patients with cirrhosis and portal hypertension, hepatic flow parameters such as apparent arterial and portal perfusion measured with MRI correlated with the severity of cirrhosis and portal pressure elevation [19]. It should be noted that considering the role of the spleen in splanchnic circulation, a significant decrease in splenic perfusion was observed in patients with chronic liver disease and the spleen perfusion is negatively correlated with wedged hepatic vein pressure measured by dynamic CT [20].

In clinical practice, portal hypertension is assessed by measuring the hepatic venous pressure gradient (HVPG) invasively by means of a hepatic vein catheter [21]. Elastographic techniques such as TE [22], acoustic radiation force impulse (ARFI) imaging [23, 24], shear wave elastography (SWE) [25], and MRE [26, 27] were introduced to noninvasively assess portal hypertension. Considering blood flow and perfusion, in patients with hepatic fibrosis and portal hypertension, shear modulus and shear wave speed measured by elastographic methods include contributions not only from fibrosis-related collagen accumulation but also from pressure and mechanical coupling between fluid phase and solid tissue. The most direct evidence came from studies investigating hepatic stiffness in patients with portal hypertension before and after transjugular intrahepatic portosystemic shunt (TIPS) implantation [26]. The TIPS procedure provides instantaneous portal decompression, leading to a reduction in HVPG with a concomitant increase in both arterial and total liver perfusions [18]. A significant reduction in hepatic stiffness after TIPS was found using both THE [28] and MRE [26]; examples of MR elastograms reflecting liver stiffness before and after TIPS placement are shown in Fig. 20.4a. The fact that in these studies, liver softening was observed shortly after TIPS implantation suggests that reduced hepatic stiffness is mostly related to changes in portal pressure and clearly underlines the pressure sensitivity of shear stiffness.

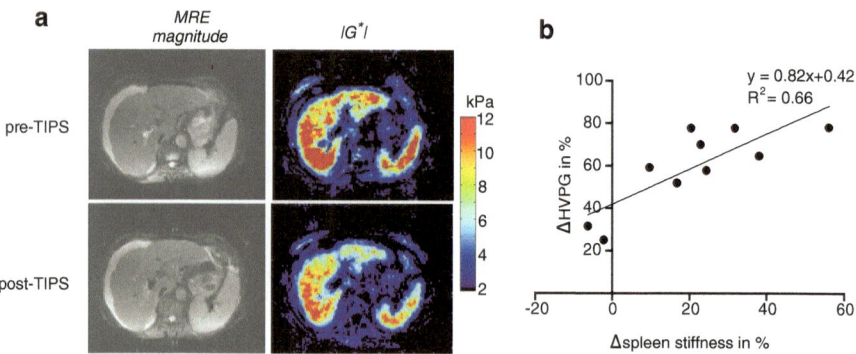

Fig. 20.4 (**a**) Magnitude of the MRE signal and reconstructed |G^*| map of a patient before and after TIPS placement. Softening of both liver and spleen after portal decompression is apparent from the |G^*| map. (**b**) The relative change in splenic shear stiffness and the relative changes in HVPG after TIPS placement are linearly correlated. The figures are taken and modified from [26] with journal permission

Portal decompression following TIPS placement also causes hematological variations in the spleen [29]. Splenomegaly is commonly observed in patients with portal hypertension as a result of congestion. In portal hypertension, the spleen becomes engorged with blood because of impaired flow through the splenic vein, which empties into the portal vein. After TIPS placement, as spleen congestion is relieved, the size of the spleen decreased [29], and splenoportal venous velocity increased [30]. A mechanical response of the spleen to the hematological variation after TIPS placement is the reduction of splenic shear stiffness, as reported in [23, 26]. The MRE study [26] found not only a significant reduction in spleen stiffness but also a linear correlation between relative changes in spleen stiffness before and after TIPS implantation with relative changes of HVPG (Fig. 20.4b). A similar finding was reported by an ARFI study, whereas a reduction in shear stiffness due to TIPS was only observed in the spleen [23]. The authors of both studies hypothesize that the less advanced fibrosis stage of the spleen makes it more responsive to portal decompression than the cirrhotic liver.

The sensitivity of shear stiffness to portal pressure is confirmed by numerous elastography studies reporting either hepatic or splenic shear stiffness or both to be correlated with HVPG. Among the MRE studies, Nedredal et al. showed, in a canine model of cholestatic chronic liver disease, that there was a direct correlation between spleen stiffness and the HVPG [31]. In pigs with acute portal hypertension, both splenic and hepatic stiffness were reported to significantly correlate with portal pressure [32]. In a recent MRE study, the loss modulus of the liver and the spleen correlated with the HVPG in 36 patients with cirrhosis [27]. In an ultrasound elastography study, SWE was used to investigate hepatic stiffness in patients with liver cirrhosis. Here, both liver stiffness and its changes after medication correlated with the HVPG [25]. A study based on real-time elastography (RTE) in patients with chronic liver damage showed that splenic elasticity correlated well with HVPG [33]. Furthermore, TE and ARFI revealed good correlation between liver stiffness and HVPG in patients with chronic liver disease and portal hypertension [22, 24].

To summarize, hepatic and splenic shear stiffness are elevated in patients with portal hypertension. Compared to hepatic stiffness, splenic stiffness seems more responsive to HVPG changes. Shear stiffness can serve as an imaging marker for the noninvasive assessment of portal pressure.

20.2.2.2 Renal Dysfunction

In patients with chronic kidney disease (CKD), the presence of sclerotic glomeruli, tubular atrophy, and peritubular fibrosis leads to a decreased blood flow in the peritubular vascular plexus. The damage of microcirculation reduces renal blood flow and perfusion, as reported in [34, 35]. Current diagnostic methods for CKD include blood testing (serum creatinine level), urinalysis, and renal biopsy, which is still the gold standard despite its invasiveness. Serum and urine markers are not specific enough, and renal biopsy has a small sampling size, limiting its ability to quantify overall renal damage [36]. Elastographic methods have been introduced to assess renal mechanical properties and renal function noninvasively [37–41]. However, renal stiffness determined by elastography reflects more than renal fibrosis alone. Animal experiments suggest that urinary and vascular pressure related to renal blood flow as well as perfusion are other influencing factors [42–44]. In a supersonic shear wave elastography study in which the kidneys of pigs were investigated in vivo, a decrease and an increase in renal elasticity could be induced by renal artery and renal vein ligation, respectively [42]. Additionally, a positive correlation of parenchymal elasticity with urinary pressure was observed in the same study. Using MRE, Warner et al. demonstrated, in a pig model, that hemodynamic variables such as renal blood flow modulate renal stiffness [44]. In this study, renal cortex stiffness decreased during acutely reduced renal blood flow, while, in pigs with chronic renal arterial stenosis (RAS), reduced renal blood flow did not translate directly into decreased renal stiffness due to a simultaneous increase in renal stiffness due to renal fibrosis in chronic RAS. A recent SWE study on ex vivo porcine kidneys also reported that an increase in perfusion pressure elevated the average shear modulus in the renal cortex [43]. An influence of perfusion pressure on renal stiffness was also reported in patients. In an ARFI elastography study, where renal stiffness was measured in healthy volunteers and patients with CKD, despite the presence of interstitial fibrosis, the patients had significantly lower shear wave velocity (SWV) compared with healthy volunteers [38]. These findings were confirmed by another ARFI study in 183 patients with CKD, where a positive correlation was found between renal SWV and estimated glomerular filtration rate (eGFR) [41]. eGFR reflects renal function and correlates positively with renal perfusion, as demonstrated in [45]. In [41], the authors hypothesized that diminished renal blood flow led to reduced kidney stiffness in patients with CKD and that variation in blood flow may affect SWV values more than progression of tissue fibrosis. This hypothesis was supported by a later ARFI study including patients with different degrees of CKD, which found a similar positive correlation between renal SWV and eGFR despite large variation among the patients included [37]. More interestingly, the authors also found that significant SWV differences only existed between patients with stage 1 or no CKD and patients with stage 4 or 5 CKD. The lack of SWV differences for the intermediate CKD stages hints at the complex interplay between hemodynamic variation and fibrosis—two major factors affecting renal stiffness.

In addition to native kidneys, hemodynamics also play a role in estimating renal stiffness of kidney transplants. MRE in patients with renal allograft showed that renal stiffness was higher in functioning allografts than in nonfunctioning ones (Fig. 20.5a). Dysfunctional allografts were characterized by a high degree of fibrosis and reduced GFR. Figure 20.5b shows that renal stiffness $|G^*|$ is positively correlated with GFR [40]. The authors concluded that reduced renal blood flow and perfusion in patients with dysfunctional renal transplants resulted in a decreased renal stiffness despite the presence of fibrosis. Similar findings were reported in another MRE study where stiffness in transplant kidneys with mild fibrosis was lower than in transplants without significant fibrosis [39]. Healthy native kidneys were also investigated and compared to renal transplants in [40]. Stiffness of native kidneys was significantly lower than in transplant recipients with normal graft function. The authors attributed this finding to denervation of sympathetic nerves in kidney transplants, a process which leads to increased water excretion and

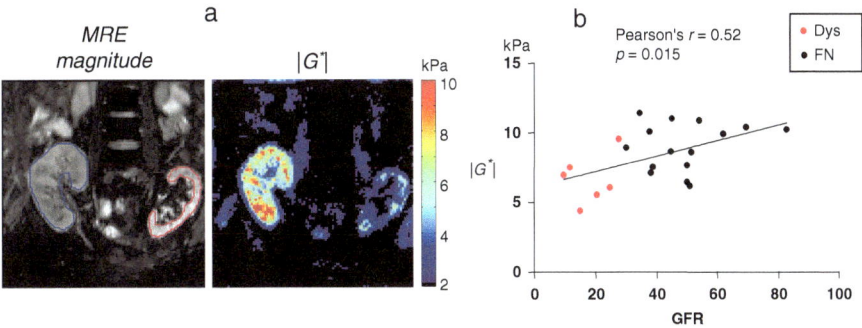

Fig. 20.5 (**a**) Magnitude of the MRE signal and reconstructed |G*| map of a patient with two transplant kidneys. In the |G*| map, the functioning transplant (blue) appears to be stiffer than the nonfunctioning one (red). (**b**) Renal stiffness |G*| is positively correlated with GFR. The figures are modified from [40] with journal permission

intratubular fluid [46, 47], resulting in an elevated renal stiffness in functional renal transplants.

To summarize, reduced renal perfusion in CKD leads to decreased renal stiffness, which could mask changes in kidney stiffness related to other renal pathologies. Therefore, one needs to take into account the effects of perfusion- and pressure-related factors in order to adequately interpret the mechanical properties obtained by renal elastography.

The relationships between abdominal shear stiffness and pressure/perfusion under different physiological and pathological conditions discussed in this section are summarized in Table 20.1.

20.3   The Influence of Perfusion on Brain Stiffness

Under resting conditions, blood flow in the brain accounts for about 15% of the cardiac output (see Fig. 20.1). Gray matter has a high rate of oxidative metabolism, which is about six times higher than that of white matter. Therefore, the brain is very sensitive to hypoxia, and 10 mins of brain ischemia leads to irreversible cell damage. The primary function of cerebral circulation is to ensure a constant supply of oxygen (O_2) to the brain parenchyma.

Regulation of blood flow in the human brain is extremely complex. There exist multiple overlapping regulatory paradigms and key structural components. The interaction of these components, as well as the components themselves, is not yet fully understood. Major categories of mechanisms discovered so far are pressure autoregulation, metabolic regulation, and neurogenic regulation [48].

For some vascular beds (e.g., renal or cutaneous), the dilatation of large arteries induced by a moderately severe level in the partial pressure of carbon dioxide ($PaCO_2$) in the blood (hypercapnia) has relatively small effects on blood flow [49]. In strong contrast, cerebral blood flow is largely dependent on $PaCO_2$ [50], making hypercapnia a very effective vasodilator in the cerebral circulation.

Table 20.1 Relationship between shear stiffness and pressure/perfusion for abdominal organs under different physiological and pathological conditions

Organ (condition)	Pressure	Perfusion	Stiffness
Liver (postprandial/ water ingestion)		Portal perfusion ↑; arterial perfusion ↓ [5]	↑ [9–11]
Liver (valsalva maneuver)	Abdominal pressure ↑	Hepatic venous flow ↓ [15]; portal flow ↔ [16]	↓ [15]
Spleen (postprandial/ water ingestion)		Splenic venous flow ↔ [6, 7]	↓ [10, 11]
Pancreas (water ingestion)			↓ [11]
Liver (portal hypertension: patients vs. controls)	Portal pressure ↑	Portal and total liver perfusion ↓ Arterial perfusion ↑ [18]	↑ [23]
Spleen (portal hypertension: patients vs. controls)	Increased pressure ↑ due to congestion	↓ [20]	↑ [23]
Liver (TIPS placement)	Portal pressure ↓	Arterial and total liver perfusion ↑ Portal perfusion ↔ [18]	↓ [26, 28]
Spleen (TIPS placement)	Congestion relieved ↓	↑ [30]	↓ [23, 26]
Kidney (patients with CKD)	↓ [34]	↓ [35]	↑ with fibrosis [37] ↓ with reduced eGFR [37, 41]
Transplant kidney (patients with renal allograft dysfunction)		↓ [40]	↑ in functioning vs. ↓ dysfunctioning transplants ↓ with reduced GFR [40]

↑ increase, ↓ decrease, ↔ no change

20.3.1 Quantification of Blood Perfusion in Brain Parenchyma

The measurement of tissue perfusion depends on the ability to repeatedly measure the concentration of a tracer agent in a target organ of interest. Basically, two tracer types can be distinguished: (a) exogenous tracers such as radiographic contrast material or radionuclides [51, 52] and (b) endogenous tracers [53, 54]. In the following, we will focus on noninvasive perfusion MRI exploiting arterial blood water as an endogenous tracer. In arterial spin labeling (ASL) perfusion MRI, water protons are magnetically labeled (or "tagged") in the inflowing arterial blood proximal to the slice of interest. Tagging is performed by application of radio-frequency (RF) pulses specially designed to invert magnetization within a thick slab. In this way, inflowing blood can be tagged intermittently or continuously [55, 56]. Continuous labeling provides twice as much signal contrast as pulsed techniques. However, continuous labeling methods produce substantially more radio-frequency pulse-induced power deposition in the subject, which—for safety reasons—limits slice coverage and increases acquisition time.

Tagged water protons are assumed to freely diffuse from the intravascular compartment into the tissue compartment. This model is similar to that used in positron emission tomography (PET) and single-photon emission computed tomography (SPECT), where a tracer is administered to then measure its regional accumulation [56] (see Chaps. 18 and 21). While offering the advantage of noninvasiveness and use of endogenous tracers, ASL suffers from signal loss due to (a) a decreasing number of labeled protons during the transit period and (b) magnetization transfer from macromolecule-bound protons in the imaging plane to freely moving protons. Such mechanisms of magnetization transfer originate from off-resonances of the labeling RF pulses (see Chap. 10). Subtracting the label images from a control image, which is acquired using a control pulse without labeling the flowing blood, removes static background signals and reduces magnetization transfer effects.

A typical ASL data processing pipeline for cerebral blood flow (CBF) consists of (i) motion correction for control and label images and coregistration of the reference label image to the reference control image, (ii) spatial smoothing, (iii) masking of extra parenchyma voxels, and (iv) CBF quantification.

The scale of perfusion relates to the functional [57] and structural status [58] of the tissue. Topology and geometry of microvessels have been postulated to significantly influence the global shear modulus of soft biological tissues [59].

20.3.2 Interrelation of Shear Stiffness, Perfusion, and Pressure in the Human Brain

To date, little is known about the relationship between shear stiffness and blood flow in the brain. A study in nine healthy volunteers showed that jugular compression can increase the stiffness of brain tissue [60]. It was shown that subjects who do not divert venous blood through extrajugular pathways during jugular compression have higher brain stiffness than those who do, likely as a result of increased neurovascular pressure. More details on regional effects of blood flow and brain stiffness were reported in a recent study where MRE and ASL were combined in 14 healthy volunteers [61]. In this study, six regions with distinct functional and structural features were analyzed (see Fig. 20.6a): the nucleus accumbens and putamen as parts of the striatal region of the basal ganglia and the hippocampus, amygdale, thalamus, and globus pallidus as non-striatal regions of deep gray matter (DGM).

For the non-striatal regions, an inverse correlation was found for the magnitude shear modulus $|G^*|$ and CBF (see Fig. 20.6b). In contrast, the striatal regions showed significantly higher $|G^*|$ and CBF values. Hence, two clusters of data exist, showing different stiffness-perfusion properties in the CBF-MRE space. While the striatum region is characterized by high CBF and high stiffness values, the remaining DGM shows reduced stiffness with increasing CBF. This distinction between striated and non-striated DGM indicates the complex relationship between mechanical and blood flow-based parameters. The striatum is characterized by a higher blood supply and a denser neuronal network than the other DGM regions. At the same time, striatal vessels are smaller (mean vessel radius of 6.2 ± 0.7 μm) than those within,

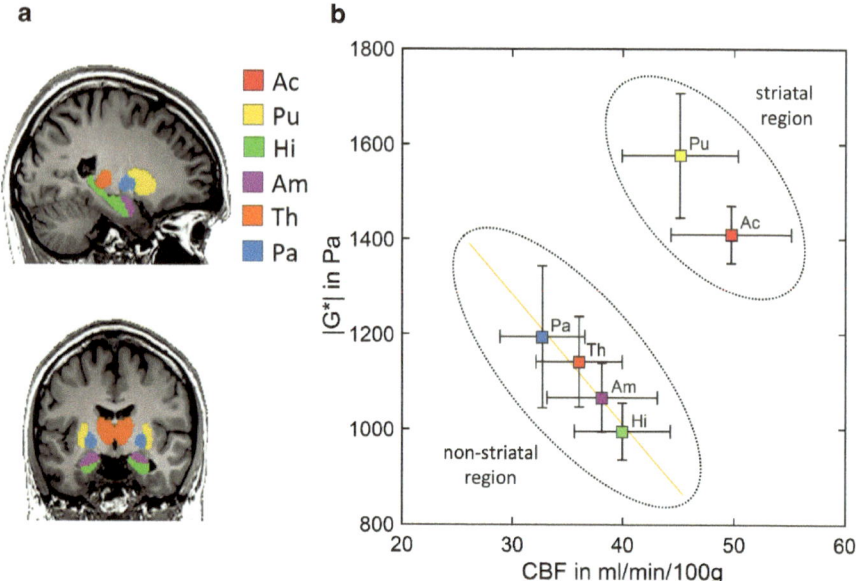

Fig. 20.6 (**a**) Analyzed deep gray matter regions in MNI space (unbiased standard magnetic resonance imaging template brain volume for normal population from the Montreal Neurological Institute). Red, nucleus accumbens (Ac); yellow, putamen (Pu); green, hippocampus (Hi); pink, amygdala (Am); orange, thalamus (Th); blue, globus pallidus (Pa). (**b**) In vivo DGM of the human brain characterized by MRE and ASL. The cluster of non-striatal DGM (Pa, Th, Am, Hi) reveals an inverse correlation between tissue stiffness ($|G^*|$) and perfusion (CBF) in these regions. Error bars reflect the standard error of the mean

e.g., the hippocampus region (mean vessel radius of $11.3 \pm 2.1\ \mu m$) or the thalamus ($8.8 \pm 1.6\ \mu m$) [62]. Thus, normalizing CBF by mean vessel area takes into account differences in the perfusion pressure gradient between the analyzed regions based on Darcy's law as explained in Chap. 3. CBF normalized to the mean vessel area represents a parameter that is related to the flux of blood through the capillary system of DGM [61]. Interestingly, such a perfusion flux parameter (q in Fig. 20.7) is linearly correlated to MRE-measured stiffness in all regions. For this reason, the distinction between striatum and other DGM regions as seen in Fig. 20.6b vanishes when considering CBF normalized to mean vessel areas, suggesting the sensitivity of MRE to the perfusion pressure gradient rather than CBF alone. More details to the model of normalized perfusion can be found in [61].

Further preliminary information is available for the effect of hypercapnia on brain stiffness in few healthy volunteers (Fig. 20.8). These data suggest that hypercapnia causes cerebral vasodilation, resulting in an increase in cerebral blood flow [65] and intracranial pressure [66]. The observed 10% increase in stiffness due to hypercapnia (Fig. 20.8) agrees with the previous observation that brain stiffness is correlated with CBF normalized to vessel area that is related to the perfusion pressure gradient. Clearly, this preliminary conclusion requires validation by more data and measurement of CBF and mean vessel areas within the same scan. Technical challenges exist for ASL and vessel size imaging with respect to long acquisition

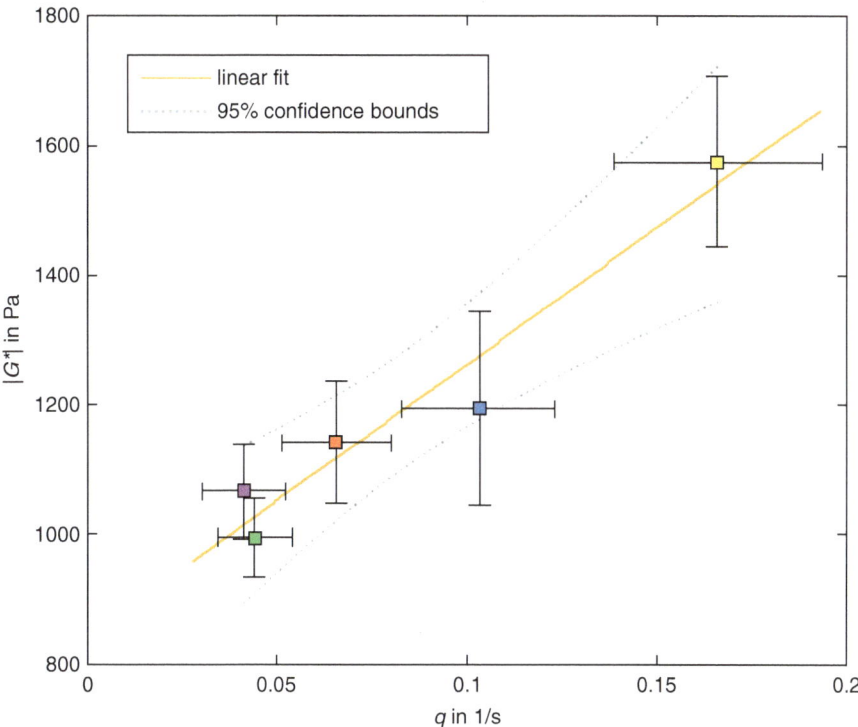

Fig. 20.7 A strong linear relationship ($R^2 = 0.92$) was observed for the perfusion flux parameter q (CBF normalized by the mean cross-sectional vessel area taken from [63, 64]) and the stiffness of DGM areas including the putamen. Error bars reflect the standard error of the mean. Colors of the symbols refer to the regions specified in Fig. 20.7

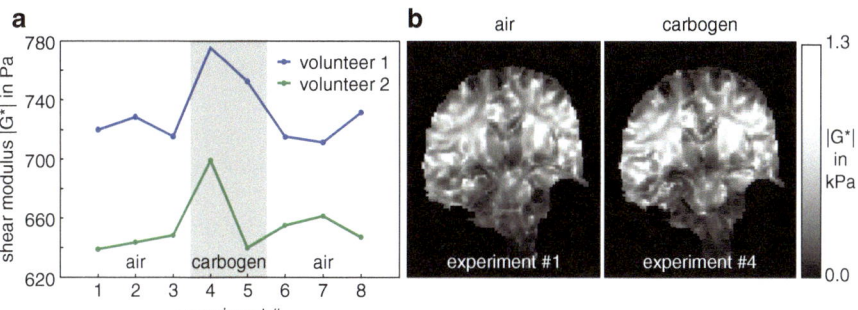

Fig. 20.8 Effect of carbogen breathing-induced hypercapnia on brain stiffness. (**a**) Time course of mean $|G^*|$ calculated from the coronal slice shown in (**b**). After initial baseline measurements (experiments 1–3), carbogen consumption (experiment 4) resulted in instantaneous stiffening of brain tissue. The decrease observed in experiment 5 under continuing carbogen breathing may be attributable to regulatory mechanisms. A return to normal values was observed for subsequent experiments (6–8) with air consumption. (**b**) High-resolution maps of $|G^*|$ for a coronal slice for air and carbogen breathing. Carbogen breathing led to an overall brightening of gray values, indicating higher brain stiffness

Table 20.2 Preliminary observations on the relationship between shear stiffness, pressure, and perfusion of brain tissue within the physiological autoregulatory regimen

Brain region (mechanism)	ICP	CBF	Stiffness
Global brain (jugular compression)	↑ [67]	CBF velocity ↑ [68]	↑ reduced venous drainage [60]
Deep gray matter (normal perfusion)		Regional differences (e.g., thalamus ↓, hippocampus ↑) [61]	Regional differences (thalamus ↑, hippocampus ↓) [61]
		Regional differences in q (e.g., thalamus ↑, hippocampus ↓) [61]	
Global brain (carbogen breathing)	↑ (ICP) [66]	↑ [65]	↑

↑ high, increase, ↓ low, decrease, *ICP* intracranial pressure, *CBF* cerebral blood flow, *q* perfusion flux rate (CBF normalized with vessel size area according to [61])

times and data consistency. MRE based on multifrequency wave inversion can be accomplished with scan times below 10 mins providing high-resolution maps of stiffness whose interpretation, however, requires more knowledge of the changes of hemodynamic parameters of the brain under autoregulation.

A summary of the preliminary results on brain stiffness versus CBF is given in Table 20.2. The basic findings suggest that brain stiffness increases with cerebral vascular pressure; however, CBF alone cannot predict regional variation in brain stiffness. To correlate CBF measured by ASL with MRE, one has to take into account that mean vessel areas vary considerably among brain regions, resulting in differences in perfusion pressure. More experiments are needed to disentangle the intricate relationship between hemodynamic parameters and effective medium mechanics of in vivo brain.

20.4   Interpretation of Contrasting Effects of Perfusion on Tissue Stiffness Observed in Different Organs

The results reviewed above demonstrate that tissue stiffness can either increase or decrease with higher blood flow and perfusion. For example, increases of stiffness due to water ingestion were reported in the liver, while at the same time, pancreatic and spleen stiffness values decreased. A simple model to explain this apparently contradictory behavior is the combination of poroelasticity and nonlinear vascular mechanics. Neglecting viscosity, shear waves in biphasic fluid-solid materials as introduced in Chap. 4 are not influenced by the vascular pressure. However, a change in vascular pressure can change the solid-fluid fraction in the biphasic material and therewith influence the effective shear modulus. In this model, an increasing fluid fraction due to enlarged vessel diameter would decrease the measured shear modulus since less solid tissue (high shear modulus) and more fluid (no shear modulus) integrate into the effective medium properties. However, considering that expansion of vessel walls can increase the stiffness of the walls by nonlinear elastic properties, the hierarchic vascular tree immersed in the tissue could yield an overall increase of tissue stiffness upon higher perfusion. This simple model of opposed effects, that is, fluid fraction vs. nonlinear vascular wall stiffness, could explain why different effects of perfusion on tissue stiffness are measured in different organs. Therefore, it is important to disentangle pathophysiological conditions and type of tissue when interpreting the response of elastography-measured stiffness to different perfusion states.

Conclusion

In the abdomen, meal or liquid intake resulted in increased hepatic stiffness and decreased splenic stiffness in both healthy subjects and patients with liver fibrosis. In patients with portal hypertension, elevated stiffness was observed in both the liver and spleen which drops after portal decompression. In patients with CKD, reduced renal stiffness was related to impaired renal perfusion. In the brain, stiffness decreased with increasing CBF. Normalization of CBF with regional vessel size in DGM yielded perfusion flux rate which was positively correlated with brain stiffness. Increased intracranial pressure caused by hypercapnia or jugular compression also resulted in elevated brain stiffness.

In both abdominal organs and cerebral tissues, the shear stiffness measured by elastography is sensitive to pressure and perfusion variations associated with changes in physiological and pathological conditions. Therefore, shear stiffness has the potential to serve as a noninvasive biomarker for the assessment of pressure and perfusion. Knowledge of the potential confounding factors and their influence on the measured tissue stiffness will improve the diagnostic accuracy of elastography.

References

1. Hall JE. Guyton and hall textbook of medical physiology. 13th ed. Oxford: Elsevier; 2015.
2. Pape HC, Kurtz A, Silbernagl S. Physiologie. Stuttgart: Thieme; 2009.
3. Burkart DJ, Johnson CD, Reading CC, Ehman RL. MR measurements of mesenteric venous flow: prospective evaluation in healthy volunteers and patients with suspected chronic mesenteric ischemia. Radiology. 1995;194(3):801–6. https://doi.org/10.1148/radiology.194.3.7862982.
4. Dauzat M, Lafortune M, Patriquin H, Pomier-Layrargues G. Meal induced changes in hepatic and splanchnic circulation: a noninvasive Doppler study in normal humans. Eur J Appl Physiol Occup Physiol. 1994;68(5):373–80.
5. Schalkx HJ, Petersen ET, Peters NH, Veldhuis WB, van Leeuwen MS, Pluim JP, van den Bosch MA, van Stralen M. Arterial and portal venous liver perfusion using selective spin labelling MRI. Eur Radiol. 2015;25(6):1529–40. https://doi.org/10.1007/s00330-014-3524-z.
6. Tsukuda T, Ito K, Koike S, Sasaki K, Shimizu A, Fujita T, Miyazaki M, Kanazawa H, Jo C, Matsunaga N. Pre- and postprandial alterations of portal venous flow: evaluation with single breath-hold three-dimensional half-Fourier fast spin-echo MR imaging and a selective inversion recovery tagging pulse. J Magn Reson Imaging. 2005;22(4):527–33. https://doi.org/10.1002/jmri.20419.
7. Pugliese D, Ohnishi K, Tsunoda T, Sabba C, Albano O. Portal hemodynamics after meal in normal subjects and in patients with chronic liver disease studied by echo-Doppler flowmeter. Am J Gastroenterol. 1987;82(10):1052–6.

8. Betal D, Hughes ML, Whitehouse GH, Roberts N. Postprandial decrease in splenic volume demonstrated by magnetic resonance imaging and stereology. Clin Anat. 2000;13(6):404–9. https://doi.org/10.1002/1098-2353(2000)13:6<404::AID-CA2>3.0.CO;2-S.

9. Ipek-Ugay S, Tzschatzsch H, Hudert C, Garcia SRM, Fischer T, Braun J, Althoff C, Sack I. Time harmonic Elastography reveals sensitivity of liver stiffness to water ingestion. Ultrasound Med Biol. 2016;42(6):1289–94. https://doi.org/10.1016/j.ultrasmedbio.2015.12.026.

10. Tzschatzsch H, Nguyen Trong M, Scheuermann T, Ipek-Ugay S, Fischer T, Schultz M, Braun J, Sack I. Two-dimensional time-harmonic Elastography of the human liver and spleen. Ultrasound Med Biol. 2016;42(11):2562–71. https://doi.org/10.1016/j.ultrasmedbio.2016.07.004.

11. Dittmann F, Tzschätzsch H, Hirsch S, Barnhill E, Braun J, Sack I, Guo J. Tomoelastography of the abdomen: tissue mechanical properties of the liver, spleen, kidney and pancreas from single MR elastography scans at different hydration states. Magn Reson Med. 2016;78:976–83. https://doi.org/10.1002/mrm.26484.

12. Yin M, Talwalkar JA, Glaser KJ, Venkatesh SK, Chen J, Manduca A, Ehman RL. Dynamic postprandial hepatic stiffness augmentation assessed with MR elastography in patients with chronic liver disease. Am J Roentgenol. 2011;197(1):64–70. https://doi.org/10.2214/AJR.10.5989.

13. Berzigotti A, De Gottardi A, Vukotic R, Siramolpiwat S, Abraldes JG, Garcia-Pagan JC, Bosch J. Effect of meal ingestion on liver stiffness in patients with cirrhosis and portal hypertension. PLoS One. 2013;8(3):e58742. https://doi.org/10.1371/journal.pone.0058742.

14. Barone M, Iannone A, Brunetti ND, Sebastiani F, Cecere O, Berardi E, Antonica G, Di Leo A. Liver stiffness and portal blood flow modifications induced by a liquid meal consumption: pathogenetic mechanisms and clinical relevance. Scand J Gastroenterol. 2015;50(5):560–6. https://doi.org/10.3109/00365521.2014.1003396.

15. Ipek-Ugay S, Tzschätzsch H, Braun J, Fischer T, Sack I. Physiological reduction of hepatic venous blood flow by Valsalva maneuver decreases liver stiffness. J Ultrasound Med. 2016;36:1305–11. https://doi.org/10.7863/ultra.16.07046.

16. Bang DH, Son Y, Lee YH, Yoon KH. Doppler ultrasonography measurement of hepatic hemodynamics during Valsalva maneuver: healthy volunteer study. Ultrasonography. 2015;34(1):32–8. 10.14366/usg.14029.

17. Walser EM, Nguyen M. Hepatic perfusion and hemodynamic effects of transjugular intrahepatic portosystemic shunts. Semin Intervent Radiol. 2005;22(4):271–7. https://doi.org/10.1055/s-2005-925553.

18. Weidekamm C, Cejna M, Kramer L, Peck-Radosavljevic M, Bader TR. Effects of TIPS on liver perfusion measured by dynamic CT. Am J Roentgenol. 2005;184(2):505–10. https://doi.org/10.2214/ajr.184.2.01840505.

19. Annet L, Materne R, Danse E, Jamart J, Horsmans Y, Van Beers BE. Hepatic flow parameters measured with MR imaging and Doppler US: correlations with degree of cirrhosis and portal hypertension. Radiology. 2003;229(2):409–14. https://doi.org/10.1148/radiol.2292021128.

20. Tsushima Y, Koizumi J, Yokoyama H, Takeda A, Kusano S. Evaluation of portal pressure by splenic perfusion measurement using dynamic CT. Am J Roentgenol. 1998;170(1):153–5. https://doi.org/10.2214/ajr.170.1.9423623.

21. Bosch J, Abraldes JG, Berzigotti A, Garcia-Pagan JC. The clinical use of HVPG measurements in chronic liver disease. Nat Rev Gastro Hepat. 2009;6(10):573–82. https://doi.org/10.1038/nrgastro.2009.149.

22. Schwabl P, Bota S, Salzl P, Mandorfer M, Payer BA, Ferlitsch A, Stift J, Wrba F, Trauner M, Peck-Radosavljevic M, Reiberger T. New reliability criteria for transient elastography increase the number of accurate measurements for screening of cirrhosis and portal hypertension. Liver Int. 2015;35(2):381–90. https://doi.org/10.1111/liv.12623.

23. Gao J, Ran HT, Ye XP, Zheng YY, Zhang DZ, Wang ZG. The stiffness of the liver and spleen on ARFI imaging pre and post TIPS placement: a preliminary observation. Clin Imaging. 2012;36(2):135–41. https://doi.org/10.1016/j.clinimag.2011.11.014.

24. Salzl P, Reiberger T, Ferlitsch M, Payer BA, Schwengerer B, Trauner M, Peck-Radosavljevic M, Ferlitsch A. Evaluation of portal hypertension and varices by acoustic radiation force

impulse imaging of the liver compared to transient elastography and AST to platelet ratio index. Ultraschall Med. 2014;35(6):528–33. https://doi.org/10.1055/s-0034-1366506.

25. Choi SY, Jeong WK, Kim Y, Kim J, Kim TY, Sohn JH. Shear-wave elastography: a noninvasive tool for monitoring changing hepatic venous pressure gradients in patients with cirrhosis. Radiology. 2014;273(3):917–26. https://doi.org/10.1148/radiol.14140008.

26. Guo J, Buning C, Schott E, Kroncke T, Braun J, Sack I, Althoff C. In vivo abdominal magnetic resonance elastography for the assessment of portal hypertension before and after transjugular intrahepatic portosystemic shunt implantation. Investig Radiol. 2015;50(5):347–51. https://doi.org/10.1097/RLI.0000000000000136.

27. Ronot M, Lambert S, Elkrief L, Doblas S, Rautou PE, Castera L, Vilgrain V, Sinkus R, Van Beers BE, Garteiser P. Assessment of portal hypertension and high-risk oesophageal varices with liver and spleen three-dimensional multifrequency MR elastography in liver cirrhosis. Eur Radiol. 2014;24(6):1394–402. https://doi.org/10.1007/s00330-014-3124-y.

28. Tzschätzsch H, Marticorena-Garcia S, Ipek-Ugay S, Braun J, Sack I, Althoff C. Time-harmonic elastography of the liver is sensitive to intrahepatic pressure gradient and liver decompression following transjugular intrahepatic portosystemic shunt (TIPS) implantation. Ultrasound Med Biol. 2016;43(3):595–600.

29. Jalan R, Redhead DN, Allan PL, Hayes PC. Prospective evaluation of haematological alterations following the transjugular intrahepatic portosystemic stent-shunt (TIPSS). Eur J Gastroenterol Hepatol. 1996;8(4):381–5.

30. Ran HT, Ye XP, Zheng YY, Zhang DZ, Wang ZG, Chen J, Madoff D, Gao J. Spleen stiffness and splenoportal venous flow: assessment before and after transjugular intrahepatic portosystemic shunt placement. J Ultrasound Med. 2013;32(2):221–8. [pii]

31. Nedredal GI, Yin M, McKenzie T, Lillegard J, Luebke-Wheeler J, Talwalkar J, Ehman R, Nyberg SL. Portal hypertension correlates with splenic stiffness as measured with MR elastography. J Magn Reson Imaging. 2011;34(1):79–87. https://doi.org/10.1002/jmri.22610.

32. Yin M, Kolipaka A, Woodrum DA, Glaser KJ, Romano AJ, Manduca A, Talwalkar JA, Araoz PA, McGee KP, Anavekar NS, Ehman RL. Hepatic and splenic stiffness augmentation assessed with MR elastography in an in vivo porcine portal hypertension model. J Magn Reson Imaging. 2013;38(4):809–15. https://doi.org/10.1002/jmri.24049.

33. Hirooka M, Ochi H, Koizumi Y, Kisaka Y, Abe M, Ikeda Y, Matsuura B, Hiasa Y, Onji M. Splenic elasticity measured with real-time tissue elastography is a marker of portal hypertension. Radiology. 2011;261(3):960–8. https://doi.org/10.1148/radiol.11110156.

34. Khatir DS, Pedersen M, Jespersen B, Buus NH. Evaluation of renal blood flow and oxygenation in CKD using magnetic resonance imaging. Am J Kidney Dis. 2015;66(3):402–11. https://doi.org/10.1053/j.ajkd.2014.11.022.

35. Rossi C, Artunc F, Martirosian P, Schlemmer HP, Schick F, Boss A. Histogram analysis of renal arterial spin labeling perfusion data reveals differences between volunteers and patients with mild chronic kidney disease. Investig Radiol. 2012;47(8):490–6. https://doi.org/10.1097/RLI.0b013e318257063a.

36. Braun JP, Lefebvre H. Kidney function and damage. In: Clinical biochemistry of domestic animals, vol. 1. London: Elsevier; 2008. p. 485–528.

37. Bob F, Bota S, Sporea I, Sirli R, Popescu A, Schiller A. Relationship between the estimated glomerular filtration rate and kidney shear wave speed values assessed by acoustic radiation force impulse elastography: a pilot study. J Ultrasound Med. 2015;34(4):649–54. https://doi.org/10.7863/ultra.34.4.649.

38. Guo LH, Xu HX, Fu HJ, Peng A, Zhang YF, Liu LN. Acoustic radiation force impulse imaging for noninvasive evaluation of renal parenchyma elasticity: preliminary findings. PLoS One. 2013;8(7):e68925. https://doi.org/10.1371/journal.pone.0068925.

39. Lee CU, Glockner JF, Glaser KJ, Yin M, Chen J, Kawashima A, Kim B, Kremers WK, Ehman RL, Gloor JM. MR elastography in renal transplant patients and correlation with renal allograft biopsy. a feasibility study Acad Radiol. 2012;19(7):834–41. https://doi.org/10.1016/j.acra.2012.03.003.

40. Marticorena Garcia SR, Fischer T, Durr M, Gultekin E, Braun J, Sack I, Guo J. Multifrequency magnetic resonance Elastography for the assessment of renal allograft function. Investig Radiol. 2016;51(9):591–5. https://doi.org/10.1097/RLI.0000000000000271.
41. Asano K, Ogata A, Tanaka K, Ide Y, Sankoda A, Kawakita C, Nishikawa M, Ohmori K, Kinomura M, Shimada N, Fukushima M. Acoustic radiation force impulse elastography of the kidneys: is shear wave velocity affected by tissue fibrosis or renal blood flow? J Ultrasound Med. 2014;33(5):793–801. https://doi.org/10.7863/ultra.33.5.793.
42. Gennisson JL, Grenier N, Combe C, Tanter M. Supersonic shear wave elastography of in vivo pig kidney: influence of blood pressure, urinary pressure and tissue anisotropy. Ultrasound Med Biol. 2012;38(9):1559–67. https://doi.org/10.1016/j.ultrasmedbio.2012.04.013.
43. Helfenstein C, Gennisson JL, Tanter M, Beillas P. Effects of pressure on the shear modulus, mass and thickness of the perfused porcine kidney. J Biomech. 2015;48(1):30–7. https://doi.org/10.1016/j.jbiomech.2014.11.011.
44. Warner L, Yin M, Glaser KJ, Woollard JA, Carrascal CA, Korsmo MJ, Crane JA, Ehman RL, Lerman LO. Noninvasive in vivo assessment of renal tissue elasticity during graded renal ischemia using MR elastography. Investig Radiol. 2011;46(8):509–14. https://doi.org/10.1097/RLI.0b013e3182183a95.
45. Heusch P, Wittsack HJ, Blondin D, Ljimani A, Nguyen-Quang M, Martirosian P, Zenginli H, Bilk P, Kropil P, Heusner TA, Antoch G, Lanzman RS. Functional evaluation of transplanted kidneys using arterial spin labeling MRI. J Magn Reson Imaging. 2014;40(1):84–9. https://doi.org/10.1002/jmri.24336.
46. DiBona GF, Sawin LL. Effect of renal denervation on dynamic autoregulation of renal blood flow. Am J Physiol-Renal. 2004;286(6):F1209–18. https://doi.org/10.1152/ajprenal.00010.2004.
47. Yoshimoto M, Sakagami T, Nagura S, Miki K. Relationship between renal sympathetic nerve activity and renal blood flow during natural behavior in rats. Am J Physiol-Reg I. 2004;286(5):R881–7. https://doi.org/10.1152/ajpregu.00105.2002.
48. Peterson EC, Wang Z, Britz G. Regulation of cerebral blood flow. Int J Vasc Med. 2011;2011:823525. https://doi.org/10.1155/2011/823525.
49. Heistad DD, Abboud FM. Dickinson W. Richards lecture: circulatory adjustments to hypoxia. Circulation. 1980;61(3):463–70.
50. Willie CK, Tzeng YC, Fisher JA, Ainslie PN. Integrative regulation of human brain blood flow. J Physiol. 2014;592(5):841–59. https://doi.org/10.1113/jphysiol.2013.268953.
51. Frackowiak RS, Lenzi GL, Jones T, Heather JD. Quantitative measurement of regional cerebral blood flow and oxygen metabolism in man using 15O and positron emission tomography: theory, procedure, and normal values. J Comput Assist Tomogr. 1980;4(6):727–36.
52. Gobbel GT, Cann CE, Fike JR. Measurement of regional cerebral blood-flow using ultrafast computed-tomography - theoretical aspects. Stroke. 1991;22(6):768–71.
53. Rosen BR, Belliveau JW, Chien D. Perfusion imaging by nuclear magnetic resonance. Magn Reson Q. 1989;5(4):263–81.
54. Wong EC. An introduction to ASL labeling techniques. J Magn Reson Imaging. 2014;40(1):1–10. https://doi.org/10.1002/jmri.24565.
55. Edelman RR, Siewert B, Darby DG, Thangaraj V, Nobre AC, Mesulam MM, Warach S. Qualitative mapping of cerebral blood flow and functional localization with echo-planar MR imaging and signal targeting with alternating radio frequency. Radiology. 1994;192(2):513–20. https://doi.org/10.1148/radiology.192.2.8029425.
56. Detre JA, Williams DS, Zhang W, Roberts DA, Leigh JS, Koretsky AP. Noninvasive perfusion MR imaging using spin labeling of arterial water. Diffusion and perfusion magnetic resonance imaging. New York: Raven; 1995.
57. Chen JJ, Jann K, Wang DJ. Characterizing resting-state brain function using arterial spin Labeling. Brain Connect. 2015;5(9):527–42. https://doi.org/10.1089/brain.2015.0344.
58. Le Bihan D. Theoretical principles of perfusion imaging. Application to magnetic resonance imaging. Investig Radiol. 1992;27(Suppl 2):S6–11.
59. Parker KJ. Experimental evaluations of the microchannel flow model. Phys Med Biol. 2015;60(11):4227–42. https://doi.org/10.1088/0031-9155/60/11/4227.

60. Hatt A, Cheng S, Tan K, Sinkus R, Bilston LE. MR Elastography can be used to measure brain stiffness changes as a result of altered cranial venous drainage during jugular compression. Am J Neuroradiol. 2015;36(10):1971–7. https://doi.org/10.3174/ajnr.A4361.
61. Hetzer S, Birr P, Fehlner A, Hirsch S, Dittmann F, Barnhill E, Braun J, Sack I. Perfusion alters stiffness of deep gray matter. J Cereb Blood Flow Metab. 2017:271678X17691530. https://doi.org/10.1177/0271678X17691530.
62. Shen Y, IM P, Ahearn T, Clemence M, Schwarzbauer C. Quantification of venous vessel size in human brain in response to hypercapnia and hyperoxia using magnetic resonance imaging. Magn Reson Med. 2013;69(6):1541–52. https://doi.org/10.1002/mrm.24258.
63. Jensen JH, Lu H, Inglese M. Microvessel density estimation in the human brain by means of dynamic contrast-enhanced echo-planar imaging. Magn Reson Med. 2006;56(5):1145–50. https://doi.org/10.1002/mrm.21052.
64. Mann DM, Eaves NR, Marcyniuk B, Yates PO. Quantitative changes in cerebral cortical microvasculature in ageing and dementia. Neurobiol Aging. 1986;7(5):321–30.
65. Pollock JM, Deibler AR, Whitlow CT, Tan H, Kraft RA, Burdette JH, Maldjian JA. Hypercapnia-induced cerebral hyperperfusion: an underrecognized clinical entity. Am J Neuroradiol. 2009;30(2):378–85. https://doi.org/10.3174/ajnr.A1316.
66. Marx P, Weinert G, Pfiester P, Kuhn H. The influence of hypercapnia and hypoxia on intra-cranial pressure and on CSF electrolyte concentrations. In: Schürmann K, Kasner M, Brock M, Reulen H-J, Voth D, editors. Brain edema/Cerebello pontine angle tumors, vol. 1. Berlin Heidelberg: Springer; 1973. p. 195–8. https://doi.org/10.1007/978-3-642-65734-4_26.
67. Hulme A, Cooper R. The effects of head position and jugular vein compression (JVC) on intracranial pressure (ICP). A clinical study. In: Beks JWF, Bosch A, Brock M, editors. Intracranial pressure III. Berlin Heidelberg: Springer; 1976. p. 259–63. https://doi.org/10.1007/978-3-642-66508-0_43.
68. Frydrychowski AF, Winklewski PJ, Guminski W. Influence of acute jugular vein compression on the cerebral blood flow velocity, pial artery pulsation and width of subarachnoid space in humans. PLoS One. 2012;7(10):e48245. https://doi.org/10.1371/journal.pone.0048245.

Radionuclide Imaging of Cerebral Blood Flow

21

Ralph Buchert

Abstract

Blood flow serves numerous important functions in the living body. It is particularly important for the brain, because of the brain's high energy demand and its lack of capacity to store energy. Impairment of cerebral blood flow plays a central role in a wide spectrum of diseases, including not only cerebrovascular diseases but also neurodegenerative diseases such as Alzheimer's disease. Thus, measurement of cerebral blood flow has many clinical and preclinical indications. This chapter describes radionuclide imaging methods for quantitative imaging of regional cerebral blood flow. After introducing the general principles of radionuclide imaging, positron-emission tomography (PET) with the freely diffusible tracer oxygen-15-labeled water and single photon emission computed tomography (SPECT) with the chemical microsphere Tc-99m-HMPAO are presented in detail. A representative clinical application is shown for both modalities. Finally, the utility of multi-pinhole small animal SPECT with Tc-99m-HMPAO for brain perfusion imaging in mice is discussed.

21.1 Cerebral Blood Flow: Some Physiological Aspects

Blood flow serves numerous important functions including supply of organs and tissues with energy sources, oxygen, and other essential substrates, (slow) signal transmission by delivery of messenger substances (hormones), and removal of carbon dioxide and metabolic waste. Among all organs, blood flow is particularly important for the brain. In humans, the brain contributes only about 2% (1.5 kg) to

R. Buchert

Department of Nuclear Medicine, Charité – Universitätsmedizin Berlin, Berlin, Germany

e-mail: ralph.buchert@charite.de

© Springer International Publishing AG 2018

I. Sack, T. Schaeffter (eds.), *Quantification of Biophysical Parameters in Medical Imaging*, https://doi.org/10.1007/978-3-319-65924-4_21

451

the total body weight, but it consumes more than 20% of the body's total glucose and oxygen requirement [1]. This 10:1 ratio clearly demonstrates that the brain is quite an "expensive organ" [2]. The primary reason for this is the brain's mode of information processing via neurotransmitter systems (see next subsection). Unlike many other organs, the brain has no relevant energy storage capacity and, therefore, strongly relies on continuous supply of glucose and oxygen by blood (metabolism of glucose is the brain's only source of energy, except after prolonged starvation [3]). To meet this demand, blood flow through the brain (cerebral blood flow, CBF) is high even at rest, on average 50 ml blood per 100 g of brain tissue per min. Furthermore, CBF is regulated independent of systemic blood pressure [4]. Reduction of CBF, either globally, that is, uniformly within the whole brain, or locally in some specific brain region can cause severe health problems. Depending on the extent and duration of CBF impairment, health problems range from acute (ischemic) stroke to mild cognitive dysfunction that might be reversible after normalization of CBF. Brain metabolism stops when CBF falls below a threshold of about 18–20 ml/100 g/min.

21.2   Cerebral Blood Flow as Biomarker of Signaling-Related Synaptic Activity

About 75% of the glucose consumption in brain gray matter occurs within the neuropil and is associated with signaling-related synaptic activity [5–7]. The remaining fraction is mainly used for maintaining neuronal and glial resting potentials [7]. The signaling-related energy consumption per g of gray matter tissue equals that in human leg muscle during marathon running [7]. It is used to replenish ATP stores which have been exhausted by Na/K-ATPase in restoring ion gradients across the cell membrane after spike activity ("heat of recovery") [6]. Interestingly, in humans, unlike rodents, the reversion of postsynaptic effects requires a larger fraction of synaptic energy consumption than the reversion of action potential-related presynaptic effects [7]. The fact that signaling-related energy consumption accounts for about 75% of total gray matter energy usage implies that gray matter glucose metabolism is proportional to the action potential frequency (spike rate) to good approximation.

There is also a close temporal and regional linkage between signaling-related synaptic activity and CBF [8, 9]. Therefore, CBF can also be considered a surrogate of synaptic activity, although the mechanisms underlying the relationship between synaptic activity and CBF are less well understood than the mechanisms underlying the relationship between synaptic activity and glucose utilization. The hypothesis of "metabolic" neurovascular coupling, according to which CBF is regulated by vasoactive metabolic products of energy consumption, has been challenged by the "neuronal" hypothesis suggesting that neuronal energy demand is communicated to the vasculature within the neurovascular unit in an anticipatory manner so that vasoreaction can occur independently of actual glucose metabolism [1, 10]. Nevertheless, CBF is a useful surrogate of synaptic activity under most conditions.

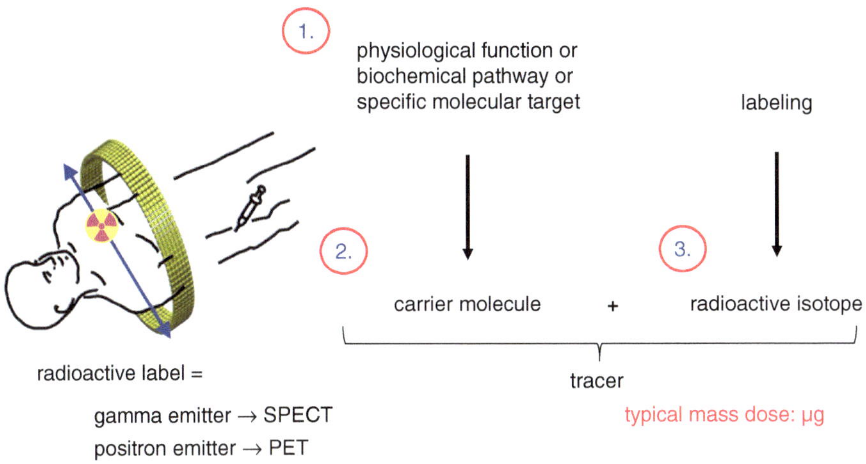

Fig. 21.1 Illustration of the basic principle of molecular imaging in nuclear medicine (radionuclide imaging). The first step is to decide which physiological function, biochemical pathway, or specific molecular target is to be investigated. Then, an appropriate carrier molecule with high affinity and selectivity for the function, pathway, or target is selected. In the third step, the carrier molecule is labeled by integrating or linking a radioactive atom. The radioactively labeled carrier molecule is called a tracer because it traces the path of the respective endogenous biomolecules. A small amount of the tracer is administered, typically by intravenous injection. In the patient's body, the tracer takes part in the physiological function or follows the biochemical pathway or binds to the molecular target for which it has been selected. The temporal and spatial distribution of the tracer within the body can be measured by detecting the decay of the radioactive label. A SPECT camera or a PET camera is used for this purpose, depending on the decay mode of the radioactive label. Tracer kinetic modeling of time-activity curves derived from sequentially acquired SPECT or PET images ("dynamic imaging") allows quantitative characterization of the function, pathway, or target of interest. For example, PET with oxygen-15 water as tracer and a dynamic acquisition protocol of 60 s total duration allows determination of regional cerebral blood flow in absolute units (mL blood per g of tissue per minute)

21.3   Basic Principle of Molecular Imaging in Nuclear Medicine (Radionuclide Imaging)

Radionuclide techniques for noninvasive imaging of physiological functions, biochemical pathways, and specific molecular targets in the living body are based on the so-called tracer principle first formulated by George de Hevesy in 1913: "it would interest me to follow the path of the cup of tea taken in through my body." A tracer for radionuclide imaging consists of two components: the carrier and the radioactive label (Fig. 21.1). The carrier is a molecule that is specific for the physiological function, biochemical pathway, or molecular target to be investigated. After administration, e.g., by intravenous injection into the blood, the carrier molecule participates in the physiological function, enters the biochemical pathway, or binds to the molecular target with high affinity and selectivity. To follow the path of the carrier molecule in the living body from outside, the molecule is labeled radioactively. The radioactive label most often is a gamma emitter (with an appropriate physical half-life in the range of several hours and primary gamma energy similar to the X-ray energy of about 100 keV in computed tomography), or it is a positron emitter (with an appropriate half-life in the range of minutes to a few hours). The spatial distribution of the tracer within the body can be computed from projection images of photons from the decay of the radioactive label leaving the body (when positrons decay, the photons originate from positron-electron annihilation following positron decay). The projection images are obtained with a gamma camera (single photon emission computed tomography, SPECT) or a camera for positron-emission tomography (PET), depending on the type of the radioactive label: SPECT in the case of a gamma emitter and PET in the case of a positron emitter.

Radioactive labeling can be done by replacing one of the atoms of the carrier molecule by an appropriate radioactive isotope of the same element. The practically most relevant example of this approach is the replacement of one stable carbon-12 atom by a positron-emitting carbon-11 atom for PET imaging (all biomolecules contain carbon atoms). The advantage of this approach is that the chemical and thus the biochemical properties of the carrier molecule are not changed. The disadvantage is the short half-life of carbon-11 of 20 min, which makes C-11-PET particularly demanding and expensive. An alternative method for radioactive labeling is to replace an atom or a group of atoms (e.g., a hydroxyl group, OH) by a radioactive isotope (e.g., F-18) or another group of atoms including a radioactive isotope. It is evident that this alters the chemical and biochemical properties of the carrier molecule so that it is mandatory to test whether the pharmacokinetics of the labeled carrier molecule is still adequate for tracing the physiological function, biochemical pathway, or specific molecular target of interest. In some rare cases, the labeling-induced changes of tracer kinetics simplify acquisition and interpretation of the radionuclide images. The most prominent example is the glucose analog F-18-fluorodeoxyglucose (FDG; see glossary) for PET imaging of glucose metabolism [11], which is widely used in oncological indications (whole-body tumor imaging) as well as for the detection and differentiation of neurodegenerative diseases such as Alzheimer's disease and frontotemporal lobar degeneration (see, e.g., Chap. 18 "Quantification of Functional Heterogeneities in Tumors by PET Imaging"). FDG is trapped within cells after hexokinase-mediated phosphorylation, while C-11-labeled glucose is further metabolized, resulting in loss of the radioactive label from the cell [12]. FDG retention measured by a single late static PET scan, after all the FDG injected into blood has been taken up and trapped within cells, is proportional to the metabolic rate of glucose to good approximation.

Radionuclide imaging has excellent sensitivity: nM to pM concentrations of the tracer are sufficient for imaging [13]. This provides considerable flexibility in the choice of carrier molecules for many physiological functions, biochemical pathways, and specific molecular targets. It allows imaging of targets of rather low density without significant side effects. Because of the very low doses required, pharmacological or toxicological issues that may prevent the use of a molecule for radionuclide imaging are rare.

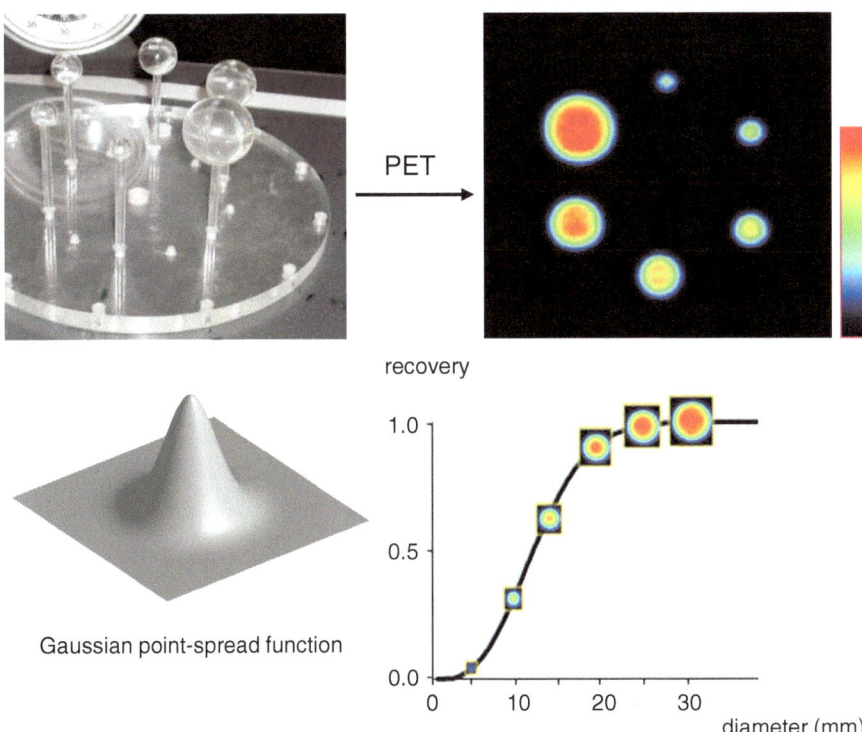

Fig. 21.2 Impact of the spatial resolution on the recovered tracer concentration in radionuclide imaging. The upper row shows a transverse PET image (*right*) through the center of the spheres (of varying diameter) of a phantom (*left*). Although all spheres were filled with the same F-18 solution (i.e., the same tracer concentration), the tracer concentration is strongly underestimated in the PET image of the smaller spheres (*right*). This so-called recovery effect is due to the limited spatial resolution of PET imaging. Assuming an isotropic three-dimensional Gaussian point-spread function (*bottom left*, two-dimensional) for PET imaging, it is straightforward to compute tracer recovery in PET for spherical lesions (*bottom right*). The recovery is scale invariant, i.e., it depends on the ratio of sphere diameter to the full width of half maximum of the Gaussian point-spread function

21.4   SPECT versus PET for Brain Radionuclide Imaging

For human applications, PET provides better spatial resolution and better count sensitivity than SPECT (for preclinical applications in small animals, see subsection 21.1.7 "Preclinical application: multi-pinhole Tc-99m-HMPAO SPECT in the mouse"). Limited spatial resolution results in underestimation of the tracer concentration in "small" brain structures [14] (Fig. 21.2). This recovery effect (see glossary) depends on the size of the brain structure relative to the spatial resolution in the reconstructed images, the latter typically specified as full width of half maximum (FWHM) of the point-spread function of the imaging procedure. The smaller the size of the structure relative to the FWHM, the smaller is the recovery of the structure's actual tracer concentration in the image. In structures whose size approaches the FWHM, tracer concentration is underestimated by about 70%. Procedures for recovery correction based on structural information, such as structural MRI, improve quantitative accuracy but are not yet available routinely. The recovery effect is more limiting in brain SPECT than in brain PET, given that spatial resolution of human brain imaging with modern cameras is about 5 mm FWHM in PET compared to about 10 mm FWHM in SPECT.

Improved count sensitivity of PET compared to SPECT allows better statistical image quality in PET compared to SPECT after administration of standard tracer doses (Fig. 21.3). This also contributes to higher power of brain PET for detecting alterations of physiological functions, biochemical pathways, and specific molecular

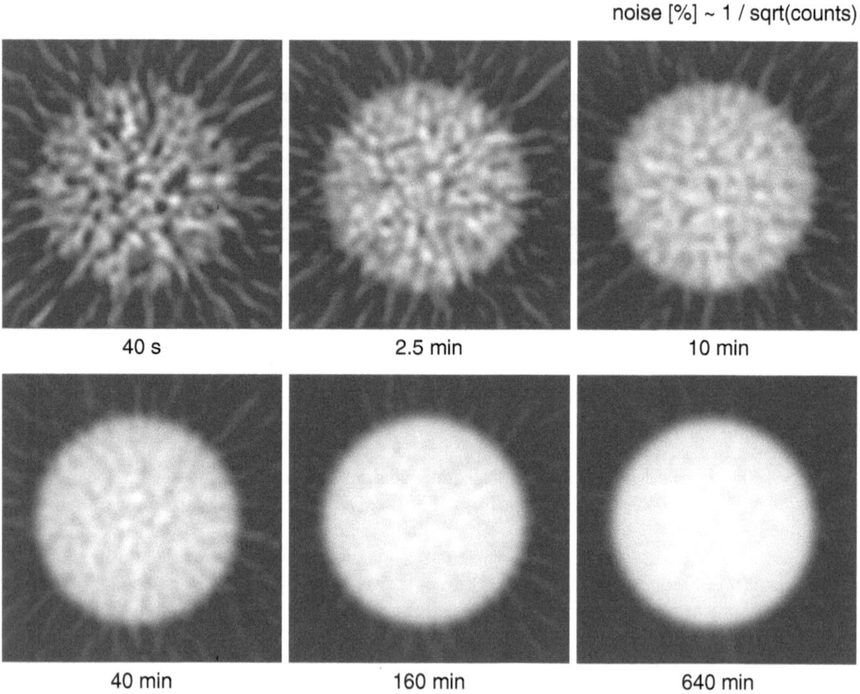

noise [%] ~ 1 / sqrt(counts)

40 s   2.5 min   10 min

40 min   160 min   640 min

Fig. 21.3 Impact of statistical noise on radionuclide images. PET images of the same transverse slice through a cylinder phantom uniformly filled with a solution of the positron emitter F-18 and acquired with different scan durations are shown. The nonuniformity of the PET images at shorter scan durations is a consequence of statistical noise associated with the small number of radioactive decays (counts) detected per picture element (pixel). Statistical image quality can be improved by increasing the scan duration and/or by increasing the radioactivity dose. Both options are limited in patients for obvious reasons. Thus, high count sensitivity of the PET (or SPECT) camera is of paramount importance for radionuclide imaging. It might be worth noting that count sensitivity and statistical noise are a much greater problem in radionuclide imaging than in X-ray CT, for example. This is due to the fact that the tracer in radionuclide imaging usually distributes throughout the body (not only to the organ of interest), so that only a fraction of the radioactive decays in the body contributes to the PET image. This is different in CT, where only X-rays in the direction of the detector are used, so that each single X-ray contributes to the image (even if it is absorbed in the body)

targets compared to brain SPECT, at least for PET and SPECT tracers with similar pharmacokinetics, which, however, rarely is the case. For CBF radionuclide imaging, the most widely used SPECT tracer hexamethyl-propyleneamine oxime (HMPAO) labeled with the gamma emitter technetium-99m [15, 16] has considerably inferior pharmacokinetic properties compared to the most widely used PET tracer oxygen-15 (O-15)-labeled water (see next subsection). It is worth noting that PET with O-15-water quite literally fulfills the vision of George de Hevesy cited above.

21.5   Tc-99m-HMPAO SPECT versus O-15-Water PET

O-15-water PET has been used for CBF measurement since the 1980s [17, 18] and is generally considered the gold standard for this purpose [19]. The pharmacokinetics of O-15-water is rather simple: O-15-water is inert, i.e., it is not metabolized in the body, and it is freely diffusible (see glossary), i.e., it enters the tissue from the blood by passing the blood-brain barrier (see glossary) during a single capillary passage and then very quickly equilibrates between tissue and venous blood. A mathematical description of the pharmacokinetics of O-15-water is given by the Kety-Schmidt model [20] illustrated by the "reversible" one-tissue compartment model shown in Fig. 21.4. The O-15-water concentration in tissue, C_t, is given by the first-order differential equation:

1–tissue compartment models

2– tissue compartment models

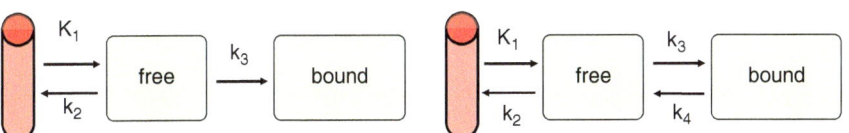

Fig. 21.4 One- and two-tissue compartment models used for tracer kinetic modeling in radionuclide imaging for quantitative characterization of physiological functions, biological pathways, or specific molecular targets. The arrows represent unidirectional transport of the tracer. The rate constants $K_1,\ldots, k_4$ for the different transport steps are assumed to be independent of time during the PET or SPECT scan. This requirement is fulfilled due to the low mass doses of the tracer, so that the function, pathway, or target under consideration is not altered by the presence of the tracer. The pictorial representation of the models is easily translated (using the law of mass action) into a single first-order differential equation with constant coefficients (one-tissue models) or a system of two such equations that can be solved even analytically

$$dC_t / dt = K_1 C_p - k_2 C_t \qquad (21.1)$$

where C_p is the concentration of O-15-water in blood (arterial plasma) and K_1 and k_2 are the rate constants for unidirectional transport of O-15-water from blood to tissue and vice versa (t is the time and d/dt is the derivative with respect to time). Equation (21.1) is easily solved to obtain

$$C_t(t) = K_1 C_p * \exp(-k_2 t) = K_1 \int_0^t C_p(s) \exp(-k_2(t-s)) ds \qquad (21.2)$$

where the symbol "asterisk" denotes the mathematical operation of convolution. Iterative minimization (e.g., using the Levenberg-Marquardt method or more specific procedures [21]) to fit the operational Eq. (21.2) to the time-activity curve C_t in the brain tissue (measured by dynamic PET imaging) and the time-activity curve C_p in blood (measured by arterial blood sampling using an automatic blood sampling device [22]) yields estimates of the rate constants, K_1 and k_2, either for specific regions of interest or in a voxel-by-voxel manner to generate parametric images of K_1 and k_2 (Fig. 21.5).

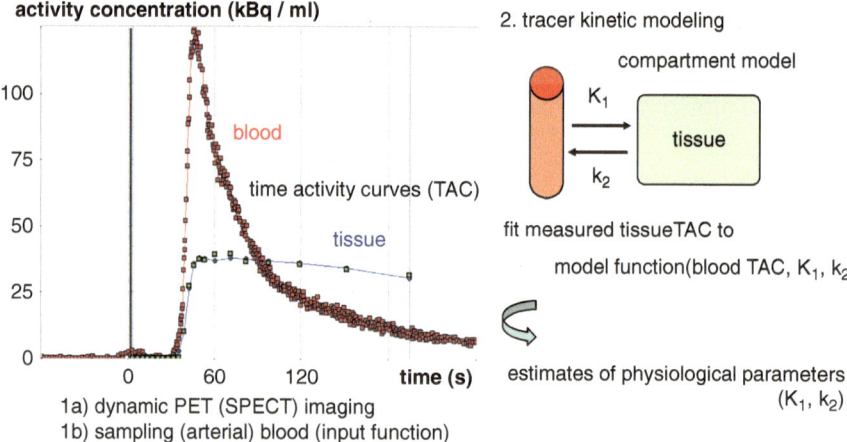

Fig. 21.5 Workflow of tracer kinetic modeling for quantitative characterization of physiological functions, biological pathways, or specific molecular targets in radionuclide imaging. When a dynamic acquisition protocol is used, a sequence of PET or SPECT scans is acquired, usually starting simultaneously with tracer administration, to generate time-activity curves of the tracer. Arterial blood is drawn throughout the scanning procedure to measure the concentration of the (unmetabolized fraction of the) tracer in arterial plasma. The resulting blood-time curve serves as input function in tracer kinetic modeling. Using the (compartment) model assumed to properly describe tracer pharmacokinetics, the measured input function, and some start values for the rate constants, a synthetic tissue time-activity curve can be computed. The synthetic tissue time-activity curve is compared to the curve measured by PET or SPECT (e.g., using the sum of squared differences). Then the rate constants are varied to obtain the best agreement between the two curves (the synthetic and the measured one). The rate constants that provide the best fit are assigned to the tissue. Tracer kinetic modeling can be performed for a region of interest (e.g., a whole organ or specific parts of it) or pixel for pixel to generate parametric volume maps of the rate constants

Now, the rate constant K_1 is the product of CBF and the single-pass extraction fraction E (see glossary), i.e.,

$$K_1 = E \times \text{CBF}. \tag{21.3}$$

According to the Renkin-Crone model [23, 24], extraction fraction E is given by

$$E = 1 - \exp(-\text{PS}/\text{CBF}), \tag{21.4}$$

where PS is the permeability surface area product characterizing the permeability of the blood-brain barrier for the tracer (see glossary), O-15-water in this case (or Tc-99m-HMPAO in the case of HMPAO SPECT; see below). The PS is assumed to be independent of CBF (e.g., no recruitment of additional capillaries [25]). It is evident from Eq. (21.4) that the extraction fraction decreases with increasing CBF. However, when the permeability surface area product is very large compared to CBF, which is a reasonable assumption for O-15-water in the physiological range of CBF, the second term of the right-hand side of Eq. (21.4) can be neglected, resulting in $E \approx 1$. Inserting this into Eq. (21.3) results in

$$K_1 \approx \text{CBF} \tag{21.5}$$

so that the estimates of K_1 obtained by tracer kinetic modeling of O-15-water PET according to the operational Eq. (21.2) can be considered estimates of CBF. Thus, quantitative estimation of CBF from O-15-water PET is rather straightforward. However, O-15-water PET is not widely available because of the short half-life of O-15 (2 min), which requires a cyclotron for O-15 production on site of the PET system [26]. In addition, O-15-water is not authorized as a medicinal product, unlike Tc-99m-HMPAO.

The use of Tc-99m-HMPAO is based on the principle of chemical microspheres, according to which (a) the tracer is fully extracted from arterial blood to tissue during a single capillary passage and then (b) is locally retained in tissue, meaning that clearance of the tracer from tissue can be neglected. When Tc-99m-HMPAO is used, fixation in tissue is due to glutathione-dependent metabolism to hydrophilic forms, followed by binding to nondiffusible cell components [27, 28]. The pharmacokinetics of (chemical) microspheres is even simpler than the pharmacokinetics of O-15-water. Mathematically, it is described by what is known as the irreversible one-tissue compartment model shown in Fig. 21.4. It is evident that the irreversible one-tissue compartment model can be considered the "$k_2 = 0$" case of the reversible one-tissue compartment model. Its operational equation, therefore, is easily obtained from Eq. (21.2) by setting $k_2 = 0$, which results in

$$C_t = K_1 \int_0^t C_p(s)\,\mathrm{d}s. \tag{21.6}$$

The assumption that the chemical microsphere is fully extracted to tissue during a single capillary passage means that $E = 1$, resulting in (see Eq. (21.3)) $K_1 = CBF$, so that from Eq. (21.6)

$$CBF = C_t(T) / \int_0^T C_p(s)\,ds \qquad (21.7)$$

where T is a late time point after intravenous injection of the tracer at which all tracer molecules have been taken up and are trapped in tissue (no tracer left in blood).

Unfortunately, the kinetics of Tc-99m-HMPAO shows considerable violation of the microsphere principle. First, glutathione-dependent metabolism is not fast enough to avoid clearance of unmetabolized Tc-99m-HMPAO from tissue (i.e., $k_2 > 0$) [29]. Second, the permeability surface area product is considerably smaller for Tc-99m-HMPAO compared to O-15-water. At normal CBF, for example, only about 70% of Tc-99m-HMPAO in blood is extracted to tissue during a single capillary passage (i.e., $E = 0.7$) [30]. The smaller permeability surface area product of Tc-99m-HMPAO limits the capability of Tc-99m-HMPAO SPECT to measure changes of CBF (see next subsection). More details on the tracer chemistry in SPECT and PET are provided in Chap. 11.

21.6   Clinical Applications

Impairment of CBF is the primary or a contributing cause of many neurological and psychiatric diseases. This explains why assessment of CBF plays an important role in clinical routine patient care in a wide spectrum of diseases including both cerebrovascular and other diseases. The spectrum of indications also includes neurodegenerative diseases such as Alzheimer's or Parkinson's disease [31] or the localization of the seizure onset area as part of presurgical workup in patients with suspected unifocal epilepsy [32]. In these noncerebrovascular diseases, CBF is typically used as a biomarker of neuronal activity, e.g., to detect or rule out neuronal dysfunction or neuronal degeneration (see Sect. 21.1.2).

In molecular imaging of the brain using exogenous tracers, e.g., for specific neurotransmitter receptors, CBF is often the most important "covariate of no primary interest" (nuisance variable), since uptake and retention of the tracer in the brain depends not only on the availability and/or function of the tracer's specific target in tissue but also on tracer delivery to the tissue after intravenous administration (see Eq. (21.4)). In these cases, reliable characterization of the specific target (for correct etiological classification of the disease or for therapy decisions) might require measurement of CBF in addition to the specific target.

In clinical studies, e.g., to evaluate the efficacy of a new drug for Alzheimer's disease, measurement of CBF can be used to assess treatment effects on neuronal function as a (secondary) endpoint. In this respect, measurement of CBF might be considered a kind of objective neuropsychological testing which is less sensitive to day-to-day variability of cognitive performance than conventional paper-and-pencil tests. Increased test-retest stability of CBF measurement compared to standard cognitive tests provides increased statistical power for the detection of the response to the novel drug (its efficacy). CBF measurement as surrogate for brain function appears particularly useful in preclinical studies with rodents, as neuropsychological testing obviously is limited in mice and rats compared to humans.

The following subsections will present two very different clinical indications of radionuclide perfusion imaging. In the first application, O-15-water PET most likely is more appropriate than Tc-99m-HMPAO SPECT, while the second application is only possible with Tc-99m-HMPAO SPECT and not with O-15-water PET.

21.6.1 O-15-Water PET for Measurement of the Cerebrovascular Reserve Capacity in Chronic Cerebrovascular Diseases

The annual risk of stroke is 5.5% per year in patients with internal carotid artery steno-occlusion and even higher in patients with occlusion-induced impairment of CBF [33–35]. Cerebrovascular reserve capacity (CVRC; see glossary) is one of the determining cerebral hemodynamic parameters in decision making for treatment of chronic cerebrovascular diseases such as moyamoya vasculopathy or atherosclerotic cerebrovascular disease [36–39]. Extra-intracranial bypass surgery reduces the risk of stroke in patients suffering from these diseases and impaired CVRC [40, 41]. Thus, early and reliable detection of CVRC is highly relevant clinically in patients with steno-occlusive cerebrovascular disease.

The CVRC can be assessed by imaging regional CBF under pharmacological challenge with the vasodilator acetazolamide compared to CBF at rest. Both perfusion SPECT with Tc-99m-HMPAO and O-15-water PET are recommended in current guidelines for moyamoya vasculopathy [37, 42, 43], despite the limitations of Tc-99m-HMPAO SPECT compared to O-15-water PET with respect to both general image quality (spatial resolution, statistical noise) and the kinetic properties of the tracer as discussed above [37, 44]. This might be explained by the fact that systematic head-to-head comparison of O-15-water PET and Tc-99m-HMPAO SPECT for CVRC assessment in the same patients has not been performed so far.

Figure 21.6 shows Tc-99m-HMPAO SPECT and O-15-water PET images obtained without and with acetazolamide challenge in the same patient. O-15-water PET was performed within 1 week after Tc-99m-HMPAO SPECT since the negative SPECT finding (i.e., no indication of CVRC impairment) had been suspected to be false negative by the referring neurosurgeon. O-15-water PET revealed strongly reduced or even missing CVRC in the left frontal lobe, in line with steno-occlusion and clinical symptoms, suggesting O-15-water PET to be true positive in this patient.

This case suggests higher sensitivity of O-15-water PET for the detection of impaired CVRC, which appears to be mostly due to better image quality (spatial resolution, statistical noise) of PET compared to SPECT but also due to superior pharmacokinetics of O-15-water compared to Tc-99m-HMPAO. The smaller permeability surface area of Tc-99m-HMPAO results in considerable underestimation of the acetazolamide-induced CBF increase from the increase in Tc-99m-HMPAO uptake. According to the Renkin-Crone model, a 30% increase in CBF (following acetazolamide challenge) results in an increase in Tc-99m-HMPAO uptake of only about 15%. Thus, the increase in HMPAO uptake underestimates the CBF increase by about 50%. It appears very plausible that the reduced size of the measured

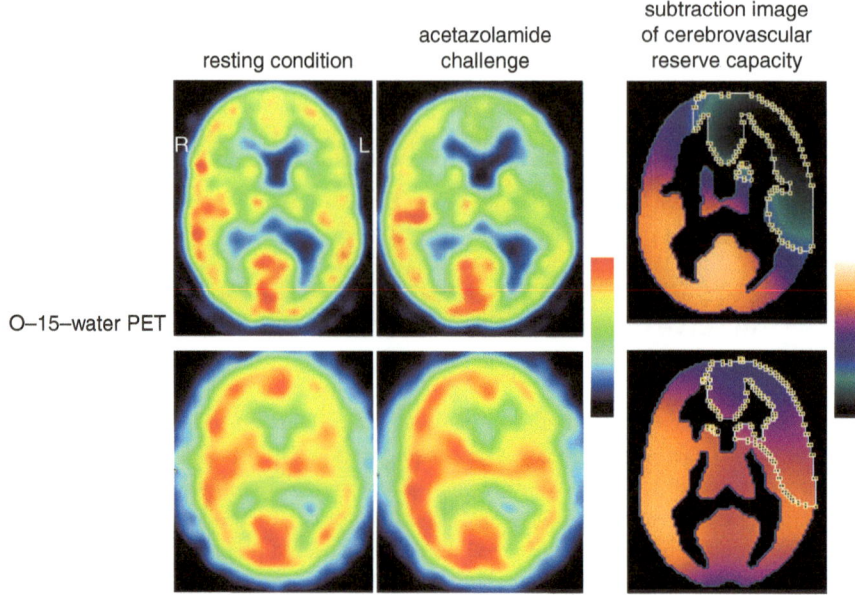

Fig. 21.6 Assessment of cerebrovascular reserve capacity (CVRC) by measuring the acetazolamide-induced change in regional cerebral blood flow using either Tc-99m-HMPAO SPECT or O-15-water PET in the same patient. Tc-99m-HMPAO SPECT was interpreted as normal (no indication of impaired CVRC). This was suspected to be false negative by the referring neurosurgeon so that O-15-water PET was performed only a few days later. O-15-water PET showed markedly reduced CVRC in the left frontal lobe, consistent with the site of steno-occlusive disease

acetazolamide effect compared to the actual effect size limits the detection of impaired CVRC by Tc-99m-HMPAO SPECT. The situation is considerably better for O-15-water, although incomplete permeability of the blood-brain barrier at high flow rates has also been reported for O-15-water [45]. However, systematic head-to-head comparison of Tc-99m-HMPAO SPECT and O-15-water PET in a (preferably prospective) study including a sufficiently large sample of patients is required to confirm the superiority of O-15-water PET for CVRC measurement.

21.6.2 Tc-99m-HMPAO SPECT in the Presurgical Evaluation of Patients with Epilepsy

Patients with unifocal epilepsy that does not well respond to pharmacotherapy may benefit from surgical removal of the brain area that triggers their seizures, the so-called seizure onset area. Depending on the localization of the seizure onset area within the brain, surgical removal can result in full cure without severe adverse effects. This requires reliable localization of the seizure onset area prior to surgery. If localization is unclear after standard diagnostic workup including neuropsychological testing, analysis of seizure semiology, MRI of the brain [46], and extracranial electroencephalography (EEG), both interictal EEG (between seizures) and ictal EEG (during a

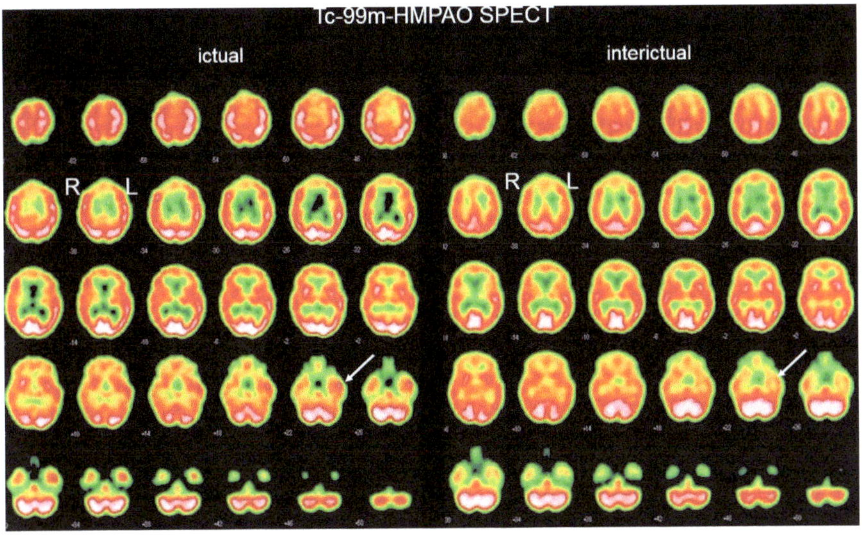

Fig. 21.7 Tc-99m-HMPAO perfusion SPECT with tracer injection 20 s after seizure onset (ictal, *left*) in a patient with epilepsy. There is distinct ictal hyperperfusion in the left temporal lobe (*arrow*), which is not seen between seizures (interictal SPECT, *right*). This strongly suggests that the seizure onset area is located in the left temporal lobe

seizure), ictal perfusion imaging may allow reliable localization of the seizure onset area. The rationale for this is that neuronal activity is strongly increased in the seizure onset area during the seizure, resulting in a strong local increase in CBF that can be measured by brain perfusion imaging. It is mandatory to measure CBF in the early phase of the seizure, not later than 20 s after seizure onset, because seizures often spread through the brain, resulting in increased CBF also outside the seizure onset area. As a consequence, there is a considerable risk of mislocalization of the seizure onset area if tracer injection is delayed. Since seizures are often associated with strong involuntary movements, it is clear that brain imaging is not possible during seizures. A quite elegant solution is provided by Tc-99m-HMPAO SPECT. Tc-99m-HMPAO is injected during the first 20 s of the seizure. It is taken up (more or less proportional to local CBF) and retained in the brain tissue within a few seconds. After that, Tc-99m-HMPAO is trapped within the brain for several hours. Thus, the distribution of Tc-99m-HMPAO in the brain presents a "frozen image" of CBF during the seizure that can be acquired with a SPECT camera up to several hours after the seizure. A representative case is shown in Fig. 21.7. O-15-water PET requiring acquisition within 60 s of O-15-water injection cannot be used for ictal CBF imaging.

21.7 Preclinical Application: Multi-Pinhole Tc-99m-HMPAO SPECT in the Mouse

Spatial resolution of radionuclide imaging is an even larger limitation in preclinical small animal experiments than in human applications (Fig. 21.8). This is due to the fact that downscaling of PET and SPECT imaging technology to small animals such

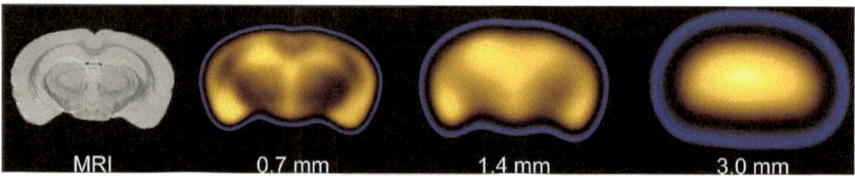

Fig. 21.8 Impact of spatial resolution in mouse brain imaging. The colored images were obtained by smoothing the MRI (left) with Gaussian filters to simulate spatial resolution of 0.7 mm (small animal pinhole SPECT), 1.4 mm (modern small animal PET), and 3.0 mm (outdated small animal PET)

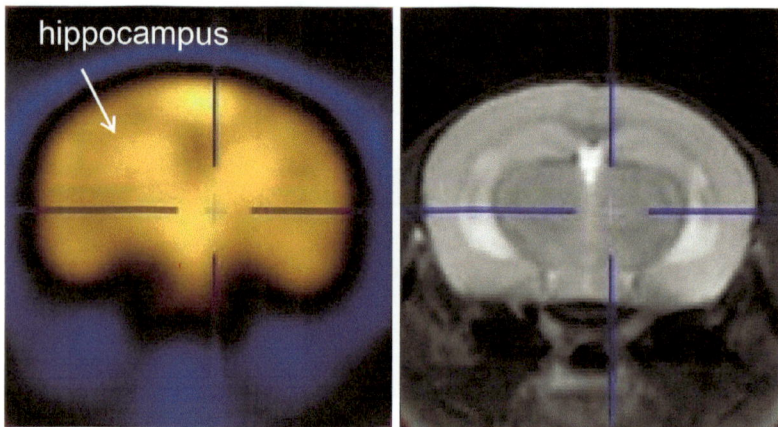

Fig. 21.9 Brain perfusion SPECT with Tc-99m-HMPAO in the mouse (*left*) and MRI of the same animal (*right*)

as the mouse is not perfect: the mouse brain (about 0.5 mL) is more than three orders of magnitude smaller than the human brain (about 1500 mL), whereas volume resolution of dedicated small animal PET systems (about 0.003 mL [47, 48]) is only less than two orders of magnitude better than the resolution of human PET systems (about 0.1 mL [49, 50]). However, small animal SPECT with multiheaded cameras equipped with multiple pinhole collimators [51–53] can provide volume resolution down to 0.0001 mL. This is achieved by using the pinhole collimators to zoom the animal onto the detectors. Thus, SPECT can provide better spatial resolution than PET in small animal imaging, in contrast to imaging in humans. Brain perfusion SPECT with Tc-99m-HMPAO in the mouse has recently been validated [54] and used as biomarker of neuronal activity in a mouse model of neurofibromatosis type 1 [55] (Fig. 21.9).

Small animal PET with O-15-water has poorer spatial resolution compared to small animal Tc-99m-HMPAO SPECT not only due to limitations of small animal PET technology but also due to the "travel distance" of the positron from the O-15 decay in tissue before it annihilates with an electron to produce the two 511 keV photons that are detected by the PET camera [56, 57]. The mean distance of 2.5 mm between the origin of the O-15 decay (=site of the tracer molecule) and the annihilation event (=site reconstructed by PET) results in considerable degradation of the spatial resolution of small animal O-15-water PET (Fig. 21.8). The relative magnitude of the effect of the positron range on spatial resolution is much smaller in human O-15-water PET because of the poorer camera resolution. The mean range of positrons from the decay of F-18 is 0.6 mm [57].

Multiple-head, multi-pinhole small animal SPECT systems also allow fast dynamic imaging for measuring time-activity curves for fully quantitative characterization of physiologic or pathologic functions or targets by tracer kinetic modeling. This is possible because these small animal SPECT systems allow "stationary" SPECT imaging, i.e., without rotation of the detector heads, while human SPECT systems require rotation of the detector heads to generate tomographic images (typically about 60 projection images in step-and-shoot mode with a two-headed SPECT system) [58].

Glossary

Blood-brain barrier: The blood-brain barrier is the border between blood and tissue (including extracellular fluid) that regulates the exchange of substrates between blood and tissue in the brain. The permeability of the blood-brain barrier is highly selective, allowing required substrates to enter the brain while blocking entry of toxic substrates.

Cerebral blood flow (CBF) or regional cerebral blood flow: Regional cerebral blood flow refers to the amount of blood (in mL) that flows through the brain tissue (in g) in a given brain region per unit time (e.g., min). Thus, the unit of regional cerebral blood flow is mL blood/g of tissue/min, often given as mL blood/100 g of tissue/min. There are various methods to measure regional cerebral blood flow which tend to provide systematically different results due to method-specific limitations. The definition of regional cerebral blood flow does not rely on a specific measurement method.

Cerebrovascular reserve capacity (CVRC): Cerebrovascular reserve capacity refers to the brain's maximum ability to increase regional cerebral blood flow when required.

F-18-fluorodeoxyglucose (FDG): Glucose analog labeled with the positron emitter F-18 (physical half-life of 110 min) for use as a tracer of glucose metabolism with positron-emission tomography (PET). FDG-PET is widely used for detection and metabolic characterization of tumors and as a surrogate of neuronal brain activity in the detection and differentiation of neurodegenerative diseases such as Alzheimer's disease and Parkinson's disease.

Freely diffusible: In the context of this section, diffusibility is the ability of the radionuclide imaging tracer to cross the blood-brain barrier. Freely diffusible means that the blood-brain barrier is infinitely permeable for the tracer so that the tracer concentration equilibrates between blood and tissue during a single capillary passage.

Permeability surface area product: In the context of this section, the permeability surface area product is a measure of the permeability of the blood-brain barrier for a given substrate. It depends not only on the substrate but also on the status (integrity) of the blood-brain barrier. Thus, it can vary between brain regions and with diseases. It is assumed not to depend on regional cerebral blood flow.

Recovery effect: The recovery effect or, more precisely, the effect of limited recovery describes the fact that the apparent tracer concentration in radionuclide images (SPECT or PET) is smaller than the actual tracer concentration. This is due to the limited spatial resolution of radionuclide imaging, which causes blurring of the actual tracer distribution and, as a consequence, underestimation of local tracer concentration (assuming that the total amount of tracer is preserved). The recovery effect is closely related to the partial volume effect that describes the mixing of signal from neighboring tissues.

Single-pass extraction fraction: The single-pass extraction fraction is the fraction of tracer that enters the brain during a single capillary transit. The single-pass extraction fraction of a freely diffusible tracer is 1.

Tc-99m-hexamethyl-propyleneamine oxime (HMPAO): Tracer for imaging regional cerebral blood flow with SPECT. It is labeled with the metastable isomer technetium-99m (Tc-99m, half-life of 6 h) of technetium-99. Tc-99m is easily made available for routine use by Mo-99/Tc-99m generators.

References

1. Mergenthaler P, Lindauer U, Dienel GA, Meisel A. Sugar for the brain: the role of glucose in physiological and pathological brain function. Trends Neurosci. 2013;36:587–97.
2. Aiello LC, Wheeler P. The expensive tissue hypothesis: the brain and the digestive system in humans and primate evolution. Curr Anthropol. 1995;36:199–221.
3. Hasselbalch SG, Knudsen GM, Jakobsen J, Hageman LP, Holm S, Paulson OB. Brain metabolism during short-term starvation in humans. J Cereb Blood Flow Metab. 1994;14:125–31.
4. Heiss WD. Cerebral blood flow: physiology, pathophysiology and pharmacological effects. Adv Otorhinolaryngol. 1981;27:26–39.
5. Kadekaro M, Crane AM, Sokoloff L. Differential effects of electrical stimulation of sciatic nerve on metabolic activity in spinal cord and dorsal root ganglion in the rat. Proc Natl Acad Sci U S A. 1985;82:6010–3.
6. Sokoloff L. Energetics of functional activation in neural tissues. Neurochem Res. 1999;24:321–9.
7. Attwell D, Laughlin SB. An energy budget for signaling in the grey matter of the brain. J Cereb Blood Flow Metab. 2001;21:1133–45.
8. Phillips AA, Chan FH, Zheng MM, Krassioukov AV, Ainslie PN. Neurovascular coupling in humans: physiology, methodological advances and clinical implications. J Cereb Blood Flow Metab. 2016;36:647–64.
9. Venkat P, Chopp M, Chen J. New insights into coupling and uncoupling of cerebral blood flow and metabolism in the brain. Croat Med J. 2016;57:223–8.
10. Attwell D, Buchan AM, Charpak S, Lauritzen M, Macvicar BA, Newman EA. Glial and neuronal control of brain blood flow. Nature. 2010;468:232–43.

11. Reivich M, Sokoloff L. Application of the 2-deoxy-D-glucose method to the coupling of cerebral metabolism and blood flow. Neurosci Res Program Bull. 1976;14:474–5.
12. Spence AM, Muzi M, Graham MM, O'Sullivan F, Krohn KA, Link JM, Lewellen TK, Lewellen B, Freeman SD, Berger MS, Ojemann GA. Glucose metabolism in human malignant gliomas measured quantitatively with PET, 1-[C-11]glucose and FDG: analysis of the FDG lumped constant. J Nucl Med. 1998;39:440–8.
13. Massoud TF, Gambhir SS. Molecular imaging in living subjects: seeing fundamental biological processes in a new light. Genes Dev. 2003;17:545–80.
14. Kessler RM, Ellis JR Jr, Eden M. Analysis of emission tomographic scan data: limitations imposed by resolution and background. J Comput Assist Tomogr. 1984;8:514–22.
15. Nowotnik DP, Canning LR, Cumming SA, Harrison RC, Higley B, Nechvatal G, Pickett RD, Piper IM, Bayne VJ, Forster AM, et al. Development of a 99Tcm-labelled radiopharmaceutical for cerebral blood flow imaging. Nucl Med Commun. 1985;6:499–506.
16. Catafau AM. Brain SPECT in clinical practice. Part I: perfusion. J Nucl Med. 2001;42:259–71.
17. Raichle ME, Martin WR, Herscovitch P, Mintun MA, Markham J. Brain blood flow measured with intravenous H2(15)O. II. Implementation and validation. J Nucl Med. 1983;24:790–8.
18. Herscovitch P, Markham J, Raichle ME. Brain blood flow measured with intravenous H2(15) O. I. Theory and error analysis. J Nucl Med. 1983;24:782–9.
19. Wintermark M, Sesay M, Barbier E, Borbely K, Dillon WP, Eastwood JD, Glenn TC, Grandin CB, Pedraza S, Soustiel JF, Nariai T, Zaharchuk G, Caille JM, Dousset V, Yonas H. Comparative overview of brain perfusion imaging techniques. Stroke. 2005;36:e83–99.
20. Kety SS, Schmidt CF. The nitrous oxide method for the quantitative determination of cerebral blood flow in man: theory, procedure and normal values. J Clin Invest. 1948;27:476–83.
21. Boellaard R, Knaapen P, Rijbroek A, Luurtsema GJ, Lammertsma AA. Evaluation of basis function and linear least squares methods for generating parametric blood flow images using 15O-water and positron emission tomography. Mol Imaging Biol. 2005;7:273–85.
22. Boellaard R, van Lingen A, van Balen SC, Hoving BG, Lammertsma AA. Characteristics of a new fully programmable blood sampling device for monitoring blood radioactivity during PET. Eur J Nucl Med. 2001;28:81–9.
23. Crone C. The permeability of capillaries in various organs as determined by use of the 'indicator diffusion' method. Acta Physiol Scand. 1963;58:292–305.
24. Renkin EM. Transport of potassium-42 from blood to tissue in isolated mammalian skeletal muscles. Am J Phys. 1959;197:1205–10.
25. Barfod C, Akgoren N, Fabricius M, Dirnagl U, Lauritzen M. Laser-Doppler measurements of concentration and velocity of moving blood cells in rat cerebral circulation. Acta Physiol Scand. 1997;160:123–32.
26. Cho BK, Tominaga T. Moyamoya disease update. Tokyo: Springer; 2010.
27. Colamussi P, Calo G, Sbrenna S, Uccelli L, Bianchi C, Cittanti C, Siniscalchi A, Giganti M, Roveri R, Piffanelli A. New insights on flow-independent mechanisms of 99mTc-HMPAO retention in nervous tissue: in vitro study. J Nucl Med. 1999;40:1556–62.
28. Neirinckx RD, Burke JF, Harrison RC, Forster AM, Andersen AR, Lassen NA. The retention mechanism of technetium-99m-HM-PAO: intracellular reaction with glutathione. J Cereb Blood Flow Metab. 1988;8:S4–12.
29. Lassen NA, Andersen AR, Friberg L, Paulson OB. The retention of [99mTc]-d,l-HM-PAO in the human brain after intracarotid bolus injection: a kinetic analysis. J Cereb Blood Flow Metab. 1988;8:S13–22.
30. Andersen AR, Friberg HH, Schmidt JF, Hasselbalch SG. Quantitative measurements of cerebral blood flow using SPECT and [99mTc]-d,l-HM-PAO compared to xenon-133. J Cereb Blood Flow Metab. 1988;8:S69–81.
31. Smolinski L, Czlonkowska A. Cerebral vasomotor reactivity in neurodegenerative diseases. Neurol Neurochir Pol. 2016;50:455–62.
32. Apostolova I, Lindenau M, Fiehler J, Heese O, Wilke F, Clausen M, Stodieck S, Buchert R. Detection of a possible epilepsy focus in a preoperated patient by perfusion SPECT and computer-aided subtraction analysis. Nuklearmedizin. 2008;47:N65–8.

33. Yonas H, Smith HA, Durham SR, Pentheny SL, Johnson DW. Increased stroke risk predicted by compromised cerebral blood flow reactivity. J Neurosurg. 1993;79:483–9.
34. Klijn CJ, Kappelle LJ, Tulleken CA, van Gijn J. Symptomatic carotid artery occlusion. A reappraisal of hemodynamic factors. Stroke. 1997;28:2084–93.
35. Eicker SO, Turowski B, Heiroth HJ, Steiger HJ, Hanggi D. A comparative study of perfusion CT and 99m Tc-HMPAO SPECT measurement to assess cerebrovascular reserve capacity in patients with internal carotid artery occlusion. Eur J Med Res. 2011;16:484–90.
36. Gibbs JM, Wise RJ, Leenders KL, Jones T. Evaluation of cerebral perfusion reserve in patients with carotid-artery occlusion. Lancet. 1984;1:310–4.
37. Lee M, Zaharchuk G, Guzman R, Achrol A, Bell-Stephens T, Steinberg GK. Quantitative hemodynamic studies in moyamoya disease: a review. Neurosurg Focus. 2009;26:E5.
38. Settakis G, Molnar C, Kerenyi L, Kollar J, Legemate D, Csiba L, Fulesdi B. Acetazolamide as a vasodilatory stimulus in cerebrovascular diseases and in conditions affecting the cerebral vasculature. Eur J Neurol. 2003;10:609–20.
39. Webster MW, Makaroun MS, Steed DL, Smith HA, Johnson DW, Yonas H. Compromised cerebral blood flow reactivity is a predictor of stroke in patients with symptomatic carotid artery occlusive disease. J Vasc Surg. 1995;21:338–44. discussion 344-335
40. Grubb RL Jr, Powers WJ, Clarke WR, Videen TO, Adams HP Jr, Derdeyn CP. Surgical results of the carotid occlusion surgery study. J Neurosurg. 2013;118:25–33.
41. Yamada S, Oki K, Itoh Y, Kuroda S, Houkin K, Tominaga T, Miyamoto S, Hashimoto N, Suzuki N, Research Committee on Spontaneous Occlusion of Circle of W. Effects of surgery and antiplatelet therapy in ten-year follow-up from the registry study of research committee on moyamoya disease in Japan. J Stroke Cerebrovasc Dis. 2016;25:340–9.
42. Research Committee on the Pathology and Treatment of Spontaneous Occlusion of the Circle of Willis; Health Labour Sciences Research Grant for Research on Measures for Infractable Diseases. Guidelines for diagnosis and treatment of moyamoya disease (spontaneous occlusion of the circle of Willis). Neurol Med Chir. 2012;52:245–66.
43. Pandey P, Steinberg GK. Neurosurgical advances in the treatment of moyamoya disease. Stroke. 2011;42:3304–10.
44. Rahmim A, Zaidi H. PET versus SPECT: strengths, limitations and challenges. Nucl Med Commun. 2008;29:193–207.
45. Bos A, Bergmann R, Strobel K, Hofheinz F, Steinbach J, den Hoff J. Cerebral blood flow quantification in the rat: a direct comparison of arterial spin labeling MRI with radioactive microsphere PET. EJNMMI Res. 2012;2:47.
46. Wellmer J, Parpaley Y, von Lehe M, Huppertz HJ. Integrating magnetic resonance imaging postprocessing results into neuronavigation for electrode implantation and resection of subtle focal cortical dysplasia in previously cryptogenic epilepsy. Neurosurgery. 2010;66:187–94. discussion 194–185
47. Kemp BJ, Hruska CB, McFarland AR, Lenox MW, Lowe VJ. NEMA NU 2-2007 performance measurements of the Siemens Inveon preclinical small animal PET system. Phys Med Biol. 2009;54:2359–76.
48. Zhang H, Bao Q, NT V, Silverman RW, Taschereau R, Berry-Pusey BN, Douraghy A, Rannou FR, Stout DB, Chatziioannou AF. Performance evaluation of PETbox: a low cost bench top preclinical PET scanner. Mol Imaging Biol. 2011;13(5):949–61.
49. Brambilla M, Secco C, Dominietto M, Matheoud R, Sacchetti G, Inglese E. Performance characteristics obtained for a new 3-dimensional lutetium oxyorthosilicate-based whole-body PET/CT scanner with the National Electrical Manufacturers Association NU 2-2001 standard. J Nucl Med. 2005;46:2083–91.
50. Mawlawi O, Podoloff DA, Kohlmyer S, Williams JJ, Stearns CW, Culp RF, Macapinlac H, National Electrical Manufacturers A. Performance characteristics of a newly developed PET/CT scanner using NEMA standards in 2D and 3D modes. J Nucl Med. 2004;45:1734–42.
51. Beekman F, van der Have F. The pinhole: gateway to ultra-high-resolution three-dimensional radionuclide imaging. Eur J Nucl Med Mol Imaging. 2007;34:151–61.

52. Sharma S, Ebadi M. SPECT neuroimaging in translational research of CNS disorders. Neurochem Int. 2008;52:352–62.
53. Branderhorst W, Vastenhouw B, van der Have F, Blezer EL, Bleeker WK, Beekman FJ. Targeted multi-pinhole SPECT. Eur J Nucl Med Mol Imaging. 2011;38(3):552–61.
54. Apostolova I, Wunder A, Dirnagl U, Michel R, Stemmer N, Lukas M, Derlin T, Gregor-Mamoudou B, Goldschmidt J, Brenner W, Buchert R. Brain perfusion SPECT in the mouse: normal pattern according to gender and age. NeuroImage. 2012;63:1807–17.
55. Apostolova I, Niedzielska D, Derlin T, Koziolek EJ, Amthauer H, Salmen B, Pahnke J, Brenner W, Mautner VF, Buchert R. Perfusion single photon emission computed tomography in a mouse model of neurofibromatosis type 1: towards a biomarker of neurologic deficits. J Cereb Blood Flow Metab. 2015;35:1304–12.
56. Jodal L, Le Loirec C, Champion C. Positron range in PET imaging: an alternative approach for assessing and correcting the blurring. Phys Med Biol. 2012;57:3931–43.
57. Partridge M, Spinelli A, Ryder W, Hindorf C. The effect of β+ energy on performance of a small animal PET camera. Nucl Instrum Methods Phys Res A. 2006;568:933–6.
58. Lange C, Apostolova I, Lukas M, Huang KP, Hofheinz F, Gregor-Mamoudou B, Brenner W, Buchert R. Performance evaluation of stationary and semi-stationary acquisition with a non-stationary small animal multi-pinhole SPECT system. Mol Imaging Biol. 2014;16:311–6.

Cardiac Perfusion MRI

22

Amedeo Chiribiri

Abstract

Coronary artery disease (CAD) is the leading cause of death and costs in the Western world. A large study has demonstrated that up to 60% of patients with chest pain and suspected CAD, who undergo expensive invasive catheterisation, might not need the procedure. Therefore, there is a strong need for a reliable diagnostic test to triage patients at intermediate risk of CAD for the appropriate treatment. Over the last decades, several clinical landmark studies have shown that accurate assessment of tissue's blood supply (perfusion) could serve as a gatekeeper for treating the right patients. Perfusion is essential for the integrity of the heart and is an early marker of the so-called ischaemic cascade that leads to non-reversible tissue damage and thus chronic heart disease.

Coronary artery disease (CAD) is the leading cause of death and costs in the Western world. A large study has demonstrated that up to 60% of patients with chest pain and suspected CAD, who undergo expensive invasive catheterisation, might not need the procedure [1]. Therefore, there is a strong need for a reliable diagnostic test to triage patients at intermediate risk of CAD for the appropriate treatment. Over the last decades, several clinical landmark studies have shown that accurate assessment of tissue's blood supply (perfusion) could serve as a gatekeeper for treating the right patients [2–4]. Perfusion is essential for the integrity of the heart and is an early marker of the so-called ischaemic cascade that leads to non-reversible tissue damage and thus chronic heart disease.

The quest to quantify myocardial perfusion non-invasively is motivated by the need to offer patients a non-invasive, observer-independent and reproducible alternative to cardiac catheterisation for assessing the presence of CAD. Assessing the perfusion status of the myocardium can also provide an insight in the complex

A. Chiribiri
Division of Imaging Sciences and Biomedical Engineering,
King's College London, London, UK
e-mail: amedeo.chiribiri@kcl.ac.uk

© Springer International Publishing AG 2018
I. Sack, T. Schaeffter (eds.), *Quantification of Biophysical Parameters in Medical Imaging*, https://doi.org/10.1007/978-3-319-65924-4_22

471

pathophysiological mechanisms leading to the onset of myocardial ischaemia. This is of particular value in consideration of the multiple mechanisms that regulate myocardial perfusion (also known as coronary autoregulation). The complexity of these mechanisms is due to the unique characteristics of the heart, both as vital organ and as a pump for the blood which generates its own perfusion pressure. The heart has very demanding energy requirements. The entire organ contains approximately 700 mg of adenosine triphosphate (ATP) [5], which has a central role in cellular bioenergetics. This quantity of ATP however is enough to power the heart for less than 10 s and requires continuous regeneration of ATP. It has been calculated that a normal heart requires on average approximately 6000 g of ATP per day to function. The regeneration of the cellular pool of ATP is almost entirely dependent on oxidative metabolism and it is maintained in the mitochondria. These are highly concentrated in the cardiac tissue, about 25 times more than in skeletal muscle, and constitute up to 35% of the cardiac mass. Myocardial oxygen consumption, which is already relatively high at rest with a heart-rate in the range of 60–70 beats/min, can increase up to sixfold during maximal exercise. As an adaptation to high oxygen demand, the heart maintains a very high oxygen extraction, so that 70–80% of the oxygen delivered by the coronary circulation is extracted by the tissue. This compares with 20–40% in the skeletal muscle and is facilitated by the high capillary density in the tissue, approaching 4000 capillaries/mm^3. Oxygen extraction in the myocardium cannot further increase to compensate for hypoxia, differently from other organs. It follows that any increase in cardiac activity and myocardial oxygen consumption must be met by a nearly simultaneous increase in oxygen availability obtained through an increase in myocardial blood perfusion. As a practical consequence of the close relationship between oxygen delivery and perfusion, accurate measurements of myocardial perfusion can be used as a surrogate measurement of myocardial oxygenation.

The need for the sensitive, non-invasive detection of myocardial perfusion can be addressed by the external detection of a flow tracer as it distributes in the tissue following its injection. The majority of myocardial perfusion scans is currently performed with single photon emission computed tomography (SPECT) [6]. However, this imaging modality has a relatively low spatial resolution (e.g. 5–10 mm) and allows for the measurement of relative perfusion (i.e. regions are compared to a reference region) only. Absolute quantification of cardiac perfusion is possible with new solid state SPECT detectors, which are currently under investigation [7]. Quantitative measurement of cardiac perfusion is possible by positron emission tomography (PET) which is still considered as a gold standard. In PET, time-resolved acquisition of the first-pass of tracer uptake and quantification of tracer concentration were developed. However, its limited availability and high scanning costs prohibit a wide clinical use.

Cardiac magnetic resonance imaging (MRI) plays an increasing role in the diagnostic and prognostic assessment of patients with suspected cardiovascular disease, justified by its high spatial resolution, tissue contrast and morphological detail as well as the ability to provide reproducible quantitative data [8, 9]. Compared with PET, magnetic resonance MRI imaging has a better temporal and spatial resolution, the absence of any radiation hazards, and the availability of highly stable and inert MRI contrast agents of low toxicity [10]. The spatial resolution of below 2 mm allows the measurement of the transmural distribution of blood flow [11]. Whilst visual assessment of cardiac perfusion MRI is one of the methods of choice by current guidelines for the evaluation of patient with suspected CAD and to guide coronary revascularisation, there is a long line of evidence proving that cardiac perfusion MRI can also be used to provide quantitative measurements of myocardial perfusion, leading to new insights in coronary physiology and potentially in a more accurate and objective assessment of patients with heart disease.

22.1 Principles of Perfusion Quantification by Cardiac Perfusion MRI

Perfusion cardiac MRI is a real-time imaging technique based on the visualisation of the first-pass of a Gadolinium-based contrast agent during its wash in across the cardiac chambers and the left ventricular myocardium. The contrast agent changes the relaxation times of the water protons in its vicinity. Since the T1-relaxation time is significantly shortened, a T1-weighted imaging sequence is applied. Usually a saturation prepulse is employed before the imaging sequence resulting in high signal ("bright pixel value") for short T1-values. The contrast agent is usually injected as a bolus in a peripheral vein and diluted with blood in the veins leading to the right heart and in the pulmonary circulation (Fig. 22.1). The contrast agent mixed with

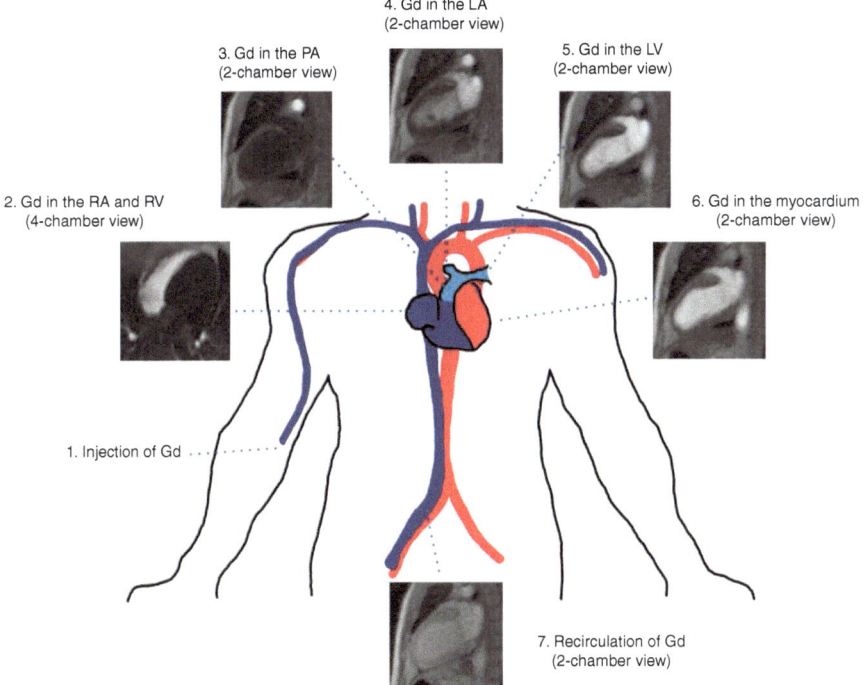

Fig. 22.1 First-pass perfusion cardiac MRI following peripheral venous injection of a Gadolinium (Gd)-based contrast agent. The contrast agent is usually injected in the forearm and it mixes with blood whilst progressing quickly in the vascular system and in the heart. Myocardial enhancement is expected in 15–20 heart beats in a normal subject, recirculation is usually complete in 30–40 heart beats

blood flows then through the left ventricle and into the aorta and the coronary arteries, causing a detectable signal increase in the myocardium, proportional to the local perfusion rate. The wash in of the contrast agent is measured with an ECG-triggered sequence to visualise the wash in of contrast agent in the cardiac chambers and in the myocardium between successive heart beats. Cardiac perfusion MRI is usually acquired twice, during vasodilatory stress and in resting conditions. Whilst suitable for visual interpretation, the resulting dynamic series of images are also suitable for quantitative analysis.

Perfusion quantification by cardiac MRI is based on the central volume principle, a corollary of the Fick's principle, also known as principle of conservation of mass. The first and simplest description of the principle was stated by Stewart for the assessment of cardiac output in 1897 [12]: *"A solution of a substance* [or indicator] *which can be easily recognised and quantitatively estimated in the blood is permitted to flow for a definite time at an approximately uniform rate into the heart. The injected substance mingles with the blood, and passes out with it into the circulation. At a convenient point of the vascular system a sample of blood is drawn off just before the injection and another during the passage of the substance; and the quantity of solution which must be added to a given volume of the first sample, in order that it may contain as much of the injected substance as the second sample, is determined. This determination, it is evident, gives us the means of estimating the extent to which the injected solution has been mixed with blood in the heart, and, therefore, knowing the quantity of the solution which has run into the heart, we can calculate the output in the given time"*. In Stewart's original description, the indicator dilution experiment was based on a constant speed of injection of the indicator, resulting in a steady concentration in the vascular system for the duration of the experiment. Whilst useful to provide a basic understanding of the central volume principle, Stewart's method is not directly transferrable to perfusion MRI experiments, due to the different method used to inject the MRI contrast agent. Gadolinium is injected via a peripheral vein over a period of time. It is therefore subjected to a process of mixing and dilution with blood, before it reaches the heart to generate an input function of significant duration and non-uniform amplitude, known as arterial input function (AIF). In order to apply the central volume principle to the quantification of myocardial perfusion we will have to use the formulation of the central volume principle by Zieler for measurements of flow and volume by sudden injection of the indicator [13]. The first-pass of the contrast agent across the cardiac cavities is here assimilated to a rapid injection of the indicator performed directly in the left ventricle or directly in the coronary arteries. The coronary circulation is assumed to be a time-invariant system, with a single input and a single output. Following the arrival of the indicator (i.e. contrast agent or tracer) at the entrance of the vascular system, the indicator molecules are free to follow a multitude of pathways in the microvascular network of the myocardium and to reach the right atrium across the coronary venous system. If the injection of tracer is performed instantaneously (i.e. approximates an ideal bolus or a Dirac "delta function"), there will be a time lag between the injection and the arrival of the tracer at the exit of the vascular system, followed by a steep rise in concentration and a slower wash out. The mean transit time (MTT) recorded at tissue level will represent the average MTT of indicator molecules across all possible vascular pathways. In the ideal case of an instantaneous injection, and if the tracer is well mixed with blood, then the wash out rate will be a monotonic exponential decay function, dependent on the flow rate divided by the volume of the vascular system, according to the central volume theorem. This theorem states that

$$V = F\bar{t}$$

where V is the vascular volume of the system, F is flow and $\bar{t}$ is the average MTT of the particles of indicator. Literally, this formulation of the central volume principle only applies to intravascular indicators and is widely used in MR-perfusion of the brain, since Gadolinium-based contrast agents are confined to the vasculature due to the blood-brain barrier.

The methods described for the quantification of myocardial perfusion do not assume an exponential wash out, since as explained the injection of the contrast agent in the coronary circulation is not instantaneous therefore the approximation to an exponential decay would result in significant quantification errors. The response of the vascular system to an arbitrary AIF is instead determined via convolution and deconvolution operations

$$C_{myo}\left(t\right) = \int_0^\infty C_{aif}\left(t-\tau\right)h\left(\tau\right)d\tau = C_{aif}\left(t\right)^* h\left(t\right) \qquad (22.1)$$

The response of the system $C_{myo}(t)$ to a given concentration of the indicator in the arterial input function $C_{AIF}(t)$ is known as the transfer function. The transfer function $h(t)$ also represents the normalised distribution of the transit times of the indicator across the vascular system and characterises the myocardial circulation that transforms the delivery of contrast agent from the input into the observed myocardial enhancement. Also the $h(t)$ has units of inverse time (1/dynamic scan interval, if the data are not rescaled in absolute units of time).

Let m units of indicator be injected at time zero into a vascular system, and measure the concentration of the indicator at the exit as a function of time, $C_{out}(t)$. The amount of indicator, dm, leaving the system during a small time interval between time t and $t + dt$ is the concentration of indicator leaving the system, $C_{out}(t)$, multiplied by the volume of fluid leaving the system during this time interval (or flow rate, F, in units of mL/min) multiplied by time.

$$dm = C_{out}\left(t\right)\cdot F \cdot dt \qquad (22.2)$$

The amount of indicator which leaves the system at time t is equal to the amount introduced into the system during the time interval between s and $s + t$ time units before t. The amount of indicator introduced during this time interval is $m_{in} \cdot ds$. The fraction of indicator eliminated per unit time at time t is $k(t)$. Therefore, the amount of indicator leaving per unit time is

$$k\left(s\right)\cdot m_{in}\cdot ds.$$

The rate at which indicator leaves the system at time t is:

$$m_{out} = \int_0^t k(s) \cdot m_{in}(t-s) \cdot ds \tag{22.3}$$

This is also equal to:

$$m_{out} = C_{out}(t) \cdot F.$$

Therefore:

$$C_{out}(t) = \frac{1}{F} \int_0^t k(s) \cdot m_{in}(t-s) \cdot ds \tag{22.4}$$

The amount of indicator at the input to the system is

$$m_{in} = C_{aif}(t) \cdot F.$$

Under assumption that flow is constant and that the vascular system is linear and stationary, the Eq. (22.4) can be written as:

$$C_{out}(t) = \int_0^t C_{aif}(t-s)k(s)ds \tag{22.5}$$

where $k(t)$ is a transfer function which describes the fraction of injected indicator leaving the system per unit of time. However, CMR allows the detection of the mass of contrast agent residing in the tissue ROI, C_{myo}, instead of measurement of ROI out flow mass. Therefore, we need to find a relation between C_{myo} and C_{aif} in order to be able to estimate the flow from CMR images.

According to indicator–dilution theory, the amount of tracer which remains in the tissue region after injection of the indicator is:

$$C_{myo} = \int_0^t F\left[C_{aif}(s) - C_{out}(s)\right]ds. \tag{22.6}$$

By inserting Eq. (22.5) in the above equation instead of $C_{out}(t)$:

$$C_{myo} = \int_0^t FC_{aif}(s)^* [(1-k(s))]ds = C_{aif}(t)^* h(t).$$

Here $h(t)$ represents the fraction of contrast agent which remains in the ROI at time t and is called tissue impulse response. The quantity of indicator remaining in the system during any time interval can never be less than zero and never be more than all of it that entered the system.

For any input function C_{aif} and for finite flow rates, we have $C_{out}(t = 0) = 0$, as tracer cannot instantaneously pass through a ROI and reach the output after injection. Therefore, assuming an instantaneous injection of contrast agent in the coronary circulation, the initial amplitude of the (tissue impulse response) will be equal to the flow rate, and the perfusion value in the myocardium ROI will be equal to flow divided by density of myocardium, in g/ml of tissue [14–16].

22.2 Violations of the Assumptions of the Central Volume Principle in Real Vascular Systems

A limitation of quantitative cardiac perfusion MRI is the fact the some of the contrast agent is extravasated in the interstitial space. Therefore, in the time frame acquired by the scanner, $\int_0^\infty h(t)\,dt < 1$.

Another limitation derives from the injection protocols that can be implemented in clinical practice. The distribution of transit time of the indicator particles must be identical to the distribution of transit time of the blood. However, this differs between stress and rest conditions and is therefore a potential source of systematic error for the determination of myocardial perfusion, depending on changes in haemodynamic conditions and cardiac output. To correct for this error, rest and stress AIF are used for quantification. Data from the literature consistently support that resulting perfusion indices correlate linearly with the actual MBF. Correction for differences in input function shape would be less critical with left atrial, left ventricular or intracoronary injection of contrast agent. In this case, the characteristics of the AIF would largely be determined by the power injector settings and less by the haemodynamic conditions. This is however unpractical in a clinical setting, where peripheral venous injections of Gadolinium are preferred.

22.3 Deconvolution Operations for Quantitative Perfusion Analysis

The task of calculating the myocardial transfer function $h(t)$ by direct deconvolution of the $C_{myo}(t)$ and $C_{AIF}(t)$ is challenging because it amounts to inverting the convolution of $h(t)$ with $C_{myo}(t)$, which is in this context an ill-posed noise sensitive inverse problem and needs regularisation. This can be obtained by discrete Fourier transform, single value decomposition and least square minimisation, usually by Tikhonov regularisation and Levenberg–Marquardt algorithm [17].

Model-independent signal deconvolution techniques have been widely used for myocardial perfusion quantification. Model-independent deconvolution was originally developed for intravascular contrast agents in the assessment of brain perfusion, where a precise measurement of the AIF and of the quantity Q of tracer injected into the vascular compartment is more easily obtained. Model-independent deconvolution has however gained popularity for myocardial perfusion assessment. A variety of different deconvolution methods (Fermi function modelling, B-spline basis, exponential basis and autoregressive moving average model—ARMA) are used to estimate the tissue specific transfer function and perfusion rate, usually adopting a forward modelling approach to overcome the intrinsic limitations of the deconvolution process [18, 19].

Fermi function modelling is the most common method used for perfusion quantification on the basis of the observed similarity between the transfer function $h(t)$ for an intravascular tracer and shape of the Fermi function [20, 21]. Fermi function modelling has proven robust to different scanning and modelling parameters and independent from spatial resolution and the signal to noise levels. This approach has been validated in a number of clinical studies in comparison with gold standards such as perfusion phantom, microspheres, position emission tomography (PET) and fractional flow reserve (FFR) [22–27]. However, the choice of the deconvolution method depends on several factors including signal to noise ratio of the data, computational burden and desired accuracy of the results [16, 28].

22.4   Model-Based Quantification

Model-based quantification is a less popular alternative to model-independent deconvolution methods. In this case, perfusion quantification is based on tracer-kinetic modelling. Assumes that a vascular system can be divided in pharmacodynamic compartments (intravascular, interstitial and intercellular), transfer constants (K_{trans}) are defined which describe the permeability of different compartments to the contrast agent. Most commonly, two compartments are considered for the purpose of myocardial perfusion quantification, including an intravascular and an interstitial extracellular compartment, since Gadolinium-based contrast agent cannot permeate cellular membranes.

A further simplification of the two-compartment model can be done by assuming a very high permeability of the vessels to the contrast agent, with fast diffusion in the extracellular space driven by the large concentration gradient between compartments during first pass, so that the exchange is only limited by the flow. In this one-compartment model, the concentration of tracer in the extracellular space is determined by the tracer concentration in plasma and no concentration difference between plasma and extracellular compartment can be measured. Therefore, model-based deconvolution relies on the quality of the estimates of the pharmacodynamic constants in use. The main limitation of model-based quantification is that the tracer-kinetic modelling does not provide myocardial flood flow measurements directly but rather provides a product of myocardial blood flow and contrast agent extraction during first pass. The calculation of myocardial blood flow is therefore based on correct estimates of the permeability surface product for the contrast agent. There are only a few reports in the literature trying to estimate this parameter based on experiments performed in canines. The direct application of these assumptions to human studies is still challenging.

An alternative for model-based deconvolution is represented by distributed models based on partial differential equations, like the blood tissue exchange (BTEX) model originally developed by Bassingthwaighte [29]. These models account for the concentration of the contrast agent in different compartments, for variations in blood flow, for the permeability surface area product for the capillaries, for the fluid volume in the capillaries and for the volume of the interstitial fluid, and for the geometry of the capillaries and the diffusion of the contrast agent.

The main advantage of the BTEX model is that it explicitly estimates the permeability surface area product which is used to calculate an extraction fraction to account for the flow dependent leakage of contrast agent from the vascular space into the interstitium.

22.5   Other Technical Factors Affecting the Accuracy of Quantitative Cardiac Perfusion MRI

The fundamental assumption at the basis of quantitative perfusion measurements is the existence of a linear correlation between the concentration of the indicator and the signal measured at voxel level, and a known relationship between signal intensity and Gadolinium concentration. In the case of cardiac MRI, several factors could affect this relationship and potentially impede accurate perfusion measurements. We will discuss here of signal saturation effects and surface coil spatial inhomogeneities.

22.5.1 Signal Saturation

Cardiac perfusion MRI is based on the acquisition of dynamic and strongly T1-weighted images (Fig. 22.2). This is commonly achieved using a saturation-recovery scheme, which destroys all longitudinal magnetisation with such prepulse. Consequently, magnetisation will recover exponentially with different rates given by the T1-relaxation time resulting in high signal for magnetisation with short T1-times (i.e. due to presence of contrast agent). The signal intensity depends on the contrast

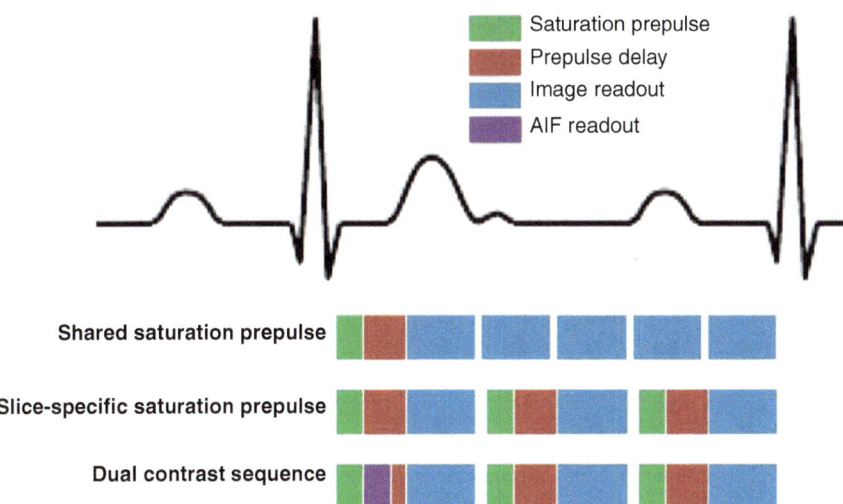

Fig. 22.2 Acquisition sequences for quantitative cardiac perfusion MRI. Sequences with shared saturation prepulse between slices enable the acquisition of 5–6 slices, with improved coverage of the left ventricle, including the apex. However, contrast is different between slices, as well as the dose-signal response, making perfusion quantification challenging. Therefore, sequences with slice-specific saturation prepulse are preferred. The readout of every slice is preceded by a saturation prepulse and by a prepulse delay of approximately 100 ms. The dual sequence is a further refinement with slice specific saturation prepulse, where an additional readout for the AIF is performed immediately after the saturation prepulse of the first slice, with the effect of avoiding full magnetisation recovery in the AIF

agent dose and the recovery delay. In a first order approximation the recovery curve can be linearised and the signal intensity is proportional to contrast agent concentration. However, this is not true for high concentration resulting in errors [30]. Therefore, a compromise in sequence parameter is required to detect lower concentration of contrast agent in the myocardium (usually a by a longer recovery delay) and high concentration in the blood pool (usually a by a shorter recovery delay). In addition, a high concentration results also in additional T2*-relaxation effects that reduces the measured signal intensity. Therefore a combination of full magnetisation recovery (high gadolinium concentration and long saturation-recovery delay) and T2* effect can cause clipping of the AIF, i.e. a reduction of the AIF-peak signal. If these saturation effects are not avoided, significant overestimation of myocardial perfusion can result.

Two approaches are currently in use to minimise signal saturation effects: the dual-bolus and the dual-contrast approach.

The dual-bolus is currently the most common approach to minimise signal saturation effects. It consists of a combination of two consecutive injections of contrast agent [31]. The first injection is performed with low dosage (for example, 0.0075 mmol/kg of body weight of Gadolinium) and is used to record a low amplitude and non-saturated AIF. The second injection is given instead with high dosage (for example, 0.075 mmol/kg of body weight of Gadolinium) and is sufficient to elicit a myocardial response suitable for visual and quantitative assessment. The dilution ratio of the pre-bolus, in the previous example 1:10, is then used during post-processing to correct the data and calculate a non-saturated AIF. When using a dual-bolus approach, a sufficient pause between the diluted pre-bolus and the main bolus is essential to avoid overlap between the input functions. In patients with normal cardiac output, a 25-s pause is usually sufficient.

The dual-contrast approach consists in the acquisition, in each cardiac cycle, of a low-resolution image to measure the AIF [32–34]. This is obtained acquiring an additional image soon after the saturation prepulse of the first slice (Fig. 22.2). The short saturation time AIF can provide more accurate measurements by avoiding full recovery of the magnetisation, which would instead be observed with longer saturation-recovery times as those used to image the myocardium. The acquisition of the AIF slice is followed by the acquisition of high-resolution images with longer saturation-recovery time to measure the myocardial signal with high contrast. Additional correction for T2* effects has been proposed [34]. The dual-sequence method has been shown to provide accurate measurement of the AIF when using high-dose and single-injection protocols and has the potential to become the preferred acquisition protocol, as it does not require changes to routine protocols in use for visual assessment.

It has recently become clear however that saturation effects, which are more commonly observed in the AIF, can also affect the myocardial signal intensity curves, resulting, in this case, in an underestimation of myocardial perfusion rate.

22.5.2 Surface Coil Spatial Inhomogeneities

Flexible phased array surface coils are regularly applied for cardiac MRI to increase the signal to noise ratio. These coils introduce a distance-to-coil-dependent signal

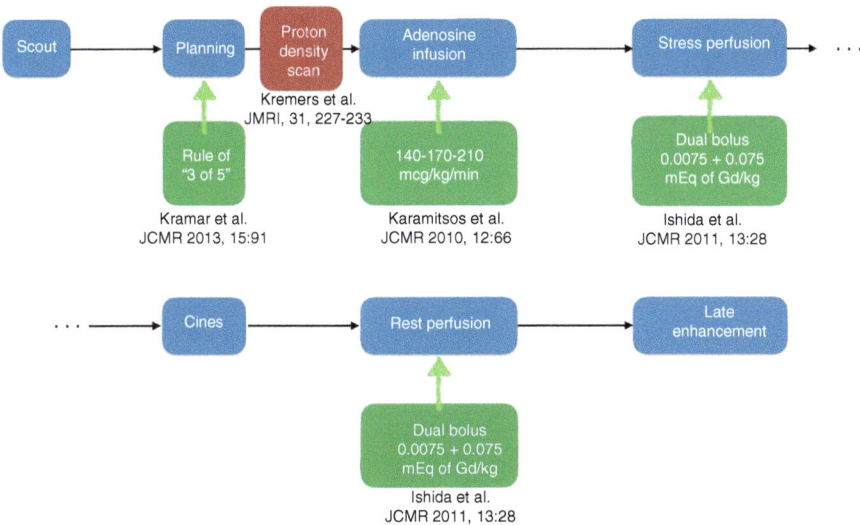

Fig. 22.3 Clinical acquisition protocol for quantitative cardiac perfusion MRI using a universal dual bolus approach. This protocol for quantitative cardiac perfusion MRI can easily be achieved in under 45 min of scanning and requires one additional breath hold to acquire a proton density image to be used for correction of the surface coil spatial inhomogeneities

intensity pattern. However, as already mentioned the deconvolution technique requires a homogeneous sensitivity over the heart, since the same AIF is applied to all myocardial segments.

Different methods have been described to correct intensity variations induced by surface coils. One approach involves the estimation of the sensitivity profile derived directly from the images, for example by using the baseline myocardial signal before the arrival of the contrast agent. Another approach involves the acquisition of additional 2D or 3D proton density (PD)-weighted images to assess the radiofrequency (RF) coil sensitivity [34–36].

22.6   Clinical Acquisition Protocols

Despite the increased complexity of the acquisition protocols required to obtain quantitative perfusion measurements, these can easily be integrated in clinical routine protocols. An example using the dual-bolus approach is shown in Fig. 22.3. Following the acquisition of scout images and planning of the perfusion sequence to cover the basal, mid-ventricular and apical segments ("rule of 3 of 5") [37], a proton density-weighted image is acquired to correct for the coil bias during post-processing. Then adenosine stress (from 140 to 210 mcg/min/kg of body weight, depending on the individual response [38]) is administered for at least 3 min, at the end of which the stress-perfusion acquisition is performed. A body weight-adjusted dose of 0.075 mmol/kg of body weight of Gadolinium is injected as main bolus, preceded by a tenfold diluted pre-bolus. The two injections are separated by a 25″

pause of the injector. After the injection of the main bolus of contrast agent, adenosine infusion can be interrupted. At least 10–15′ wash out is allowed after the first injection. During this time, cine sequences can be acquired, followed by rest perfusion imaging. This is acquired using the same dual-bolus scheme and the same sequence as the stress injection. Rest data are useful in post-processing to identify imaging artefacts and to provide quantitative rest data to be used to calculate myocardial perfusion reserve (MPR = stress/rest myocardial perfusion rate). Rest perfusion is followed by late gadolinium enhancement imaging. This protocol can be easily achieved in a clinical setting in less than 45′ of scanning.

22.7   High-Resolution Quantification

One of the main advantages of cardiac perfusion MRI, in comparison with other imaging techniques, is the elevated in-plane spatial resolution. Regardless of the specific technical parameters used for the acquisition of the images, cardiac

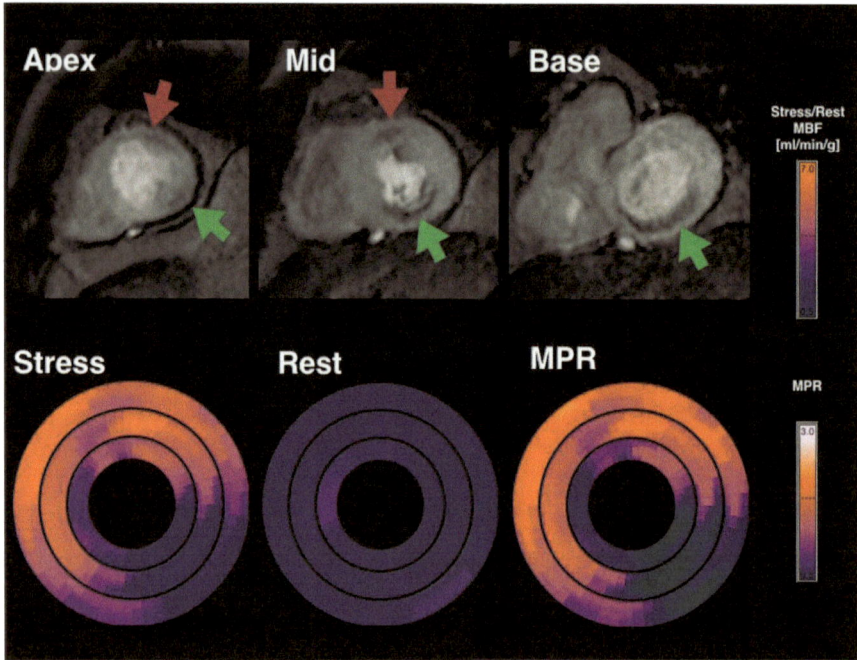

Fig. 22.4 Example of positive stress perfusion cardiac MRI. The top row shows the apical, mid-ventricular and basal left ventricular segments at peak enhancement during first-pass of Gadolinium during adenosine stress. The images demonstrate significant subendocardial perfusion abnormalities in the perfusion territory of the left anterior descending coronary artery (LAD; red arrows) and more severe abnormalities in the perfusion territory of the right coronary artery (green arrows). The bottom row shows bull's-eye plots obtained with high-resolution quantitative cardiac perfusion MRI at stress and rest, and an MPR map. Note the more extensive perfusion abnormalities on quantitative perfusion maps

perfusion MRI allows a spatial resolution comparable or superior to nuclear medicine techniques. This is considered the key factor explaining the superiority of cardiac perfusion MRI over nuclear perfusion imaging and SPECT in particular. Due to the complex interactions between coronary vasculature and myocardium, myocardial ischaemia arises from and affects more severely the subendocardial layers of the left ventricular myocardium and less the outer epicardial layers. Cardiac perfusion MRI is capable of an independent visualisation of multiple left ventricular layers, and allows a sensitive identification of ischaemia at its onset from the endocardium. This might conversely be missed by lower spatial resolution techniques due to partial volume effects.

High-resolution [25, 39] pixel-wise cardiac perfusion quantification has recently been proposed to preserve the spatial detail enabled by MRI and generate myocardial perfusion maps (Fig. 22.4). Validation studies in phantom, animal models and patients have shown high accuracy and good sensitivity for the detection of subendocardial ischaemia.

References

1. Patel MR, Peterson ED, Dai D, Brennan JM, Redberg RF, Anderson HV, Brindis RG, Douglas PS. Low diagnostic yield of elective coronary angiography. N Engl J Med. 2010;362:886–95.
2. Shaw LJ, Berman DS, Maron DJ, Mancini GBJ, Hayes SW, Hartigan PM, et al. Optimal medical therapy with or without percutaneous coronary intervention to reduce ischemic burden: results from the clinical outcomes utilizing revascularization and aggressive drug evaluation (COURAGE) trial nuclear substudy. Circulation. 2008;117(10):1283–91. https://doi.org/10.1161/CIRCULATIONAHA.107.743963.
3. Tonino PA, De Bruyne B, Pijls NH, Siebert U, Ikeno F, van' t Veer M, et al. Fractional flow reserve versus angiography for guiding percutaneous coronary intervention. N Engl J Med. 2009;360(3):213–24. https://doi.org/10.1056/NEJMoa0807611.
4. van Nunen LX, Zimmermann FM, Tonino PAL, Barbato E, Baumbach A, Engstrøm T, et al. Fractional flow reserve versus angiography for guidance of PCI in patients with multivessel coronary artery disease (FAME): 5-year follow-up of a randomised controlled trial. Lancet. 2015;386:1853. https://doi.org/10.1016/S0140-6736(15)00057-4.
5. Duncker DJ, Bache RJ. Regulation of coronary blood flow during exercise. Physiol Rev. 2008;88(3):1009–86. https://doi.org/10.1152/physrev.00045.2006.
6. Hachamovitch R, Hayes SW, Friedman JD, Cohen I, Berman DS. Comparison of the short-term survival benefit associated with revascularization compared with medical therapy in patients with no prior coronary artery disease undergoing stress myocardial perfusion single photon emission computed tomography. Circulation. 2003;107(23):2900–7. https://doi.org/10.1161/01.CIR.0000072790.23090.41.
7. Pazhenkottil AP, Nkoulou R, Kuest S, et al. Absolute coronary blood flow and coronary flow reserve assessed by gated SPECT with cadmium-zinc-telluride detectors: a direct comparison with 13N-ammonia PET. J Am Coll Cardiol. 2013;61:E1005.
8. Greenwood JP, Maredia N, Younger JF, Brown JM, Nixon J, Everett CC, et al. Cardiovascular magnetic resonance and single-photon emission computed tomography for diagnosis of coronary heart disease (CE-MARC): a prospective trial. Lancet. 2012;379(9814):453–60. https://doi.org/10.1016/S0140-6736(11)61335-4.
9. Schwitter J, Wacker CM, van Rossum AC, Lombardi M, Al-Saadi N, Ahlstrom H, et al. MR-IMPACT: comparison of perfusion-cardiac magnetic resonance with single-photon emission computed tomography for the detection of coronary artery disease in a multicentre, multivendor, randomized trial. Eur Heart J. 2008;29(4):480–9. https://doi.org/10.1093/eurheartj/ehm617.

10. Bettencourt N, Chiribiri A, Schuster A, Ferreira N, Sampaio F, Duarte R, et al. Cardiac magnetic resonance myocardial perfusion imaging for detection of functionally significant obstructive coronary artery disease: a prospective study. Int J Cardiol. 2013;168(2):765–73. https://doi.org/10.1016/j.ijcard.2012.09.231.

11. Ismail TF, Hsu L-Y, Greve AM, Gonçalves C, Jabbour A, Gulati A, et al. Coronary microvascular ischemia in hypertrophic cardiomyopathy - a pixel-wise quantitative cardiovascular magnetic resonance perfusion study. J Cardiovasc Magn Reson. 2014;16(1):49. https://doi.org/10.1186/s12968-014-0049-1.

12. Stewart GN. Researches on the circulation time and on the influences which affect it. J Physiol. 1897;22(3):159–83.

13. Zierler KL. Theoretical basis of indicator-dilution methods for measuring flow and volume. Circ Res. 1962;10(3):393–407.

14. Zarinabad N. Advanced quantification of myocardial perfusion. A dissertation submitted to graduate school of King's College London in partial fulfilment of the requirements for the degree of Doctorate of philosophy. King's College London. 2016.

15. Zarinabad N, Chiribiri A, Breeuwer M. Myocardial blood flow quantification from MRI – an image analysis perspective. Curr Cardiovasc Imaging Rep. 2013;7(1):9246. https://doi.org/10.1007/s12410-013-9246-9.

16. Zarinabad N, Chiribiri A, Hautvast G, Shuster A, Sinclair M, van den Wijngaard JPHM, et al. Modelling parameter role on accuracy of cardiac perfusion quantification. In: Functional imaging and modeling of the heart, Lecture notes in computer science, vol. 7945. Berlin: Springer; 2013. p. 370.

17. Sammut E, Zarinabad N, Vianello PF, Chiribiri A. Quantitative assessment of perfusion – where are we now? Curr Cardiovasc Imaging Rep. 2014;7(7):9278. https://doi.org/10.1007/s12410-014-9278-9.

18. Pack NA, DiBella EVR. Comparison of myocardial perfusion estimates from dynamic contrast-enhanced magnetic resonance imaging with four quantitative analysis methods. Magn Reson Med. 2010;64(1):125–37. https://doi.org/10.1002/mrm.22282.

19. Zarinabad N, Chiribiri A, Hautvast GL, Ishida M, Schuster A, Cvetkovic Z, et al. Voxel-wise quantification of myocardial perfusion by cardiac magnetic resonance. Feasibility and methods comparison. Magn Reson Med. 2012;68(6):1994–2004. https://doi.org/10.1002/mrm.24195.

20. Jerosch-Herold M, Wilke N, Stillman AE. Magnetic resonance quantification of the myocardial perfusion reserve with a Fermi function model for constrained deconvolution. Med Phys. 1998;25(1):73–84.

21. Wilke N, Jerosch-Herold M, Wang Y, Huang Y, Christensen BV, Stillman AE, et al. Myocardial perfusion reserve: assessment with multisection, quantitative, first-pass MR imaging. Radiology. 1997;204(2):373–84.

22. Chiribiri A, Schuster A, Ishida M, Hautvast G, Zarinabad N, Morton G, et al. Perfusion phantom: an efficient and reproducible method to simulate myocardial first-pass perfusion measurements with cardiovascular magnetic resonance. Magn Reson Med. 2013;69(3):698–707. https://doi.org/10.1002/mrm.24299.

23. Christian TF, Rettmann DW, Aletras AH, Liao SL, Taylor JL, Balaban RS, Arai AE. Absolute myocardial perfusion in canines measured by using dual-bolus first-pass MR imaging. Radiology. 2004;232(3):677–84. https://doi.org/10.1148/radiol.2323030573.

24. Lockie T, Ishida M, Perera D, Chiribiri A, De Silva K, Kozerke S, et al. High-resolution magnetic resonance myocardial perfusion imaging at 3.0-Tesla to detect hemodynamically significant coronary stenoses as determined by fractional flow reserve. J Am Coll Cardiol. 2011;57(1):70–5. https://doi.org/10.1016/j.jacc.2010.09.019.

25. Lee DC, Simonetti OP, Harris KR, Holly TA, Judd RM, Wu E, Klocke FJ. Magnetic resonance versus radionuclide pharmacological stress perfusion imaging for flow-limiting stenoses of varying severity. Circulation. 2004;110(1):58–65. https://doi.org/10.1161/01.CIR.0000133389.48487.B6.

26. Morton G, Chiribiri A, Ishida M, Hussain ST, Schuster A, Indermuehle A, et al. Quantification of absolute myocardial perfusion in patients with coronary artery disease: comparison between

cardiovascular magnetic resonance and positron emission tomography. J Am Coll Cardiol. 2012;60(16):1546–55. https://doi.org/10.1016/j.jacc.2012.05.052.

27. Schuster A, Sinclair M, Zarinabad N, Ishida M, van den Wijngaard JPHM, Paul M, et al. A quantitative high resolution voxel-wise assessment of myocardial blood flow from contrast-enhanced first-pass magnetic resonance perfusion imaging: microsphere validation in a magnetic resonance compatible free beating explanted pig heart model. Eur Heart J Cardiovasc Imaging. 2015;16(10):1082–92. https://doi.org/10.1093/ehjci/jev023.

28. Biglands JD, Magee DR, Sourbron SP, Plein S, Greenwood JP, Radjenovic A. Comparison of the diagnostic performance of four quantitative myocardial perfusion estimation methods used in cardiac MR imaging: CE-MARC substudy. Radiology. 2015;275(2):393–402. https://doi.org/10.1148/radiol.14140433.

29. Bassingthwaighte JB, Wang CY, Chan IS. Blood-tissue exchange via transport and transformation by capillary endothelial cells. Circ Res. 1989;65:997–1020.

30. Ichihara T, Ishida M, Kitagawa K, Ichikawa Y, Natsume T, Yamaki N, et al. Quantitative analysis of first-pass contrast-enhanced myocardial perfusion MRI using a patlak plot method and blood saturation correction. Magn Reson Med. 2009;62:373. https://doi.org/10.1002/mrm.22018.

31. Ishida M, Schuster A, Morton G, Chiribiri A, Hussain S, Paul M, et al. Development of a universal dual-bolus injection scheme for the quantitative assessment of myocardial perfusion cardiovascular magnetic resonance. J Cardiovasc Magn Reson. 2011;13(1):28. https://doi.org/10.1186/1532-429X-13-28.

32. Gatehouse PD, Elkington AG, Ablitt NA, Yang G-Z, Pennell DJ, Firmin DN. Accurate assessment of the arterial input function during high-dose myocardial perfusion cardiovascular magnetic resonance. J Magn Reson Imaging. 2004;20(1):39–45. https://doi.org/10.1002/jmri.20054.

33. Sánchez-González J, Fernandez-Jiménez R, Nothnagel ND, López-Martín G, Fuster V, Ibañez B. Optimization of dual-saturation single bolus acquisition for quantitative cardiac perfusion and myocardial blood flow maps. J Cardiovasc Magn Reson. 2015;17(1):329–12. https://doi.org/10.1186/s12968-015-0116-2.

34. Kellman P, Hansen MS, Nielles-Vallespin S, Nickander J, Themudo R, Ugander M, Xue H. Myocardial perfusion cardiovascular magnetic resonance: optimized dual sequence and reconstruction for quantification. J Cardiovasc Magn Reson. 2017;19:43. https://doi.org/10.1186/s12968-017-0355-5.

35. Kremers FP, Hofman MB, Groothuis JG, Jerosch-Herold M, Beek AM, Zuehlsdorff S, et al. Improved correction of spatial inhomogeneities of surface coils in quantitative analysis of first-pass myocardial perfusion imaging. J Magn Reson Imaging. 2009;31(1):227–33. https://doi.org/10.1002/jmri.21998.

36. Murakami JW, Hayes CE, Weinberger E. Intensity correction of phased-array surface coil images. Magn Reson Med. 1996;35(4):585–90.

37. Kramer CM, Barkhausen JR, Flamm SD, Kim RJ, Nagel E. Standardized cardiovascular magnetic resonance (CMR) protocols 2013 update. J Cardiovasc Magn Reson. 2013;15(1):1–1. https://doi.org/10.1186/1532-429X-15-91.

38. Karamitsos TD, Ntusi NA, Francis JM, Holloway CJ, Myerson SG, Neubauer S. Feasibility and safety of high-dose adenosine perfusion cardiovascular magnetic resonance. J Cardiovasc Magn Reson. 2010;12(1):66. https://doi.org/10.1186/1532-429X-12-66.

39. Villa ADM, Sammut E, Zarinabad N, Carr-White G, Lee J, Bettencourt N, et al. Microvascular ischemia in hypertrophic cardiomyopathy: new insights from high-resolution combined quantification of perfusion and late gadolinium enhancement. J Cardiovasc Magn Reson. 2016;18:1–11. https://doi.org/10.1186/s12968-016-0223-8.

Marc Dewey and Marc Kachelrieß

Abstract

Quantification of myocardial perfusion is the holy grail of cardiovascular imaging. Computed tomography angiography (CTA) is the most accurate noninvasive diagnostic test to diagnose obstructive coronary artery disease but lacks the ability to quantify the functional relevance of coronary artery stenosis. Using myocardial CT perfusion might enable comprehensive assessment of coronary artery disease by quantification of myocardial blood flow. The rather high radiation dose and the complicated analysis of 4D CT are the main challenges for achieving this goal. This chapter summarizes the current status of myocardial perfusion imaging and describes potential technical and clinical solutions for myocardial perfusion assessment by CT.

23.1 Technical Principles and Clinical Challenges

23.1.1 Current Clinical Status and Paradigm Shift

Anatomic assessment of coronary artery disease (CAD) is still most commonly used in deciding about coronary revascularization [1], although including the functional component of CAD has been shown to improve cardiovascular outcome [2].

M. Dewey (✉)
Department of Radiology, Charité - Universitätsmedizin Berlin, Berlin, Germany
e-mail: marc.dewey@charite.de

M. Kachelrieß
Medical Physics in Radiology, German Cancer Research Center (DKFZ),
Heidelberg, Germany
e-mail: marc.kachelriess@dkfz.de

© Springer International Publishing AG 2018   487
I. Sack, T. Schaeffter (eds.), *Quantification of Biophysical Parameters in Medical Imaging*, https://doi.org/10.1007/978-3-319-65924-4_23

Thus, there is a clinical need for a paradigm shift by going beyond morphology and including physiology in diagnosis, prognosis and therapy of CAD [3]. The widespread use of physiological information on CAD in clinical practice could translate into improved health in millions of Europeans, help reduce the annual 2 million unnecessary invasive coronary angiographies in Europe [4], and improve cost-effectiveness [5].

23.1.2 Potential and Limitations of Conventional 3D CT

With their high resolution of 30 voxels/mm^3, high-end volumetric (3D) computed tomography (CT) scanners are excellent tools for the noninvasive evaluation of the coronary arteries (Fig. 23.1). However, while 3D CT angiography (CTA) is very accurate for identifying and characterizing obstructive CAD, it remains a test to investigate morphology, and the specificity and accuracy of 3D CTA greatly drops compared with measures of physiology, such as fractional flow reserve allowing to assess the functional relevance of coronary obstructions during catheterization (Fig. 23.2). Moreover, CT does not allow quantifying the functional relevance of anatomic CAD with high accuracy and low radiation dose [6, 7]. Thus, CT suffers from significant limitations and currently lacks several key functions that would be required of a one-stop modality for CAD diagnosis.

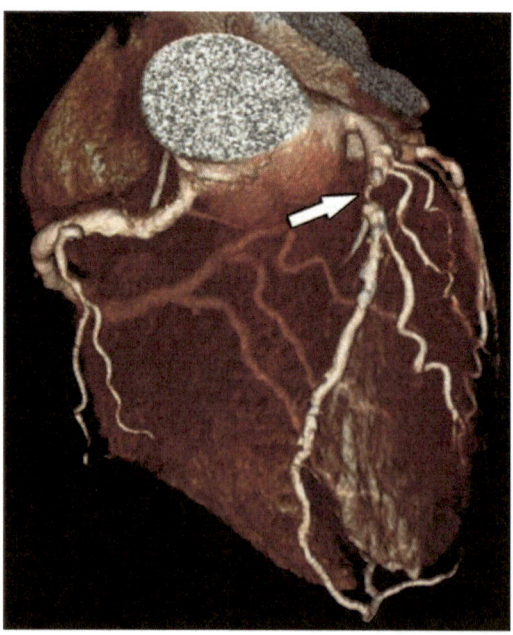

Fig. 23.1 3D CTA provides noninvasively derived images for the detection of coronary stenosis and plaques (arrow) with highest spatial resolution of 30 voxels per mm^3

Fig. 23.2 3D CTA has high diagnostic accuracy in an evaluation of its sensitivity and specificity using summary receiver operating characteristic curves versus morphology (invasive angiography) but reduced accuracy and specificity versus physiology (fractional flow reserve)

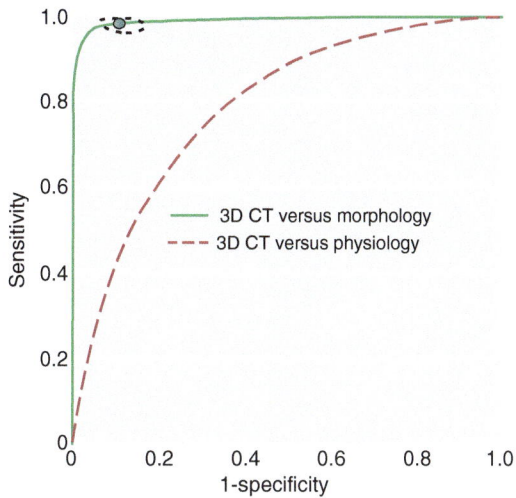

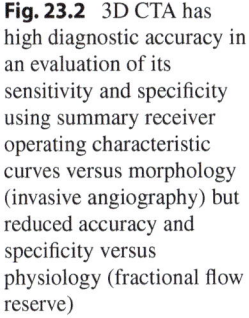

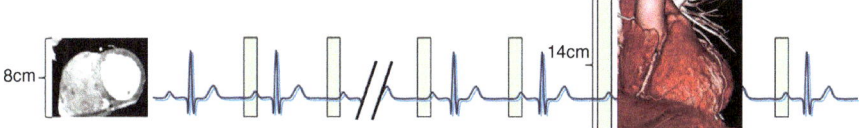

Fig. 23.3 4DCT covering the heart (shown in short axis) every beat, while coronary CT is integrated during a single beat with larger coverage. Adapted from the CTP and CT-FFR chapter in *Cardiac CT* [31]

23.1.3 Major Clinical Challenge: Radiation Dose and Data Analysis

The major challenge for myocardial blood flow (MBF) quantification is that at least two 3D CT volumes need to be acquired (one during rest and one during stress conditions) which, in principle, doubles the radiation exposure of a typical standard CT examination. With the repetitive acquisition of 3D volume of the entire heart during every heart beat (4D CT) during the first-pass inflow of the diffusible contrast agent into the myocardial tissue (Fig. 23.3), radiation dose further increases greatly. Moreover, data analysis of 4D CT becomes challenging considering that such a 4D dataset encompasses 2 to 3 billion voxels. Temporal averaging of 3D volumes acquired in different heart beats within the 4D CT dataset may help to better depict myocardial perfusion deficits (Fig. 23.4) and smoothen the resulting time-attenuation curves provided image registration is properly done (Fig. 23.5).

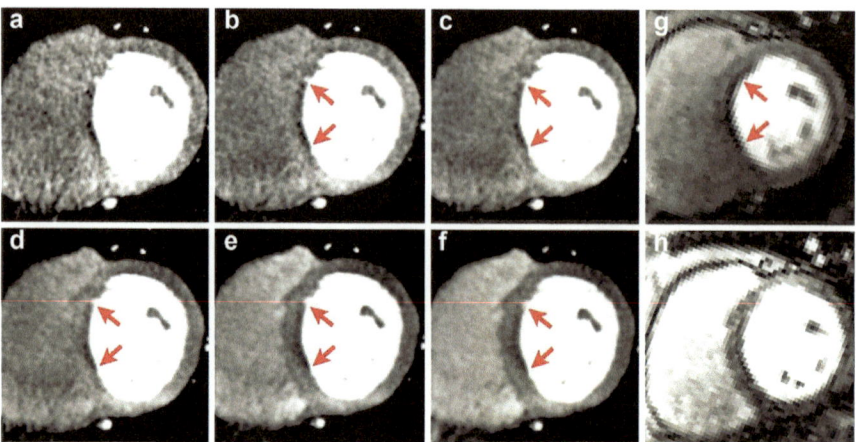

Fig. 23.4 Effect of temporal averaging of 3D volumes from a dynamic 4D CT dataset on image quality and depiction of perfusion deficits (arrows). Basal cardiac short axis slice of a 63 year old male with typical angina pectoris and hyperlipidemia. Panels (**a**)–(**f**) show the combination of one, two, three, four, six, and eight CTP 3D datasets from consecutive heart beats for temporal averaging. Panels (**g, h**) show stress and rest MR imaging, respectively. The 4D CTP datasets were read to have no relevant motion and good image quality. Please note the decrease in noise from (**a–f**). From (**b–f**), the subendocardial perfusion defect in the septal wall (arrows) is demarcated very well, while it was read as false negative in the dynamic dataset (**a**). The stress MR imaging as the reference standard confirmed the septal subendocardial perfusion defect (arrows in **g**), which was not visible on the rest images (**h**). Figure from [32]

23.1.4 Clinical Potential

Our CARS-320 and CORE-320 trials showed the potential of adding physiology to CAD assessment already by use of nonquantitative CT perfusion (CTP) [8, 9]. Despite the above major challenge of radiation dose for clinical implementation, both 3D and 4D of myocardial CT perfusion offer enormous potential to replace existing inaccurate and expensive approaches to assessing the functional relevance of CAD in clinical practice. Currently, a single comprehensive and noninvasive imaging tool for CAD morphology and perfusion is missing [10]. The primary clinical potential of cardiac CT perfusion of the myocardium would be to allow for a personalized assessment of MBF and improved individualized prediction of most appropriate subsequent therapies (Fig. 23.6).

23.1.5 Current Status of Imaging

Definitive diagnosis of CAD in patients with high pretest probability of disease is recommended using invasive coronary angiography (ICA) [11, 12]. Whether

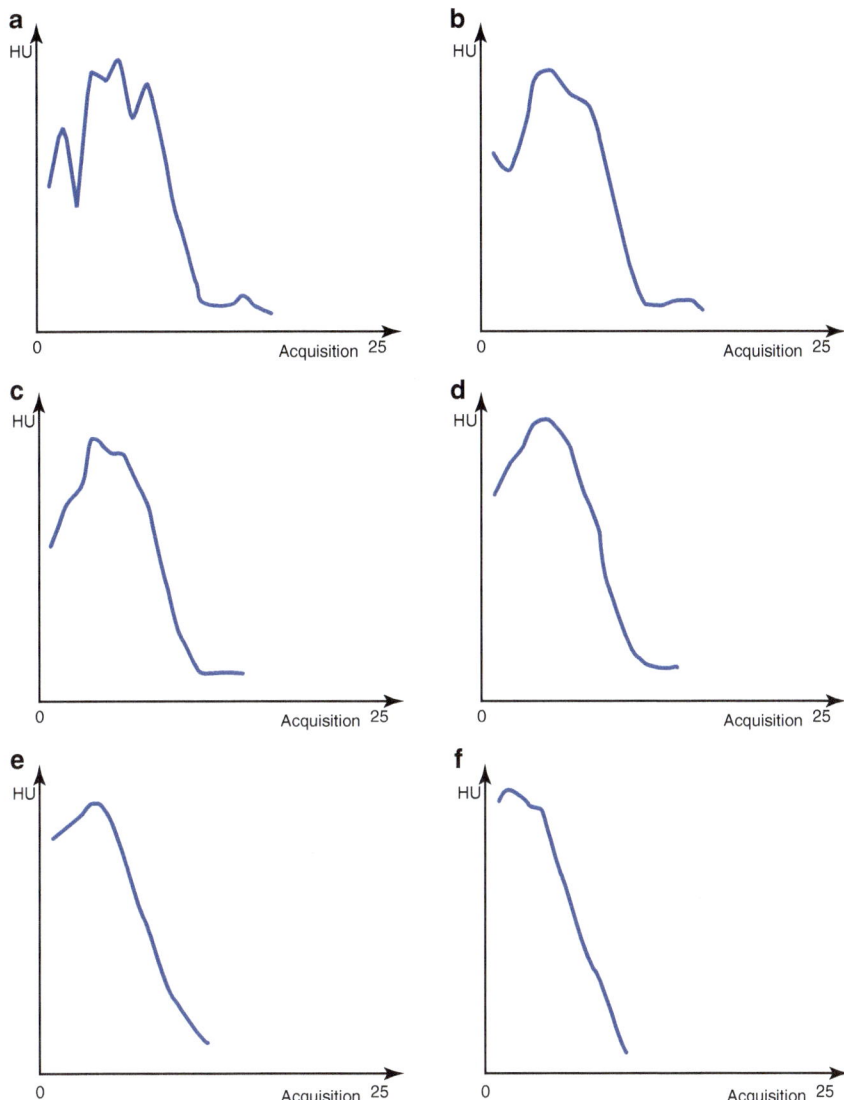

Fig. 23.5 Effect of temporal averaging of 3D volumes from 4D CT perfusion on the time-attenu-ation curves (TAC) within the left ventricle. TACs calculated from a measurement point in the left ventricle. The y axis demonstrates Hounsield Units values, while the x axis shows different acqui-sition phases. (**a**) Input TAC without temporal averaging. (**b**) Temporal averaging of two consecu-tive 3D datasets. (**c**) Temporal averaging of three consecutive 3D datasets. (**d**) Temporal averaging of five consecutive 3D datasets. (**e**) Temporal averaging of seven consecutive 3D datasets and (**f**) temporal averaging of eight consecutive 3D datasets. With increasing temporal averaging levels the TAC becomes smoother, but also shorter and less accurate as oversmoothing caused by the temporal average occurs which can be seen at the lower peak. Figure from [32]

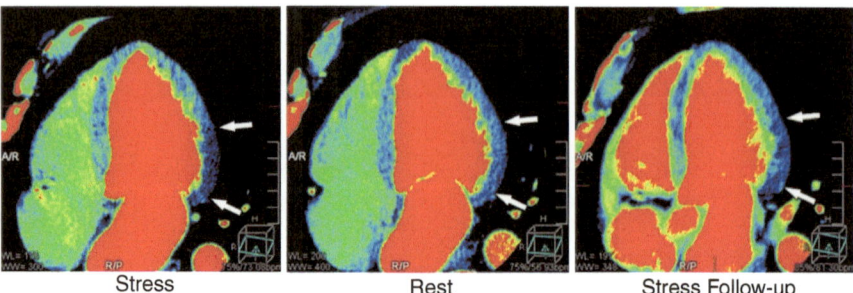

Stress   Rest   Stress Follow-up

Fig. 23.6 Cardiac CT perfusion results at baseline and over time. Stress and rest cardiac CT perfusion show a reversible perfusion defect ("ischemia") in the lateral myocardial wall (arrows), i.e., the perfusion defect is only present after infusion of adenosine (see Chap. 14, Fig. 14.1) and perfusion becomes normal during rest (arrows). Importantly, rest perfusion results are able to predict in this patient the improved stress CT perfusion results on follow-up of the patient after successful interventional treatment of his obstructive coronary artery disease. Thus, cardiac CT perfusion may not only act as an imaging biomarker but may also allow personalized therapy planning by enabling individual prediction of therapy response

intermediate coronary stenosis is relevant can be assessed by invasively determining the fractional flow reserve (FFR) [2]. FFR-guided management has been reported to reduce major adverse cardiovascular events [2] and is thus recommended in the guidelines [11, 12]. It would be a pivotal advancement if the ischemic relevance of coronary stenosis could be assessed noninvasively [10]. Magnetic resonance imaging (MRI, see Chap. 22) agrees well with positron emission tomography (PET) for absolute MBF [13, 14] but does not reach the accuracy of CTA for anatomic disease [15, 16]. Single-photon emission computed tomography using dedicated cameras also holds promise for MBF [17]. PET is more commonly used for quantification and is the research reference standard for MBF and absolute coronary flow reserve (CFR) [18–20]. Further advantages of PET over FFR are that the microvascular component is captured [20], the prognostic power is incremental to ICA [21], and CTA is enabled by hybrid imaging [22]. CTA alone is widely available, less costly, and most accurate for stenosis detection and noninvasive plaque characterization [15, 23, 24]. Thus, CTA has become the clinical reference standard for 15–50% pretest probability of CAD in the European stable chest pain guidelines [11, 12]. However, CT is greatly limited by its rather high and nonpersonalized radiation exposure. The first and comprehensive European Commission report on medical radiation dose shows that CT accounts for only 8.7% of all examinations but the CT-related radiation dose is far higher with about 57% [25], demonstrating the great need to further reduce CT-related dose. The unique potential of CTP, most importantly the optimal linear tracer–signal relationship [26], has not been exploited so far for accurately capturing MBF [7].

23.2   Technical Challenges and Potential Solutions

23.2.1 Technical Issue

The major technical issue for the above clinical challenge of too high radiation exposure for clinical implementation is to reduce dose as much as possible to achieve clinically acceptable dose levels without compromising the ability to quantify MBF during rest and stress.

23.2.2 Primary Obstacle and The ALARA Principle

The primary obstacle to CTP is the level of radiation imparted by dynamic scanning during 4D CT, which can result in effective doses of up to 54 mSv [6]. This needs to be lowered significantly to a tolerable level to achieve clinical acceptance of CTP. Currently, vendor-provided packages cater neither for the specific demands of the 4D examination nor the individual patient. Still the implementation of the "as low as reasonably achievable" (ALARA) principle is required for clinical implementation by novel technical developments.

23.2.3 Potential Solutions

The following technical approaches are potential solutions for the above challenges and require further analysis towards reduced radiation dose during cardiac CT perfusion without limiting the ability to quantify perfusion. Compressed sensing with reduced acquisition of line integrals during certain heart beats but also omitting certain heart beats during 4D CT acquisition approaches at all by using prior information (Fig. 23.7).

Another promising approach to better understanding myocardial perfusion is fractal analysis of imaging (Fig. 23.8). The concept of self-similarity was first described by Leibniz in 1695, and Benoit Mandelbrot coined the term fractal in 1975 [27]. Fractal analysis is a rather new concept in radiological and nuclear perfusion imaging [28] and has only recently been discussed for use in cardiac medicine [29]. We patented fractal analysis of transition regions in perfusion imaging and showed that this is a promising approach to myocardial perfusion imaging for clinical differentiation of coronary artery disease and microvascular dysfunction (Fig. 23.9) [30].

Acknowledgment Professor Dewey would like to thank his group members S. Feger, S. Lukas, F. Michallek, M. Rief, and E. Zimmermann for an exciting collaboration and excellent work on myocardial perfusion imaging. Professor Kachelrieß would like to thank F. Pisana for support.

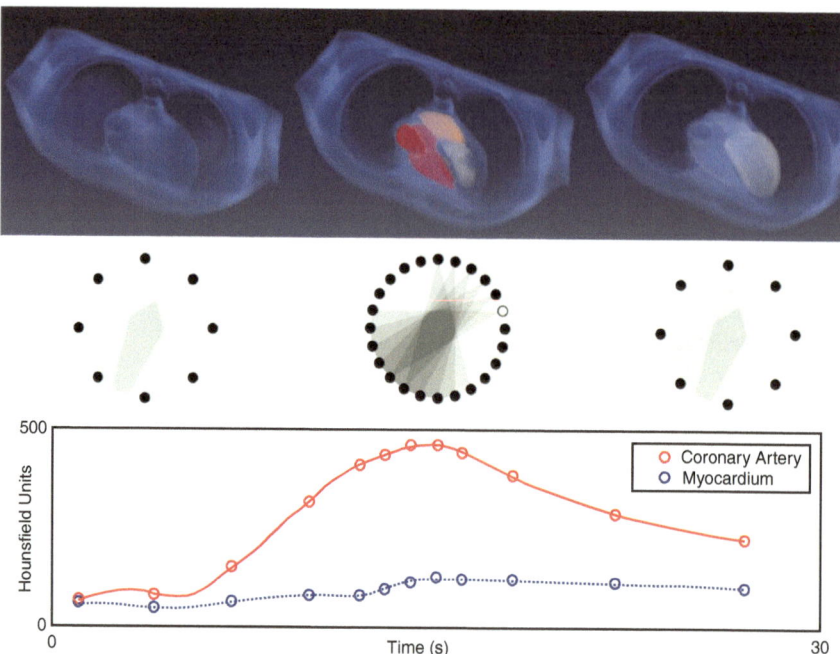

Fig. 23.7 Dose reduction in 4D CTP (top row) to about 1 mSv by limiting the number of angular projections and reducing tube current at certain heartbeats (middle row) and restricting the number of acquired time points as visualized in the time-attenuation curve for the coronary artery and the myocardium (bottom row)

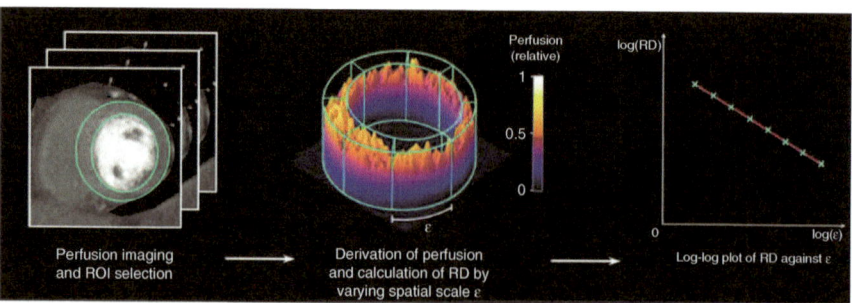

Fig. 23.8 Principle of fractal analysis using the relative dispersion (RD) algorithm for improved analysis of any type of perfusion imaging. Figure from [28]

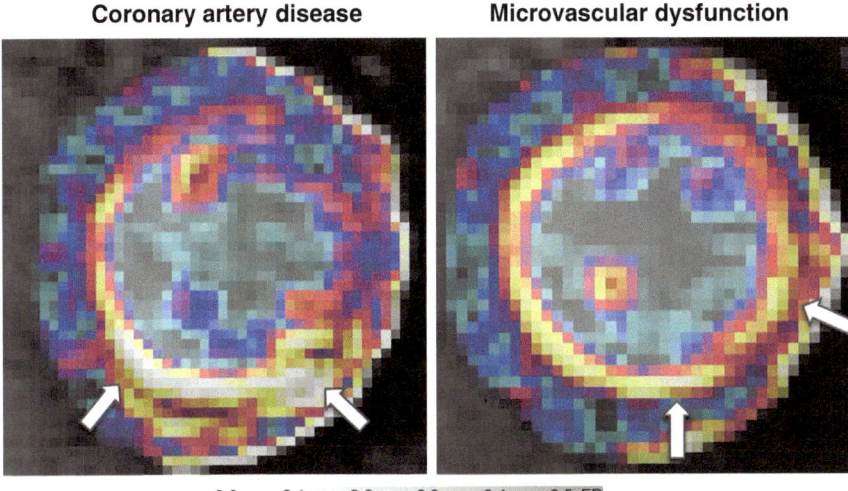

Fig. 23.9 Principle of fractal analysis for improved analysis of myocardial perfusion imaging that allows for the first the noninvasive differentiation of myocardial perfusion deficits (arrows) that arise from obstructive coronary artery disease versus those that are due to coronary microvascular dysfunction. Figure from [30]

References

1. Krone RJ, Shaw RE, Klein LW, Blankenship JC, Weintraub WS, American College of Cardiology - National Cardiovascular Data Registry. Ad hoc percutaneous coronary interventions in patients with stable coronary artery disease--a study of prevalence, safety, and variation in use from the American College of Cardiology National Cardiovascular Data Registry (ACC-NCDR). Catheter Cardiovasc Interv. 2006;68:696–703.
2. De Bruyne B, Fearon WF, Pijls NH, et al. Fractional flow reserve-guided PCI for stable coronary artery disease. N Engl J Med. 2014;371:1208–17.
3. Arai AE. Computed tomography perfusion to assess physiological significance of coronary stenosis in the post-FAME era (Fractional Flow Reserve versus Angiography for Multivessel Evaluation). J Am Coll Cardiol. 2013;62:1486–7.
4. Moschovitis A, Cook S, Meier B. Percutaneous coronary interventions in Europe in 2006. EuroIntervention. 2010;6:189–94.
5. Dewey M, Hamm B. Cost effectiveness of coronary angiography and calcium scoring using CT and stress MRI for diagnosis of coronary artery disease. Eur Radiol. 2007;17:1301–9.
6. So A, Hsieh J, Imai Y, et al. Prospectively ECG-triggered rapid kV-switching dual-energy CT for quantitative imaging of myocardial perfusion. JACC Cardiovasc Imaging. 2012;5:829–36.

7. Ishida M, Kitagawa K, Ichihara T, et al. Underestimation of myocardial blood flow by dynamic perfusion CT: explanations by two-compartment model analysis and limited temporal sampling of dynamic CT. J Cardiovasc Comput Tomogr. 2016;10:207–14.
8. Rief M, Zimmermann E, Stenzel F, et al. Computed tomography angiography and myocardial computed tomography perfusion in patients with coronary stents: prospective intraindividual comparison with conventional coronary angiography. J Am Coll Cardiol. 2013;62:1476–85.
9. Rochitte CE, George RT, Chen MY, et al. Computed tomography angiography and perfusion to assess coronary artery stenosis causing perfusion defects by single photon emission computed tomography: the CORE320 study. Eur Heart J. 2014;35:1120–30.
10. Loewe C, Stadler A. Computed tomography assessment of hemodynamic significance of coronary artery disease: CT perfusion, contrast gradients by coronary CTA, and fractional flow reserve review. J Thorac Imaging. 2014;29:163–72.
11. Montalescot G, Sechtem U, Achenbach S, et al. 2013 ESC guidelines on the management of stable coronary artery disease: the Task Force on the management of stable coronary artery disease of the European Society of Cardiology. Eur Heart J. 2013;34:2949–3003.
12. Windecker S, Kolh P, Alfonso F, et al. 2014 ESC/EACTS guidelines on myocardial revascularization: the task force on myocardial revascularization of the European Society of Cardiology (ESC) and the European Association for Cardio-Thoracic Surgery (EACTS) developed with the special contribution of the European Association of Percutaneous Cardiovascular Interventions (EAPCI). Eur Heart J. 2014;35:2541.
13. Qayyum AA, Hasbak P, Larsson HB, et al. Quantification of myocardial perfusion using cardiac magnetic resonance imaging correlates significantly to rubidium-82 positron emission tomography in patients with severe coronary artery disease: a preliminary study. Eur J Radiol. 2014;83:1120–8.
14. Miller CA, Naish JH, Ainslie MP, et al. Voxel-wise quantification of myocardial blood flow with cardiovascular magnetic resonance: effect of variations in methodology and validation with positron emission tomography. J Cardiovasc Magn Reson. 2014;16:11.
15. Schuetz GM, Zacharopoulou NM, Schlattmann P, Dewey M. Meta-analysis: noninvasive coronary angiography using computed tomography versus magnetic resonance imaging. Ann Intern Med. 2010;152:167–77.
16. Morton G, Plein S, Nagel E. Noninvasive coronary angiography using computed tomography versus magnetic resonance imaging. Ann Intern Med. 2010;152:827–8. author reply 8-9
17. Petretta M, Storto G, Pellegrino T, Bonaduce D, Cuocolo A. Quantitative assessment of myocardial blood flow with SPECT. Prog Cardiovasc Dis. 2015;57:607–14.
18. Gould KL, Johnson NP, Bateman TM, et al. Anatomic versus physiologic assessment of coronary artery disease. Role of coronary flow reserve, fractional flow reserve, and positron emission tomography imaging in revascularization decision-making. J Am Coll Cardiol. 2013;62:1639–53.
19. Bengel FM, Higuchi T, Javadi MS, Lautamaki R. Cardiac positron emission tomography. J Am Coll Cardiol. 2009;54:1–15.
20. Saraste A, Kajander S, Han C, Nesterov SV, Knuuti J. PET: is myocardial flow quantification a clinical reality? J Nucl Cardiol. 2012;19:1044–59.
21. Taqueti VR, Hachamovitch R, Murthy VL, et al. Global coronary flow reserve is associated with adverse cardiovascular events independently of luminal angiographic severity and modifies the effect of early revascularization. Circulation. 2015;131:19–27.
22. Kajander S, Joutsiniemi E, Saraste M, et al. Cardiac positron emission tomography/computed tomography imaging accurately detects anatomically and functionally significant coronary artery disease. Circulation. 2010;122:603–13.
23. Schuetz GM, Schlattmann P, Dewey M. Avoiding overestimation of clinical performance by applying a 3x2 table with an intention-to-diagnose approach: an exemplary meta-analytical evaluation of coronary CT angiography studies. BMJ. 2012;345:e6717. https://doi.org10.1136/bmj.e6717

24. Hulten EA, Carbonaro S, Petrillo SP, Mitchell JD, Villines TC. Prognostic value of cardiac computed tomography angiography: a systematic review and meta-analysis. J Am Coll Cardiol. 2011;57:1237–47.
25. EC. European Commission. Radiation Protection N° 180. Medical radiation exposure of the european population. Part 1/2. Brussels, BE: Directorate-General for Energy. Directorate D — Nuclear Safety & Fuel Cycle. Unit D3 — Radiation Protection. 2014.
26. Kalender W. Computed tomography. Hoboken: Wiley; 2011.
27. Mandelbrot B. The fractal geometry of nature. San Francisco: W. H. Freeman; 1982.
28. Michallek F, Dewey M. Fractal analysis in radiological and nuclear medicine perfusion imaging: a systematic review. Eur Radiol. 2014;24:60–9.
29. Captur G, Karperien AL, Hughes AD, Francis DP, Moon JC. The fractal heart - embracing mathematics in the cardiology clinic. Nat Rev Cardiol. 2017;14:56–64.
30. Michallek F, Dewey M. Fractal analysis of the ischemic transition region in chronic ischemic heart disease using magnetic resonance imaging. Eur Radiol. 2017;27(4):1537–46.
31. Kitagawa K, Erglis A, Dewey M. Ch. 19. Cardiac CT2014.
32. Feger S, Shaban A, Lukas S, et al. Temporal averaging for analysis of four-dimensional whole-heart computed tomography perfusion of the myocardium: proof-of-concept study. Int J Cardiovasc Imaging. 2017;33:371.